MW00843334

INTEGRATED DESIGN OF WATER TREATMENT FACILITIES

Preliminary Studies to Plant Start-up

INTEGRATED DESIGN OF WATER TREATMENT FACILITIES

Susumu Kawamura

A WILEY-INTERSCIENCE PUBLICATION

JOHN WILEY & SONS, INC.

New York · Chichester · Brisbane · Toronto · Singapore

Copyright © 1991 by John Wiley & Sons, Inc.

All rights reserved. Published simultaneously in Canada.

Reproduction or translation of any part of this work
beyond that permitted by Section 107 or 108 of the
1976 United States Copyright Act without the permission
of the copyright owner is unlawful. Requests for
permission or further information should be addressed to
the Permissions Department, John Wiley & Sons, Inc.

Library of Congress Cataloging in Publication Data:

Kawamura, Susumu.
    Integrated design of water treatment facilities / Susumu Kawamura.
      p. cm.
    Includes bibliographical references.
    ISBN 0-471-61591-9

    1. Water treatment plants—Design and construction. I. Title.
TH4538.K39 1991
628.1'62—dc20   90-40995
                                                         CIP

Printed in the United States of America

10 9 8 7 6 5 4 3 2

This book is dedicated to my wife, Reiko, and my brother,
Dr. Shigeru Tanaka. Without their encouragement and understanding
this book would never have been written.
I also wish to express my deepest appreciation to my daughters
for their help: especially to Mika for her devoted efforts in editing
and typing the manuscript and to Midori for her support.

◼◼◼◼ PREFACE

There are many books relating to water treatment on the market; however, only few are capable of acting as a guide for project control as well as for the design of actual water treatment processes and the operational aspects of the plant. A few books have tried to cover a broad spectrum of the treatment plant design projects and have ended up with typical handbook styles. These books are mere compilations of the thesis with little or no coordination among individual subjects.

Integrated Design of Water Treatment Facilities is written by a single author who has more than 35 years of experience in all phases of the design project including research pilot studies, preliminary design studies, the design, construction management, and actual plant operation. This book is especially geared for professional engineers and college students who seek emphasis on the practical rather than principle, method over methodology, and wish to have a useful guide and reference instead of a useless book that occupies space on a bookshelf.

Unlike other published works, this book covers an entire project sequence. It is the first book to describe not only basic design criteria for each process but also to show how to design each process in order to maximize overall process efficiency while minimizing operation and maintenance costs. It is for these reasons that this book is entitled *Integrated Design of Water Treatment Facilities*.

The first chapter describes the management of design projects and covers typical forms of contracts and the roles of professional engineers. Project management is a subject with which most new engineers are not familiar; however, it is of vital importance to be knowledgeable about this topic since the successful completion of a project on time as well as within budget can only be accomplished through a firm project management process.

Chapter 2 deals with the preliminary design studies. This chapter covers the issues that design engineers need to explore when working on actual design projects. Bench scale and pilot studies have become important parts of predesign studies in recent years because the EPA's latest stringent drinking water standards, effective 1990, have made the ordinary, conventional treatment process used in the past obsolete and incapable of meeting these new standards. This chapter not only discusses the pilot studies but also describes the concept of a basic plant layout, items to be covered by geotechnical evaluation of the site, plant hydraulics, and project management planning in the design phase.

Chapter 3 handles the design of basic treatment plant components. Most of the existing books address each process separately and are often written by different authors, resulting in little integration among the unit processes. These other books simply illustrate how theory can be applied to actual process design and do not adequately consider topics such as viability of construction, operation, maintenance, and other practical details. This book illustrates how each process should be designed using an integrated approach. Many sample calculations are included to show how the actual plant design should be accomplished.

Needless to say, the heart of a water treatment process train is the filtration process; thus, the pretreatment processes should be designed to maximize filter efficiency. Pretreatment process designs, especially the flocculation process, should be different for conventional rapid sand filters as opposed to high-rate filters with reverse graded filter bed under direct filtration process trains. If the filtration process is a membrane process rather than one of a granular medium, the difference in pretreatment requirements is pronounced. Many inexperienced engineers do not realize these basic principles because their educational backgrounds are gained from textbooks that simply address how to maximize each process efficiency and fail to discuss the important basic denominator. This chapter also touches on the process control and plant instrumentation, which are topics generally not covered in most water treatment books.

Chapter 4 details the subordinate plant facilities such as chemical feed systems, including ozonation, waste washwater handling, sludge handling and disposal, the intakes, grit chamber, and operations building design. In reality, design engineers are always seeking practical design guides and criteria for these facilities.

Chapter 5 gives consideration to and design guides on plant components including flow metering, liquid level measuring, valve selection, and standby power supply systems. These are all vital plant design issues that should be included in the plant design.

Plant hydraulics, specifications, cost estimates, and incidental works such as geotechnical studies, surveying, and design considerations in cold weather are briefly examined in Chapter 6.

Chapter 7 shows special water treatment processes that go beyond the conventional complete water treatment process. Lime–soda ash softening, iron and manganese removal, taste and odor control, control of THMs and VOCs, fluoridation and defluoridation, corrosion control, GAC adsorption, ion exchange and membrane separation processes, and the removal of heavy metals and inorganic contaminants are all subjects of discussion.

Chapter 8 deals with the management of procurement and plant construction. This chapter describes the key issues in the procurement stage, construction management activities, design considerations, and the preparation of a practical "operation and management manual" for treatment plants. There are also several other good reference books available that specifically cover construction management. Since many disputes and lawsuits have occurred because of poor construction management, this is an important subject for all design engineers. Engineers are also encouraged to take on the construction management task since they will learn so much from this activity.

Chapter 9 presents guidelines for writing a practical operations and maintenance manual for the water treatment plant and for setting up plant operator training programs before plant start-up.

The last chapter outlines plant start-up and necessary follow-up activities after the plant goes into operation. These topics are totally neglected by other existing books despite the fact that every treatment plant design must undergo this phase of operation after plant construction. It is an unfortunate fact that a significant number of engineers repeat the same design errors because they have not obtained any feedback from plant operational staffs due to a lack of proper follow-ups.

The sum total of this book covers practical design as well as construction procedures which are found in actual water treatment design and construction practices. The text is written at the level of professional engineers but is sufficiently basic so that it is useful to engineering students at the university level.

English units are used in this book since they are still the standard measure of units in the United States. However, the new SI units (metric units) are also shown in parenthesis and some of the calculation examples are deliberately presented in metric units not only for the benefit of engineers in the rest of the world but also for American engineers who may wish to familiarize themselves with metric units.

It should be clearly noted that this book is not written by James M. Montgomery Consulting Engineers, Inc. The entire text is written by me on my own time. This achievement is based on my beliefs and on nearly 40 years of experience in various aspects of water treatment engineering both in the United States and overseas.

In my professional career, I have been privileged to be associated with outstanding colleagues and predecessors at various locations: particularly at J. M. Montgomery Consulting Engineers in California and Hanshin Water Work Bureau in Japan. Their intellectual stimulation and the working experience gained through their interaction strongly influenced my thinking and this is reflected throughout the book.

Particular appreciation is given to Richard S. Holmgren, Jr., President of J. M. Montgomery Consulting Engineers, for his encouragement and collaboration.

SUSUMU KAWAMURA

San Gabriel, California
July 1990

CONTENTS

INTEGRATED DESIGN OF WATER TREATMENT FACILITIES

CHAPTER 1

Management of a Design Project

1.1 ROLE OF THE PROFESSIONAL ENGINEER

For any project the owner must prepare a budget to finance the operation and a time schedule by which it should be completed. The design phase of most water treatment plants, for municipalities, typically costs 5–10% of the estimated construction cost. The time required to design and construct a treatment plant is approximately 3 years, including a 9–12 month design period as practiced in the United States. If the owner of the water treatment plant wishes to restrict both the cost and the time required to complete the project to these figures, he must hire well qualified consulting engineers or equipment manufacturers and depend on their experience, judgment, and ingenuity.

Most European and Asian countries generally rely on either equipment manufacturers or contractors offering packaged proprietary processes to design, construct, and start the operation of the water treatment plant. The owner or hired consultant often establishes the plant site and only a few basic design criteria, such as the hydraulic loading rates for the basins and the filters and the finished water quality standard. The successful bidder then furnishes the major plant process facilities and all ancillary facilities required to complete the plant, based on the given criteria.

In the European and Asian method, tenders are invited to bid on the project after only a few design criteria have been established. The tender has the freedom to choose the process configurations, the plant siting, the type of process units, the number and dimensions of the unit process, the mode of the control systems, and the building materials. The hydraulic loadings of the treatment processes are occasionally upgraded from those dictated by the owners by submitting the results of pilot studies; those that meet the criteria established for the finished water quality. The bids are then evaluated by the owner and judged according to the quality and adequacy of the proposals and the estimated design and construction costs.

A few very large public water purveyors in both Europe and Asia are capable of performing all phases of the project, including detailed design, construction management, and plant start-up, using their own staff. However, the detailed predesign studies, such as evaluation of alternative processes and site selection, tend to be less thorough than work performed by consultants in the United States.

In regard to municipal and public owned water treatment plants, the standard practice in the United States is generally quite different from the European, except in the cases of small plants processing less than 10 mgd (0.4 m³/s) of municipal water, any size plant used in the industrial field, or water treatment plants owned by investors such as land developers. In the United States it is customary for the professional engineer to complete 80–85% of the detailed engineering design prior to bidding. In contrast, the European and Asian approach completes the bulk of the detailed design (70–80%) after bidding by the contractor.

In the United States professional engineers are required to prepare a proposal that is complete with a series of drawings and a set of specifications prior to being invited to bid on a project. The drawings themselves may total 250–400 sheets of civil, architectural, structural, mechanical, and electrical design, as well as drawings of the instrumentation and landscaping. The printed specifications may be 1–3 in. thick prior to bidding and must present the various design criteria established by the professional engineers: that is, the hydraulic profile across the plant, the various process trains, and the site plan. This document should also include a detailed design that defines the exact number, configuration, and size of each unit process, the yard piping scheme, the civil works, and the necessary ancillary facilities. Once the bid has been awarded, the engineer will merely review the shop drawings submitted by the manufacturers and manage the general construction work.

In the United States the consulting engineer is defined as a professional who is experienced in applying scientific principles to engineering problems. As a professional the consulting engineer has a duty to the public, in addition to the client. Thus, the engineer should be registered, that is, pass the required state examinations. The American Society of Civil Engineers (ASCE) publishes a guide on how to engage the engineering services of consulting engineers; refer to manual Number 45 (1981).

1.2   SELECTION OF CONSULTANTS

In principle, a consultant does not competitively bid his/her engineering services. If consultants are selected on the basis of price the owner risks hiring incompetent and inadequate services. It is therefore important for the owner to hire the best professional service that is available by paying an appropriate fee; the fee structure is discussed in Section 1.3.

Owners of water treatment plants should follow the normal procedure of selecting a professional consultant as listed in the following order.

1. Issue a "request for qualification" (RFQ) for the project and review the qualifications submitted by each firm.

2. Compile a short list of three to five firms based on their experience, knowledge, and ability to undertake the project.

3. Issue a "request for proposal" (RFP) to each of the selected firms. The RFP should ask for a detailed presentation of their qualifications and their ability to undertake the project. The proposal must include information on the size of the firm, the number of staff members, the availability of qualified personnel to be assigned to the project, and the experience of these engineers in similar lines of work.

4. Select the most qualified firm and a backup firm in case a contract cannot be negotiated with the first choice.

5. Notify the firm of its selection and begin negotiating the fee for the project and a detailed scope of the professional services that are to be rendered.

There are several important points that should be evaluated during the selection of the consultant: the technical qualifications of the firm; the personality and administrative skills of key engineers such as the project manager and project engineer; the existing workload of the firm (i.e., the ability of the firm to absorb the additional workload in relation to its capacity); the experience, reputation, and past accomplishments of the firm in similar lines of work; and the financial stability of the firm.

1.3 FEE STRUCTURE FOR SERVICES

Although most engineers assigned to a project are not involved in negotiating the fee structure of a project, since this is in the realm of the project engineer or his/her superiors, it is still important for all engineers to become familiar with this subject. After all, the design engineers are responsible for both the fiscal and technical aspects of the project.

The fees charged by the firm are based on the acceptable compensation curves or yardsticks and the prevailing standards of the industry. For more detail on this subject refer to the *Manual and Report on Engineering Practice, Number 45*, published by the ASCE, and the *Compilation of Fees*, by the American Consulting Engineers Council.

There are four basic methods by which fees are determined: (1) salary cost times by a multiplier, plus other direct costs; (2) cost plus a fixed fee; (3) fixed lump sum; and (4) a percentage of the construction cost. Most civil engineering contracts are based on the first two methods.

1. Salary Cost Times a Multiplier, Plus Other Direct Costs. This type of contract is commonly used when the scope of work cannot be accurately defined. In this method the reimbursement to the consultant is calculated by multiplying the salary cost, payroll factor, and a multiplier. Salary cost is defined as the direct payroll plus the fringe benefits given to the team members of the project; the benefits total about 130–140% of the actual salary. The multiplier applied to the salary cost is a factor that compensates for the overhead and supplies, a reasonable margin for contingencies, interest on invested capital, insurance, and profit. This multiplier varies with many factors, including geographic area, and is a negotiable item. The multiplier is approximately 2.3–2.5 times the salary cost but is subject to periodic reevaluation. For instance, if the average salary rate of a project team is $25 per hour, the payroll factor is 130% and the multiplier is 2.4. The payment provision in the contract should stipulate that reimbursement to the consultant, for the designer's time, is at $78 per hour:

$$(\$25 \times 1.3) \times 2.4 = \$78 \quad \text{or} \quad \$25 \times 3.12 = \$78$$

Other identifiable costs or direct nonsalary expenses include travel expenses, telephone calls, the subcontractor's charge, printing and binding costs, and a processing fee for executing the work. These expenses must be reimbursed by the client at actual cost plus a service charge of 10–15%.

This particular method is not limited to projects with an undefined scope of work but may also be applied to situations where the scope is fairly well defined. With this type of contract the consultants will stand a good chance of not losing money.

2. Cost Plus a Fixed Fee. This type of contract is used in cases where the scope of the work cannot be readily defined. Yet, it is important to define the scope of the work as completely as possible, for only then will the owner agree to reimburse the consultant for direct costs plus a fee. The reimbursable costs include the technical payroll and actual expenditures that are directly incurred for the project. The fixed fee contains the profit and nonallowable costs and allowable costs. The nonallowable costs include contingencies, interest on invested capital, and the availability of the consulting team. Allowable costs are defined as the direct labor, direct project costs, and indirect costs that are incurred by the labor base.

The fixed fee varies with the size of the project as well as the complexity and scope of the job. It is traditionally calculated as a percentage of the engineering costs and varies from 10 to 25%. Due to the nature of this type of contract, it is important to eliminate any increases in the engineering fee unless the scope of the engineering services is revised.

3. Fixed Lump Sum Fee. This method is frequently used for investigations and studies or design projects with a well defined scope and complexity. The fixed fee is computed by estimating the work-hours required to execute the project and the anticipated cost for rendering the service. For water treatment plant design projects, any experienced professional engineer can estimate the cost for services by analyzing the demands of the project and the total number of required drawings.

The fixed fee may also be calculated as a percentage of the estimated construction cost. However, this particular method is not generally recommended because the figures tend to be inaccurate, especially with respect to plant expansion projects.

The fixed lump sum fee contract has the client pay the firm on a monthly basis during execution of the work, unless otherwise specified. It is therefore important that the contract include a time limit for the professional services and a provision for adjusting the fee, in case the project is delayed by factors that are beyond the control of the engineer.

4. Percentage of Construction Cost. This method is a variation of the fixed lump sum fee. It was very popular in the past when the economy was stable and the available technology was not rapidly changing. Yet, this type of contract is still used for some design projects. This method of assessing fees relies heavily on the engineering reputation of the consulting firm and the customary percentage by the industry; refer to the compensation curves published in ASCE Manual No. 45 (Figure 1.3-1).

This method is not recommended for plant expansion projects because the design work is generally much more costly than for new plants. This is because the existing plant must be extensively evaluated, down to every detail, so that the new parts are compatible with the existing parts. The engineer therefore loses freedom in design and must spend more time on the project.

Note: Final agreement on the fee (by both parties) should always be negotiated after a detailed discussion of the scope of the work and the elements of the engineering costs.

1.4 PROJECT CONTROL

Professional engineers manage a broad range of activities including technical investigation and analysis, environmental and preliminary studies, planning, design, construction, compilation of the operation and maintenance manual, and plant start-up, as well as providing advice and general consultation. Regardless of the type of service, each project must be completed on schedule, within budget, and with a minimal number of errors and omissions—ideally error free. A rigorous control measure is therefore vital for any project.

In order to execute effective control over a project, a solid system of control must be established. However, if the firm already has a standard procedure, all projects should follow this method. An example of a design project control system is as follows:

1. Upon receipt of a signed contract, assign a job number to the project, including the year if possible. This number should be used on all work including calculations, drawings, and correspondence. All costs and charges pertaining to the project should be identified by this number.

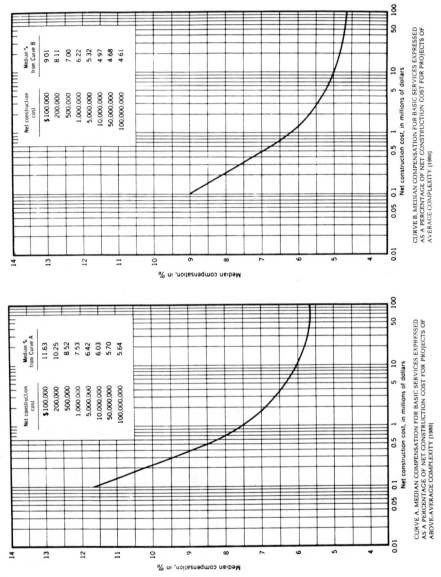

Figure 1.3-1 Compensation curves. (From ASCE Manual No. 45.)

CURVE A. MEDIAN COMPENSATION FOR BASIC SERVICES EXPRESSED
AS A PERCENTAGE OF NET CONSTRUCTION COST FOR PROJECTS OF
ABOVE-AVERAGE COMPLEXITY (1980)

Net construction cost	Median % from Curve A
$100,000	11.63
200,000	10.25
500,000	8.52
1,000,000	7.53
5,000,000	6.42
10,000,000	6.03
50,000,000	5.70
100,000,000	5.64

CURVE B. MEDIAN COMPENSATION FOR BASIC SERVICES EXPRESSED
AS A PERCENTAGE OF NET CONSTRUCTION COST FOR PROJECTS OF
AVERAGE COMPLEXITY (1980)

Net construction cost	Median % from Curve B
$100,000	9.01
200,000	8.11
500,000	7.00
1,000,000	6.22
5,000,000	5.32
10,000,000	4.97
50,000,000	4.68
100,000,000	4.61

TABLE 1.4-1a   Contract Brief

Client Name _____

Job Number _____ Profit Center _____ Date Opened _____

A.	Contract Information

C _____ For contracts already in our database, provide contract number, and proceed to Section B.

Complete remainder of Section A for new contracts which are not in our database

Contract Description _____

Profit Center _____ Proposal Number _____ Contract Amount $_____

Contract Manager Name _____ Employee Number _____

Client Address _____

_____

City _____ State _____ Zip _____

Client Contact: First Name _____ Middle Initial____ Last Name_____

Title _____ Phone Number __(____)_____

Department _____ Division _____

Referred by Client Number _____ Parent Client Number _____

____ Notice to proceed required?_____ Standard Fees & Conditions attached? Form No. _____

B.	Contract Financial Element Information

E _____ For elements already in our database, provide financial element number, and proceed to Section C.

Complete remainder of Section B for new elements which are not in our database

Element Description _____

Project Manager Name _____ Employee Number_____

Profit Center _____ Contract Type _____ Element Amount: $_____

Client Authorization:

In Accordance with: Source of Funds:

____ Our proposal countersigned by Client dated_____   First Source _____

____ Our agreement dated _____   Second Source _____

____ Letter of authorization dated _____   Third Source_____

____ Client's contract No. _____ dated _____

____ Client's P.O. No. _____ dated _____

C.	Job Information

Computer Job Description _____ _____

Job Description _____

Project Engineer Name _____ Employee Number _____

Business Element _____ Project Control Element_____ _____

Job Site Location: City _____ State _____ Zip_____

____ QA/QC Job   Percent of fees _____ %   CCM scheduled: date _____ location_____

Requested CCM members: Name _____ Emp. No.:_____ Name_____ Emp. No.:_____

Name _____ Emp. No.: _____ Name _____ Emp. No.:_____

(attach additional sheet if nessary)

Subcontractors: _____ _____ _____

_____ _____ _____

(attach additional sheet if nessary)

D.	Approvals

Project Manager _____ Date_____ Contract Manager _____ Date _____

Dept. Mgr. or BOM _____ Date_____ Group Manager _____ Date _____

Verified by Billing_____ Date_____ Verified by Corp. Counsel_____ Date_____

CA-12(3/89)

2. Prepare a contract brief, billing summary, and a budget worksheet for the project. Refer to Tables 1.4-1a, 1.4-1b, and 1.4-2 for examples of the forms.

3. Create project files.

4. Prepare a control schedule for the project, including the period of activity, the budget for each discipline of the project team, meeting

TABLE 1.4-1b Billing Summary

E.	Job Upper Limit		
Fees $ _____ _____ Firm _____ Estimated Budgeted Cost $_____ _____ Profit ____ % (Budget Worksheets are not required on Contracts with less than $10,000 in fees)			

F.	Invoicing Information		
Bill on same invoice as job no. _____ _____ Show jobs separately on invoice When to bill: _____ monthly _____ in stages _____ upon completion			

G.	Contract Type (select one)			
	G-1. Salary Cost overhead/mark-up %	G-2. Cost + Fixed Fee overhead/mark-up %	G-3. Hourly Rate mark-up %	G-4. Lump Sum _____ check here if fixed lump sum
Billing Group				Period Amt. $_____
Office Labor	_____ %	_____ %		
Field Labor	_____ %	_____ %		
Overseas Labor	_____ %	_____ %		
Rented Labor	_____ %	_____ %	_____ %	
Outside Services	_____ %	_____ %	_____ %	
Other Charges (ODC)	_____ %	_____ %	_____ %	
	Salary Cost _____ %	_____ Prov. Overhead rate Fixed Fee $_____	____ Current Std.- code _____ ____ Historical Std.- code _____ ____ Other (attach rate sch.)- code _____	

H.	Other Information
Retention _____ % Mileage Rate: _____ JMM Standard or _____ ¢ Interest Rate _____ % After No. of Days _____ Backup (only if required by client) _____ Labor _____ ODC	

I.	Special Instructions
_____ _____ _____	

J.	Billing Contact Information (complete only if different from address on reverse side)
Street Address _____ City _____ State _____ Zip_____ County _____ Country _____ First Name _____ Middle Initial _____ Last Name _____ Title _____ Phone Number (_____) _____	

K.	Invoice Description
Professional Services Rendered during the Period of: Start Date _____ To _____ _____ _____ _____	

CA-12(3/89)

Source: J.M. Montgomery, Consulting Engineers.

dates for coordinating each section of the project, a "final check" date, a date for obtaining the final authorized signature, and a date for printing and binding the proposal. Although several methods are used in scheduling the workforce requirements during the design phase, the simplest and most common are the bar chart, a graphic representation of each discipline of the job plotted against time, the Critical Path Method (CPM), the Program Evaluation and Review Technique

TABLE 1.4-2 Project Budget Worksheet

CLIENT _____ JOB NO. ⬚⬚⬚⬚⬚⬚⬚

DESCRIPTION _____

FEE ESTIMATE (see reverse):										Amount

1 FEE BASIS _____, UPPER LIMIT _____

	FF	= Fixed Fee			SC +	= Salary Cost Plus _____	%
	HR	= Hourly Rate (Schedule) _____)			% C	= Percent of Construction, ASCE Curve _____	
	PR +	= Payroll Cost Plus _____ %			O	= Other (_____)	

2 ESTIMATE OF FEE (show calcs on back) $

COST ESTIMATE:				Hours			Amount		
Item	Labor Class	Actual Salary ($/hr)	Dom.	O/S	Total	Domestic	Overseas	Total
3 Senior Company Officers	011							
4 Principal Professional	012							
5 Supervising Professional	013							
6 Senior Professional	014							
7 Professional	015							
8 Associate Professional	016							
9 Senior Designer	021							
10 Designer	022							
11 Drafter	023							
12 Temporary Labor	098							
13 Supervising Administrator	151							
14 Senior Administrator	152							
15 Administrator	153							
16 Secretarial	154							
17 Typist	155							
18 Data Processor	156							
19 Reproduction Technician	157							
20 Clerk	158							
21 SUBTOTAL (3 to 20) OFFICE DIRECT LABOR*						$	$	$
22 Senior Resident Engineer	031							
23 Resident Engineer-Inspector	032							
24 Senior Surveyor	041							
25 Surveyor	042							
26 SUBTOTAL (22 to 25) FIELD DIRECT LABOR						$	$	$
27 TOTAL DIRECT LABOR (21 and 26)						$	$	$
28 Payroll Cost (% of Total 27)	201							
29 Office Burden (% of Dom. 21)	202							
30 Domestic Field Burden (% Dom. 26)	202							
31 Overseas Burden (% of O/S 27)	202							
32 TOTAL OF 28 to 31	202					$	$	$
33 Overseas Differential	213							
34 Outside Professional & Consulting Services	203							
35 Services Division (see reverse)	204							
36 Laboratory	210							
37 Computer	207							
38 Travel and Subsistence	208							
39 Misc. (phone, sup., post., equip. rental)	205							
40 TOTAL OF 33 to 39						$	$	$
41 TOTAL ESTIMATE OF COST (27 + 32 + 40)						$	$	$

PROFIT:		
42 ESTIMATE OF PROFIT (2 LESS TOTAL 41)		$_____
43 PROFIT AS PERCENT OF FEE (42 ÷ 2)		_____ %

By: _____   _____   Approved: _____   _____
   Project Engineer   Date   Department Engineer   Date

*Use Form AC-66 for detail of Items 3 through 20 by Discipline.   Circulated by Contracts Administration _____

(PERT), and the Integrated Budget and Schedule Monitoring technique (IBSM). Refer to Figure 1.4-1.

5. Investigate all requirements established by local, state, and federal agencies.

6. Organize a project team.

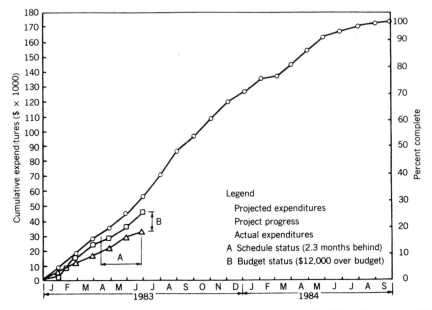

Figure 1.4-1 Integrated Budget and Schedule Monitoring (IBSM). (Courtesy of J. M. Montgomery Consulting Engineers.)

7. Arrange for all necessary outside services such as soil analysis and site survey.
8. Select members for the technical advisory committee and the value engineering team.
9. Determine the number of technical advisory meetings that will be held during the project. The first meeting should take place after approximately 5–10% of the project is completed.
10. Prepare a memorandum after each meeting.
11. Encourage active input from the client and keep the client informed on the progress of the project by holding periodic meetings.
12. Review the cost of the project at the end of each month.
13. Prepare construction specifications based on the Standard Specifications and edit them to make the job more specific.
14. Have the completed drawings and specifications checked by the project engineer and an independent checker. Each sheet should be red/yellow lined.
15. Edit the bid documents and submit the preliminary drawings and specifications to the client for review.
16. Arrange an estimate for the construction costs.
17. Schedule the production of construction documents.

18. Obtain signatures from the company officer and the client on the final tracings.
19. Present the final drawings and specifications to the client and the appropriate governmental agency (local, state, or federal) as required.
20. Arrange for the advertising of bids and bid openings.

It is imperative that the calculations, drawings, and specifications be checked prior to finalizing the bid documents. Failure to do so may result in errors that could potentially have disastrous effects on plant construction and operation or even lead to a lawsuit. Since many specifications are similar among various projects, in both outline and technical provisions, it is strongly advised that firms standardize their specifications as a means of project control.

Lastly, it is essential to schedule the total workload of a firm, as well as individual projects. Without proper scheduling, a firm will operate inefficiently and will have problems in effectively controlling ongoing projects.

CHAPTER 2

Preliminary Studies

2.1 FEASIBILITY STUDY

The feasibility study is the starting point for all water treatment design projects. It is often conducted as part of the master development plan for the region or regions. Since this subject is not the focal point of this book, it is discussed only briefly.

2.1.1 Planning Period

A span of 10–20 years is considered to be a reasonable period for planning a water supply system, including the capacity of the water treatment plant. It is important to note that some regions, such as southern California and the Bay area of northern California, have already exceeded the predictions for water demand made in the early 1980s. Yet, some cities in the United States are well below the predicted levels due to declining industrial activities and severe weather conditions. However, a span of 15 years is generally agreed upon as an acceptable planning period for an immediate system design and construction.

2.1.2 Water Supply Areas

When determining the areas to be served by the water supply system, engineers should limit their evaluation to the areas included in the planning system: those considered to receive water during the design period. However, adjacent areas and the potential amalgamation of neighboring districts should also be included in the overall design.

2.1.3 Future Population

There are several methods by which to predict the growth or decline of the future population. Since most college textbooks cover this subject in great detail, it is not discussed here.

2.1.4 Maximum Daily Water Demand

The first task in the feasibility study is the estimation of the maximum daily per capita water demand. For a new system, this estimation is generally derived from data acquired from communities sharing similar backgrounds, characteristics, and trends in development. In the case of plant expansion, the best estimation is provided by the trends of the existing water supply system. Once the maximum daily demand per capita is determined, which is the basis for evaluating the water treatment capacity, the maximum daily demand may be calculated as the product of the maximum daily per capita demand and the estimated population to be served within the planning period. The average annual rate of use in gallons per capita per day (gpcd) of 100 (80 to 130) is common in the United States and the maximum daily demand is usually 150% of the average annual rate.

2.1.5 Evaluation and Selection of the Water Source

The source of the raw water may be a river, lake, artificial reservoir, ground-water, and, in some cases, reclaimed sewage or seawater. The evaluation and selection of the proper water source should be based on the following issues: (1) quantity of required water, (2) quality of the raw water, (3) climatic condition (i.e., icing), (4) potential difficulties in constructing the intake, (5) operator safety, (6) providing minimal operations and maintenance costs for the treatment plant, (7) possibility of future contamination of the water source, and (8) the ease of enlarging the intake if required at a future date.

For small communities, underground water, including water collected by an infiltration gallery, should seriously be considered as a water source because it can potentially save a significant amount of capital as well as operations and maintenance costs. The quality of the underground water is generally superior to that of surface water and there are no problems associated with icing during the winter months—an added benefit for cold weather regions.

2.1.6 Size of the Water Treatment Plant

The capacity of a water treatment plant is determined by the maximum daily demand placed on the system. However, the future water demand of the area being served should also be taken into account when calculating the size of the plant and the site area included. Additional issues that affect plant size are the reliability of the water supply and the cost effectiveness of supplying water from one large plant versus two or even three medium sized plants at different locations and different elevations. As a rule of thumb, the required available site area for a conventional plant may be estimated by the following formula:

$$A \geq Q^{0.6}$$

where A = area in acres,
Q = ultimate plant capacity in mgd.

2.1.7 Treatment Plant Site

Evaluation of the treatment plant site is primarily based on the distance from the intake, the layout of the treatment process units, the environmental impact of the treatment plant, and the method of water distribution (gravity or by pumping). Moreover, the following items must be included in the evaluation of the treatment plant site: (1) geographical location, (2) information obtained from the geological study, (3) availability of electric power and utilities, (4) accessibility to major highways, (5) history of flooding or the presence of earthquake faults, (6) construction costs, (7) site maintenance costs, (8) operator safety and the safety of neighboring homes, and (9) provisions for future plant expansion.

2.1.8 Financing

The preliminary site evaluation must present various means by which to fund the project. Several methods are used to obtain necessary funding for constructing water treatment plants: revenue bonds, general obligation bonds, special assessment bonds, state and federal aid funds, operating revenues, contributions by customers, and private funds. The most common method is the issuance of revenue bonds since water treatment plants are revenue-producing facilities.

The first step in obtaining financing requires the engineer to provide an estimate of the plant construction costs. Then a team of consultants—engineering, financial, and legal—must study the proposal. It is imperative that the team include a financial specialist and legal expert so that the utility can properly decide if it can issue revenue bonds or if the community should be asked to issue general obligation bonds: a bond attorney is required to draw up the proceedings and to issue the approving opinion, without which the bonds cannot be marketed. Finally, the preliminary engineering, financial, and legal studies must be submitted to the city council and should disclose whether the project can be accomplished at a reasonable cost through acceptable financing methods within the existing statutory limitations. These studies often save the governing board of the utility from unnecessary embarrassment and keeps the project from becoming a political football.

In recent years, "privatization," which involves private financing and turnkey type projects, has emerged. In the public water–utility industry the term turnkey construction refers to having a single entity design, construct, and deliver a properly functioning facility that is ready for service. This type of contract has the advantages of maintaining engineering continuity and quality control.

Privatization of the water treatment industry has allowed organizations from the public sector to enter into a market that has traditionally been dominated by public funding. Private companies may now finance, own, construct, and operate the facilities. Since these facilities are run as a business enterprise, their design, construction, and operation are generally more cost efficient than their public counterparts. However, local governments have the option of purchasing these types of facility at fair market value at the end of the lease terms. Despite the tax reform bill passed by the House of Representatives in December 1986—which removed a large part of the tax exemption benefits for the investor and owners—the tax exempt interest of this income, for the investor or lessor, should still attract private investors.

BIBLIOGRAPHY

Bray, F. A., "Municipal Utility Bond Financing," *J. AWWA*, 62(8):468(August 1970).

Doctor, R. D., "Private Sector Financing for Water System," *J. AWWA*, 78(2):47(February 1986).

Earl, T. E., and Greenstein, S. A., "Economic Considerations in Facility Planning," *J. AWWA*, 62(11):705(November 1970).

Joint Discussion, "Pros and Cons of Turn-key Construction," *J. AWWA*, 65(12):766(December 1973).

Seader, D., "Privatization: An Emerging Management and Financing Trend," *Water/ Engineering & Management*, p. 44(March 1984).

2.2 BENCH SCALE AND PILOT STUDIES

The predesign efforts must include bench scale and pilot studies whenever possible. These studies allow the engineer to obtain valuable basic design data and to either optimize or confirm the proposed raw water and filter waste wash treatment processes. The bench scale study may be conducted in a short period of time and with a limited budget. However, a pilot study requires a minimum of 6 months to 1 year in order for it to yield meaningful and reliable data. Consequently, these studies are quite expensive.

2.2.1 Bench Scale Studies

A proper bench scale study addresses the following objectives: optimization of chemical coagulants, the chemical application sequence, confirmation of the proper mixing conditions for flocculation, estimation of the hydraulic surface loading for the sedimentation process through the measurement of floc settling velocities, the total level of trihalomethane that the water can potentially produce, and the control of taste- and odor-producing compounds through the use of oxidants or activated carbon. The Phipps and Bird jar tester is the instrument that is primarily employed for these studies and a minimum of 200 work-hours are generally required to complete the bench scale studies, which includes the time required to write the report.

The coagulant optimization study involves selecting the most effective type of coagulant and the appropriate dosage that will produce good settled water quality, as well as optimal conditions for the filtration process. Engineers should realize that a conventional water treatment process is comprised of many unit processes: coagulation, flocculation, sedimentation, and filtration, and that these subsystems must be interrelated and compatible if the sequence is to produce good quality water. It is recommended that the effectiveness of coagulants, such as alum, ferric salts, and cationic polymers, be evaluated both individually and as a combination of alum and polymer. The recommended jar testing procedure is presented in Section 3.2.6 (Example 1, Filtration Process) and Appendix 7.

The suggested coagulant dosages for the first raw water jar test run are as follows:

Alum or ferric chloride dosage (mg/L) 3, 6, 12, 20, 40, 60

Cationic polymer alone (mg/L) 0.25, 0.5, 1, 2, 4, 8

$$\frac{\text{Alum}}{\text{Cationic polymer}} \text{ (together)} \quad \frac{\text{(mg/L)} \quad 6 \quad 12 \quad 20 \quad 6 \quad 12 \quad 20}{\text{(mg/L)} \quad 0.25 \quad 0.25 \quad 0.25 \quad 0.5 \quad 0.5 \quad 0.5}$$

See Table A7-1 of Appendix 7 for sample jar test results.

The flocculant optimization test should be designed to determine the most effective type of polymer or additives (such as bentonite clay), the optimum chemical dosage, and the proper application sequence for both coagulant and chemical flocculation aids. The most effective flocculant aids are usually anionic polymers or nonionic polymers because they can greatly increase both floc size and settling velocity. However, the timing of their application is very important; they should be fed when pinpoint floc have formed, which is approximately 5 min after alum flocculation begins. Table 3.2.3-1 clearly illustrates this issue. The sequence of chemical application is a critical consideration for other water treatment chemicals such as lime and alum, chlorine and polymer, and chlorine and powdered activated carbon (PAC).

The mixing conditions for the flocculation process may be optimized by a jar test and data obtained in this manner are useful for the plant design. The jar test allows the engineer to determine the optimum level of the mixing energy input (G-value), the mixing time, and the energy input pathway (tapered mixing requirement). Although most existing conventional treatment plants use an average G-value of about 35 with a mixing time of 30 min, a wide range of numbers should be tested at optimum coagulant and flocculant dosages in order to maximize the conditions. Tapered mixing—the use of high-intensity mixing at the beginning of flocculation, followed by a gradual reduction toward the end—helps promote the formation of large floc and should therefore be studied as well.

Four major parameters that significantly influence the optimum mixing condition are water temperature, raw water quality, type of coagulant, and coagulant dosage. Thus, it is recommended that bench scale tests be performed during the winter, summer, and either spring or autumn to better comprehend the need to tailor coagulant dosages to the changing raw water characteristics and to estimate the seasonal sludge production rate.

The fourth objective of the bench scale studies—floc settling velocities— must be evaluated at optimal levels of coagulant and mixing conditions. The test should be conducted during both cold and hot weather months for reasons previously stated.

It is relatively simple to produce a settling velocity distribution curve from a jar test. The turbidity of the settled water is measured for samples drawn from 10 cm below the surface of a 2 liter beaker at 1, 2, 4, 8, and 16 min after the jar test has been initiated. The samples collected at these times represent the respective settling velocities of 10, 5, 2.5, 1.25, and 0.625 cm/ min. A settling velocity of 4 cm/min is equivalent to a surface loading of 1 gpm/ft^2. The measurement of floc settling velocities is discussed in detail by T. R. Camp and H. E. Hudson. Engineers must note that the bench scale test results may not be relevant if the turbidity is artificially produced or if the water is not fresh, more than 1–2 days old, since the test results do not accurately portray actual water treatment conditions.

Refer to Section 7.4 for a discussion of the total potential THM formation and Section 7.3 for taste and odor control by oxidants or activated carbon.

2.2.2 Pilot Plant Studies

In recent years, engineers have been forced to evaluate nonconventional treatment processes due to new EPA drinking water regulations, public concern over the safety of the drinking water, the ever increasing costs of plant construction, and the emergence of new water treatment technologies. Consequently, the use of pilot plant studies, to evaluate the proposed treatment processes, is becoming an increasingly necessary part of the design process.

Although pilot studies can potentially result in significant cost savings, in plant construction and operations, the cost of the pilot studies themselves is rather large: a good pilot study may cost anywhere from $100,000 to $1,000,000. Thus, if meaningful results are to be obtained from a pilot study, the most appropriate type and best manufactured equipment must be selected. Moreover, the pilot plant must be operated by highly qualified personnel for a period of at least 6–12 months.

Pilot plant studies may be necessary to achieve some or all of the following objectives:

1. Obtain permits from regulatory agencies for nonconventional processes.
2. Evaluate the practicability of a new treatment process.
3. Compare the effectiveness of alternative processes.
4. Obtain a guide for process design criteria and operating costs.
5. Improve existing processes.
6. Investigate the cause of problems.
7. Confirm the effectiveness of the proposed treatment process.
8. Discover unforeseen problems resulting from the implementation of the proposed treatment process.

Although the pilot plant studies are a valuable guide in establishing both the process design criteria and the operational parameters of the plant, they also have their limitations. The major problems are (1) difficulty in testing the raw water on a year-round basis; (2) use of an improper type of clay when simulating abnormal raw water conditions (high turbidity); (3) using raw water that has been stored for over 1–2 days; (4) differences in operational conditions, including the technical and operational knowledge of the plant operators—that is, pilot plant versus actual treatment plant; (5) problems encountered in scaling-up; (6) failure to foresee long-term effects of the new process; and (7) arriving at conclusions biased by personal expectations.

Pilot studies may be classified into two basic categories: study of hydraulic characteristics and study of unit process performance. The first includes the study of flow characteristics in clearwells or sedimentation tanks and the second includes pilot studies on corrosion and on the processes of filtration, flocculation, GAC adsorption, and ozonation.

Once a pilot study is planned, the engineer must determine the variables to be studied and what their characteristics should be. Since flocculation, sedimentation, and preozonation are filtration pretreatment processes, it is very important to select a combination of these unit processes that will yield both maximum efficiency and optimum economic benefits for the overall treatment system.

From a practical standpoint, a pilot study usually cannot examine all the relevant factors due to time and budgetary constraints. Thus, it is necessary to reduce the number of variables by selecting only a few very important items. However, engineers must first establish the following issues during the initial stages of the study: (1) the purpose of the study, (2) the duration and cost of the experiment, (3) the availability of equipment and technical staff, and (4) the important variables of the study.

A basic pilot plant has a coagulation, flocculation, clarification, filtration, and final disinfection process. The study should first optimize the dosage of the coagulant and other required chemicals: that is, determine the dosage that yields the best filter performance for the selected filter bed configuration. The physical variables of the pretreatment process, such as the mixing intensity, detention time, and hydraulic loading, should then be evaluated using the optimum coagulant dosage. The optimum dosage is based on results obtained from jar test studies. However, the use of ordinary pilot plants is virtually useless in predicting the settled water quality or in optimizing horizontal flow sedimentation tanks. This subject is discussed later.

The piloting filter, including the GAC adsorption process, yields meaningful data and is relatively easy to monitor. Filter performance may be evaluated effectively through the use of small diameter columns (2–4 in.). However, if filter washing is conducted for study purposes, large sized columns should be used to minimize the "side wall" effect. A rule of thumb is to use a column diameter that is 1000 times the average size of the granular filter medium. For example, if the average size of the filter sand is 0.8 mm, the size of the filter column should be 80 cm (30 in.) in diameter. In the case of filtration studies, the size of the filter column should be 100 times the effective size of the filter medium.

The pilot filter study should address the following issues: optimization of the filter bed and filter rate, evaluation of the filter washing conditions, and control of initial and terminal turbidity breakthrough. In some cases, the declining rate versus constant rate filtration may be included as a study item. To achieve these objectives, a minimum of three filter columns is absolutely necessary, and five to six columns are preferred, with one of the units as a control column.

Media optimization involves the evaluation of different types of medium materials, determination of the size to depth relationship, and the comparison of the effectiveness of monomedium versus multimedia. Basic information in regard to size, depth, and material of filter beds is presented in Table 3.2.6-1.

One of the most popular filter beds in the last half of the 20th century is a dual media filter bed with a depth of 30 in. (0.75 m). The bed is composed of a layer of anthracite coal which is 20 in. (0.5 m) thick (1 mm effective size) and 10 in. (0.25 m) of filter sand (0.5 mm effective size). This bed should be used as a control for comparing the performance of other types of bed.

Engineers are advised to use a constant-rate filtration process for the pilot filter study. Filtration rates of 3, 6, and 9 gpm/ft^2 (7.5, 15, and 22.5 m/h) should be evaluated since a filtration rate of 6 gpm/ft^2 (15 m/h) is generally considered to be an acceptable filtration rate for high-rate filtration using dual media beds. Figure 2.2.2-1 is a schematic diagram of the recommended standard pilot plant configuration.

The quality of the filter effluent obtained from pilot filters is often significantly below that produced by actual operational filters. Under identical conditions, the turbidity of the pilot filter effluent is often twice as high as the actual filter; 0.2 NTU for pilot plant filters versus 0.1 NTU for actual filters. Engineers should also be aware that the initial turbidity breakthrough,

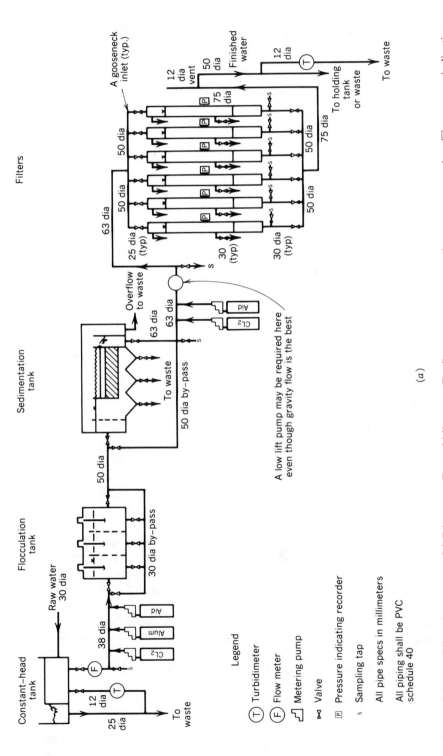

Figure 2.2.2-1 Pilot plant set-up. (a) Diagram. T, turbidimeter; F, flow meter; , metering pump; , valve; □, pressure indicating recorder; S, sampling tap. All pipe specs are in millimeters; all piping shall be PVC Schedule 40. (b) Filter columns, 800 mm in diameter, for both backwash and filtration performance evaluation study. (c) Filter columns, 75 mm in diameter, for filtration performance study only. (From S. Kawamura, *AWWA Proceedings* No. 20164, May 1982.)

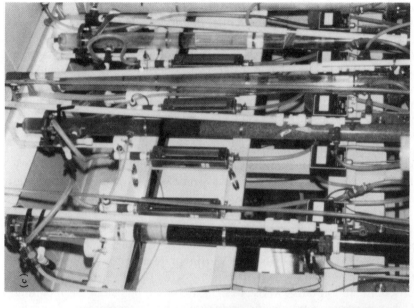

(c)

(b)

Figure 2.2.2-1 (Continued)

at the beginning of the filtration cycle, is usually quite severe for pilot plants. This may be attributed to the degree of ripening or maturation of the filter bed. Unlike operational treatment plants, the filters of the pilot plant are always thoroughly washed. Consequently, the filter beds are seldom allowed to become well ripened. The reasons for the frequent washings are (1) minimizing the conditions present in the previous pilot filter tests and (2) lack of operational knowledge by the pilot filter operators who are usually engineers or scientists. These individuals generally wash the filter beds until the waste wash water becomes clear, in contrast to the practice of actual plant operators, who allow a turbidity of 5–10 NTU to remain in the filter waste wash.

As previously mentioned, pilot tests for horizontal sedimentation tanks have limitations. If the purpose of the piloting is to study the hydraulic characteristics of the tank, a hydraulic scale model that applies Froude's similarity law should be adopted. However, if the object of the study is to evaluate the ability of the tank to remove turbidity or other constituents, it is difficult to scale down the horizontal sedimentation tanks due to problems associated with adjusting the physical characteristics of the floc, even if the hydraulic flow characteristics have properly been established. Consequently, traditional design criteria, such as the hydraulic overflow rate and detention time, are used to size the pilot units based on a manageable (small) flow rate.

Pilot sedimentation tanks whose designs are based on actual basin parameters always have geometrically distorted shapes, for example, very deep and short. Furthermore, unlike their prototypes, these basins fail to produce effluent of acceptable quality and are incapable of producing a good sludge deposition profile. The way to produce an acceptable quality effluent is by improving the floc settleability through the application of an appropriate type of polymer (as a flocculant aid) in conjunction with alum, in addition to installing a high-rate settler module, such as a tube settler, in the pilot tank. The addition of the high-rate settler module provides an overflow rate of approximately 1 to 2 gpm/ft² (2.5 to 5.0 m/h) for the surface area of the tank covered by the settler module.

Based on data obtained from numerous pilot scale upflow sludge blanket clarifiers, it may be stated that the overflow rate of the prototype pilot scale unit may be used, to an acceptable degree, to model the upflow sludge blanket clarifiers and the high-rate settler modules.

The relationship between the upward flow velocity and the floc settling velocity is the major clarification mechanism for both upflow reactor clarifiers and the high-rate settlers. This simple relationship enables the pilot study to be possible, provided that there is no distinct flow short-circuiting within the unit.

Hydraulic scale model studies are not frequently conducted even for large scale projects because they are always tailored for specific projects and therefore cannot easily be used for other projects, making them very costly. However, the hydraulic scale model study is the only positive means of studying the flow characteristics of a sedimentation tank, including the selection of an

TABLE 2.2.2-1   Relationships Between Hydraulic Scale Model and Prototype

Scale	Relation[a]	Example (1/25 scale model)
Length	$1{:}\alpha$	$1{:}\frac{1}{25}$
Area	$1{:}\alpha^2$	$1{:}\frac{1}{625}$
Volume	$1{:}\alpha^3$	$1{:}\frac{1}{15,625}$
Time	$1{:}\alpha^{0.5}$	$1{:}\frac{1}{5}$
Velocity	$1{:}\alpha^{0.5}$	$1{:}\frac{1}{5}$
Discharge	$1{:}\alpha^{2.5}$	$1{:}3125$
Acceleration	$1{:}\alpha^0 = 1$	$1{:}1$
Force	$1{:}\alpha^3$	$1{:}\frac{1}{15,625}$
Work	$1{:}\alpha^4$	$1{:}\frac{1}{390,625}$

[a]α is the size reduction ratio for the hydraulic scale model.

effective flow diffuser and baffle walls. A few academic and research institutes do conduct hydraulic scale model studies but purely for academic interests.

Froude's similarity law should be applied to all scale model studies:

$$\mathrm{Fr} = \frac{v^2}{Lg}$$

where v = characteristic velocity of the system,
$\quad L$ = characteristic linear dimension,
$\quad g$ = acceleration of gravity.

Froude's law expresses the condition for similarity between the forces of gravity and inertia. In other words, when the Froude number is the same for both the model and the prototype, the gravity force to inertial force ratios are the same; the paths of flow are similar. The fundamental relationships for models of structure, based on Froude's law, are presented in Table 2.2.2-1.

BIBLIOGRAPHY

AWWA, *Design of Pilot-Plant Studies*, AWWA Seminar Proceedings No. 20164 (May 1982).

Camp, T. R., "Sedimentation and the Design of Settling Tanks," *ASCE Trans.*, 3:895 (1946).

Hickox, G. H., "Hydraulic Models," in *Handbook of Applied Hydraulics*, 2nd ed., C. V. Davic, Editor-in-chief McGraw-Hill, New York, 1952.

Hudson, H. E. Jr., *Water Clarification Processes*, Van Nostrand Reinhold, New York, 1981.

Kawamura, S., "Hydraulic Scale-Model Simulation of the Sedimentation Process," *J. AWWA* 73(7):417 (July 1981).

2.3 PRELIMINARY ENGINEERING STUDY

Once the owner has decided to proceed with the construction phase of a project based on the feasibility study, a preliminary engineering study, including planning, must be initiated. This study should address the following issues:

1. Quality and treatability of the raw water.
2. Objectives for the finished water quality.
3. Additional goals and objectives.
4. Restrictions and constraints on plant design.
5. Alternative treatment processes.
6. Preliminary design criteria for treatment unit processes.
7. Hydraulic grade across the plant.
8. Geotechnical considerations.
9. Structural design conditions and criteria.
10. Plant waste handling and disposal.
11. Various schemes for instrumentation and control.
12. Preliminary cost estimates.
13. Recommended treatment processes.
14. Plant layout and architectural design.
15. Management planning of the design phase.
16. Environmental impact statement (EIS) for the new plant.

The preliminary study must conclude with concrete recommendations with respect to the proper treatment process train, hydraulic loading values for each unit process, and all major equipment and appurtenant facilities. At this stage of the project, proper procedure dictates that each of the recommendations be discussed with the appropriate regulatory agencies to solicit their opinions.

2.3.1 Quality and Treatability of the Raw Water

Data on the surface water quality, taken over a sufficient period of time (5–10 years), should be both reviewed and evaluated in order to assess the physical, chemical, microbiological, and radiological characteristics of the raw water. A risk assessment must also be made in regard to the possible contamination of the water supply by chemical spills or radioactive wastes. Moreover, the degree of present and future land development in the watershed must be studied.

If groundwater is selected as the source of the process water, the same considerations associated with surface water also apply. Groundwater as a raw water source necessitates additional studies, such as the geological conditions, water table, the drawdown of the water table as the result of pumping, problems associated with seawater intrusion, and the potential leaching of industrial wastes, domestic wastes, agricultural chemicals, and fertilizers into the groundwater.

For the reasons just mentioned, data analysis is a very important aspect of evaluating the quality of a water source. Although arithmetic averages are often presented in many instances, a statistical presentation is generally more meaningful. The statistical presentation may be constructed by plotting data on an arithmetic or a probability scale. The two most important aspects of data analysis are that it is a means of describing the central tendency of the data and it allows engineers to determine the degree of variability in the data. Those components in the raw water whose maximum concentration levels are limited by the Drinking Water Standards must be carefully analyzed and evaluated.

The treatability of the raw water may be evaluated through the use of bench scale tests and a pilot study. These issues may be found in Section 2.1. If there is an existing water treatment plant in the vicinity of the proposed plant site, the design engineers should consult the operational data of the existing plant because it will provide valuable information on the treatability of the raw water.

The 1986 Amendment to the Safe Drinking Water Act (SDWA-PL-99-339) required the EPA to promulgate a national primary drinking water regulation (NPDWR) by December 19, 1987. This amendment specified the criteria by which filtration was ordered as a standard treatment technique for all public water systems supplied by any source of surface water.

2.3.2 Objectives for the Finished Water Quality

The objective of a public water supply system is to provide safe and aesthetically appealing water to consumers without interruption and at a reasonable cost. The required quality is measured against the standards set by state and federal governments; both regulate drinking water quality. Since only the national regulations are discussed in this section, engineers are strongly advised to contact local regulatory agencies for further restrictions.

In 1974 the EPA established the National Interim Primary Drinking Water Regulations (NIPDWR) and thereby set the maximum contaminant levels (MCLs) for a variety of inorganic and organic chemicals, as well as physical, microbiological, and radioactive contaminants. These primary standards have been in effect since June 1977 and are designed to protect the public health. Compliance with these standards is therefore mandatory. The National Interim Secondary Drinking Water Regulations generally relate to the aesthetic quality of a water supply and thus are recommended goals.

As previously mentioned, the EPA also issued the 1986 Amendment and the National Primary Drinking Water Regulation (NPDWR). Furthermore, the EPA plans to increase the MCLs from the present rate of 49 contaminants (late 1980s) to almost 190 by the year 2000.

A few of the more significant regulations are as follows:

1. Requirements for filtered water turbidity. The filtered water must have less than 0.5 NTU for 95% of the time.
2. Disinfection requirements. A minimum of 99.9% of *Giardia lamblia* cysts must be removed and/or inactivated and at least 99.99% (4 logs) of enteric viruses must be removed and/or inactivated. The EPA proposes 100% removal of total coliform (1988) and a plate count of less than 10 heterotrophic bacteria per milliliter of potable water for finished water (1989).
3. Proposed restrictions on MCLs for disinfection by-products (DBPs), including chloroform, THMs, haloacetic acids, haloketones, haloacetonitriles, and aldehydes.
4. Proposed restrictions on MCLs for volatile organic compounds (VOCs).
5. Proposed MCLs for synthetic organic compounds (SOCs) and inorganic compounds (IOCs) such as pesticides, PCBs, water treatment chemicals (acrylamide, epichlorohydria, and styrene), as well as nitrate, nitrite, and asbestos.
6. Proposed MCLs for corrosion by-products such as lead (5 µg/L) and copper (1.3 mg/L) and a control requirement for the pH of potable water.

The drinking water criteria act as the basis for the design criteria of all water treatment processes. These criteria, however, will change with time. With the knowledge that available raw water sources could deteriorate further, new water treatment plants must be designed in a manner that will allow for flexibility in the treatment processes, a comfortable cushion for potential increases in the available headloss across the plant, and adequate space to accommodate the addition of new processes and/or future plant expansion.

The International Standards for Drinking Water, as set by the World Health Organization (WHO) in 1980, are widely accepted by many developing countries as their drinking water standard. Tables 2.3.2-1a, 2.3.2-1b, and 2.3.2-2 list the MCLs set by the National Primary Drinking Water Regulations in November 1985 and some of the more updated proposed MCLs, for SOCs, IOCs, and SMCLs (EPA, 1988).

2.3.3 Additional Goals and Objectives

Beyond the finished water quality objectives, the goals of the owner, for the water treatment plant, must be identified and established. The goals of the overall treatment plant should state what the plant is to accomplish and how these goals affect the water distribution systems and the environment. Once

TABLE 2.3.2-1a   Proposed/Recommended Primary Standards (1985)

Proposed MCLs for Volatile Organic Chemicals

Compound	Proposed MCL (mg/L)
Trichloroethylene	0.005
Carbon tetrachloride	0.005
Vinyl chloride	0.001
1,2-Dichloroethane	0.005
Benzene	0.005
1,1-Dichloroethylene	0.007
1,1,1-Trichloroethane	0.200
p-Dichlorobenzene	0.750

Proposed RMCLs for Microbiological Contaminants

Contaminant	Proposed RMCL
Giardia	0
Viruses	0
Total coliforms	0
Turbidity	0.5 ntu

Proposed RMCLs for Synthetic Organic Chemicals

Synthetic Organic Chemical	RMCL (mg/L)
Acrylamide	0
Alachlor	0
Aldicarb, aldicarb sulfoxide, and aldicarb sulfone	0.009
Carbofuran	0.036
Chlordane	0
cis-1,2-Dichloroethylene	0.07
DBCP	0
1,2-Dichloropropane	0.006
o-Dichlorobenzene	0.62
2,4-D	0.07
EDB	0
Epichlorohydrin	0
Ethylbenzene	0.68
Heptachlor	0
Heptachlor epoxide	0
Lindane	0.0002
Methoxychlor	0.34
Monochlorobenzene	0.06
PCBs	0
Pentachlorophenol	0.22
Styrene	0.14
Toluene	2.0
2,4,5-TP	0.052
Toxaphene	0
trans-1,2-Dichloroethylene	0.07
Xylene	0.44

TABLE 2.3.2-1a (Continued)

Proposed RMCLs for Inorganic Chemicals

Contaminant	Proposed RMCL (mg/L)
Arsenic	0.050
Barium	1.5
Cadmium	0.005
Chromium	0.12
Copper	1.3
Lead	0.020
Mercury	0.003
Nitrate	10
Nitrite	1
Selenium	0.045
Asbestos	7.1 million long fibers/L

this is established, the required function of each unit process (of the treatment process train) must be identified and the objectives of each of these units should be defined.

Design engineers have traditionally considered each unit process as an individual item and have optimized each unit separately on the premise that this would result in the most efficient overall plant design. However, such "suboptimization" usually does not result in the optimization of the total plant design. Additionally, objectives such as optimization of initial costs, life cycle costs, and construction time should be evaluated in the preliminary engineering study.

2.3.4 Restriction and Constraints on Plant Design

Project engineers are rarely, if at all, given complete freedom in their designs. Thus, the restrictions and constraints must clearly be defined at the beginning of the design phase. Restrictions may be due to economic, physical, chemical, temporal, climatic, geological, sociological, legal, or aesthetic considerations and may be imposed by local, state, or federal agencies.

Constraints are defined as restrictions placed on the acceptable values of design variables—properties of the system which are controlled by the designers. Examples of constraints are building codes, zoning laws, and OSHA regulations and standards. For instance, a 33 in. (84 cm) diameter steel pipe is not standard size, nor is a 60 horsepower (45 kW) electric motor a standard item. These restrictions and constraints may fix the component properties or establish a range in which they must fall. Although these restrictions are generally made for reasons of standardization and safety, serious design problems often arise when designing treatment plants for developing countries

TABLE 2.3.2-1b Proposed/Recommended Primary Standards (1988–1989)

MCLGs/MCLs TO BE PROMULGATED by March 1989

Organics

SOC	NIPDWR (mg/l)	MCLG (mg/l)	MCL (mg/l)
acrylamide	—	0	treatment tech.
alachlor	—	0	0.002
aldicarb	—	0.01	0.01
aldicarb sulfone+	—	0.04	0.04
aldicarb sulfoxide	—	0.01	0.01
atrazine	—	0.002	0.002
carbofuran	—	0.04	0.04
chlordane	—	0	0.002
chlorobenzene	—	0.1	0.1
2,4-D	0.1	0.07	0.07
dibromochloropropane (DBCP)	—	0	0.0002
o-dichlorobenzene (o-DCB)	—	0.6	0.6
cis-1,2-dichloroethylene	—	0.07	0.07
trans-1,2-dichloroethylene	—	0.07	0.07
1,2-dichloropropane	—	0	0.005
epichlorohydrin	—	0	treatment tech.
ethylbenzene	—	0.7	0.7
ethylene dibromide (EDB)	—	0	0.00005
heptachlor	—	0	0.0004
heptachlor epoxide	—	0	0.0002
lindane	0.004	0.0002	0.0002
methoxychlor	0.1	0.4	0.4
PCBs	—	0	0.0005
pentachlorophenol	—	0.2	0.2
styrene	—	0	0.005
tetrachloroethylene	—	0	0.005
toluene	—	2	2
toxaphene	0.005	0	0.005
2,4,5-TP	0.01	0.05	0.05
xylene*	—	10	10

* MCLG/MCL for sum of o-, m-, and p-xylene isomers.
+ MCLG/MCL for aldicarb sulfone in the presence of aldicarb sulfone or aldicarb sulfoxide = 0.01 mg/l.

Inorganics

IOC	NIPDWR (mg/l)	MCLG (mg/l)	MCL (mg/l)
arsenic	0.05	0	0.03
asbestos*	—	7 MFL	7 MFL
barium	1.0	5	5
cadmium	0.010	0.005	0.005
chromium	0.05	0.1	0.1
mercury	—	0.002	0.002
nitrate (as N)**	10	10.0	10.0
nitrite (as N)	—	1.0	1.0
selenium	0.01	0.05	0.05

NOTES:

* MCLG/MCL applies to fibers > 10 μm in length. MFL - million fibers per liter.
** MCLG/MCL for total nitrate and nitrite = 10 mg/l (as N).

PROPOSED SMCLs

The EPA also proposed secondary maximum contaminant levels (SMCLs) for 11 contaminants. An SMCL is defined in § 1401(2) of the SDWA as a regulation which applies to public water systems and which specifies the maximum level which is required to protect the public welfare. An SMCL is not an enforceable standard. It may be set for a contaminant which adversely affects the taste or odor or other "appearance" of the water.

SOC/IOC	SMCL mg/l	SOC/IOC	SMCL mg/l
chlorobenzene	0.1	toluene	0.04
o-dichlorobenzene	0.01	xylene	0.02
p-dichlorobenzene	0.005	pentachlorophenol	0.03
1,2-dichloropropane	0.005	aluminum	0.05
ethylbenzene	0.03	silver	0.09
styrene	0.01		

June 1, 1988

TABLE 2.3.2-2 EPA Proposed Secondary Drinking Water Regulations with Optional Enforcement by States (1979)

Contaminant	Maximum Contaminant Level
Chloride	250 mg/L
Color	15 color units
Copper	1 mg/L
Corrosivity	Noncorrosive
Foaming agents	0.5 mg/L
Hydrogen sulfide	0.05 mg/L
Iron	0.3 mg/L
Manganese	0.05 mg/L
Odor	3 threshold odor number
pH	6.5–8.5
Sulfate	250 mg/L
Total dissolved solids	500 mg/L
Zinc	5 mg/L

due to the limited number of available components, materials, technology, and qualified personnel, in addition to the bureaucratic red tape associated with the importation of goods.

2.3.5 Treatment Process Selection: Alternatives

The basis for selecting treatment process alternatives is established by the characteristics of the raw water and the finished water quality goals. Consideration must be given to the future implementation of more stringent EPA drinking water quality standards and to possible changes and variability in the raw water quality. Thus, the goals and objectives, as well as the restrictions and constraints defined in the previous section, all bear upon the selection of alternative processes. Furthermore, the availability of major equipment, postinstallation services, and the capability of operators and maintenance personnel, as well as the waste handling requirements and the availability and cost of water treatment chemicals, all greatly affect the selection of the water treatment process, especially in remote regions and developing countries.

Final selection of the most appropriate treatment process scheme must be based on reliability, constructability, ease of operation, simple maintenance, and, most importantly, cost. Value analysis may be practiced during the selection process, with respect to the whole system and the individual unit processes. A cost analysis of the various alternatives should be based on two criteria—present worth and useful life.

Bench scale studies and pilot studies, conducted during either the feasibility study or preliminary engineering study phase, are quite valuable in confirming the effectiveness of the selected alternative processes if the engineer is unable to draw on previous experiences in treating the raw water. Experience acquired from existing water treatment plants (treating the same source of water) and plant scale simulation tests provide important guides in selecting both the treatment process scheme and the hydraulic loading of each unit process.

There are three basic water treatment purification processes and two modified schemes: (1) conventional complete treatment, (2) direct filtration, and (3) in-line filtration. The modified versions are high-level complete and two-stage filtration. Figures 2.3.5-1a and 2.3.5-1b illustrate these five basic processes and Table 2.3.5-1 shows the applicable raw water qualities for the basic treatment processes.

2.3.6 Preliminary Design Criteria for Treatment Unit Processes

In preparation for the cost comparisons and site plans used in evaluating site adequacy, the preliminary sizes of all major treatment unit processes must be established. This preliminary report should consider the coagulation, flocculation, sedimentation, and filtration processes; disinfection process including type of disinfectant, disinfecting condition, and byproducts; the filter washing, waste-wash handling, sludge handling, and chemical feed and storage facilities; the clearwell; and the control building. The ozonation and GAC adsorption processes should also be included in the preliminary design report if they are integrated into the treatment plant. Table 2.3.6-1 is an example of preliminary design criteria.

2.3.7 Hydraulic Grade Across the Plant

It is important to establish the hydraulic grade line across the plant early in the preliminary engineering study because both the proper selection of the plant site elevation and the suitability of the site—for executing balanced cuts and fills to accommodate all process units requiring certain water elevations and depths of structures—depend on this consideration.

Most conventional water treatment plants require 16–17 ft (4.9–5.2 m) of headloss across the plant. This means that a difference of 16–17 ft must exist between the water level at the head of the plant and the high water level in the clearwell, which is the tail end of the treatment plant process train. Modern treatment plants using preozonation, as well as postozonation and granular activated carbon adsorption processes, require almost 25 ft (7.6 m) of available head across the plant. Under these circumstances, if the plant site is flat, the following criteria must be met: the high water level in the clearwell must be set at ground level because of the groundwater table; the water level at the head of the process train must be 25 ft above the ground level; and the majority of the unit processes in the first half of the process train must necessarily be in the air. It is therefore obvious that a flat and level site is not the best choice for this type of treatment plant. The ideal plant site will have a 3–5% one-

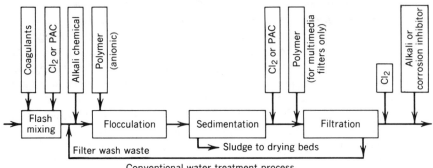

Conventional water treatment process

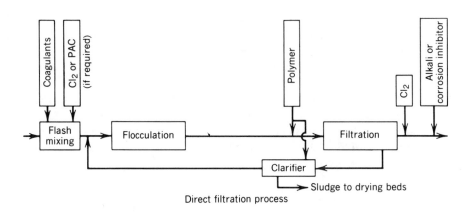

Direct filtration process

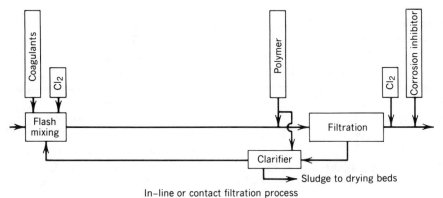

In-line or contact filtration process

(a)

Figure 2.3.5-1 Basic treatment processes.

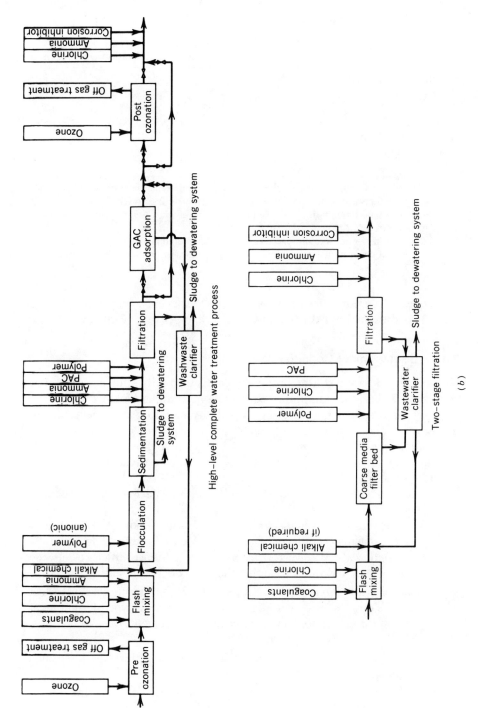

High-level complete water treatment process

Two-stage filtration

(b)

Figure 2.3.5-1 (Continued)

33

TABLE 2.3.5-1 Applicable Raw Water Quality for the Basic Treatment Processes

	Conventional Complete	Two-Stage Filtration	Direct Filtration	In-Line Filtration
Turbidity (NTU)	<5000	<50	<15	< 5
Color (apparent)	<3000	<50	<20	<15
Coliform (#/mL)	$<10^7$	$<10^5$	$<10^3$	$<10^3$
Algae (ASU/mL)	$<10^5$	$<5 \times 10^3$	$<5 \times 10^2$	$<10^2$
Asbestos fiber (#/mL)	$<10^{10}$	$<10^8$	$<10^7$	$<10^7$
Taste and odor (TON)	<30	<10	< 3	< 3

Notes: (1) The criteria shown are a general condition. Exceptions may be possible under certain conditions and pilot studies may be used to evaluate their feasibility.
(2) Slow sand filtration is applicable in cases where the raw water quality is acceptable for direct filtration.
(3) If the raw water turbidity exceeds 1000 NTU, a presedimentation process is required for all conventional complete treatment processes in order to assure good quality filtered water.

way slope and a ground elevation that satisfies the necessary elevations. Figure 6.1.2-2 is an example of the hydraulic grade line across a plant.

2.3.8 Geotechnical Considerations

Ground considerations of the proposed plant site greatly influence the entire construction cost, the duration of construction, and plant maintenance. The items of major importance are the availability of information necessary to design the foundations of the treatment facilities, ground characteristics that could affect construction, and soil characteristics that allow engineers to provide measures against cut and fill slopes as well as corrosion of pipes and concrete structures.

The following items can be obtained only from a geotechnical investigation.

1. *Soil Pressure*. Allowable bearing pressures under footings, lateral earth pressure against cantilever walls, lateral earth pressure against fully restrained walls, passive resistance force, friction coefficient between footing bases and the soil, and anticipated degree of settlement.

2. *Data on Excavation and Fill*. Cut slope for excavation, permanent cut and fill slopes, shoring cut angle for pipe trenches, type of equipment required for excavation, and compaction requirements under structures, around structures, and under roadways.

3. *Groundwater Level*. Expected high groundwater level and seasonal fluctuations.

4. *Site Seismicity*. Both the peak ground acceleration force and the seismic surcharge earth load against the walls of buried structures must be used in the structural design.

TABLE 2.3.6-1 Preliminary Design Criteria: Lake Ray Roberts Water Treatment Plant

Item	Units	Initial	Interim	Ultimate	Comments
Plant Capacity	mgd	10	50	100	The interim capacity is based on the dependable yield of 19.76 mgd available to the City of Denton from Lake Ray Roberts multiplied by 2.22 which is the maximum day drought demand factor. Ultimate capacity is for a regional plant.
Raw Water Pumping Station					
Number of Pumps	no.	2	3	5	Minimum number with at least one standby pump at each capacity
Capacity (each)	mgd	10	25	25	
Influent Flow Meter					
Type	-	-	-	-	Universal Venturi Tube
Number	no.	1	1	2	
Size	in	20	42	42	
Capacity	mgd	3-14	15-70	30-140	
Ozone System					
Ozone Dosage	mg/l	3	3	3	Maximum dose
Ozone Production					
Design Requirement	lb/day	250	1.250	2.500	
Equipment Capacity	lb/day	330	1.670	3.330	Equipment operates at a maximum of 75% of capacity
Ozone Feed Gas	-	Air	Air	Oxygen	
Ozone Generators					
Type	-	Medium Frequency			
Number	no.	2	6	6	Switching the feed gas from air to oxygen doubles the generation capacity of each generator
Capacity (each)	lb/day	330	330	670	
Total Capacity	lb/day	670	2.000	4.000	A backup generator is supplied for reliability
Ozone Contact Basins					
Type	-	3-stage countercurrent			
Number	no.	1	5	10	
Length (each)	ft	38	38	38	
Width (each)	ft	14	14	14	
Water Depth	ft	18	18	18	
Volume (each)	gal	71,600	71,600	71,600	
Total Volume	gal	71,600	358.000	716.000	
Design Detention Time	min	10	10	10	The first phase basin will be designed to operate at a 5 minute contact time also.
Flash Mix					
Type	-	-	-	-	Pumped diffusion
Number	no.	1	1	2	
G-value	sec⁻¹	750	750	750	
Mixing Time	sec	2.5	0.5	0.5	
Pump Capacity	gpm	400	400	400	
Motor Horsepower	hp	5	5	5	
Flocculation Basins					
Type	-	-	-	-	Horizontal flow, baffled and compartmentalized for plug flow characteristics
Number of Basins	no.	1	5	10	
Size (each basin)	ftxft	37x64	37x64	37x64	

TABLE 2.3.6-1 (Continued)

Item	Units	Initial	Interim	Ultimate	Comments
Water Depth	ft	12	12	12	
Stages	no.	4	4	4	Stages separated by diffuser walls
Volume (each basin)	ft³	27,650	27,650	27,650	
Volume (total)	ft³	27,650	138,250	276,500	
Detention Time	min	30	30	30	
Flocculators:					
Type	-	-	-	-	Vertical shaft (two speed in first two stages, variable speed in last two stages)
Number (each basin)	no.	8	8	8	Two in each stage
Number (total)	no.	8	40	80	
Mixing Energy (G, sec⁻¹):					
1st Stage	sec⁻¹	60/30	60/30	60/30	
2nd Stage	sec⁻¹	60/30	60/30	60/30	
3rd Stage	sec⁻¹	40 to 5	40 to 5	40 to 5	
4th Stage	sec⁻¹	40 to 5	40 to 5	40 to 5	

Sedimentation Basins

Item	Units	Initial	Interim	Ultimate	Comments
Type	-	-	-	-	Rectangular, plug flow
Number	no.	1	5	10	
Size (each basin)	ftxft	37x250	37x250	37x250	
Water Depth	ft	12	12	12	
Volume (each basin)	ft³	111,720	111,720	111,720	
Volume (total)	ft³	111,720	558,600	1,111,720	
Detention Time	min	120	120	120	
Surface Loading Rate	gpd/ft²	1,078	1,078	1,078	State requirement is a maximum of 800 gpd/ft²
Horizontal Flow Velocity	fpm	2	2	2	
Weir Overflow Rate:					
End Weir (Length)	ft	37	37	37	
Overflow Rate	gpd/ft	270,000	270,000	270,000	State requirement is 20,000 gpd/ft maximum

Sludge Collectors:

Item	Units	Initial	Interim	Ultimate	Comments
Type	-	-	-	-	Chain and Flight (with small chain and flight cross collector)
Number (each basin)	no.	2	2	2	
Number (total)	no.	2	10	20	
Speed	fpm	2	2	2	

Filters

Item	Units	Initial	Interim	Ultimate	Comments
Type	-	-	-	-	Gravity, equal loading self-backwashing
Number	no.	4	16	30	
Size	ftxft	12x24	12x24	12x24	
Area (each)	ft²	290	290	290	
Area (total)	ft²	1,160	4,640	8,700	
Filtration Rate					
with all filters	gpm/ft²	6	7.5	8	
with one filter out	gpm/ft²	8	8	8.3	
Backwash Rate					
Maximum	gpm/ft²	22	22	22	
Average	gpm/ft²	18	18	18	Supplemental backwash pump from clearwell may be necessary at initial capacity

Media (dual)
Anthracite (20") d, = 1.1 to 1.20 mm, u.c. = 1.4 or less s.g. = 1.6 to 1.65
Sand (10") d, = 0.53 to 0.6 mm, u.c. = 1.4 or less s.g. = 2.63
Gravel (12") graded from 1/16" to 2.5"
Underdrain Leopold or Concrete Teepee
Surface Wash

Item	Units	Initial	Interim	Ultimate	Comments
Type	-		Fixed Grid		
Rate	gpm/ft²	3	3	3	

Filter Ripening Addition of polymer to backwash

Waste Washwater Reclamation

Item	Units	Initial	Interim	Ultimate	Comments
Volume per backwash	gallons	44,660	44,660	44,660	22 gpm/ft² at 7 minutes
Maximum filter wash cycles per day	no.	3	3	3	Average of one wash per filter per day
Waste production rate (maximum)	mgd	0.54	2.1	4.0	Worst case, 3 backwashes per filter per day
Washwater Holding Tank:					
Number	no.	1	1	2	
Size (each)	ftxft	40x40	40x40	40x40	Sized for two backwash volumes each
Effective water depth	ft	8	8	8	
Effective volume	gallons	96,000	96,000	192,000	

TABLE 2.3.6-1 (Continued)

Item	Units	Initial	Interim	Ultimate	Comments
Recycle Pumps:					
Number	no.	2	2	4	
Capacity (each)	gpm	300	1-600	2-600	
			1-1,000	2-1,000	
Capacity (total)	gpm	600	1,600	3,200	Total capacity sized for worst case; one 300 gpm pump would operate on average initially, one 600 gpm pump would operate on average at interim capacity, and one 1,000 gpm pump would operate on average at ultimate capacity
Horsepower (each)	hp	5	10 & 15	10 & 15	
Sludge Drying Lagoons					
Estimated Production (dry solids)					
Average	lb/day	1,109	5,545	11,090	10 NTU and 10 mg/l ferric sulfate
Critical	lb/day	4,860	24,290	48,580	50 NTU and 25 mg/l ferric sulfate Critical period assumed at one month per year
Volume of Sludge Produced at 4% solids (64 lbs/ft³)					
Average	ft³/day	430	2,170	4,340	
Critical	ft³/day	1,900	9,490	18,980	
Lagoon Sizing					
Number	no.	4	4	8	Designed to handle critical month at ultimate capacity
Effective Depth	ft	6	6	6	
Area (each)	ft²	12,000	12,000	12,000	
Volume (each)	ft³	72,000	72,000	72,000	
Volume (total)	ft³	288,000	288,000	576,000	
Days of Storage					
Average	day	660	131	131	
Critical	day	150	30	30	
Clearwell					
Number	no.	1	1	2	
Capacity (total)	gal	1x10⁶	1x10⁶	2x10⁶	
Water Depth (minimum)	ft	10	10	10	
Size (each)	ftxft	115x115	115x115	115x115	
High Service Pump Station					
Number of Pumps	no.	2	3	5	Minimum number with one standby pump at each capacity
Capacity (each)	mgd	10	25	25	

	Dose (mg/l)		
Chemicals	Average	Maximum	Comments
Ferric Sulfate	5 to 10	25	Lower end of average can be used with preozonation
Cationic Polymer	0.5 to 1	2	Lower end of average can be used with preozonation
Anionic/Nonionic Polymer	0.05	0.3	Use as filter aid and/or for filter ripening
Hydrogen Peroxide	0.2	0.6	Supplemental taste and odor control chemical performing as ozone catalyst
Caustic Soda (NaOH)	2 to 5	10	
Chlorine	1 to 2	5	Average dose will depend on the final disinfectant and disinfection byproduct regulations
Aqueous Ammonia	0.3 to 0.6	1.5	Maintain a minimum chlorine to ammonia ratio of 3 to 1
Hydrofluosilic Acid			To be determined in design
Spare Chemical			For example, corrosion control chemical, determined in design

J.M. Montgomery, Consulting Engineers.

2.3.9 Structural Design Conditions and Criteria

The preliminary engineering study includes both general and specific structural design considerations and criteria for the proposed plant site. Generally, all structures must be structurally sound, that is, capable of withstanding dead weight, live loads, water pressure, earth pressure, various forces resulting from earthquakes, vibrations produced by moving mechanical equipment, loads arising form the snow pack, ice pressure, wind pressure, and the anticipated loading and impact during construction. Since a majority of the structures are water bearing (tanks and basins), they must be leakproof. Moreover, structures holding filtered and finished water must not be polluted by untreated water or waste water. Lastly, structures situated in areas with high underground water levels or where the underground water levels become high during the rainy seasons should be designed to resist buoyancy, especially when the tanks are empty. When all these requirements are considered, the minimum thickness of the reinforced concrete walls and slabs of water-bearing structures should be 8 in. (200 mm), regardless of the loading conditions.

With the exception of tropical regions, the ambient and water temperatures change significantly with the seasons. Thus, the plant design must account for thermal stresses and the expansion and contraction of the structures. The local weather conditions determine the temperature range and this in turn strictly dictates the spacing of the expansion joints for both concrete and metal structures.

The expansion joints for reinforced concrete structures are usually spaced every 33–65 ft (10–20 m), depending on factors such as if the structure is exposed to sunlight, filled with water, or located underground. These joints should provide a space of 0.5–1 in. (12–25 mm) to allow movement of the structures. Each expansion joint must have a water stop, joint filler material and sealant, and special measures that allow the reinforcing concrete bars to move at the joint (e.g., a reinforcing bar in a pipe sleeve filled with grease).

With respect to concrete and reinforced concrete structures, the water/cement ratio is a critical factor in both the strength of the concrete, corrosion of concrete surface, and the degree of shrinkage during the curing period. To minimize the formation of cracks during shrinkage, this ratio is generally limited to less than 0.5. This figure also maintains the compression strength of the walls and slabs at a minimum of 4000 psi (280 kg/cm^2). The allowable shrinkage rate of water treatment concrete tanks is usually limited to 0.04–0.05%. When designing treatment plants for regions experiencing earthquakes, engineers must use the proper seismic acceleration coefficients—those recommended by the local building codes and geotechnical consultants—in all structural design computations.

Should a plant site have high levels of sulfate in the soil and/or underground water, Type 2 or Type 5 cement may be considered. General purpose cement Type 1 is usually not capable of withstanding the corrosive conditions.

2.3.10 Plant Waste Handling and Disposal

Water treatment plants yield two basic types of waste: recoverable wastes and nonrecoverable wastes. The first type are the filter wash wastes, waste water from the "filter-to-waste" process, the supernatant of the sludge drying beds, and plant overflow. The latter include sludge from both the clarifiers and the filter wash-waste holding tanks, sanitary and chemical wastes, and wastes produced by other unit processes such as the sludge press or ion exchanger.

The acceptable modes of waste handling and processing are determined by the Water Pollution Control Act Amendment of 1972 and the National Primary Drinking Water Regulation (1986 Amendments). Thus, the preliminary engineering study must be conducted with these regulations in mind.

In the past, the recoverable wastes were collected in a holding tank, then recycled to the head of the plant without any further treatment, except in cases where natural plain sedimentation was employed. However, the new EPA drinking water treatment guidelines now consider this practice as unacceptable. Since the wastes usually contain high levels of hazardous materials, such as microorganisms, SOCs, DPBs, and heavy metals, the wastes must now be treated (flocculation, sedimentation, and disinfection) prior to recycling. The bottom line is that the quality of the recycled water must be better than or equally as good as the raw water entering the plant. However, the Water Pollution Control Act also allows the treated recoverable wastes to be discharged to a nearby water course as long as a permit is acquired from the Regional Administrator.

Handling of nonrecoverable wastes is a more complicated issue. Several alternatives exist. The simplest is discharge into the sewer system. However, this method is often impractical because of the expensive charges levied by wastewater departments, particularly when they are not located in the same city as the treatment plant.

The fundamental methods of processing and handling nonrecoverable wastes are concentration of the wastes, treatment and disposal of separated liquids, disposal of concentrated materials, and the recovery of by-products from the wastes. By-products that may be recovered are aluminum sulfate and calcium oxide; the wastes also yield inert compounds that can be used to produce bricks and aggregates for road construction. The most common handling and processing practices are gravity thickening, physical and chemical separation, heat treatment, and other types of chemical engineering process.

Generally, only a limited number of alternatives exist for the disposal of the final residual wastes. Thus, the engineer must carefully research the availability of appropriate disposal sites, requirements for disposal, methods of transporting the final residue, and the costs associated with final disposal.

The preliminary engineering study must evaluate all feasible alternatives for the treatment, handling, and ultimate disposal of all waste produced by the water treatment plant. Furthermore, the study must present a realistic recommendation based on cost considerations and the effectiveness of the proposed waste processing and handling scheme.

2.3.11 Instrumentation and Control Systems

The general motives for installing a sophisticated instrumentation and control system for a treatment plant and water supply system are (1) the continuous production and supply of safe drinking water, (2) the automatic execution of corrective measures and automatic response to potentially disastrous situations, (3) minimizing the potential for human error, (4) the capability to quickly solve analytic problems, and (5) the ability to diagnose problems in remotely located equipment before a malfunction occurs.

Because of cost constraints, managers of all utilities, as well as those in the water supply business, are seeking means to cut both the operation and maintenance costs; the reduction of personnel is usually their target. Moreover, managers of water production and supply facilities are troubled by the new stringent water quality standards set by the EPA, including the requirement for frequent or possibly continuous monitoring of water quality for certain parameters.

Modern water treatment and supply control systems are commonly referred to as Supervisory Control and Data Acquisition (SCADA) systems. Although the SCADA systems have powerful features, it is important to be certain that these features are truly beneficial to the operation and maintenance of the facilities. SCADA systems that are poorly designed or those with no previous applications, such as a newly designed system, can potentially cause numerous problems and the expenditure of unnecessary amounts of money. Engineers should not be swayed by the promotional sales pitch to acquire state-of-the-art equipment because these systems may become obsolete within a period of 10 years, thus making the replacement parts hard to find or even unavailable.

Another consideration is the complexity of the water treatment and supply systems. In contrast to most industrial facilities, the water system is relatively simple and the qualifications to become a plant operator are less stringent. In light of this fact, the real need for a high-level SCADA system is often questionable; the only appropriate situation would be if the facility is very large and has an exorbitant amount of data that must be processed quickly. Although the new computer-controlled SCADA technology is exciting, project engineers must avoid the potential pitfalls.

Four basic types of plant instrumentation and control system are currently available. They are manual, semiautomatic, automatic, and supervisory. Their respective definitions are as follows: manual initiation of a function by the operator; manual initiation of an automatic function; the use of sensors, limit switches, timers, analytical instruments, controller, and control logic devices such as relays or programmable controllers; remote control over plant operations while on the plant site or from locations away from the plant site.

The preliminary engineering study should evaluate all four levels of the system based on the local conditions, size and complexity of the treatment and water supply systems, the management philosophy of the owner, and the anticipated financial constraints. The final recommendation should be somewhat based on owner preference. However, final selection must also be based on sound engineering judgment.

2.3.12 Preliminary Cost Estimates

The best system among the various water treatment alternatives may be selected on the basis of cost effective construction and the costs associated with plant maintenance and operation. The preliminary cost estimate of the treatment plant is also necessary to assess the capital that must be raised to construct the plant.

The construction cost estimate is comprised of all costs associated with the construction contract, the overhead and profit of the contractor, but excluding the engineering and legal expenses. This estimate should also assume competitive bidding contracts.

A common method in formulating preliminary cost estimates is to use the cost estimation curves developed by the EPA (*Estimating Water Treatment Costs*, EPA 600/2-79-162b, August 1979). Other methods employ construction cost estimation data, such as Figure 2.3.12-1, supplied by reputable consulting

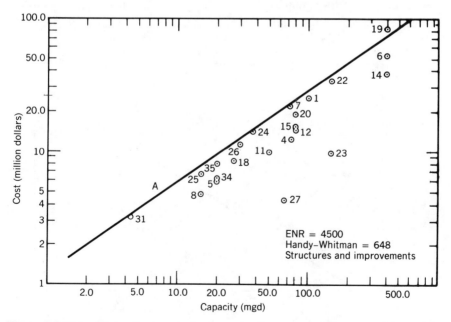

Figure 2.3.12-1 Water treatment plant construction curves. (Courtesy of James M. Montgomery, Consulting Engineers.)

engineering firms. The cost figure obtained from the curves may be adjusted to a geographical area in the United States and to current standards through the application of a special cost index. The two best recognized cost indexes are the Engineering News-Record Construction Index and the Handy–Whitman Index of Water Utility Construction Costs. Figure 2.3.12-1 is based on the ENR Index of 4500. This index was created in 1931 and was developed for 20 different cities in the United States and two Canadian cities, each representing a local geographic area. This index is designated as 100 and appears in the *Engineering News-Record*, a magazine published two to four times a year by McGraw-Hill Book Co.

It is important that engineers realize that the preliminary estimates cannot account for all factors unique to the plant site. The preliminary estimates should therefore be considered as a budget estimate with an expected accuracy of approximately $+30\%$ to -15%.

Cost estimates prepared for the purpose of comparing two or three alternative systems must take into account the useful life, salvage value, and annual revenue of the system, in addition to the initial construction costs. This can be achieved either by converting all costs to equivalent, uniform annual costs and income or by converting all costs and revenues to present worth at time zero. Present worth is defined as the yield attained at a later date of money invested at time zero times the required costs and revenues at a specified interest rate.

If the cost effectiveness of a proposed treatment system for a community is to be evaluated, the engineer must also perform a value analysis and appraise the treatment plant and other related facilities—for example, intake,

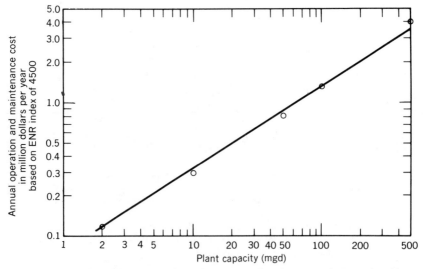

Figure 2.3.12-2 Operation and maintenance cost estimation curve (conventional process with a good raw water quality).

low service pumping facility, and high service pumping stations. Furthermore, the location of the water distribution reservoirs and the water distribution pipe system must be analyzed.

The operation and maintenance costs of a water treatment plant typically include costs associated with (1) labor, supervision, and administration; (2) chemicals; (3) power; (4) maintenance and repair; and (5) miscellaneous supplies and services. Additional factors that affect the actual operation and maintenance costs are the policy of the owner, the complexity of the system, and the local environment and weather. These costs rise on an annual basis due to the continuing inflationary trends in labor, power, and equipment. Figure 2.3.12-2 is a cost estimation curve for expenses associated with operations and maintenance. The curves are established on data obtained from a basic, conventional water treatment plant processing raw water of fairly good quality.

2.3.13 Recommended Treatment Processes

After the first 11 steps of the preliminary engineering study have been completed and thoroughly evaluated, the project engineer is ready to make a recommendation to the owner in regard to the most appropriate type of water treatment system. The steps discussed up to this point may be grouped into four basic procedures: data collection and problem formulation; synthesis and analysis of various subsystems and system alternatives; value analysis; and appraisal of the chosen alternatives. The recommended treatment system must be effective, reliable, simple in both design and operation, cost effective, and a proven type of system with minimal operation and maintenance costs.

2.3.14 Plant Layout and Architectural Design

The establishment of the plant layout is one of the final steps of the preliminary design study. Once the alternative treatment processes are narrowed down to one or two, efforts to define the plant layout for the proposed site or sites should begin. The engineer must attempt to obtain input from qualified architects for aesthetic reasons, from superintendents of local waterworks for their operational preferences, and from the owner for his particular preferences on this issue. The use of a computer-aided design and drafting system (CADD) is the most effective method for conducting this task.

Although the layout is greatly influenced by the general site topography and the extent of future plant expansions, there are three basic plant layouts: cluster, satellite or college campus, and an intermediary form of the cluster and satellite. Regardless of the layout pattern, the water should flow throughout the entire process train by means of gravity, unless pumping between the unit processes is absolutely necessary.

From an engineering standpoint, there are eight important characteristics that must be considered during work on the layout. The first is the minimization of costs associated with the civil works. However, the design engineer may not delete the careful study of the site conditions: site topography, the existence of geographical faults, soil conditions, natural water course, cut and fill requirements, access roads to the site, the slope of these roads, and the drainage and protection of the proposed site against flooding.

The second item to be considered is the ease of construction. This item addresses issues such as providing a simple arrangement of water treatment units, providing only the minimum number of required units, allowing easy access to and between the units for construction equipment, avoiding the design of yard piping that is too deep or that cross over each other, and selection of simple and uncomplicated individual structures.

Automatic, equal hydraulic loading to each tank or filter is the third consideration. This goal can only be attained through a proper layout of the units. Over 70% of all existing water treatment plants have a significant imbalance in their hydraulic loadings from one treatment unit to another, primarily as the result of improper design of the layout.

The fourth item is the centralization of control and operation. Veteran plant operators unanimously prefer this type of layout because it allows for easier and more effective plant supervision and control. Experience has shown that it is more effective to arrange subsystems requiring close supervision, such as filters, chemical feeders, and the flocculation process, in close proximity to the control/administration building rather than in a widely scattered pattern.

The fifth consideration is the physical separation of the major unit process structures. Massive concrete structures of different depths and weight must be physically separated to prevent the formation of serious cracks due to uneven subsidence and to facilitate easy access by construction equipment for the installation and repair of the plant equipment. Physical separation is very important for plant sites that have soft ground or frequent earthquake activity. The majority of plant sites have significant differential ground subsidence between existing and newly constructed structures; this fact must be given careful consideration.

The development of a master plan for the ultimate plant layout and yard piping scheme is the sixth consideration. Engineers must allocate areas for future process units in order to minimize any interference and inconvenience to the existing treatment system and daily operational activities. From a safety standpoint, two independent means of access and two alternative roads within the site should be provided to the plant. This scheme is beneficial for both future plant expansion and the transportation of chemicals in large trailer trucks during normal plant operation.

The seventh significant consideration is the climatic conditions that exist at the plant site. For example, treatment plants located in very cold regions must house all unit processes to protect them from freezing. Consequently, the layout must be a compact scheme that avoids large separations between the major process units to minimize the size of the housing structures. If the unit processes must be separated, underground passage tunnels should be considered; it therefore follows that the process units must be arranged in a way that minimizes tunnel construction costs.

The eighth and last, but not least, consideration is architectural design. The aesthetic features of the plant layout and building and the landscaping of the plant site are superficial but very important issues. Most water treatment plants in the United States are publicly owned. They are frequently visited by public groups, school children, and engineers from other municipalities, states, or countries. Furthermore, the plant sites are often near residential areas or areas of natural preservation. Thus, the proposed treatment plant may meet resistance from groups of local citizens or other organizations. The plant must therefore be functional and safe and must also blend into the surrounding environment, as well as be aesthetically pleasing when viewed from the surrounding neighborhoods. It is very important to consider spending at least 2–3% of the total design cost on improving the aesthetics of the plant.

2.3.15 Preliminary Planning of the Project

In order for the design of a water treatment system to be effectively and successfully executed, a balanced team of highly qualified specialists must be assembled. Preliminary selection of experts and management planning of the project must be accomplished at the end of the preliminary engineering study phase.

One tool widely used in project management is the bar chart. However, the Program Evaluation and Review Technique (PERT), the Integrated System of Project Management (ISPM), and the Critical Path Method (CPM) have also been used in the management of complex projects: these methods may also be used in project scheduling. PERT emphasizes the control phase of project management; CPM defines the duration of a project and thereby minimizes the cost, whereas, ISPM stresses the aspect of project cost control. Figure 2.3.15-1 is an example of a project control schedule.

2.3.16 Environmental Analysis Report

In accordance with the National Environmental Protection Act of 1969, all water treatment and supply projects are required to file an Environmental Impact Statement (EIS) prior to implementation. The EIS must include detailed studies and an analysis of the environmental impact of the facility. A "nonbuild" alternative is also required.

Preparation of an EIS requires a team of many specialists (a biologist, a hydrologist, archaeologists, and economists) to both develop and complete the studies. Projects that improve the existing environment or those with a negative impact statement, indicating no environmental impact by the proposed project, can proceed to the design stage rather quickly. However, under certain circumstances, an EIS may require a significant amount of time and

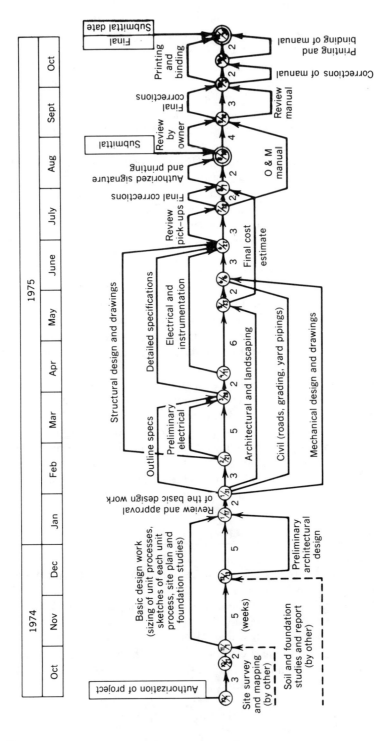

Figure 2.3.15-1 An example of the project control schedule.

money. The findings of the EIS may also have a tremendous impact on the design, construction schedule, and total cost of the project.

BIBLIOGRAPHY

AWWA, *Water Treatment Plant Design*, AWWA, New York, 1969.

George, C. E., "Effective Project Scheduling," *Water & Sewage Works*, p.386 (December 1971).

Grant, E. L., *Principles of Engineering Economy*, 3rd ed. Ronald Press, New York, 1950.

Kawamura, S., "Pilot Studies of Mixing and Settling," *AWWA Seminar Proceeding—Design of Pilot Plant Studies*, p.25 (1982).

Merritt, F. S., *Standard Handbook for Civil Engineers*, 3rd ed., McGraw-Hill, New York, 1983.

Montgomery, J. M., Consulting Engineers, *Water Treatment—Principles and Design*, Wiley, New York, 1985.

Olson, R. V., "Critical Path Method of Work Scheduling," *J. AWWA*, 61(9):447 (September 1969).

CHAPTER 3

Design of Basic Treatment Process Units

3.1 BASIC APPROACH

A water treatment plant should be designed to produce a continuous supply of safe drinking water regardless of the raw water characteristics and the environmental conditions. Thus, the application of space-age technology in the process design and subsystem selection is not always the best approach. The ultimate plant design has a system and subsystem that are proved to be simple, effective, reliable, durable, and cost effective. This basic philosophy is applicable to all circumstances and cases. Design engineers should study the following basic rules prior to design work:

1. Carefully evaluate the local conditions.
2. Create a reasonably conservative design that is cost effective to construct.
3. Apply the best knowledge and skill to the design.
4. Design a plant that is easy and safe to construct, as well as simple and safe to operate.
5. Allow for maximum operational flexibility and minimal operations and maintenance costs.
6. Design a plant that is aesthetically pleasing and does not have a negative impact on the environment.
7. Acquire a plant site that adequately satisfies the basic criteria and is free of all potential disasters.
8. Do not be embarrassed to obtain help from qualified associates, consultants, or specialists; an engineer should only perform services that are in his/her area of competence.
9. Be certain that the project meets all pertinent legal requirements and engineering standards.
10. Respect the wishes of the owner.

A water treatment process is an assemblage of unit processes that is created to effectively produce an abundant supply of safe drinking water. The treatment system generally consists of four or more interrelated and compatible components. Since the unit processes must be interrelated, the operation of each component affects the performance of the others, that is, the entire system. Most water treatment textbooks and design guide manuals presently on the market tend to view each component as an individual unit. Consequently, readers may erroneously believe that the most efficient overall plant performance is obtained by optimizing the design of the individual components. This type of suboptimization generally does not achieve optimum total plant design. Since the pivotal purification processes is filtration, all pretreatment unit processes should be designed and operated to maximize the efficiency of the filtration process and the subsequent process of disinfection.

As previously mentioned, the design engineer should only work in the area of his/her competence. The design team is usually composed of civil, structural, mechanical, electrical, and instrumentation engineers and engineers specializing in computers, soil foundation, and hydraulics, as well as an architect. The team may also include a chemical engineer if necessary. It is very important that both the project engineer and project manager be familiar with the outlines of work performed by each of the disciplines since they must coordinate the progress of the various sections on a daily basis (ideally).

Design engineers must remember to utilize commercially available components whenever feasible because this helps to keep the project costs down. Components such as gates, valves, pumps, pipes, and heating and ventilation equipment, in addition to certain equipment such as flow meters, chemical feeders, mixers, sludge collection units, and instrumentation and control units, should be evaluated for substitution by commercially available units. The key to creating a good water treatment plant design is to pay attention to important details and avoid spending time on insignificant items.

3.1.1 Design Procedure

Efficient and successfull design work can only be accomplished by strictly following the standard procedures for the project as established by the organization. Section 1.4 presents a detailed example of project control procedures. The key activities are as follows:

1. Assign job numbers.
2. Establish budget and project schedules (see Figure 2.3.15-1).
3. List all required drawings.
4. Organize a design team.
5. Allocate a budget for each discipline of the project team.
6. Firmly establish the process and P&I diagrams during the early stage of the project.
7. Perform all necessary calculations and establish a reasonably detailed

basic sketch for each process unit. These should be circulated to each of the disciplines (refer to Figure 3.1.1-1).

8. Establish the details of all subordinate plant facilities, including the chemical feed system and the waste and sludge handling system.

9. Discuss and finalize item numbers 6 through 8 with the heads of each discipline prior to releasing them to the design teams.

10. Begin design work based on the project schedule.

11. Establish a tight coordination among the various disciplines of the project.

12. Maintain control of the overall project at all times.

13. Edit all specifications written by other members of the team. Use the standardized specifications of the firm whenever possible.

14. Carefully review the design calculations, drawings, and specifications with the help of experts. Do not sign the documents until each item has been checked and corrected.

15. Estimate the construction cost and arrange to have the construction documents produced.

The design project will be completed successfully if the following control methods are understood and followed.

TABLE 3.1.1-1 Anticipated Number of Drawings for a Water Treatment Plant Design

Discipline	Percentage of Total	Number of Drawings if Total is 200 Drawings
General	5	10
Civil	10	20
Architectural	8	16
Structural	30	60
Mechanical	20	40
Electrical and instrumentation	25	50
Landscaping	2	4

Notes: (1) The total number of drawings may vary depending on the conditions of the plant site, local weather conditions, the complexity of the design, and the plant size.

(2) Small conventional plants, those processing less than 10 mgd, may have 70–120 total drawings depending on the degree to which the proprietary treatment units are specified.

(3) Medium sized conventional plants (10–50 mgd) require anywhere from 120 to 200 drawings, but generally over 150.

(4) Large sized conventional plants (over 50 mgd) require 200–350 sheets of drawings.

(5) If an intake pumping station, high service pumping station, or ozonation system is included in the plant design, the total number of drawings will be increased approximately by 7%, 7%, and 9%, respectively.

1. The project engineer must have a clear understanding of the project objectives and the project goal.
2. A good and balanced design team must be established.
3. An adequate budget, reasonable design period, and sufficient personnel must be available to complete the project.
4. The project engineer must coordinate the various design disciplines.
5. Changes to and deviations from the originally established design concept and sketches should be kept at an absolute minimum.
6. The project control procedure must be strictly followed.
7. Both the project engineer and project manager must be knowledgeable and experienced and willing to devote their time to the project and to bring a positive attitude to the design team.

Tables 3.1.1-1 to 3.1.1-3 and Figure 3.1.1-1 may aid inexperienced engineers in grasping some of the basic design requirements and may ultimately help them to create realistic project control schemes. Table 3.1.1-1 gives a

TABLE 3.1.1-2 Normal Design Time per Drawing

Discipline	Total Hours/Drawing (Design and Drafting)	Breakdown of the Hours	
Project engineering	19	Specifications	6 h
		Cost estimate	3 h
		Engineering	10 h
Civil	85	Engineering	45 h
		Drafting	40 h
Architectural	80	Engineering	35 h
		Drafting	45 h
Structural	70	Engineering	35 h
		Drafting	35 h
Mechanical	75	Engineering	30 h
		Drafting	35 h
Electrical and instrumentation	60	Engineering	20 h
		Drafting	40 h
Landscaping	40	Engineering	15 h
		Drafting	25 h

Notes: (1) The engineering time for all disciplines includes the hours required to check the design calculations and drawings and the time required to coordinate efforts with other disciplines of the project.
(2) If a good design efficiency is to be maintained, the weighted average hours per drawing, for the engineering and drafting phase, are 80 h.
(3) The cost of one sheet of drawing ranges from $3000 to $7000 (1989 cost) depending on the nature of the drawing, the required design time, the complexity of the drawing, and the number of corrections that must be made.

TABLE 3.1.1-3 Breakdown of Water Treatment Plant Construction Costs (Approximate)

Civil work	7.0%
(earthwork, grading, paving, fencing)	
Yard pipings	8.0%
Landscaping and irrigation	1.0%
Operations building	10.0%
(chemical feed system included)	
Flocculation and sedimentation basins	17.0%
Filters	20.0%
Clearwell	8.0%
Pumping facilities	7.0%
Meter vaults (L.S.)	2.0%
Filter washwaste holding and recycling	3.0%
Sludge drying beds	2.0%
Miscellaneous items	0.3%
Chemical storage facilities	1.0%
Electrical and instrumentation works	12.0%
Testing and disinfecting works	0.2%
Move on and move off	1.5%
(contractors)	

Notes: (1) The table does not include the overhead and profit of the contractor; these are generally 20% of the total cost shown above.
(2) The above figures are based on a high-rate conventional process.

rough idea of the total number of drawings required for certain sized plants using a conventional complete treatment process. Table 3.1.1-2 lists the normal design time and work-hours required to complete a sheet of drawings for each of the various disciplines. Table 3.1.1-3 is a very basic breakdown of the construction costs. Figure 3.1.1-1 presents the basic design sketches of each unit process: the basis for the detailed design by each of the disciplines of the design team.

3.2 COMMON WATER TREATMENT PROCESSES

3.2.1 Introduction

The phrase "common water treatment processes" connotes a system that combines the processes of coagulation, flocculation, sedimentation, filtration, and disinfection with the necessary process control and instrumentation measures. Until the mid-20th century the design of water treatment plants was handled exclusively by civil engineers and emphasis was placed on the hy-

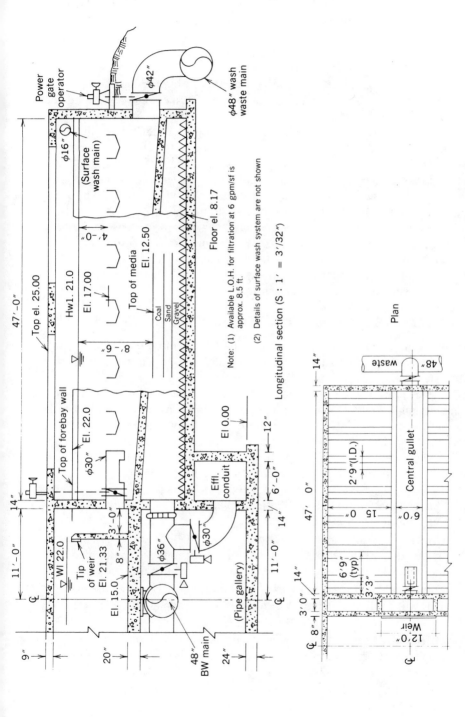

Figure 3.1.1-1 Dual media bed alternative (16 filters) equal loading, constant-level control type filters (surface wash by water jets).

53

draulic, foundation, and structural aspects of design. It is therefore understandable why these early treatment plants tend to lack chemical engineering, microbiological, electrical, mechanical, and architectural considerations. Consequently, the performances of these plants do not always meet the expectations of the design and are not always aesthetically pleasing. The present trend is to have sanitary engineers design the treatment processes and the overall plant design because they are familiar with a broader range of subjects.

The efficient and successful execution of a modern water treatment design is dependent on the concerted efforts of various specialists. The team members must therefore pool their knowledge and skills so that the performance and interaction of each component are designed in a complementary manner. This type of communication is essential because the design of each component affects the design of the entire treatment system.

The cost effectiveness of the design phase may be improved by establishing a value engineering team, in addition to the design team. The project members may also opt to include both construction experts and experienced plant operators (of plants similar to the plant being designed). Figure 3.2.1-1 presents the members of a typical design team.

3.2.2 General Considerations

When designing basic treatment process units, the most important considerations are the fundamental design philosophy, the design procedures, and control over the design phase of the project; these issues are discussed in Section 3.1. However, certain items must be established during the early stage of any design project: plant layout, process diagrams, hydraulic profile across the plant, process trains, and the issue of modifying and/or expanding the existing plant. The project engineer and manager are responsible for establishing these items based on some basic rules and considerations. A few of the more important general design concepts are listed below.

- Create a master plan of the treatment plant site, including future process units and major yard piping.
- Use the module expansion concept if the plant is to be expanded in stages. In order to avoid any operational confusion, each module to have the same size and elevation and ideally have identical processes.
- Provide adequate access roads to each unit and module of the process. This will facilitate simple plant maintenance and allow easy installation of future modules.
- Special attention must be given to the hydraulics of the plant. For example, provide equal hydraulic loading to each tank and every process unit and maintain a uniform backwash flow distribution to each filter.
- Study the plant design to ensure that the safety of both plant operators and the environment has been addressed.
- Research the flood level of the plant site, carefully plan the access roads to the plant site, and establish the various utility hookups.

- Establish an adequate area for the plant site by calculating the area required by the sludge handling facilities and the area required to install new facilities, such as GAC adsorption bed and related facilities, at a future date.

Plant Layout The plant layout is primarily dictated by the topography of the plant site. Thus, poor site selection severely restricts the choice of alternative layouts. The design phase should only refine the preliminary layout, which was established in the predesign stage. The design phase focuses on the anticipated traffic flow and operator movement, on establishing hydraulic balance among the process units and basins, on the aesthetics of the plant, and the location of the main control building; the main control building should be situated as near as possible to the pretreatment process, filters, and chemical feed system.

Two important considerations must be addressed during work on the plant layout. The first is to provide a single chemical application point for all pretreatment process units. The second is to provide a single chemical feed point to the filter influent for all the filters and to use this same scheme for the combined filter water prior to the clearwells. Although special attention must be given to the yard pipings and the layout of the process units, the single-point application scheme is recommended because it significantly simplifies the chemical feed system and provides uniform water quality control. If the chemicals are divided to more than two points, the design will require additional feeders or flow splitting devices such as "rotameters" and the feed rate to each process unit must also be paced to the actual process flow rates.

Figure 3.2.2-1 illustrates the three basic plant layouts. Miscellaneous engineering considerations with regard to the layout are described in Section 2.3.14.

Process Diagram The process diagram is the fundamental basis for the design work. This is especially true for the design of the instrumentation and control systems and the subsystems. A basic process diagram should be established during the preliminary design phase and refined at the beginning of the final design phase. The instrumentation and control specialist and the mechanical specialist may then commence with their design work.

The process diagram should include the following items: (1) all unit processes in the correct sequence; (2) all major pipe connections with the flow directions; (3) all chemicals that are to be used and the application point(s) of each; (4) all major water sampling points that are necessary to maintain quality control; (5) the location and size of all major flow meters, valves, and connecting pipes in the process train; (6) the location of all major pumps, blowers, screens, and other such items of the process train; and (7) the control points for the pressure, water level, flow rate, and water quality of the process. All these items are essential for designing the piping and instrumentation diagram (P&ID). Details of the P&ID are discussed in Section 3.2.9. Refer to Figure 3.2.2-2 for an example of a water treatment plant process diagram.

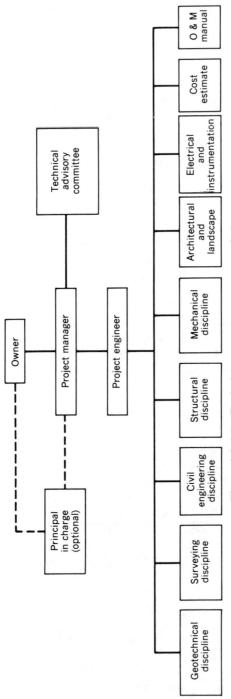

Figure 3.2.1-1 Typical water treatment system design team.

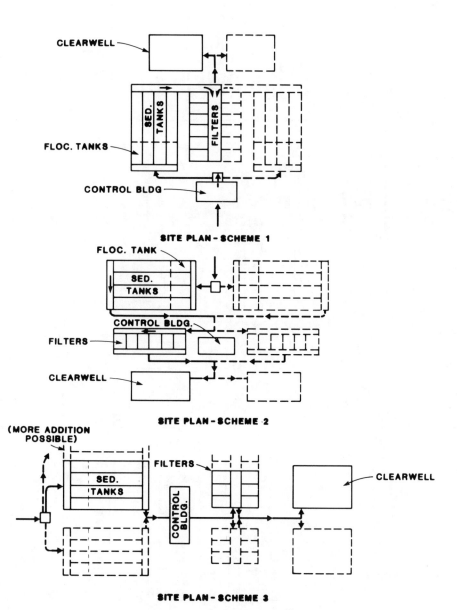

Figure 3.2.2-1 Three basic plant layouts.

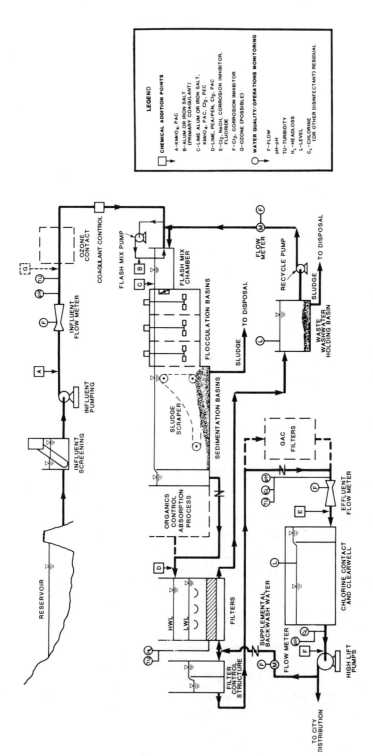

Figure 3.2.2-2 An example of process diagram.

Preliminary Hydraulic Grade Line Across the Plant   Once the process flow diagram is finalized, the design engineer should produce a diagram of the hydraulic grade line across the treatment process train. This diagram must show all unit processes, in addition to the elevations of the walkway (the top of the structure), the water level, the bottom elevation, the invert, and possibly the crown levels of all connecting pipes and channels. Moreover, the finished ground level against each structure of the process unit must be shown in order to facilitate the work of the remaining disciplines of the design team, especially the civil, structural, and mechanical engineers. It must be emphasized that this diagram is a key component of the entire design project because it provides the basic information for the design team. Any significant modifications to this diagram, during the middle or latter stages of the design period, will have a negative impact on design efforts, especially with respect to the cost and deadline of the project.

Design of the Basic Processes   The detailed design work of each process should commence as early in the design phase as possible. The basic design criteria of each process are based on the hydraulic loading, which should have already been established during the preliminary design phase. Thus, the size of each process unit and the type of major equipment are loosely established. However, at this stage all the details, including the dimensions and shape of each process, must be firmly set. Moreover, all major equipment must be selected to ensure that they will fit comfortably into each unit structure. The detailed design phase should therefore begin with the chief engineer of the mechanical discipline. This is particularly important with respect to process structures that house numerous pieces of equipment.

The process structures must be carefully reviewed because certain types of unit require special foundations, a space around the unit, temperature control, dust control, complete isolation, a large capacity of water and power, or an explosion-proof room. All equipment, particularly the mechanical and electrical units, require servicing or may need to be replaced. Consequently, adequate spacing must be provided between the units. There have been a few cases where the valves in the filter pipe gallery were so inaccessible, due to a complicated piping scheme, that it was a major undertaking to try to service or replace them.

Design engineers should also design the unit process structures based on the largest and heaviest available units on the market because the exact manufacturer of the installed equipment will not be known until after the plant construction bid is awarded to a contractor; contractors always select the most cost effective unit that meets the design specifications. There have been several embarrassing cases where the equipment was delivered during construction but could not be installed in the structure because the door size was too small or where the replacement unit could not be installed due to the lack of adequate spacing between the equipment. These types of problem should be discovered and corrected during the checking procedure of the latter stages of the design phase. However, this requires the expenditure of more time and money. It is therefore extremely important to sort out these issues at the beginning of the design phase.

The project manager and the project engineer should also set up a good system of design documentation, including a format for the calculation sheets and filing system, and disseminate this program among the members of the design team. The design calculations and the record of equipment selection are essential in case of an accident and/or lawsuit.

The project engineer should be capable of handling all the basic design calculations and the selection of the principal equipment. Yet, he/she should only handle the area of his/her competence. Qualified experts should be consulted for the other areas; the project engineer must clearly convey the objectives and limitations of the system and subsystems. Table 3.2.2-1 contains some basic design criteria used to size certain unit processes and pipelines.

Modification of the Existing Plant   An existing plant may be modified to expand the plant capacity and/or rehabilitate the old plant. The approach and considerations used in this situation are quite different from those used in designing a new treatment plant since the modification must be based on the original plant design. Difficult problems may arise if the original plant was designed by another firm many years ago, when governmental regulations were more lenient.

The modification of existing treatment plants is basically approached in three phases: first, evaluate the existing plant based on available documents and field trips; second, perform field testing of the raw water quality, the actual unit process efficiency, and study the local conditions; and third, evaluate and finalize the plant modification scheme based on the regulatory requirements of the local and national agencies.

The following steps must be performed when evaluating an existing plant:

1. Collect, analyze, and evaluate all available documents concerning the design, operation, and historical raw and finished water quality of the plant.
2. Interview the chief operator of the plant regarding operational experiences and obtain a list of both positive and negative comments and changes that he/she would like to have implemented.
3. Obtain field measurements of the dimensions of all major process units and measure the elevations of the water surface level at selected points in the process train, at the designed flow rate of the plant, and at a different flow rate. Do not rely on the "as-built" drawings since the plant may have been modified beyond what the "as-built" drawings show.
4. Evaluate all existing major equipment, conduct an inventory of the major equipment, and evaluate their condition and utility.

TABLE 3.2.2-1   Basic Design Criteria

GUIDE ON DESIGN FLOW RATE FOR WATERWORKS

Water Treatment Processes including Intake Facility: $Q_{max\text{-}day}$

Plant Hydraulic Capacity: 1.25 to 1.5 × $Q_{max\text{-}day}$

Clearwell Capacity: Equal or Larger than 0.2 × $Q_{max\text{-}day}$

                            Or Fire Fighting Capacity (Local Code)

High Service Pump Station: $Q_{max\text{-}day}$

Water Distribution Reservoir: $Q_{max\text{-}hour}$
(In the City)

Notes: $Q_{ave\text{-}day}$ is an average annual daily flow rate

$Q_{max\text{-}day} = 1.5 \times Q_{ave\text{-}day}$

$Q_{max\text{-}hour} = 1.5$ to $2.0 \times Q_{max\text{-}day}$

$Q_{min\text{-}day} = 0.25 \times Q_{ave\text{-}day}$

GUIDE ON PIPE SIZE SELECTION FOR WATERWORKS

Raw Water Main   :   6 to 7 fps

Flocculated Water Line :   1.0 to 1.5 fps for Conventional Rapid Sand filter
                             with Alum floc.

                          :   3.0 fps for Direct Filtration Process or Filters
                             with Reverse Graded media Bed

Filter Influent Line   :   2.5 to 3.5 fps in general
                             3.5 to 4.5 fps for Polymer Fed Filter Influent

Filter Effluent Line   :   5.0 to 6.0 fps

Filter Wash Water Main :   8.0 to 9.0 fps

Wash Waste Main   :   6.0 fps

Distribution Main   :   6.0 fps

Pump Suction Line   :   4.0 to 6.0 fps

Pump Discharge Line   :   7.0 to 9.0 fps

5. Conduct an inventory of and evaluate the electrical, instrumentation, and control systems.
6. Inspect all major structures for structural integrity and evaluate their life expectancy.

Field testing is the next phase in modifying an existing plant. The following five steps must be performed:

1. Evaluate the treatability of the raw water by conducting bench scale tests. Be sure to evaluate the type of coagulant, optimum coagulant dosage, optimum mixing conditions ($G \times t$), floc settling rate, and the filterability of the settled and flocculated water.
2. Evaluate the effectiveness of the existing filter washing system by obtaining core samples of the filter bed both before and after filter washing. Also analyze the actual backwash rate, effective sizes, uniformity coefficients, and specific gravities of the filter media. Furthermore, the mud deposition profile across the bed depth should be assessed.
3. Evaluate the actual flow-through time of the flocculation and sedimentation processes and the clearwell by conducting tracer tests.
4. Survey and investigate the geotechnical aspect of the locations where the additional process units will be constructed.
5. Verify the hydraulic bottlenecks and the performance of the unit processes at higher than designed flow rates by performing high-flow simulation tests on an isolated number of basins and filters.

The final phase in modifying an existing plant is evaluating the results obtained from the first two phases and establishing a plant modification scheme. Some of the more important considerations are as follows:

1. Evaluate and establish a balanced flow rate and process performance between the existing facilities and the new additions.
2. Try to use the existing filter washing system by specifying the same size filters at the same elevation. This practice will also avoid any confusion for the plant operators.
3. Check to see that the design criteria of all the unit processes meet the present requirements set by the state, local, and federal agencies and those of the foreseeable future.
4. Verify that the owner's requests are incorporated in the modification scheme.
5. If any part of the old system is salvageable, try to integrate the old instrumentation and control system into the new scheme.
6. Establish a basic plant modification scheme so that the existing plant operations are minimally disrupted by the construction.

BIBLIOGRAPHY

Kawamura, S., et al., "Improving Water Supply in Thailand," *J. AWWA*, 80(5):59(June 1988).

Merritt, F. S., *Standard Handbook for Civil Engineers*, 3rd ed., McGraw-Hill, New York, 1983.

3.2.3 Coagulation and Chemical Application Considerations

Coagulation is defined as the destabilization of charge on colloids and suspended solids, including bacteria and viruses, by a coagulant. Flash mixing is an integral part of coagulation. This chapter discusses the role of chemical application and flash mixing in effective water treatment.

Purpose The purpose of flash mixing is to quickly and uniformly disperse water treating chemicals throughout the process water. Effective flash mixing is especially important when using metal coagulants such as alum and ferric chloride, since their hydrolysis occurs within a second and subsequent adsorption to colloidal particles is almost immediate. In practical design, the dispersion of metal coagulants should therefore be completed within 1–2 s, although theoretically, it should be completed in a fraction of a second. The time requirement for other chemicals, such as polymers (polyelectrolytes), chlorine, alkali chemicals, ozone, and potassium permanganate, is not as critical since they do not undergo hydrolytic reactions. Thus, from a practical point of view, the dispersion time of these chemicals may be completed within several seconds or less. Although flash mixing is an important unit process in water treatment, most conventional treatment plants are capable of producing quality water without the use of a flash mixer provided that coagulant in excess of 30–40% is used. Many existing water treatment plants have proved this point, contrary to some literature which claims that flash mixing is absolutely essential for water treatment. It should be noted that flash mixing of coagulant is very different from the normal concept of liquid–liquid mixing used in the chemical industries. Under normal conditions the ratio of liquid alum volume to raw water volume is 1:50,000 (in the water treatment field). Thus, the instantaneous dispersal of such a small amount of alum into the raw water is a very unique situation.

Considerations The following items should carefully be evaluated when designing an effective flash mixing system: (1) type of coagulant to be employed, (2) number of chemicals to be fed and the characteristics of each, (3) local conditions, (4) raw water characteristics, (4) type of chemical diffusers, (5) available headloss for the flash mixer, (6) variations in the plant flow rate, (7) type of subsequent process, (8) cost, and (9) other miscellaneous items.

Type of Coagulant The selection of coagulant is extremely important in two respects: to set design criteria for the flash mixing system and for effective flocculation and clarification. The most commonly used coagulants are the metal salt coagulants: aluminum sulfate, ferric chloride, and ferric sulfate. Synthetic polymers such as polydiallyl dimethyl ammonium (PDADMA) and natural cationic polymers such as chitosan (made from crustacean shells) are also employed.

The difference between metal salt coagulants and cationic polymers is their hydrolytic reaction with water. Metal salts undergo hydrolysis when they are fed to the process water; yet, all polymers do not. This hydrolytic reaction produces hydroxocomplexes, such as $Al(H_2)_6^{3+}$, $Fe(H_2O)_3^{3+}$, $AlOH^{2+}$, and $Fe(OH)^{2+}$, in a pH range found in the process trains of most water treatment plants. The formation of the hydrolytic products occurs in a very short period

TABLE 3.2.3-1 Importance of Time Delay Sequencing for Anionic Polymers as Flocculant Aids

A. Jar Test of Surface Water for Turbidity Removal

Raw water: Turbidity = 18 Tu, pH = 7.9, Alkalinity = 84 mg/L

Time of polymer addition with respect to alum		Alum Dosage (mg/L)			
		0	10	13	16
No polymer addition		18 Tu	10.5 Tu	6 Tu	4 Tu
Anionic polymer (8181) 0.2 mg/L	5 min before	—	—	7.5	4.5
	same time	—	—	6.5	3.5
	5 min after	—	—	2.5	1.2
	10 min after	—	—	1.2	0.7

B. Jar Test of Iron Removal by Chlorine Oxidation for Well Water

Raw water: pH = 6.7, Alkalinity = 198 mg/L, Hardness = 188 mg/L, Fe = 4.4 mg/L

Chemical dosages: Cl_2 = 5 mg/L, NaOH = 15 mg/L

Time of polymer addition with respect to chlorine		Turbidity (Tu)	Iron (mg/L)	pH
No polymer addition		5	2.5	8.0
Anionic polymer (7763) 0.2 mg/L	30 sec after	5	2.2	8.0
	2.5 min after	3	1.6	8.0
	5 min after	2	1.4	8.0

of time: less than 1 second. These products readily adsorb to colloid particles and cause destabilization of their electrical charge. In turn, the hydrolytic products are quickly polymerized through hydrolytic reactions.

In summary, flash mixing or instantaneous mixing is important because (1) hydrolysis and polymerization are very fast reactions; (2) a uniform supply of coagulant and a uniform process water pH are essential for the production of hydrolytic products; and (3) adsorption of these species to colloidal particles occurs quickly.

When cationic polymers are used as coagulants, instantaneous mixing is not critical because hydrolytic reactions do not occur; the rate of colloid adsorption is much slower because of their large physical size. Thus, from a practical point of view, the mixing time for polymers may be 2–5 s.

Number of Chemicals and Their Characteristics   More than two or three chemicals are commonly fed into the flash mixer, which is located at the head of the treatment plant. The most frequently used chemicals are alum, cationic polymers, potassium permanganate, chlorine, powdered activated carbon (PAC), ammonia, lime or caustic soda, and anionic and nonionic polymers. Design engineers should evaluate the characteristics of each of these chemicals and classify them into two or three categories based on the following criteria: necessity of instantaneous mixing, potential chemical reactions that prevent the process chemicals from acting individually, and the proper chemical addition sequence (to the raw water) which yields the most effective results.

The proper chemical application sequence is very important because it produces remarkable results and also reduces the total chemical cost. For example, when anionic polymers are fed to the process water during pinpoint floc formation, the size and weight of the alum floc are dramatically improved. Generally, a 5–10 min lag time is essential for optimum floc formation. Table 3.2.3-1 illustrates this point.

The selection of an appropriate chemical application sequence is also vital in cases where the raw water does not have sufficient alkalinity. In this type of situation, alkali chemicals such as lime or caustic soda are fed to the water in conjunction with alum.

The most reasonable chemical application sequence is to first lower the pH of the raw water through the addition of alum, thereby allowing divalent and trivalent aluminum hydroxocomplexes to form; these complexes effectively reduce colloid charge. The next step is to adjust the pH of the water to the range of minimum aluminum solubility in order to allow aluminum hydroxide (floc) formation. Figure 3.2.3-1 presents the solubility equilibria of aluminum and iron hydroxides, while Figure 3.2.3-2 illustrates the steps involved in a proper coagulation and flocculation process.

Local Conditions   The local conditions that should be investigated during the planning of the process design are the availability of equipment and parts,

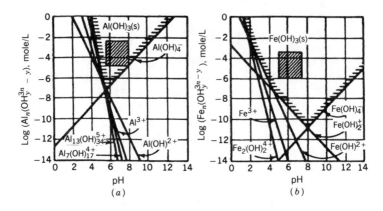

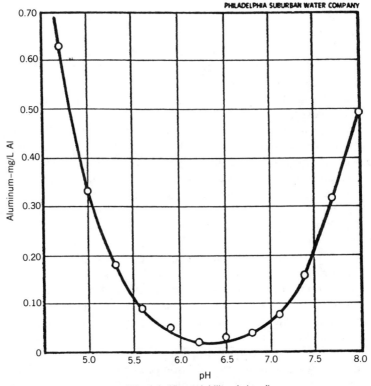

Effect of pH on solubility of alum floc
(constant alum dosage of 1.0 g.½gal(17.1 mg/L).
pH adjusted with acetic acid and sodium hydroxide
(c)

Figure 3.2.3-1 Solubility of common floc versus pH. (a,b) Equilibrium composition of solutions in contact with freshly precipitated $Al(OH)_3$ and $Fe(OH)_3$. Calculated using representative values for the equilibrium constants for solubility and hydrolysis equilibria. Shaded areas are approximate operating regions in water treatment practice; coagulation in these systems occurs under conditions of oversaturation with respect to the metal hydroxide. (From Stumm and O'Melia, 1968.) (c) Effect of pH on solubility of alum floc. [Constant alum dosage of 1.0 g/gal (17.1 mg/L). pH is adjusted with acetic acid and sodium hydroxide.]

the availability of services within the area, and the reliability of the power supply.

Other conditions that influence the choice and design of the flash mixing system are the frequency of sand storms and the winter temperatures. For instance, in remote areas and developing countries, the use of hydraulic jumps and in-line static mixers is the most appropriate choice. If the system is located aboveground, it should be protected by a building structure. However, if the plant is located in an area experiencing frequent sand storms, severe winter weather, in a quiet residential neighborhood, or an area with a high rate of vandalism and, in some cases, the potential for military attack, the flash mixing system must be located in an underground vault.

Raw Water Characteristics   The design of the coagulation process is determined by the raw water characteristics. The most common problem associated with the process of coagulation is clogging of the chemical diffuser orifices and feed lines. This is directly related to the hardness of the process water and, in part, to dissolved and suspended solids. If the water hardness exceeds 30 mg/L, the metal salt and alkali chemical feed lines and the diffuser orifices will become clogged within several months or less. The design of the system should therefore facilitate simple cleaning of the feed lines, injection nozzles, and orifices. Although low water temperatures have an adverse effect on chemical dispersion and reaction rates, under normal conditions this influence is minimal. However, raw water alkalinity and pH are critical factors in both coagulation and flocculation. Appropriate pH ranges for metal salt coagulants are illustrated in Figure 3.2.3-1.

Type of Chemical Diffuser   The most commonly used water treatment chemicals may be categorized into three basic groups: gases, liquids, and solids. Most of these chemicals are converted into solution prior to being fed to the raw water, with the exception of ozone, carbon dioxide, lime, and powdered activated carbon (PAC). Both ozone and carbon dioxide are generally manufactured on site; these gases contain only a small percentage of the essential components and their solubility is very low. They are therefore fed to the process water through deep channel contact tanks and are released from the tank bottom as fine bubbles to ensure efficient gas transfer. Further discussion of gas mixing is not covered in this chapter since flash mixing usually does not effectively achieve this result.

Lime and PAC are generally fed to raw water in the form of a slurry. This is dispite the fact that for small plants lime may be fed to the raw water as saturated calcium hydroxide water, provided that a lime saturator is used.

In summary, water treatment chemicals are fed to the flash mixing system in either of two forms—solution or slurry. For both pipe and channel flow application, the chemical diffuser (for solutions) is most often a pipe with multiple orifices. If the chemicals are in slurry form, two types of diffuser may be used: for pipe application the open end of the diffuser pipe acts as

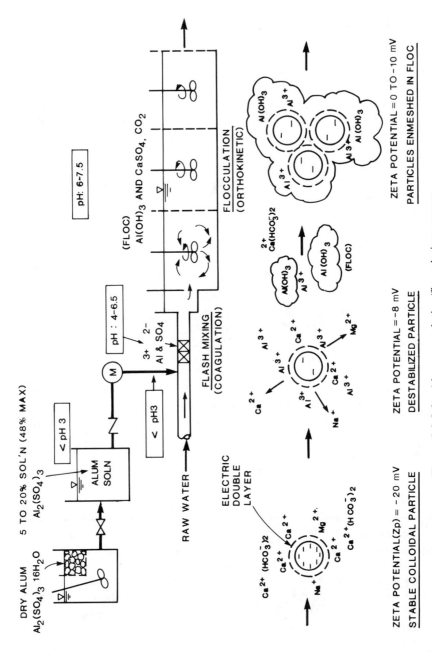

Figure 3.2.3-2 Alum coagulation/flocculation process.

the diffuser, while channel flow application uses a trough with notches or holes at the bottom.

Special attention must be given to chemical solutions that produce scale since they can clog the diffuser orifices. Metal salt coagulants, such as alum or ferric chloride, and alkali chemicals (caustic soda, soda ash, lime saturated solutions, and ammonia solutions) are scale-forming chemicals. When these chemicals are employed, a provision for periodic cleaning should be included in the design of the feed system; an additional diffuser must be specified so that plant operations are not interrupted when one diffuser is being cleaned. Figure 3.2.3-3 depicts the most commonly used chemical diffusers. Figure 3.2.3-4 illustrates the basic design criteria for multi-orifice diffusers.

Available HeadLoss for the Flash Mixer Water treatment plants may occasionally have excess pressure in the raw water line. This pressure must be reduced by means of a special valve or electric power generator. When headloss can be afforded, a hydraulic mixing system such as an in-line static mixer, weir, or hydraulic jump should be considered. In any case, a headloss of approximately 2 ft is generally required to ensure good mixing. However, if a plant has tight hydraulic conditions that cannot afford headloss, a special system such as pump diffusion should be employed. If instantaneous mixing is not absolutely necessary, a properly designed pipe diffuser system is the correct choice.

Since mechanical mixers do not produce significant headloss, they may be considered as a viable alternative to the flash mixer. However, many studies indicate that most mechanical flash mixing systems designed are ineffective due to flow short-circuiting and frequent maintenance and repairs. In recent years, several articles have stressed the inability of mechanical mixing systems to achieve instantaneous mixing in tanks lacking the proper type of baffling.

Variation in Plant Flow Rate Variations in water demand cause fluctuations in the plant flow rate on daily and seasonal bases. Since variations in plant flow rate can easily exceed the 1:2 ratio, the mixing energy of the hydraulic flash mixing system must also change in proportion to the plant flow rate. Recent studies in effective flash mixing have shown that the values for the product of mixing energy (G-value expressed in seconds^{-1}) and mixing time (t in seconds) should range from approximately 300 to 1600. Hydraulic mixing can be used if $G \times t$ falls within this range. If the seasonal plant flow fluctuation is too great to satisfy the proper $G \times t$ range, the design must provide two parallel hydraulic flash mixing systems: the design criteria will be satisfied if only one of the two systems is used during a low flow season.

Due to variations in mixing time, plant flow rate variation becomes an important design factor in other mixing systems such as pump diffusion and mechanical mixers. The mixing energy must therefore be adjusted to satisfy the proper $G \times t$ range. The pump diffusion system uses a flow-regulating valve to control the pump discharge flow, while mechanical flash mixers employ a variable mixing speed control unit.

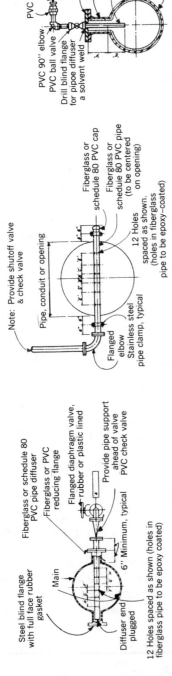

Steel blind flange
with full face rubber
gasket

Fiberglass or schedule 80
PVC pipe diffuser

Fiberglass or PVC
reducing flange

Flanged diaphragm valve,
rubber or plastic lined

Provide pipe support
ahead of valve

6" Minimum, typical

PVC check valve

Main

Diffuser end
plugged

12 Holes spaced as shown (holes in
fiberglass pipe to be epoxy coated)

Type A: In-line Diffuser

Note: Provide shutoff valve
& check valve

Fiberglass or
schedule 90 PVC pipe

Pipe, conduit
or openings

Flanged
elbow

Stainless steel
pipe clamp, typ.

Type D: In a Tank Inlet or in a Structure

Note: Provide shutoff valve
& check valve

Pipe, conduit or opening

Fiberglass or
schedule 80 PVC cap

Fiberglass or
schedule 80 PVC pipe
(to be centered
on opening)

12 Holes
spaced as shown
(holes in fiberglass
pipe to be epoxy-coated)

Flanged
elbow

Stainless steel
pipe clamp, typical

Type B: In a Tank Inlet or in a Channel

Lime diffuser
480 dia. asbestos
cement pipe cut
in half

Seal ends with
concrete (typ)

Stainless stl.
187 x 187.8
with concrete
anchors
(typ)

12 Dia. holes
200 D.C. (in mm)

Type E: Gravity Feed from a Trough Above
Water Surface

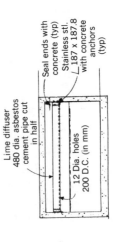

PVC
PVC union   check valve

PVC diaphragm
valve, see specs

PVC blind flange

150 lb. Flanged outlet

PVC pipe diffuser

Main

PVC 90" elbow

PVC ball valve

Drill blind flange
for pipoe diffuser
a solvent weld

Type C: In-line (Removable)

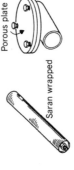

Porous plate

Saran wrapped

Type F: Fine–bubble Gas Diffusers

Figure 3.2.3-3 Chemical diffusers.

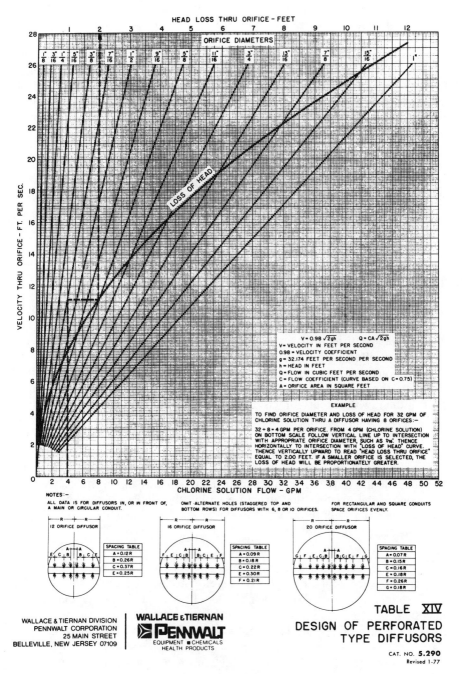

Figure 3.2.3-4 Diffuser pipe design guide.

Type of Subsequent Process   The design engineer must carefully consider the type of treatment processes that will be subsequent to the flash mixing system since the coagulant dosage and the degree of pretreatment can be quite different. Aside from sludge conditioning and waste washwater reclamation, four basic process trains are used in modern water treatment: in-line filtration, direct filtration, two-stage filtration, and complete conventional treatment. Lime softening and high-level complete treatment processes, such as ozonation and granular activated carbon adsorption, are considered to be modified complete conventional treatments. Iron and manganese removal (through oxidation) may be achieved through either direct filtration or the conventional complete process train.

The in-line, direct, and two-stage filtration processes have two distinct characteristics; they have no sedimentation process and require very low levels of coagulant (2–6 mg/L). Inadequate flash mixing of coagulant results in higher filtered water turbidity because the small amount of coagulant fails to disperse uniformly.

In contrast, the conventional complete process always includes sedimentation and thereby allows the formation of good settleable floc, provided that an adequate dosage of alum is applied (15–30 mg/L). Hence, if sufficient alum is fed to the process water, the role of flash mixing may not be obvious. Yet, an effective flash mixing system requires 30–40% less alum. This reduction was confirmed by several operational treatment plants after the installation of effective flash mixers.

If the conventional complete process is used in treating raw water with low turbidity, the benefits of the flash mixing system may not be apparent. However, if the raw water turbidity is high, use of an effective flash mixing system dramatically improves the turbidity of the settled water (Table 3.2.3-2).

Sludge conditioning and wash waste clarification must be treated differently from ordinary water treatment because of the high concentration of solids. The coagulant used in alum sludge conditioning is a polymer, not alum. Polymers do not require instant mixing. In fact, intense mixing may adversely disintegrate the existing floc structure within the sludge. Unfavorable conditions will not arise if the G-value of sludge conditioning is $100–150 \text{ s}^{-1}$ with a mixing time of 1 min or less.

Cost   Cost effectiveness is always an important consideration; it is particularly crucial when designing plants for developing countries. With this in mind, the flash mixing system should be reliable, effective, and easy to operate and maintain. The mechanism of the flash mixing system should be simple so that its service and repair can easily be accomplished by the operational staff. Exprience has shown that plant operators will not use the flash mixing system if it is overly complicated, requires frequent maintenance, or generates too much noise. Also, the operators will not use the flash mixer if they feel that it does not produce a discernible difference in the treatment process. For these reasons, the cost of the flash mixing system should be approximately 0.3% of the construction cost, unless there are some justifiable reasons.

TABLE 3.2.3-2 Effectiveness of Flash Mixing

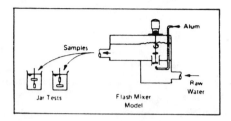

Value of the flash mixer was approximately 1,000 s⁻¹; mixing propeller rotation was set to give upward flow.

Effect of Flash Mixing by a Bench-Type Test*

Factor	Test No.										Average
	1	2	3	4	5	6	7	8	9	10	
Quality of raw water											
Turbidity—*Jtu*	8	28	30	35	38	70	85	500	550	600	195
pH	6.7	7.0	7.1	6.9	6.9	6.8	6.8	6.8	6.6	6.5	6.8
Alkalinity—*mg/l*											
(as CaCO₃)	17	25	26	22	21	20	21	15	14	14	19.5
Alum dosage—*mg/l*	10	11	10	9	10	15	20	25	20	30	16
Time required to form visible floc—*min*											
Case 1	3.5	1.5	1.5	1.5	1.5	2.7	1.5	1.3	1	1.3	1.7
Case 2	3.5	1.5	2.5	1.5	1.5	3	2.5	2.3	3	2	2.3
Turbidity of supernatant —*Jtu*											
Case 1	1.0	1.5	2	9	7	3	1.8	12	8	6	5.1
Case 2	1.0	2	3	7	5	4	3.0	17	14	13	6.1

*Case 1 is with flash mixing and case 2 is without flash mixing.

Effect of Flash Mixing Tested on the Actual Plant—Kabutoyama Plant,* Kobe, Japan

Factor	Test 1	Test 2	Test 3
Quality of raw water			
Turbidity—*Jtu*	9.0	7.5	6.5
pH	6.6	6.65	6.75
Alkalinity—*mg/l*			
(as CaCO₃)	16.5	19	19
Alum dosage—*mg/l*	10	10	12
Turbidity of flocculating water —*Jtu*			
Flash mixer on	5.0	3.0	1.5
Flash mixer off	5.5	5.0	2.5
Turbidity of settling tank effluent—*Jtu*			
Flash mixer on	1.0	0.8	0.3
Flash mixer off	1.0	1.0	0.5

*1958—40 mgd capacity.

Source: S. Kawamura, *J. AWWA*, 65(6):417(1973).

Miscellaneous Items Although some articles emphasize the adverse effect of back mixing during the flash mixing process, this theory is somewhat debatable. Back-mixing implies the blending of chemicals to the mixture of raw water and recycled chemical-laden raw water. A typical example of back-mixing is a tank with a mechanical mixer on center (for mixing alum), providing 1–2 min detention time.

It is important to mention that operators of many plants using mechanical mixers as the flash mixing system observe no measurable benefits by operating the mixer. In fact, mechanical problems, such as noise and energy waste, and a high maintenance cost are often cited as reasons not to run the mixer.

Flash mixing by means of compressed air has been tried. However, this method has never become popular because there is some doubt that it can achieve instantaneous and uniform dispersion of coagulant and because of the noise generated by the compressor.

Type and Selection Guide This section presents commonly used alternatives in the flash mixing process and guidelines for making the proper selections.

Available Alternatives Flash mixing may be achieved by means of a hydraulic, mechanical, or pump diffusion system. Figure 3.2.3-5 shows the most common types of flash mixer. They are as follows:

- Mechanical mixer
- Hydraulic mixer
- In-line mechanical mixer
- In-line static mixer
- Diffusion mixing by pressured water jets
- Miscellaneous—diffusion by pipe grid

Selection of the flash mixing system should be based on the considerations discussed in the early part of this section.

When establishing the design of a flash mixing process, the order of preference is based on effectiveness, reliability, minimal maintenance, and cost. Thus, the order of choice is as follows:

1. Diffusion mixing by pressured water jets.
2. In-line static mixing.
3. In-line mechanical mixing.
4. Hydraulic mixing.
5. Mechanical mixing.
6. Diffusion by pipe grid.

Discussion of Alternatives

DIFFUSION MIXING BY PRESSURED WATER JETS For various reasons, diffusion mixing by pressured water jets is the first choice among the alternative flash mixing processes. The advantages of this type of flash mixing system are that there is essentially no additional headloss by the mixer, it is very effective, it has a controllable degree of mixing, its power consumption is less than half that of a mechanical mixer system, it is cost effective, and all parts of the system are "off-the-shelf" items.

The source of the pressured water may be either the plant utility water line, the discharge of the high service pumps, or pumped process water. The raw water should not be used with dilute and diffuse coagulant unless the turbidity of the raw water is continually low (less than 5 NTU) and virtually free of any suspended solids, including fish and freshwater clams. If the above criteria are not met, serious nozzle clogging problems will occur unless a basket strainer is provided in the pump suction line. Ideally, settled or filtered water should be pumped. The pressure used to diffuse the chemicals must be a minimum of 10 psi (0.7 kg/cm^2).

One disadvantage of this alternative is the potential for coagulant and debris present in the pumped water to clog the nozzles. A second disadvantage is the difficulty in applying it to a system with extra large pipes or channels, for example, those having a diameter greater than 100 in. or 2500 mm. In the extra large system other means of dispersion, such as multiple nozzles or a pipe grid with injection nozzles, should be employed. In order to avoid clogging the injection nozzles with metal hydroxide scale, it is extremely important to ensure that the metal salt coagulants are not excessively diluted prior to injection. In general, alum solutions should not be diluted beyond 1% (5% for ferric chloride). The dilutions may be adjusted if the dilution water is very soft and if the high level of dilution does not shift the pH of the coagulant into the range of floc formation. These metal coagulants are best fed to the water jetting from the nozzle as neat solution by means of a separate alum feed line (see Figure 3.2.3-7 and photo pictures).

IN-LINE STATIC MIXERS In-line static mixers are also known as motionless static mixers. Half a dozen different types of in-line mixer are currently available and if properly selected and applied, they are quite effective in the coagulation process. The primary application of these units is in industry because of their effectiveness. The advantages of these mixers are (1) the lack of moving parts, (2) no requirement of external energy to be input into the system, and (3) fewer clogging problems than the pump diffusion type of mixers. However, two disadvantages exist (1) the degree of mixing and the mixing time are a function of flow rate; and (2) the units are proprietary items and the design engineer must therefore rely on the performance claims of the manufacturer. From a practical point of view, a mixing time of 1–3 s and a maximum headloss of 2 ft (600 mm) across the unit are normally quite acceptable for water treatment. The treatment plant design should have an

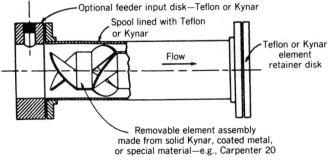

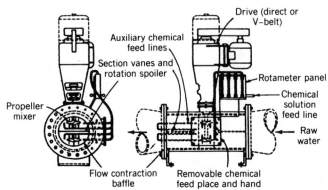

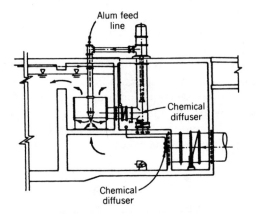

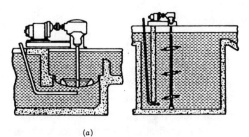

(a)

Figure 3.2.3-5 (a) Various types of flash mixers. (b) Clogged up 8-inch diluted alum. solution feed line in spite of 8 fps flow velocity (alum solution strength was 0.01 to 0.02 percent). (c) Modified alum solution dispersion (pump injection system).

Figure 3.2.3-5 (Continued)

intake screen upstream from the static mixer so that large debris will not obstruct the static mixer. It is also important to specify the mixing elements as removable in order to facilitate the necessary cleaning of debris and scale.

HYDRAULIC MIXING Parshall flumes, Venturi type meters, and weirs are categorized as hydraulic mixing devices. When raw water flows into a plant through an open channel, the plant flow can be metered by either a Parshall or Palmer–Bowles flume since they provide hydraulic jumps and large-scale turbulence. However, if the raw water is piped to the plant, Venturi type or orifice meters should be placed within the influent pipeline; these meters produce differential head and result in a certain degree of mixing—more so with the orifice meter. Weirs are seldom used as flow-measuring devices in modern treatment plants; nevertheless, they also produce downstream turbulence.

These hydraulic flow-measuring devices may also be used for chemical flash mixing; they have been used in this capacity, particularly in developing countries. However, the drawback to this procedure is that the degree of turbulence is a function of plant flow rate, and thus, there is no positive control over the degree of mixing.

MECHANICAL FLASH MIXING   The mechanical flash mixer is the most frequently used type of flash mixing unit in the water treatment industry. It consists of a tank or channel with one or more mechanical mixers. The common design parameters are as follows: $G = 300$ s^{-1}, a mixing time of 10–30 s, and a power requirement of 0.25–1.0 horsepower per million gallons per day.

Recent studies have shown that this type of rapid mixing is not preferred for use in a continuous flow process because of (1) the lack of instantaneous mixing characteristics, (2) the production of significant flow short-circuiting, (3) a mixing period that is too long for metal salt coagulants, and (4) backmixing, which may adversely affect coagulation. Furthermore, there is some difficulty in analyzing the various forces acting on the mixer shaft and impeller; this has resulted in shaft problems and gear drive failures in numerous installations, especially large-scale plants. If marginal equipment is employed, it may create relatively high operation and maintenance costs and noise pollution. For the above reasons, the mechanical flash mixers rank low in system preference.

DIFFUSION BY PIPE GRID   This type of flash mixer depends on the wake turbulence created by the pipe grids; coagulant or other chemicals are added to the flow stream through injection orifices in the grid. Kaufman et al. (1972) have suggested that a minimum density of one orifice per square inch is required to yield satisfactory coagulation and flocculation results. However, the majority of actual installations employing this criterion experience problems with scale clogging the orifices after several months to a year of plant operation. Therefore, under normal conditions, it is not recommended that this parameter be included in plant design, except in pilot scale installations.

Design Criteria   The selection of an effective, yet trouble-free flash mixing system is not a simple task, especially when designing large-scale treatment plants. The appropriate design criteria for the flash mixer process must carefully be selected. Although the velocity gradient (G-value) has been used as a design guide, it is somewhat ambiguous in determining the process because the effectiveness of coagulation is greatly affected by the following criteria: shape of the reactor tank, degree of flow short-circuiting, type of mixing element, degree of energy imput, and effective mixing time. Appropriate design parameters should be based on (1) chemical diffusion rates, (2) the degree of turbulence associated with the inertial forces (characterized by the power number), and (3) the Reynolds number, which represents the viscous forces and flow characteristics. Unfortunately, most practical design criteria are not based on these parameters.

The widely accepted general design criterion for flash mixing is the velocity gradient G:

$$G = (P/\mu V)^{0.5}$$

where P = power imput,
μ = absolute viscosity of water,
V = volume of the mixing zone.

The most effective velocity gradient, G (in s^{-1}), and mixing time, t (in s), is

$$G \times t = 300-1600$$

If pump diffusion is used as the flash mixing system, the following design criteria should be used:

1. $G \times t = 400-1600$ (1000 average).
2. Mixing jet velocity of 20–25 fps (6–7.6 m/s) at the orifice.
3. If neat alum solution is diluted prior to application to the process water, the pH of the diluted alum should be lower than 3.0. If ferric salts are used as coagulant the pH should be less than 2.0.

For in-line static mixers, the following design criteria should be used.

1. $G \times t = 350-1700$ (1000 average).
2. $t = 1-5$ s.

The above conditions may be satisfied with a single in-line static mixer if the plant flow rate variation is less than a 1:5 ratio.

Chemical Diffusers   Chemical diffusers are a component of all flash mixing systems which most design engineers fail to note. Thus, many chemical diffusers become obstructed for various reasons. Metal salt coagulants, such as alum and ferric sulfate, and alkaline chemicals, such as lime, caustic soda, and ammonia, will clog small orifices within a short period of time unless the process water is extremely soft and has very low turbidity.

The selection of the proper chemical diffuser is a function of the coagulant and the type of process employed (refer to Figure 3.2.3-3 for the following discussion). For metal salt coagulants and alkaline chemicals, the correct type of chemical diffuser is generally a Type D diffuser, since it minimizes the problem of clogging; Type D may also be used for aqua ammonia in cases where the area of chemical application is not easily accessible and if fumes are not allowed to escape into the environment due to the installation of submerged curtain walls both upstream and downstream. Type A and B diffusers may be used if mixing is achieved solely by diffusion and not by means of positive flash mixing, since the diffusers may easily be disconnected. Type A and B diffusers may also be used with chlorine, polymers, potassium permanganate, and fluoride solutions because these chemicals usually do not produce scale at the orifices. Type E diffusers may be used with lime or PAC for channel flow application upstream of the flash mixer. Type F diffusers are used exclusively for gaseous chemicals, such as ozone and carbon dioxide. Figure 3.2.3-4 illustrates the proper design of multi-orifice diffusers.

Example Design Calculations

Example 1 Mechanical Flash Mixing System

Given

Plant flow rate	1.85 m³/s (42 mgd) maximum per day
$G \times t$	1000
Mixing time	Approximately 1 s at 1.85 m³/s
Average water temperature	20° C

Determine

  (i) The size of the tank.

 (ii) The horsepower of the mixer and type of mixing blades.

(iii) The application point of alum.

(iv) The size of the inlet channel.

Solution

  (i) Provide a flash mixing tank that is 10 ft square and 12 ft deep at the head of the plant. An intermediate slab containing a 4.5 ft (1.37 m) diameter hole at midheight should be provided to create an effective flash mixing zone. Attach a 4 ft (1.22 m) length of pipe (Figure 3.2.3-6) at the opening to act as a mixing reactor tube.

Flow velocity through the hole:

$$1.85 \div \left(\frac{\pi \times (1.37)^2}{4} \right) = 1.25 \text{ m/s}$$

Mixing time: $1.22 \text{ m} \div 1.25 \text{ m/s} = 0.98 \text{ s}$

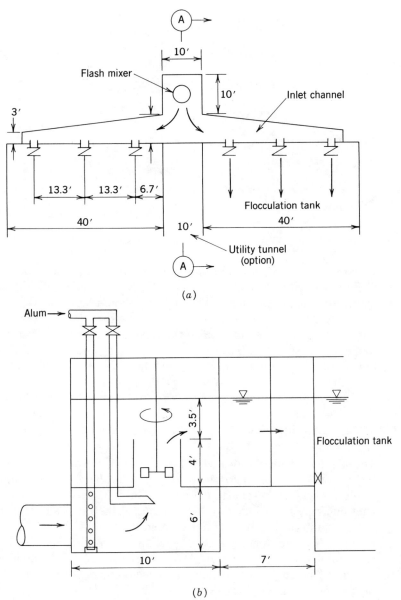

Figure 3.2.3-6 Flash mixing tank and inlet channel to the flocculation tank. (a) Plan. (b) Section A-A.

(ii) The power requirement for the mixer horsepower and mixing blades is $P = G^2 V \mu$.

When $G = 1000 \text{ s}^{-1}$

$$V = \frac{\pi \times (1.37)^2}{4} \times 1.22 = 1.8 \text{ m}^3$$

$\mu = 0.001 \text{ N·s/m}^2$ at 20° C

$P = (1000)^2 \times 1.8 \times (1 \times 10^{-3}) = 1800 \text{ J/s}$

Since 1 kW is 1000 J/s,

$$P = 1.8 \text{ kW} \quad \text{or} \quad 2.4 \text{ hp}$$

Assuming an overall efficiency of 80%,

$$P = 2.4 \div 0.8 = 3 \text{ hp}$$

which is the minimum required mixer horsepower.

Four 45° pitched blade turbines (PBT) with an overall impeller diameter of approximately 1.5 ft (0.76 m) should be selected for the mixing blades. The blades are located at the bottom of a 4.5 ft diameter flash mixing pipe. The interior of the mixing reactor tube must have four vertical stator baffles, each approximately 3 in. in width. The mixer shaft must rotate in such a manner that the water is lifted, minimizing the loss of head through the flash mixing system; shaft run-out should be limited to $\frac{1}{8}$ in. and an underwater shaft bearing should not be provided at the end of the shaft. The details of the mixer design, including shaft rotation speed, should be referred to a reputable mixer manufacturer (based on the above criteria).

(iii) The alum feed line is located below the mixing blade as shown in Figure 3.2.3-6. In order to prevent clogging of the feed line, employ a solution with a concentration greater than 5%. An additional alum feed line may be included at the entrance of the raw water line into the tank; this acts as an alternative application point should the main feed line become clogged or damaged. The end of the main chemical feed line has a simple 45° cut (Type D, Figure 3.2.3-3) to minimize clogging. The end of the auxiliary feed line is equipped with a diffuser pipe across the inlet pipe (Type B, Figure 3.2.3-3).

(iv) Assume that there are two flocculation tanks and that flash mixing occurs in the center of the two tanks. An influent channel is located between the flash mixer tank and the flocculation tanks as shown in Figure 3.2.3-6.

To assess the size of the influent channel leading to the flocculation tanks, the rule of thumb is to maintain less than 0.3 in. (7.5 mm) of the velocity head existing within the influent channel when the system undergoes maximum flow rate. This practice ensures that the water surface elevation remains fairly constant across the entire channel. it is important to note that if one of the two flocculation tanks is isolated for maintenance or repair, a flow of 1.85 m³/s may travel through one side of the channel.

Select a width of 7 ft for the midpoint of the channel and 3 ft for each end. The channels should not be designed with a width less than 2.5 ft since this causes difficulties in construction. The average water depth is assumed to be 7.5 ft.

LOCATION WITHIN THE CHANNEL

Item	Center of Entrance	Midpoint	Endpoint
Q	65 cts	43 cts	21.7 cts
	(1.85 m³/s)	(1.23 m³/s)	(0.62 m³/s)
A	52.5 ft²	37.5 ft²	22.5 ft²
v	1.24 fps	1.15 fps	0.96 fps
$v^2/2g$	0.024 ft	0.02 ft	0.014 ft
	(7.3 mm)	(6.1 mm)	(4.3 mm)

The maximum velocity head of the channel is 0.29 in. (7.3 mm); therefore, the rule of thumb criteron is met. When the two flocculation tanks are on-line, the flow velocity at the channel entrance is 0.62 fps and the velocity head is 0.07 in. (1.8 mm); the inlet condition to the flocculation tank is excellent.

Example 2 Pump Diffusion as the Flash Mixer

Given

Plant flow rate	1.5 m³/s (34 mgd) maximum per day
$G \times t$	Approximately 1000 ($G = 750$ s^{-1})
Pipe diameter	1000 mm (40 in.)
Diffusion jet velocity	6–7.5 m/s (20–25 fps)
Minimum water temperature	10° C
Alum dosage	8–50 mg/L

Determine

(i) The mixing time.
(ii) The required water horsepower.
(ii) The pump capacity.
(ii) The required jet velocity and size of orifice.
(iv) The type of nozzle.
(v) The horsepower of the pump motor.
(vi) The location of the alum feed line.

Solution

(i) The assumed length (L) of the mixing zone is

$$L = 1.5D = 1.5 \times 1 = 1.5 \text{ m}$$
$$V = 0.785 \times D^2 \times L = 0.785 \times 1^2 \times 1.5 = 1.18 \text{ m}^3$$

The mixing time at plant flow rate (t) is

$$t = 1.18 \text{ m}^3 \div 1.5 \text{ m}^3/\text{s} = 0.79 \text{ s}$$

(ii) The required water horsepower is found as follows:

$$P = G^2 \mu V$$
$$\mu = 1.336 \times 10^{-3} \text{ N·s/m}^2 \text{ at } 10°C$$
$$P = 750^2 \times (1.336 \times 10^{-3}) \times 1.18$$
$$= 887 \text{ J/s} = 0.89 \text{ kW} = 1.1 \text{ hp}$$

(ii) In order to achieve satisfactory diffusion of the primary coagulant in a short period of time, the rule of thumb is to design the pump flow rate to be 2–5% of the plant flow rate:

$$1.5 \text{ m}^3/\text{s} \times 0.03 = 0.045 \text{ m}^3/\text{s} \quad \text{or} \quad 710 \text{ gpm}$$

(iv) The velocity head of the flow issued from the blending orifice is used in computing the mixing energy. The required mixing jet velocity is calculated by using the basic pump formula:

$$\text{kW} = \frac{9.81 \times Q \times H}{e}$$

where kW = horsepower of the pump; in this case it is water horsepower,

Q = pumping rate (m³/s),

H = total dynamic head (m),

e = efficiency—in this case 100% since the theoretical water
horsepower is used to determine orifice size.

The mixing jet velocity is based on the water horsepower required
to produce the given mixing energy of 0.89 kW and a pumping capacity
of 45 L/s.

Therefore,

$$0.89 = 9.81 \times 0.045 \times H$$

$$H = 2 \text{ m} \quad \text{or} \quad 6.6 \text{ ft}$$

Since the velocity head $H = v^2/2g$,

$$v = [2 \times (2 \times 9.81)]^{0.5} = 6.26 \text{ m/s} \quad \text{or} \quad 20 \text{ fps}$$

Multiple orifices provide better efficiency of mixing. However, due
to clogging problems caused by debris, use only one nozzle that sprays
a full cone pattern of 90°. The diameter of the orifice is 3.75 in.:

$$d = \left(\frac{Q}{0.785 \times v}\right)^{0.5} = \left(\frac{0.045}{0.785 \times 6.26}\right)^{0.5} = 0.095 \text{ m}$$

(v) Model 8HF700, a full jet nozzle, manufactured by Spraying System
Company is selected. This nozzle has 92° full cone spraying, 316
stainless steel construction, a $3\frac{19}{32}$ in. (92 mm) orifice diameter, a
minimum free passage of $1\frac{7}{8}$ in. (48 mm) solids, 8 in. (200 mm) flange
pipe connection, and a 700 gpm (0.044 m³/s) capacity at 7 psi (0.5
kg/cm²) back pressure. Figure 3.2.3-7 illustrates this system and the
nozzle.

We now check the energy imput:

$$P \text{ (hp)} = (0.97 \, C_d a v^3) \div 550$$

where P = water horsepower,

C_d = coefficient of discharge = 0.75,

a = orifice area (ft²),

v = jet velocity (fps).

$$P = [0.97 \times 0.75 \times 0.07 \times (22.3)^3] \div 550 = 1.02 \text{ hp}$$

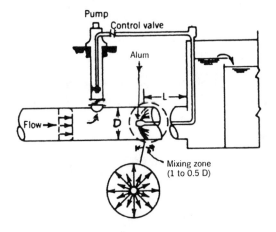

**FULL CONE
SPRAY PATTERN**

**FULL CONE
SPRAY PATTERN**

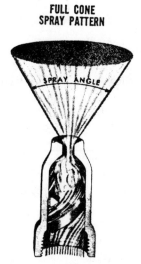

Figure 3.2.3-7 Pump diffusion system and nozzle.

Based on the *G*-value, the required horsepower is computed to be 1.1 hp. The energy imput is a little less but nonetheless acceptable.

(vi) We now determine the pump motor horsepower:

$$\text{TDH} = \text{required pressure at nozzle} + \text{pipe friction loss}$$

Note: The static head for the pump is zero due to the siphon effect (see Figure 3.2.3-6) and the pipe friction loss is assumed to be 1 ft, but the actual value should be calculated.

$$\text{TDH} = (7 \times 2.31) + 1 = 17 \text{ ft} \quad \text{or} \quad 5.2 \text{ m}$$

$$\text{BHP} = \frac{Q \times H}{3960 \times e} = \frac{700 \times 17}{3960 \times 0.8} = 3.76 \text{ hp} = 2.8 \text{ kW}$$

Therefore, use a 5 hp (3.75 kW) motor.

(vii) We now determine the location of the alum feed line.

FEED FLOW RATE OF ALUM SOLUTION

Maximum = 50 mg/L × 8.34 × 34 mgd $\quad$ = 14,178 lb/day

Average $\,$ = 15 mg/L × 8.34 × (34 ÷ 1.5) $\quad$ = 2835 lb/day

Minimum = 8 mg/L $\,$ × 8.34 × [34/(1.5 × 4)] = 378 lb/day

Note: Assume that $Q_{max} = 1.5 \times Q_{ave}$
$$Q_{mim} = 0.25 \times Q_{ave}$$

Since commercial liquid alum contains 5.4 lb of dry alum per gallon, the actual feed flow rate is 2625, 525, and 70 gal/day for the maximum, average, and minimum flow rates, respectively.

Dosage	Alum Feed Rate in gpm (1)	Pumping Rate in gpm (2)	(1)/(2) × 100	Resulting Alum Solution Strength
Maximum	1.82	700	0.26	0.13%
Average	0.36	700	0.05	0.026%
Minimum	0.05	700	0.007	0.0035%

For most cases the strength of the alum solution should be greater than 1%, upstream of the injection nozzle, and the pH of the solution should be less than 3.0. If these recommendations are followed, the nozzle and pipe may not become clogged by aluminum hydroxide scaling. Experience indicates that if these conditions are not met, heavy caking by alum hydroxides begins approximately 10–12 in. downstream of where the alum feed line terminates into the pump discharge line. Therefore, the end of the alum feed line should be located at the flange of the injection nozzle. Alternatively, the feed line exit may be located just outside the nozzle. However, the dispersion of alum is not as effective as the first choice. Note that sludge caking in the discharge line clearly confirms that alum coagulation is completed within 0.1 s after application.

Example 3   Static mixer as the Flash Mixer

Given

Plant flow rate	15 mgd (0.63 m³/s) maximum per day
$G \times t$	Approximately 1000
Pipe diameter	24 in. (600 mm)
Maximum head loss	1.5 ft at 15 mgd
Minimum water temperature	10°C

Determine

(i) The type of in-line static mixer.

(ii) The number of mixing elements.

(iii) The energy imput ($G \times t$).

Solution

(i) Several types of in-line static mixer are available for water treatment application, but a simple unit with low clogging characteristics should be selected. A provision for easy disassembly of the mixing baffles (for cleaning) must also be considered. Therefore, a motionless mixer, such as the one manufactured by Komax Systems, Inc., is preferred.

(ii) Under normal flow conditions the practical mixing time is limited to approximately 2 s; 3 s for pipes with diameters larger than 5 ft. Thus, the total length of the mixing unit can be approximated by two times the flow velocity in the pipeline. The flow velocity for a plant influent line generally ranges from 6 to 8 ft/s. The total mixer length therefore ranges from 12 to 16 ft. A rule of thumb in estimating the length of one element is to designate the length as 1.5–2.5 times the pipe diameter. Based on this criterion, the length of one element is in the range of 3–5 ft. Thus, two or three elements may be used.

(iii) We now determine the energy input:

$$G = \left(\frac{P}{\mu V}\right)^{0.5} \quad \text{and} \quad P = Qwh$$

where P = power input (lb·lb/s),
Q = pumping rate (cfs),
w = unit weight of water (62.4 lb/ft³),
h = pressure drop (ft).

If the total length of the mixing unit is assumed to be 8 ft, the volume of the unit is

$$[0.785 \times (2)^2] \times 8 = 25.12 \text{ ft}^3$$

According to the manufacturer of the static mixer unit, the pressure drop is computed by the following formula:

$$\Delta P = \left(\frac{0.007 \times Q}{D^{4.4}}\right) N$$

where ΔP = pressure drop (psi),
Q = flow rate (gpm),
D = pipe diameter (in.),
N = number of mixing elements.

Based on this information, a static mixer with two mixing elements would give the following results:

Flow Rate (mgd)	Head loss (ft)	G at 10°C (s⁻¹)	t (s)	$G \times t$
5	0.17	346	3.2	1107
10	1.34	1340	1.6	2144
15	3.0	2520	1.1	2770

However, the desired $G \times t$ value should be below 2000 at the maximum plant flow rate and approximately 1300 with an average flow rate. For this reason, the pressure drop characteristic of the element should be based on the desired $G \times t$ values; the manufacturer can reduce the pressure drop across each element by removing a portion of the baffle plates. The following table demonstrates how the desired pressure drop is achieved:

Flow Rate (mgd)	Approximate Head loss Across Mixer (ft)	G at 10°C (s⁻¹)	t (s)	$G \times t$
5	0.028	140	3.2	450
10	0.46	806	1.6	1290
15	1.5	1782	1.1	1960

Example 4   Chemical Application

Given   The following water treatment chemicals are selected to be fed at the head of the plant: alum, cationic polymers, anionic polymers, nonionic polymers, sodium hydroxide, chlorine, ammonia, PAC, and potassium permanganate.

Determine

(i) The chemical application sequence.

(ii) The number of potential application points.

(iii) The type of mixing system.

Solution

(i) The chemicals that potentially interact if fed together are alum and anionic polymers, alum and sodium hydroxide, cationic and anionic polymers, chlorine and PAC, ammonia with chlorine and alum, and chlorine with certain cationic polymers. Alum is a chemical that requires instantaneous dispersion. PAC and potassium permanganate must be fed ahead of the coagulant in order to enhance their effectiveness.

(ii) Based on the various considerations discussed earlier, the following scheme is established:

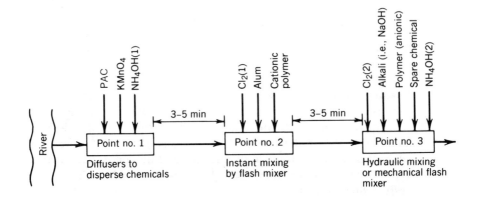

Ideally, each application point should be separated by a time lag of at least 5 min, but this is often difficult to attain. The first application point should therefore be as far upstream from the plant as practical. The second point is located at the head of the plant. It is also difficult to place the third application point so that it meets the time lag criterion; thus, it should be situated as far as possible from the second point. Anionic and nonionic polymers are good aids in floc formation if added when pinpoint flocs are formed by alum. This process generally requires 5 min. For this reason, the application points for anionic and nonionic polymers are best located in the second stage of flocculation.

(iii) As illustrated in the previous diagram, there are three application points:

Point 1: Dispersion of chemicals by diffusers.

Point 2: Instant mixing by static mixers or pump diffusion.

Point 3: Mixing by mechanical or hydraulic mixer.

Operation and Maintenance

The most important operation and maintenance aspect of the initial flash mixing process is the selection of the proper chemicals and the application of the correct quantities to an ever changing raw water quality and plant flow rate. Jar tests have been used effectively in optimizing coagulant dosage for conventional processes. However, they do not appear to supply the plant operator with the proper coagulant dosage for direct filtration and in-line filtration processes—hence the development of glass tube pilot filters. Although the glass tube pilot filter is widely used, approximately 50% of plants equipped with this coagulant control system do not use it because the system requires a great deal of operator judgment and attention.

Zeta potential measurements have also been tried, but with minimal success. In recent years, streaming current detector (SCD) units have improved upon the zeta potential measurement and appear to be successful in coagulant control. SCD units can be used for any type of water treatment process using coagulant. Yet, it still requires a significant amount of operator attention.

The second topic that must be stressed is the constant monitoring of the chemical feed and flash mixing systems. The most common water treatment problems arise from the clogging of chemical feed lines and diffusers. Malfunctioning of mechanical mixers can easily be detected by the plant operator. However, clogged static mixers, broken underwater chemical feed lines, and clogged diffuser orifices are not as easily detected unless the water is drained. Thus, scheduled plant shutdowns must be exercised during periods of low water demand.

The third consideration in proper operation and maintenance is adjustment of the mixing energy. Mechanical mixers and pump diffusion flash mixers must have their mixing energy altered with seasonal changes in raw water quality. Cold water is more viscous and theoretically requires a higher energy input to achieve the same mixing effect as obtained during the summer months. Extra power is also required to disperse coagulant when processing raw water that is highly turbid or colored.

Treatment plants that are equipped with mechanical flash mixers should have the following items inspected on a regular basis: the motor and speed reducer, mixer shaft run-out, vibrations, and noises. The lubrication oil should also be changed at certain time intervals, or by seasons, based on the recommendations of the manufacturer.

It must be emphasized that coagulation, flocculation, and clarification processes are pretreatments to filtration. Plant operators generally tend to produce large flocs and attempt to make the settled water as clear as possible by applying rather large amounts of coagulant. Consequently, the effect of the flash mixer is often undetected and is therefore turned off.

The role of the flash mixer may or may not be evident, depending on the type of process. The conventional water treatment process uses a large amount of coagulant that is more readily dispersed, forming large flocs and clear settled water. However, most modern high-rate filtration processes use only 20–30% less than the conventional alum dosage; thus, vigorous flash mixing is essential for instant dispersion. Since direct filtration processes use such small amounts of alum (4 to 6 mg/L), compared to the conventional processes, they produce very smaller floc and higher settled water turbidity. Most modern high-rate filters are capable of filtering settled water with turbidity greater than 1–2 NTU, and in many cases 3–6 NTU are considered to be optimum.

The primary objective of flash mixing is to effectively coagulate colloidal matter present in raw water by applying the proper amount of coagulant and through good flash mixing, prior to the flocculation process. This is particularly important for direct and in-line filtration processes.

Coagulation Control Test The jar test is the most widely practiced and proven method of evaluating, as well as controlling, the coagulation, flocculation, and clarification processes. Although widely used, the jar tests provide the operator with little information as to the mode of filtration (direct or in-line filtration) or whether the tests are being performed improperly. When correctly designed and performed, the jar tests may yield very meaningful data which can assist engineers in designing new treatment plants or in modifying existing plants. Plant operators also benefit from the jar tests since they will be able to optimize the efficiency of the treatment process. The recommended jar test procedure is presented in Appendix 7.

The jar tests have traditionally been performed under fixed test conditions to determine the relation between coagulant dosage and settled water turbidity. However, it is very important to evaluate the filterability of the settled water by filter paper filtration (simulated filter) since water with the lowest turbidity is not necessarily the best filter influent with respect to filter efficiency.

There are four basic objectives for the jar test:

1. Optimization of coagulant.
2. Optimization of the chemical application sequence.
3. Optimization of the mixing energy and time ($G \times t$).
4. Evaluation of clarifier and filtration performance.
5. Evaluation of Corrosive Characteristics of settled water.

OPTIMIZATION OF COAGULANT   In order to optimize the coagulation process, operators must choose the most appropriate type of coagulant (alum, ferric chloride, cationic polymers) and evaluate the optimum dosage under standard mixing conditions (to be described later). Coagulant solution strengths should also be considered since they may affect the decision. For example, by using a combination of alum and cationic polymers, the amount of sludge production is often reduced. However, cationic polymers cost 10–15 times more than alum, and the polymer dosage is therefore usually limited to 2 mg/L as a matter of economics.

OPTIMIZATION OF THE CHEMICAL APPLICATION SEQUENCE   Optimization of the chemical application sequence is a very important consideration; yet, little attention has been given to this matter. For instance, the sequence of application is extremely important when using alkali chemicals and anionic or nonionic polymers (in conjunction with alum) as flocculation aids. As described earlier, these polymers help produce excellent floc effectively only if they are fed a few minutes after alum is added to the process water.

OPTIMIZATION OF MIXING ENERGY, MIXING PATTERN, AND MIXING TIME   The optimum mixing condition depends on the type of overall treatment process train, the type of filter bed of the plant, and the raw water characteristics. In general, a tapered mixing—the use of high-intensity mixing at the beginning, followed by a gradual reduction toward the end—helps promote better flocculation, under any conditions.

Under actual plant conditions, it is common for the conventional treatment process to have a velocity gradient path of 50 s^{-1} at the beginning and 10 s^{-1} at the end of 30–40 min of mixing: $G \times t = 4 \times 10^4$ to 5×10^4. For the direct filtration process, the accepted range for the velocity gradient is from 65 to 15 s^{-1}; for 15–30 min of mixing, $G \times t = 4 \times 10^4$ to 6×10^4. Thus, these ranges should be kept in mind when conducting the jar test. Due to excessive shear force, flocculation times that are too long (over 45 min) generally produce floc types that are less than optimum for settling.

EVALUATION OF CLARIFIER AND FILTRATION PERFORMANCE   The design criteria for new sedimentation tanks or the performance of existing clarifier units may be evaluated by measuring the floc settling velocity through the use of the jar test. A settling velocity of 40 mm/min (1.5 in./min) is roughly equivalent to a hydraulic surface loading of 1 gpm/ft^2 (2.5 m/h), which is the maximum loading rate of ordinary sedimentation tanks, for alum floc, during the summer months. If the settling velocity of alum floc is much slower than the corresponding surface loading rate of the existing clarifier unit, the floc characteristics may be improved by either adding anionic polymer (0.15–0.3 mg/L) to the alum flocculation process or by substituting ferric chloride for alum; these must be evaluated by a series of jar tests. Refer to Camp (1946), Hudson (1981), and Singley (1981) for discussions on floc settling velocity measurement (see Bibliography at end of this section).

Filter performance may be evaluated by passing the supernatant through Whatman No. 1 filter paper. The best filter pretreatment is one that results in a filtrate turbidity that is at or below the turbidity goal (0.2 NTU) within the shortest filtration time.

As a rule of thumb, the following items should be recorded during the jar test:

1. Time at which pinpoint floc appears.
2. Size of the floc at the end of the flocculation period.
3. Turbidity, color (optional), pH, alkalinity, and temperature of the raw water and of selected settled water.
4. Turbidity of filtered water and the time required to filter through 50 ml of selected settled water.

Figure A7-2 in Appendix 7 presents a scale for floc size identification. Table A7-1 is an example of jar test results.

BIBLIOGRAPHY

ASCE, AWWA, and CSSE, *Water Treatment Plant Design*, AWWA, New York, 1969.

Camp, T. R., "Sedimentation and the Design of Settling Tanks," *ASCE Trans.*, 3:895(1946).

Chao, J. L., and Stone, B. G., "Initial Mixing by Jet Injection Blending," *J. AWWA*, 71:570(October 1979).

Fair, G. M., et al., *Elements of Water Supply and Disposal*, Wiley, New York, 1971.

Guven, O., and Benefield, L., "The Design of In-line Jet Injection Blenders," *J. AWWA*, 75:357(July 1983).

Hudson, H. E. Jr., *Water Clarification Process*, Van Nostrand Reinhold, New York, 1981.

Kaufman, W. J., et al., "Initial Mixing and Coagulation Process," Report No. 72-2, UC Berkley (February 1972).

Kawamura, S., "Coagulation Considerations," *J. AWWA*, 65:417(June 1973).

Kawamura, S., "Considerations for Improving Flocculation," *J. AWWA*, 68:328(June 1976).

Montgomery, J. M., Consulting Engineers, *Water Treatment Principles and Design*, Wiley, New York, 1985.

Sanks, R. L., *Water Treatment Plant Design for Practicing Engineers*, Ann Arbor Science Publishers, Ann Arbor, MI, 1979.

Singley, J. E., "Coagulation Control Using Jar Tests," AWWA Proceeding, No.20155, St. Louis, 1981.

Stumm, W., and O'Melia, C. R., "Stoichiometry of Coagulation," *J. AWWA*, 60:514(May 1968).

Weber, W. J., *Physicochemical Processes*, Wiley-Interscience, New York, 1972.

3.2.4 Flocculation Process

Purpose Flocculation is the gentle mixing phase that follows the rapid dispersion of coagulant by the flash mixing unit. Its purpose is to accelerate the rate of particle collisions, causing the agglomeration of electrolytically destabilized colloidal particles into settleable and filterable sizes.

The terms coagulation and flocculation are sometimes used interchangeably in technical literature. However, the aggregation of particulate material is actually a two-step process. The initial step involves the addition of coagulant, which reduces or eliminates the interparticulate forces that are responsible for the stability of the particulates; this process is called coagulation. The subsequent particulate collisions and enmeshment of particulates into flocs occur as the result of molecular motion and the physical mixing of the liquid; this is flocculation.

Considerations The following factors must be taken into consideration when designing a proper flocculation process: (1) raw water quality and flocculation characteristics, (2) treatment process and finished water quality goals, (3) available hydraulic headloss and plant flow variations, (4) local conditions, (5) cost, (6) relation to existing treatment facilities, and (7) miscellaneous factors. These items are discussed in detail.

Raw Water Quality and Flocculation Characteristics These are the first considerations when designing a flocculation facility. The design engineer should ideally have access to data on the year-round quality of the raw water, spanning a period of 5 years or even longer. The seven important water quality aspects are turbidity, total organic compound, pH, alkalinity, color, algae counts, and temperature. The nature of the colloids, particularly the colloidal organic compounds, and the particle size distribution on the turbidity are characteristics that are preferably known since these factors greatly affect the flocculation characteristics. The magnitude of the raw water turbidity alone is not likely to be a surrogate to coagulant dosage requirements (see Figures 3.2.4-1 and 3.2.4-2).

If historic water quality data are not available, an immediate sampling should be established prior to or as close as possible to the beginning of the predesign study period. Data on the seven basic water quality parameters should then be collected for a period of 1 year. However, if this schedule cannot be implemented, an attempt should be made to at least obtain the water quality data in both dry and wet seasons.

Flocculation characteristics may be evaluated by the jar test—a process commonly practiced in water treatment plants. The jar test must be conducted in a systematic fashion so that it will yield the necessary basic data for the design of the flocculation process as previously discussed in Section 3.2.3. The duration of the test should be as long as can be afforded. Yet, as a minimum, repeatability of the results must be demonstrated.

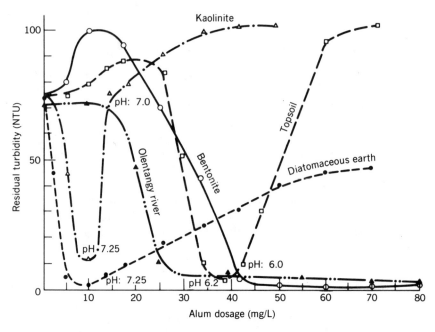

	Turbidity NTU	pH	Alkalinity meq/L
Bentonite	75	8.3	0.4
Diatomaceous	75	9.0	0.4
Kaolinite	75	8.7	0.4
Topsoil	75	6.7	0.4
Olentangy R.W.	65	8.02	3.26

(a)

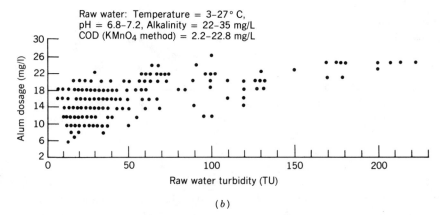

(b)

Figure 3.2.4-1 (a) Nature of turbidity and alum dosages on flocculation. (Adapted from S. Kawamura, *J. AWWA*, 68(6):328 (June 1976).) (b) Relation between raw water turbidity and optimum alum dosage for Yodo River, Japan. (Adapted from S. Kawamura, *J. Jpn. WWA*, 239:17 (September 1954).)

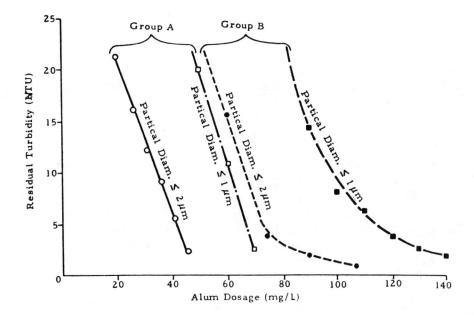

Figure 3.2.4-2   Effect of particle size on flocculation. (S. Kawamura, A Thesis, Ohio State Univ.) 1961

Treatment Process and Finished Water Quality Goals   The overall treatment process and finished water goals are the second most important considerations because both affect flocculation. For instance, in a direct filtration process, the flocculation tanks should not produce large settleable flocs since a sedimentation process is not involved. Instead, the flocculation basins should produce an appropriate type of floc so that "in-depth" filtration can be achieved by a reverse graded filter bed. In contrast, the conventional complete and the lime softening processes require thorough flocculation to produce good settleable floc with a filter influent turbidity of less than 5 NTU, regardless of the raw water quality. This is particularly true for single sand rapid filters that are characterized by "surface" filtration.

The finished water quality goals also influence the degree of flocculation. Excessive amounts of the following substances in the raw water may effectively be removed by improving the flocculation and sedimentation steps of the process train: color as a THM precursor, algae as a source of taste and odor, asbestos fibers (limitations), and certain toxic metals and compounds.

Available Hydraulic Headloss and Plant Flow Variations   Other considerations in the design of the flocculation process are the allowable headloss across the flocculation process and the magnitude of plant flow variation. If the allowable headloss is limited, hydraulic flocculation methods are ruled out. Thus, the design engineer is forced to use mechanical methods.

Similarly, plant flow rate fluctuation is an important consideration. If the fluctuation is relatively minor year-round, that is, ±50% variation from the daily average flow rate, hydraulic flocculation is applicable and will perform effectively.

Local Conditions   Local conditions should also be analyzed when selecting an appropriate type of flocculation process. The five major factors that should be evaluated are site topography, climatic conditions, availability of services, capability of the operational personnel, and level of the local technology.

Cost   Cost is always an important consideration. Both the initial cost (capital) and the operation and maintenance expenses should be evaluated.

Relation to Existing Treatment Facilities   In the case of plant expansion, the relation of the new flocculation process to the existing is a realistic problem. The basic rule is to make all the flocculation units identical so that uniform performance and operation and maintenance procedures can be maintained.

Miscellaneous   Miscellaneous considerations, such as the hydraulic characteristics of the flocculation tank and scum and sludge removal, must be incorporated during the design phase. The prevention of flow short-circuiting, the creation of effective eddies and turbulence within the tank, and the minimization of high shear forces created by the mixing blades are all important hydraulic problems. Important operational problems include the ease of scum removal from the water surface and of sludge and silt from the bottom of the basin.

Some chemists also emphasize the effect of ion concentrations, found within the raw water, on flocculation since the salt levels of the water influence the coagulant dosage requirements. Since the drinking water quality standards for the TDS of raw water already limits the presence of excessive amounts of ions, this issue is not of great concern for most design engineers.

In recent years, the process of ozonation has become popular due to the stringent limitations on trihalomethanes in drinking water. Ozone is capable of both oxidizing substances and flocculating certain organic and inorganic substances. Therefore, the flocculation process is generally more effective if it is preceded by ozonation.

Type and Selection Guide   This section describes the commonly available alternatives to the flocculation process, as well as guidelines for making the proper selections.

Available Alternatives   Flocculation mixing may be provided by either mechanical mixers or baffles. These units reduce the flow short-circuiting and induce the collision and agglomeration of particles. Figures 3.2.4-3 and 3.2.4-4 present the most common types of mechanical mixer and baffle used in the field of water treatment. The typical categories of mixing systems are listed below.

Mechanical Mixing

- Vertical shaft with turbine or propeller type blades
- Paddle type with either horizontal or vertical shafts
- Proprietary units such as Walking Beam, Flocsilator, and NU-treat.

Baffled Channel Basins

- Horizontally baffled channels
- Vertically baffled channel

Reactor Clarifier Proprietary Systems

Contact Flocculation (Gravel Packed Filter)

Diffused Air or Water Jet Agitation

Selection Criteria   Selection of the flocculation process should be based on the following criteria:

- Type of treatment process, for example, conventional, direct filtration, softening, or sludge conditioning
- Raw water quality, for example, turbidity color, and temperature
- Flocculation characteristics in response to changes in mixing intensity and mixing times

The following criteria should be used when selecting the type of mixing:

- Local conditions such as high wind, ice buildup, or increased viscosity of the gear oil in extremely cold periods
- Available headloss across the plant
- Shape and depth of the basin
- Capital and O&M costs

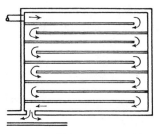

Plan of round the
end baffled mixing basin.

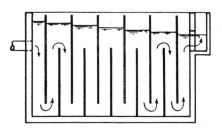

Section through over-and-under
baffled mixing chamber.

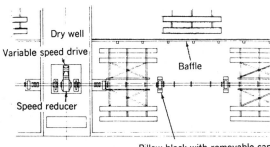

Floculator of
horizontal shaft
with mixing paddles

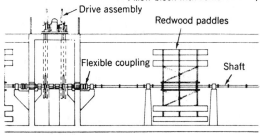

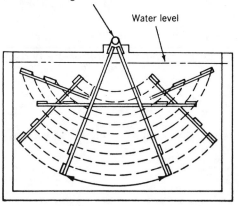

Swinging paddle type
flocculator

Figure 3.2.4-3 Baffled channels and mechanical flocculators.

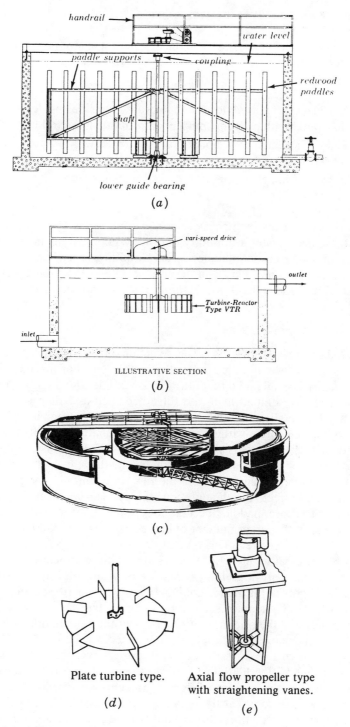

Figure 3.2.4-4 Mechanical flocculators. (a) Paddle wheel on vertical shaft. (b) Turbine mixer. (c) Moving lattice type. (d) Plate turbine type. (e) Small size vertical shaft type.

ORDER OF PREFERENCE   Based on the design of the most recent installations, the order of preference, for the selection of the flocculation process equipment, is as follows.

1. Vertical shaft flocculators, with hydrofoil type long blades, in horizontal flow tanks with proper compartmentalization in each flocculation stage.
2. Paddle flocculators in horizontal flow tanks with proper compartmentalization in each stage.
3. Baffled channel flocculation tanks for plants with fairly constant flow rates.

Reactor clarifiers and contact flocculation systems are in a category of their own; these are usually proprietary items of equipment manufacturers. Both systems may be considered under certain conditions based on plant size (usually small), local conditions, and the type of process (i.e., reactor clarifier for lime softening and contact flocculation for direct filtration with multimedia).

Discussion of Alternatives

Mechanical Mixing Systems   The selection of a proper flocculation unit depends on the overall unit process that is selected. For instance, if the raw water quality is good and direct filtration is feasible, the filter bed most likely will be a reverse graded high-rate filter bed such as a dual media type. In this situation, the floc should be small, yet physically strong in order to resist the high shearing forces within the bed. Vertical shaft high-energy flocculators produce a floc that meets these requirements. By contrast, when the raw water quality is poor (i.e., polluted or has medium to high turbidity), the suspended matter is removed from the raw water through the production of good floc and effective sedimentation; this is especially true for cold water regions. In this case, paddle flocculators generally produce larger and heavier floc because of the larger number of blades, larger total blade perimeter, and larger blade surface area. This type of flocculator rotates at slow speeds and is capable of creating more localized eddies with less shearing forces. However, a vertical shaft flocculator is capable of producing large and heavy floc if an anionic polymer is used in conjunction with alum in a specific application sequence. These examples serve only to stress the importance of the relationship between the selection of process and the flocculation equipment.

When a mechanical mixer is selected as a flocculator, the mixer should have the following characteristics: (1) it must deliver the G-values specified for each stage of flocculation; (2) it must provide sufficient eddies and turbulence so that the required velocity gradients are produced; (3) it must provide low shear forces at the edges of the mixing blades, especially during the last two stages of flocculation; and (4) it should have low maintenance and operational costs. Certain types of mixer produce a relatively uniform flow motion or a rotational motion because of their rapidly rotating blades.

Yet, the localized shear forces at the tips of the blades are high. Therefore, this type of mixing is not suited for the actual flocculation process.

Regardless of the type of flocculator, tapered mixing across the flocculation tank is always an important consideration. The last stage of the flocculation process often requires a delicate and gentle mixing condition to promote the growth of floc size. For this reason, it is best to specify an infinitely variable speed control unit (frequency control) for the final stage flocculators. Examination of records from various plants reveals that the first and second stages do not require frequent changes in speed. Therefore, a two-speed motor control mixer may be used.

Another aspect that should be considered is the reliability of a flocculation system. Vertical shaft flocculation tanks contain a large number of mixer units. Failure of one or two of the flocculator drive units usually does not significantly affect the overall efficiency of the system. Conversely, the horizontal shaft paddle flocculators usually have only one drive unit for several paddles; failure of one drive unit will terminate the movement of several flocculator paddles. It is therefore important to specify an extremely reliable drive unit when using the horizontal shaft flocculators.

In summary, all other factors being equal, vertical shaft flocculators are usually the first choice for the following reasons: (1) minimal maintenance, (2) operational flexibility, (3) very little headloss across the tank, (4) easy control of mixing intensity, (5) effectiveness, and (6) minimal impact to the overall performance if one unit malfunctions. The drive units should have hydrofoil type mixing blades and all shaft bearings should be located above the water surface of the basin. Independent of the type of mechanical mixer are the requirements for sufficient walkway space (4 ft) around each mixing unit for the control panels, for power connections, and for the installation, removal, and regular maintenance of the equipment.

Baffled Channel System   The main factors in determining the applicability of the baffled channel flocculation process are the local conditions and the presence of a moderate amount of allowable headloss across the tank. For example, many developing countries commonly have problems with properly maintaining the equipment. Since the baffled channel flocculation process does not have any mechanical parts, it should seriously be considered in this type of situation.

When properly designed, both the horizontally baffled (around-the-end flow) and the vertically baffled (over-and-under flow) channels perform well and exhibit good plug flow characterstics. However, the baffled channels have two disadvantages; there is a significant headloss of 1–2 ft (0.3–0.6 m) across the tank and the mixing intensity is a function of plant flow rate.

This discussion is not meant to imply that the use of the baffled channel flocculation process is restricted to developing countries. If the conditions of flow rate variation and available headloss can be met, this process should also be considered for use in developed countries. Some of the newer plants in both the United States and Japan use baffled channel flocculation tanks.

Contact Flocculation   Contact flocculation (gravel packed filter) is another design that requires minimal maintenance. Although it is a relatively new concept, contact flocculation is similar to the in-line filtration process. Some equipment manufacturers such as Neptune Microfloc and Culligan have long been promoting this system in their packaged water treatment units.

The advantage of contact flocculation is its compact size and its lack of moving parts. Hydraulic loadings of 10–15 gpm/ft^2 are possible and the system requires an empty bed contact time of only 3–5 min to produce a floc with the proper characteristics.

This type of flocculation system depends on plant flow rate and requires minimal variations in flow rate and water temperature for adequate mixing intensity. Contact flocculation also requires frequent backwashing with air scour.

Diffused Air or Water Jet Agitation   Of all the processes, the least preferred are diffused air and water jet flocculation. The diffused air system is characterized by a high rate of energy consumption and is also an inefficient flocculation process. Therefore, this system should serve as an auxiliary or temporary measure. Use of a water jet mixing system is not recommended for flocculation because the high shearing force on the jet path restricts the growth of floc size.

Design Criteria   When designing a flocculation process, the selection of the mode of mixing and the determination of the physical relations and characteristics of the flocculation tanks and clarifiers (sedimentation tanks) are among the first decisions to be made. As previously described, either hydraulic mixing or mechanical mixing may be chosen. In evaluating the physical characteristics, a decision must be made as to whether the flocculation tank will be built as part of the clarifier or as an independent structure. After establishing these two issues, the design may be refined: establish the number of tanks; consider the total number of tanks when the plant capacity reaches the ultimate capacity; define the number of mixing stages and the mixing energy level in each stage; and determine the type of baffle that will be used to minimize flow short-circuiting.

One of the most efficient and economical flocculation tank designs incorporates the basin into the clarifier unit. The majority of equipment manufacturers exclusively design their units in this manner. For custom design, the flocculation basin may also be combined with the clarifier by placing it at the influent end of a rectangular sedimentation tank. This arrangement offers many advantages, such as minimal land requirements, minimal floc breakage between the two processes, simplicity of design, and, above all, simple and economical construction.

The design of the flocculation tank is commonly based on only two criteria: detention time and mixing energy level. The energy level is the *G*-value or velocity gradient and is defined by Camp (1943) as

$$G = \left(\frac{P}{\mu V}\right)^{0.5}$$

where P = power input to the fluid,
  V = volume of the tank,
  μ = absolute viscosity of the fluid.

The flocculation process must be designed to provide maximum interparticle contact. These contacts are a function of the velocity gradients within the fluid; the velocity gradients are produced by hydraulic or mechanical mixing (orthokinetic flocculation). The number of particle contacts is expressed as

$$N = n_1 n_2 (G/6)(d_1 + d_2)^3$$

where N is the number of contacts between n_1 particles of diameter d_1 and n_2 particles of diameter d_2 in a unit of time. Therefore, the rate of flocculation increases with the number and size of the particles and with the power input, but decreases with the viscosity of the fluid.

The mean velocity gradient (G-value) of certain types of flocculation system may be computed as

Baffled channel: $\quad G = (gh/vt)^{0.5}$

where v = kinematic viscosity of fluid,
  t = mean detention time,
  g = gravity acceleration constant,
  h = headloss across the tank.

Mechanical mixers with paddles: $\quad G = (C_D A v^3/2 v V)^{0.5}$

where C_D = a drag coefficient that depends on the paddle shape and flow conditions (see Figure 3.2.4-5),
  A = cross-sectional area of the paddles,
  v = relative velocity of the paddle with respect to the fluid,
  v = kinematic viscosity of the fluid,
  V = volume of the flocculation tank.

It is estimated that v is from 0.5 to 0.75 of the peripheral velocity of the paddle.

The proper application of the parameters presented above is described in the following section.

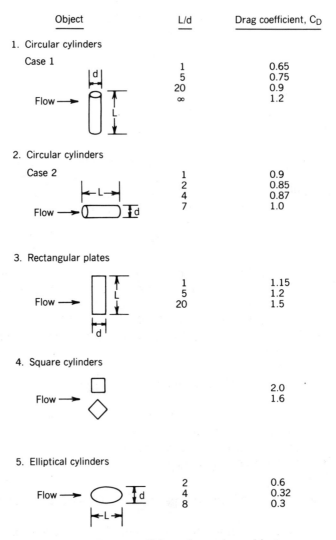

Object	L/d	Drag coefficient, C_D
1. Circular cylinders		
Case 1	1	0.65
	5	0.75
	20	0.9
	∞	1.2
2. Circular cylinders		
Case 2	1	0.9
	2	0.85
	4	0.87
	7	1.0
3. Rectangular plates		
	1	1.15
	5	1.2
	20	1.5
4. Square cylinders		
		2.0
		1.6
5. Elliptical cylinders		
	2	0.6
	4	0.32
	8	0.3

Figure 3.2.4-5   Drag coefficients for various objects.

General   The general design criteria for a basic rectangular flocculation tank are as follows:

Energy imput	$G \times t = 1 \times 10^4$ to 1×10^5
	t is in seconds (5×10^4 s^{-1} average)
	or $G = 10$–70 s^{-1} (30 s^{-1} average)
Detention times	20–30 min at maximum daily flow rate
Water depth	10–15 ft (3–4.5 m)
Flocculation stages	Two to six stages (commonly three to four)

The following factors will determine the number of flocculation stages to be included in the design.

- Type of subsequent treatment unit (sedimentation, filter, or sludge drying bed) and the type of overall process (direct filtration or complete process)
- Quality and treatability of the raw water: turbidity levels, the nature of the turbidity, and water temperature during the cold months
- Degree of flow short-circuiting across the flocculation basin (with or without baffles) and the type of baffles
- Local conditions: regulatory requirements and the operation and maintenance of the flocculation process

Vertical Shaft Flocculators   In a mechanical mixer the diameter of the mixing blade (D), in relation to the tank diameter (equivalent diameter T), is important and D/T should be greater than 0.35. Another item to watch is the maximum flow velocity induced by the mixing blade: it should be less than 8 fps (2.5 m/s) in the first stage and less than 2 fps (0.6 m/s) in the last stage of the flocculation process if good and settleable floc formation is to be expected. Mixing equipment manufacturers have established the design criteria for mechanical mixers based on their research and experience. However, their main objectives are for liquid–liquid, liquid–gas, and solid–liquid mixing as a batch operation. Thus, the manufacturers often do not have a good understanding of flocculator design for a continuous flow mixing process and the production of fragile flocs. When properly produced, the floc should have the characteristics and appearance of snow flakes.

There are two basic philosophies in specifying the mechanical flocculators. The first is simply to specify the function and a few physical characteristics such as G-value, tip speed, and the D/T ratio. The other is to write detailed specifications for the equipment, the installation procedures, and the exact power requirements including the dimensions and the elevation of the mixing blades. It appears that the first approach is better for the majority of engineers since the proper design of vertical shaft flocculators is somewhat state-of-the art.

Figure 3.2.4-6 shows the energy transmittal efficiency for some common mixing blades when used as flocculator units.

Horizontal Shaft Flocculators   The advantage of the horizontal shaft unit is that one shaft can operate a number of agitators and thereby reduce the number of drive units. However, its reliability is low because failure of one shaft results in a 25–33% loss in mixing capability. The horizontal shaft units can be equipped with either paddles or turbine blades. The basic design criteria for horizontal shaft flocculators are the same as those for the vertical type, except for the type of paddle units.

- The total paddle area should be 10–25% of the tank cross-sectional area. If the values are over 30%, the paddles will produce a distinct rotating

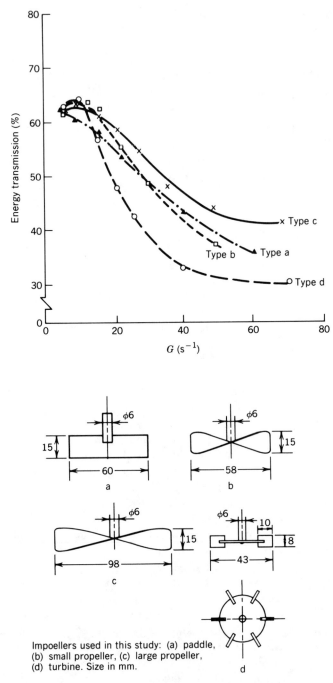

Impoellers used in this study: (a) paddle, (b) small propeller, (c) large propeller, (d) turbine. Size in mm.

Figure 3.2.4-6 Mixing efficiency as a function of G-value and type of impeller. (After Leentraar)

motion in the water, causing a reduction in local eddies and turbulence along the edge of the paddles.

- Each arm should have a minimum of three paddles so that dead space, especially near the shaft, will be minimized.
- The peripheral speed of the paddles should be between 0.5 and 3.3 fps (0.15–1 m/s).
- The designed G-value is lower than that of vertical shaft flocculators; it should initially be 50 s^{-1}, then reduced to 10 or even 5 s^{-1} in the last stage of the flocculation tank.
- For a direct filtration process, which does not have clarifiers, the flocculation tanks may be designed with "around-the-end" channels. This type of design minimizes flow short-circuiting for horizontal shaft flocculators. If the flocculation tank is part of a rectangular clarifier tank, the flow direction for both processes should be the same in order to avoid tangential inflow into the clarifier.

Tables 3.2.4-1 and 3.2.4-2 present the general design criteria and compare the major characteristics among baffled channel, horizontal shaft, and vertical shaft flocculators.

BAFFLED WALLS   When mechanical mixers are selected for the flocculation process, the use of baffles or compartmentalization improves flocculation. The shape and location of the baffles are determined by the design engineer because there are no practical design guides.

Design engineers are strongly advised to compartmentalize each stage of the mechanical mixer flocculation process with a proper baffled wall. Each baffle should have orifices that are 4–6 in. (0.1–0.15 m) in diameter uniformly distributed across its vertical surface and a velocity of 1.2–1.8 fps (0.35–0.55 m/s) should be produced through the orifice holes of each baffle at maximum flow rate; the higher velocity of 1.8 fps should be applied to the first baffle and the lower velocity of 1.2 fps to the last baffle.

In general, each baffle is placed across the tank perpendicular to the flow path. The top of the baffled wall is slightly submerged (0.5 in.) to allow the scum to flow over, and the bottom of the wall should have a space of 0.5 in. for easy drainage and sludge removal. To prevent flow short-circuiting, engineers should not design a large passageway; the passage may be greater if it is blocked off during normal operation or if a portion of the diffuser wall is capable of swinging open (by hinges) only during the cleaning operation. If timber is used to construct the baffle, it will tend to warp if the span is too long (over 10 ft) or if it is too thin (less than 2 in. net).

If the flocculation tank is designed as part of the sedimentation tank, the baffled wall located between the tanks is called the diffuser wall. In this case, the baffled wall should produce a flow rate velocity of no more than 0.8 fps (0.25 m/s) in order to avoid excessive breakage of the developed floc. Another

TABLE 3.2.4-1 General Design Criteria

	BAFFLED CHANNELS	MECHANICAL FLOCCULATORS	
		HORIZ. SHAFT WITH PADDLES	VERT. SHAFT WITH BLADES
G (SEC-1)	50 TO 5 TAPERED	50 TO 10 TAPERED	70 TO 10 TAPERED
T (MIN)	30 TO 45	30 TO 40	20 TO 40
FLOCC. STAGES (NO. OF CHANNELS OR NO. OF HORIZ. SHAFTS)	6 TO 10	3 TO 6	2 TO 4
MIX ENERGY CONTROL	FLOW PASSAGE VARIATION	VARIABLE MIXING SPEED	VARIABLE MIXING SPEED
MAX. FLOW VELOCITY OR MIXER TIP SPEED	3FPS	3FPS	6 TO 9 FPS
BLADE AREA/TANK AREA	—	5 TO 20%	0.1 TO 0.2%
BLADE: D/T	—	0.5 TO 0.75	0.2 TO 0.4
SHAFT RPM	—	1 TO 5	8 TO 25
MAJOR APPLICATION	CONV. COMPLETE TREAT.	CONV. COMPLETE TREAT.	DIRECT FILTRATION & CONV. COMPLETE TREAT.

TABLE 3.2.4-2 Comparison of Three Basic Flocculation Processes

| | BAFFLED CHANNEL | MECHANICAL FLOCCULATORS | |
		HORIZ. SHAFT WITH PADDLES	VERT. SHAFT WITH BLADES
FLOCCULATION	GOOD TO EXCELLENT	GOOD TO EXCELLENT	FAIR TO GOOD
RELIABILITY	GOOD	FAIR TO GOOD	GOOD
OPER. FLEXIBILITY	MODERATE TO POOR	GOOD	GOOD
CAPITAL COST	LOW	MODERATE TO HIGH	MODERATE TO HIGH
CONSTRUCTION	EASY	MODERATE	EASY TO MODERATE
MAINTENANCE	LOW	MODERATE	LOW TO MODERATE
FLOW CONDITION	NEAR PLUG FLOW	SHORT CIRCUITING	SHORT CIRCUITING
MAJOR ADVANTAGES	1) SIMPLE & EFFECTIVE	1) GOOD FLOC FORMATION	1) HIGH ENERGY INPUT
	2) LOW O & M COSTS	2) EFFECTIVE MIXING SYSTEM WITH GOOD TURBULENCE	2) LOW MAINTENANCE
	3) NO MOVING PARTS	3) NO HEAD LOSS	3) NO HEAD LOSS
MAJ. DISADVANTAGES	1) MIXING ENERGY IS A FUNCTION OF FLOW RATE	1) PRECISE INSTALLATION REQUIRED	1) MANY UNITS REQUIRED
	2) APPROX. 2 TO 3 FT HEAD LOSS	2) LIMITED HIGH ENERGY INPUT	2) HIGH SHEAR AT BLADES
		3) REGULAR MAINTENANCE REQUIRED	3) INADEQUATE AMOUNT OF TURBULENCE & EDDY

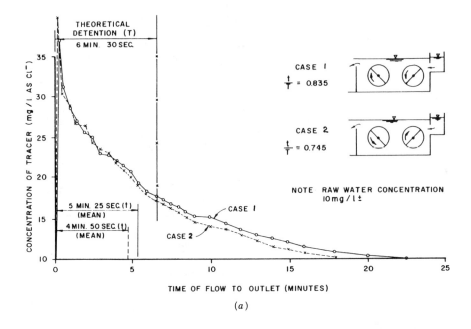

(a)

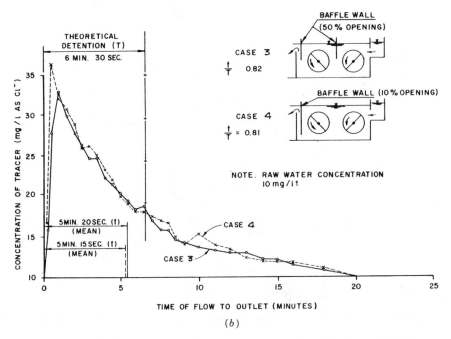

(b)

Figure 3.2.4-7 (a,c) Flocculator arrangements versus detention time. (b,d) Flocculators with baffles versus detention time.

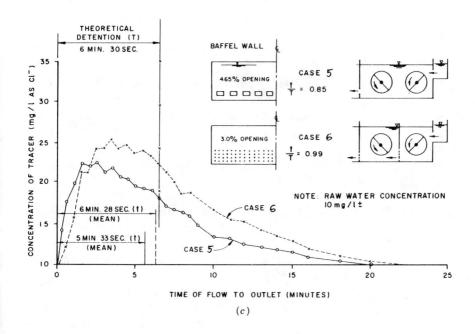

(c)

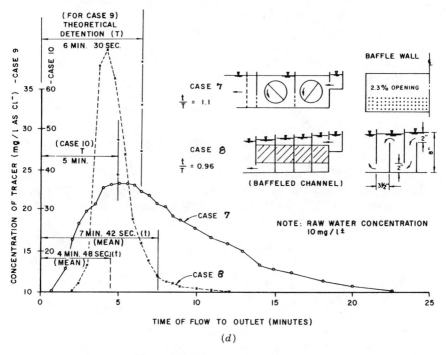

(d)

Figure 3.2.4-7 (Continued)

advantage of this type of baffled wall is the production of additional mixing energy. Baffles that are based on the above criteria provide a G-value of 5–25 s^{-1}, depending on the flow velocity through the baffles. Figure 3.2.4-7 illustrates the results of a scale model hydraulic study on the determination of an effective baffled wall system.

STATOR BAFFLES   The efficiency of vertical shaft mixers can be improved through the installation of stator baffles. The purpose of stator baffles is to increase the effective turbulence, as well as to prevent rotational fluid motion within the tank. A proper stator baffle design specifies the installation of two or four pairs of baffles on opposite sides of the compartment. The width of each individual stator baffle should be $\frac{1}{8}$ to $\frac{1}{12}$ the distance across the individual mixer compartments.

Baffled Channel (Hydraulic) Flocculators   The around-the-end baffles channel and the over-and-under baffled channel are the two basic types of baffled channel flocculator. The around-the-end baffled channel is considered to be more practical since the latter has more problems in regard to scum and silt/grit accumulation behind the baffle.

When baffles channels are designed to provide uniform flow velocity through each stage of channels (plug flow type), the headloss per 180° bend is approximately 3.2–3.5 times the velocity head in the adjacent channels. However, if the baffled channels are designed to produce the required flow velocity at only the narrowed paths, such as slits or submerged ports, the headloss will be approximately 1.5 times the velocity head of the restricted area; this type of baffled channel is called helicoidal or tangential flow because the stream of water enters tangentially into each chamber, producing a helicoidal flow pattern toward the outlet downstream of each turn. In order to facilitate easy construction and cleaning, the distance between the baffles should be a minimum of 2.5 ft (0.75 m). The minimum water depth should be 3.3 ft (1 m).

In general, a minimum of six channels should be provided and the headloss across the flocculation tank must be approximately 1–2 ft so that an average G-value of 30–40 s^{-1} will be established. The process should have a tapered mixing pattern and should have a minimum residence time of 20 min at maximum plant flow. The G-value of the baffled channel flocculation process may quickly be estimated by the following formula (using English units):

$$G = 178(h/t)^{0.5} \quad \text{at 4° C,}$$

where t = residence time in minutes,
h = headloss in feet,
G = velocity gradient in seconds^{-1}.

Tracer tests have indicated that there is little flow short-circuiting in baffled channels. Case 8 in Figure 3.2.4-7 illustrates this condition.

Example Design Calculations

Example 1   Mechanical Mixer (Vertical Shaft Flocculators)

Given

Plant flow rate	20 mgd (0.88 m^3/s) daily average
Detention time	20 min at maximum daily flow rate
Number of tanks	Two rectangular tanks
Stage of flocculation	Three stages
Energy input	60–20 s^{-1} for both stages 1 and 2
	$G = 30$–10 s^{-1} for the last stage
Type of flocculator	Vertical shaft with either 45° pitched turbine blade (PTB) or hydrofoil type propeller blades; variable speed drive control
Minimum water temperature	50°F (10°C)

Determine

(i) The size of each flocculation basin, including water depth.
(ii) The total number of flocculators.
(iii) The motor horsepower of each flocculator.
(iv) The details of the diffuser walls.

Solution

(i) Assume that the maximum daily flow rate is 150% of the daily average flow rate. The volume required for each tank is

$$\frac{1.5 \times 0.88}{2} \times 60 \times 20 = 792 \text{ m}^3 \quad \text{or} \quad 27,960 \text{ ft}^3$$

Make the average water depth 4.25 m (14 ft). The total tank area required is

$$792 \div 4.25 = 186.4 \text{ m}^2 \quad \text{or} \quad 2005 \text{ ft}^2$$

The size (shape) of each basin is as follows:

Alternative 1: 8 m × 24 m (*W* × *L*) or 26 ft × 78 ft
Alternative 2: 11 m × 16.5 m (*W* × *L*) or 36 ft × 54 ft
Alternative 3: 11 m × 18.5 m (*W* × *L*) or 36 ft × 61 ft

Alternative 1 is the most economical design but it is not the most effective tank. Alternative 2 is not as cost effective to build as Alter-

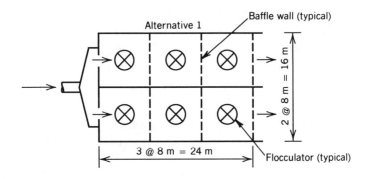

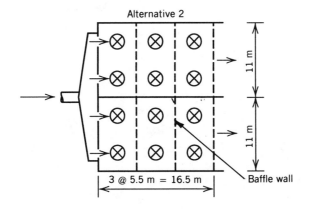

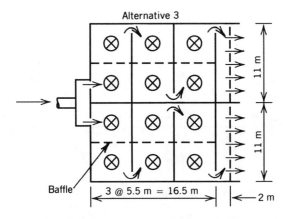

native 1. However, the larger number of flocculators gives it a greater safety factor and more effective flocculation. Therefore, adopt Alternative 2. Alternative 3 is composed of a series of complete mixing reactors that minimize the problem of flow short-circuiting. Even though this is the best type of flocculation tank this scheme requires a flow

distribution channel to be constructed, leading to the sedimentation tank. The flow direction in this channel is not ideal because it is perpendicular to the flow direction in the clarifier. However, the use of a clarifier inlet diffuser wall, which is described in Chapter 4, can produce a reasonably good inlet flow condition.

(ii) The total number of flocculators is 12 since Alternative 2 has been selected because of the cost effectiveness.

(iii) Motor horsepower of the flocculators. For all flocculators in stages 1 and 2

$$P = G^2 \mu V$$

where $\mu = 0.0013$ N·s/m^2 $(2.73 \times 10^{-5}$ lb·s/ft^2,
$\quad$ N = Newton = kg-m/s^2.

$$P = (60)^2 \times (1.3 \times 10^{-3}) \times [(5.5)^2 \times 4.25]$$
$$= 602 \text{ J/s} \quad \text{per flocculator}$$

Since 1 kW = 1000 J/s, $P = 0.6$ kW or 0.8 hp (1 hp = 0.746 kW). Assuming that the overall efficiency of a mechanical mixer is 75%, the minimum required motor horsepower for each flocculator in stages 1 and 2 is

$$0.6 \div 0.75 = 0.8 \text{ kW} \quad (1 \text{ hp})$$

For the flocculators in stage 3, the required motor horsepower can be computed in the same way.

$$P = (30)^2 \times (1.3 \times 10^{-3}) \times [(5.5)^2 \times 4.25] = 150 \text{ J/s}$$

Therefore, $P = 0.15$ kW $\div 0.75 = 0.2$ kW ($\frac{1}{4}$ hp), minimum for each flocculator in stage 3.

(iv) There will be three baffles: call them No. 1, No. 2, and No. 3 baffles, counting from the inlet side. The flow velocity through each baffle will be set as shown and the total orifice area required can be computed.

	Baffled Wall No. 1	Baffled Wall No. 2	Baffled Wall No. 3
Maximum flow velocity (m/s)	0.55	0.45	0.35
Total orifice area for both tanks (m^2)	2.4	2.93	3.77

The size of a single orifice should be 0.1–0.15 m in diameter to prevent clogging by algae or suspended matter. A reasonable distance must also be kept between the orifices so that the structural integrity of the wall is maintained. Thus, select 0.15 m as the diameter for each orifice. The following table lists the total number of orifices per tank, the total orifice area, and the headloss per baffle for both the maximum and average daily plant flow rates.

	B.W. No. 1	B.W. No. 2	B.W. No. 3
Total number of orifices	140	168	216
Total orifice area (m²)	2.47	2.97	3.82
Velocity at Q_{max} (m/s)	0.53	0.44	0.35
Velocity at Q_{ave} (m/s)	0.36	0.30	0.23
Headloss at Q_{max} (m)	0.045	0.031	0.020
Headloss at Q_{ave} (m)	0.021	0.014	0.008

Figure 3.2.4-8 shows the details of the three baffled walls in each tank. The ratio of the orifice area to the cross-sectional area of the tank is 2.5%, 3.2%, and 4.1% for walls No. 1, No. 2, and No. 3, respectively. The headloss through the orifice is computed from the formula $Q = CA (2gh)^{0.5}$, where $C = 0.8$.

Example 2 Horizontal Shaft with Paddle Type Flocculator

Given

Plant flow rate	1.85 m³/s (42 mgd) maximum per day
Number of tanks	Three rectangular tanks
Stage of flocculation	Two stages due to site limitations
Detention time	32 min at 1.85 m³/s
Shape of each tank	5 m (*W*) × 24 m (*L*) × 5.0 m (W.D.)
Type of flocculator	Horizontal shaft with paddles with a variable speed control
Energy input	(1) $G = 50$ s^{-1} maximum (2) Displacement factor* = 30 maximum
Minimum water temperature	10° C

* Beam, E. L., "Study of Physical Factors Affecting Flocculation," *W.W.E.*, p. 33 (January 1953).

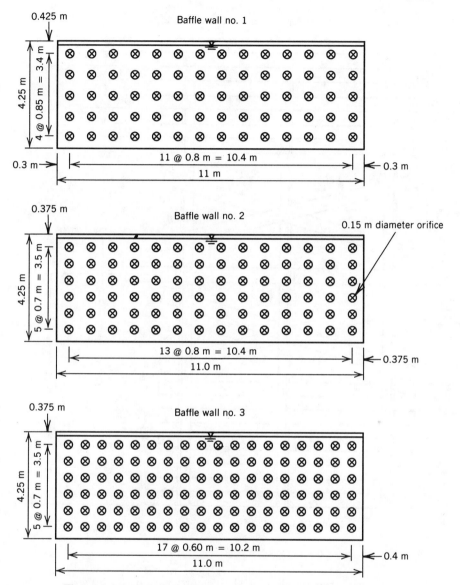

Figure 3.2.4-8 Details of the three baffled walls of each tank.

Determine

(i) The number of paddles per shaft.
(ii) The details of the paddle.
(iii) The maximum number of shaft revolutions per minute.
(iv) The required horsepower of each drive unit.

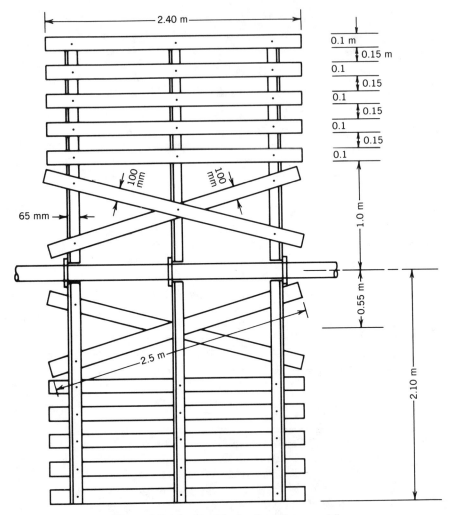

Figure 3.2.4-9 Detail of a flocculator paddle.

Solution

(i) Considering the number of shaft supports and the integrity of the paddle, a reasonable figure would be 8 paddles per shaft or a total of 48 paddles for this flocculation process.

(ii) Details of the paddle. After a few trials the shape and dimensions shown in Figure 3.2.4-9 are adapted.

Notes

1. Since there are only two stages of flocculation, each stage is compartmentalized with baffled walls similar to baffled walls No. 1 and No. 2 of the previous example (see Figure 3.2.4-8).

2. The number of paddles used is as many as is practical because the drag force and eddies produced by the paddles are proportional to the total length of the paddle edges.
3. The paddles are arranged diagonally in order to minimize flow short-circuiting; the travel speed of the paddle is minimal near the shaft.
4. Each paddle wheel contains 3.97 m² of the paddle area. This is just under 25% of the cross-sectional area of the tank.

(iii) We determine the maximum rotational speed necessary to satisfy the energy input, based on Camp's G-value formula for the paddle flocculator.

$$G = C_D(Av^3/2vV)^{0.5}$$

where $A = 3.97$ m² $\times 48 = 190.5$ m²,
$V = [2 \times (5 \times 24 \times 5)] \times 3 = 3600$ m³,
$v = 1.31 \times 10^{-6}$ m²/s at 10° C,
$C_D = 1.8$.

Try 0.55 cm/s as the peripheral speed of the paddle wheel. Incidentally, the RPM of the paddle at a velocity of 0.55 m/s is $(0.55 \times 60) \div (2\pi \times 2.1) = 2.5$.

$$G = \left(\frac{1.8 \times 190.5 \times (0.75 \times 0.55)^3}{2 \times (1.31 \times 10^{-6}) \times 3600} \right)^{0.5}$$

$$= 50.7 \text{ s}^{-1} > 50 \text{ s}^{-1}$$

Thus, the maximum RPM of the shaft must be 2.5.

$$G \times t = 50.7 \times (32 \times 60) = 9.7 \times 10^4$$

Based on Beam's design criteria of the displacement factor, $D/F \geq 30$

where $D = \Sigma d$,
$d = a \times s$
$S = 2\pi r \times$ RPM
and $d =$ displacement (pumping) capacity of each paddle in ft³/min,
$a =$ area of each paddle (ft²)
$s =$ traveling distance of the paddle area A (fpm)
$r =$ distance (ft) between the center of gravity of the paddle area (a) and the center of the shaft,
$F =$ flow rate (ft³/min).

Since 2.5 RPM as the paddle speed satisfies the G-value, use this same speed to check the displacement factor.

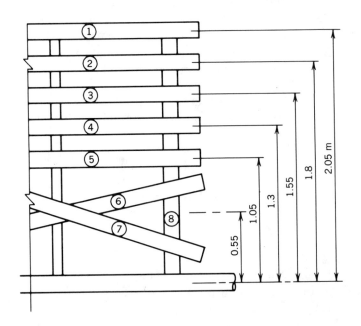

$$d_1 = 0.24 \times 2\pi \cdot 2.5 \times 2.05 = 7.74 \text{ m}^3/\text{min}$$

$$d_2 = 0.24 \times 2\pi \cdot 2.5 \times 1.8 \ \ = 6.79$$

$$d_3 = 0.24 \times 2\pi \cdot 2.5 \times 1.55 = 5.85$$

$$d_4 = 0.24 \times 2\pi \cdot 2.5 \times 1.3 \ \ = 4.91$$

$$d_5 = 0.24 \times 2\pi \cdot 2.5 \times 1.05 = 3.96$$

$$d_6 = 0.25 \times 2\pi \cdot 2.5 \times 0.55 = 2.16$$

$$d_7 = 0.25 \times 2\pi \cdot 2.5 \times 0.55 = 1.08$$

$$d_8 = 3 \times [(0.065 \times 2.1) \times (2\pi \cdot 2.5 \times 1.05)] = 6.73$$

$$2 \times \sum_{i=1}^{i=8} d_i = 2 \times 39.22 = 78.44 \text{ m}^3/\text{min} \quad \text{for one paddle}$$

Since there are 48 paddles in this flocculation process,

$$D = 78.44 \times 48 = 3760 \text{ m}^3/\text{min}$$

$$F = \ \ 1.85 \times 60 = 111 \text{ m}^3/\text{min}$$

Thus, $D/F = 3760 \div 111 = 34 > 30$. Assuming that the overall equipment efficiency is 70%, we find

$$P = 12 \div 0.7 = 17 \text{ kW} \quad (23 \text{ hp})$$

There are eight paddles mounted per shaft. A large initial torque is created when the motor is switched on. For this reason, a 22 kW (30 hp) motor is selected for each of the six drive units.

Note: Example 2 is based on an actual plant design. The flocculation system was constructed as outlined above and has been producing excellent floc since 1957.

Example 3 Baffled Channel Flocculation Process

Given

Plant flow rate	1 m³/s (23 mgd) maximum daily
Detention time	25 minutes at 1 m³/s
Number of tanks	Two tanks
Stage of flocculation	Three stages
Energy input (at 1 m³/s)	$G = 70 \text{ s}^{-1}$ for stage 1
	$G = 34 \text{ s}^{-1}$ for stage 2
	$G = 20 \text{ s}^{-1}$ for stage 3
Type of channel	Around-the-end type
Minimum water temperature	10°C

Determine

(1) The size of the tank and the number of channels.
(ii) The size of the channels and the arrangement of baffles.
(iii) The headloss and the openings at each baffle.

Solution

(i) The total volume required is $(1 \times 60) \times 25 = 1500 \text{ m}^3$. Since there are two tanks, the volume of one tank is 750 m³. Establish the average water depth as 2 m, and provide six channels (2.2 m wide) per tank. The length of each channel should be

$$750 \div [(2.2 \times 6) \times 2] = 28.4 \text{ m}$$

Make the length 28.5 m.

(ii) Figure 3.2.4-10 illustrates all the information necessary for the design: size of the channels and arrangement of baffles.

(iii) Stage 1 flocculation channels require that G be 70 s^{-1} at 0.5 m^3/s per tank. Therefore, the headloss across the stage 1 channels is

$$h_1 = G^2 v\ V/gQ = [70^2\ (1.3 \times 10^{-6}) \times 250] \div (9.81 \times 0.5)$$

$$= 0.325 \text{ m}$$

Stage 2 channels require that G be 35 s^{-1} at 0.5 m^3/s. Therefore,

$$h_2 = [35^2 \times (1.3 \times 10^{-6}) \times 250] \div (9.81 \times 0.5) = 0.081 \text{ m}$$

Stage 3 channels require that G be 20 s^{-1}. Thus,

$$h_3 = [20^2 \times (1.3 \times 10^{-6}) \times 250] \div (9.81 \times 0.5) = 0.027 \text{ m}$$

The total headloss across the tank is

$$\sum_{i=1}^{i=3} h_i = 0.433 \text{ m}$$

The opening dimension at each baffle of the stage 1 channels may be computed from the required headloss in each turn. This particular

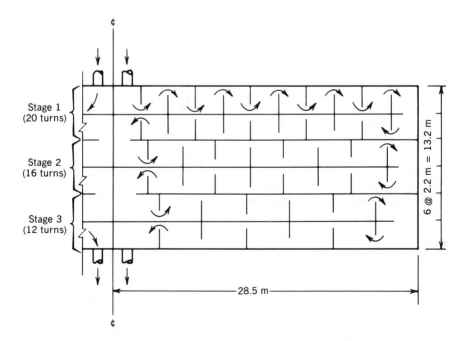

Figure 3.2.4-10 A baffled channel flocculation system.

example contains a total of 20 turns, thereby producing a headloss of 0.325 m (a headloss of 0.0163 m per turn). The headloss per turn is computed as

$$h = K(v^2/2g)$$

where $K = 1.5$ because the mean flow velocity of each channel is only 0.114 m/s and the flow enters tangentially into the next chamber; the flow enters in such a manner that a helicoidal flow pattern is developed and a clear 180° plug flow bend does not develop. The velocity required at each slit in the stage 1 channels is

$$v_1 = (2gh_1/1.5)^{0.5} = (19.62 \times 0.00163/1.5)^{0.5}$$

$$= 0.462 \text{ m/s}$$

The required width for each slit in the stage 1 channels is calculated to be

$$w_1 = Q/vH = 0.5/0.462 \times 2 = 0.54 \text{ m}$$

In stage 2 flocculation there is a total of 16 turns and the total headloss across the stage is 0.081 m. Thus,

$$v_2 = 0.258 \text{ m/s}$$

and the width of each slit is $w_2 = 0.97$ m. Stage 3 flocculation has 12 baffles and must produce a total headloss of 0.027 m; 0.00223 m at each baffle. Therefore, $v_3 = 0.17$ m/s and the width of the slit should be 1.47 m. The following table summarizes the characteristics of the baffled channel hydraulic flocculation process.

	1 m³/s (max. day)			0.667 m³/s (ave. day)		
	Stage 1	Stage 2	Stage 3	Stage 1	Stage 2	Stage 3
Slit size (m)	0.54	0.97	1.47	0.54	0.97	1.47
Flow velocity at slit (m/s)	0.47	0.26	0.17	0.31	0.17	0.11
Headloss (m) per slit	0.00163	0.0051	0.0022	0.0073	0.0023	0.001
Number of slits/stage	20	16	12	20	16	12
G-value/stage (s^{-1})	70	35	20	38	19	11
Headloss (m) across process		0.43			0.19	

Note: The design presented in Example 3 is based on the assumption of 100% efficiency. The actual design should allow for 80% efficiency. Therefore, stages 2 and 3 should have 10 turns per channel not 8 and 6 as shown. The total number of turns for the entire tank becomes 60, not 48 as shown.

Operation and Maintenance   The three basic operational procedures common to all flocculation systems are checking floc size by way of visual observation, removal of scum from the water surface, and control of algae growth on the tank walls and baffles. If mechanical mixing is provided, the mixer speed control should also be inspected.

Floc observation during the night shift requires a specific type of lighting. The regular lamps located at the walkway are inadequate because of light reflection so that proper floc observation requires a strong spotlight to be aimed toward the tank floor. This light should be located slightly above the water line. If these requirements are met, the floc can clearly be visualized due to the Tyndall effect. However, the preferred setup is to have an underwater lamp placed 12–18 in. below the water level.

The majority of water treatment plants experience problems with scum accumulation, unless the flocculation tank has been designed to avoid this. Although the scum does not have a significant effect on the water treatment process, it is clearly visible and is unsightly to most visitors.

Another problem common to many plants is the growth of algae on the tank walls and baffles. Algal growth is unsightly and can give the water an objectionable odor. Algal growth may be prevented through the use of prechlorination: a constant level of 0.3 mg/L or more of residual chlorine should be present at all times. However, for many treatment plants in the United States the use of prechlorination is limited due to restrictions placed on THM formation.

The growth of algae is most severe in the setting basin. One solution to this problem is the periodic application of a high dose of chlorine during the night for a short duration (shock treatment).

In regard to maintenance, the baffled channel process only requires an annual dewatering of the tank (every 12 months), during which any necessary repairs can be performed and sludge removal can be accomplished. Sludge removal is easier with the horizontal turning (around-the-end) baffled channels than with the vertical turning (over-and-under) channels.

The mechanical mixer flocculation units are significantly more maintenance intensive than the baffled channel type. These mixers typically require the lubrication oil of the speed-reducing units to be changed on a seasonal basis, repair of any oil leaks, and the repair and/or adjustment of the underwater shaft bearings whenever necessary.

A survey of the operational staffs for 30 plants revealed that most operators did not frequently change the flocculation speed. At the majority of the plants surveyed, the mixing speed was not even adjusted with changes in the season, despite an obvious change in water temperature. The main reason given by the operators was that the efficiency of the process did not appear to improve even when attempts were made to optimize the speed. Consequently, certain mechanical speed-reducing units were often "frozen" at one speed and were no longer capable of changing speeds.

The infinite range speed-reducing units used for the flocculators require the most maintenance. Consequently, a significant number of plants have replaced them with two-speed motor control units. However, some recent infinite range speed reducers are more reliable as well as cost effective. A variable frequency control system for the motor speed adjustment is one of these units.

BIBLIOGRAPHY

Amirtharajah, A. "Design of Flocculation Systems," in *Water Treatment Plant Design for the Practicing Engineer*, R. L. Sacks, ed., Ann Arbor Science Publishers, Ann Arbor, MI, 1979, pp. 195–229.

ASCE, *Water Treatment Plant Design*, ASCE Manual of Engineering Practice No. 19 (1952).

Brown, G. G., *Unit Operations*, Chapter 34, "Agitation," Wiley, New York, 1953.

Camp, T. R., et al., "Velocity Gradients and Internal Work in Fluid Motion," *J. Boston Soc. Civil Eng.*, 30:219(1943).

Camp, T. R., "Flocculation and Flocculation Basins," *Trans. ASCE*, 20:1(1955).

Davis, C. V., *Handbook of Applied Hydraulics*, 3rd ed., Chapter 38, "Water Treatment," McGraw-Hill, New York, 1973.

Dempsey, B. A., et al., "Polyaluminum Chloride and Alum Coagulation of Clay–Fulvic Acid Suspensions," *J. AWWA*, 77(3):74(March 1985).

Dempsey, B. A., et al., "The Coagulation of Humic Substances by Means of Aluminum Salts," *J. AWWA*, 76(4):141(April 1984).

Dentel, S. K., et al., "Using Streaming Current Detectors in Water Treatment," *J. AWWA*, 81(3):85(March 1989).

Edwards, G. A., et al., "Removing Color Caused by Humic Acids," *J. AWWA*, 77(3):50(March 1985).

Fair, G. M., and Geyer, I. C., *Water Supply and Wastewater Disposal*, Wiley, New York, 1954.

Fair, G. M., et al., *Water and Wastewater Engineering*, Vol. 2, Wiley, New York, 1968.

Hudson, Herbert E. Jr. *Water Clarification—Processes, Practical Design, and Evaluation*, Van Nostrand Reinhold, New York, 1981, pp. 40–64, 54–74, 75–100, 111–122.

Hundt, T. R., and O'Melia, C. R., "Aluminum–Fulvic Acid Interactions: Mechanics and Applications," *J. AWWA*, 80(4):176(April 1988).

James, C. R., et al., "Considering Sludge Production in Selection of Coagulants," *J. AWWA*, 74(3):148(March 1982).

Kawamura, S., "Application of Colloid Titration Technique to Flocculation Control," *J. AWWA*, 59(8):1003(August 1967).

Kawamura, S., "Coagulation Considerations," *J. AWWA*, 65(6):417(June 1973).

Kawamura, S., "Removal of Color by Alum Coagulation" (Part 1 & 2), *Water & Sewage Works*, p. 282 (August 1967), p. 324 (September 1967).

Kawamura, S., "The Fundamentals of Alum Flocculation as Applied to Water Purification," Thesis, Ohio State University, 1961.

Leentvaar, J., et al., "Some Dimensionless Parameters of Impeller Power in Coagulation–Flocculation Process," *Water Res.*, 14:140(1980).

Montgomery, J. M., Consulting Engineers, *Water Treatment Principles and Design*, Wiley, New York, 1985, pp. 116–134, 514–523.

Oldshue, J. Y., "Flocculator Impellers: A Comparison," *Chem. Eng. Prog.*, p. 72 (1983).

Olshue, J. Y., *Fluid Mixing Technology*, McGraw-Hill, New York, 1983.

Tekippe, R. J., and Ham, R. K., "Velocity Gradient Paths in Coagulation," *J. AWWA*, 63:429(1971).

Von Essen, J. A., "Energy Saving with Exotic Vessel and Impeller Design," in Proceeding of the 9th Engineering Foundation Conference on Mixing at New England College.

Walker, J. D., "High Energy Flocculation Unit," *J. AWWA*, 60(11):1271(November 1968).

3.2.5 Sedimentation (Clarification) Process

Purpose One of the most common water and wastewater treatment unit processes is sedimentation, also known as clarification. Sedimentation is broadly defined as the separation of a suspension into a clarified fluid and a more concentrated suspension.

The sedimentation process is designed to remove a majority of the settleable solids by gravitational settling, thereby maximizing downstream unit processes such as filtration. Floatation also separates liquids from solids within a suspension. However, buoyancy is the key to this process. Since buoyancy is used exclusively in wastewater treatment and seldom in the field of water treatment in the States, the floatation process is not discussed in this book.

The sedimentation process is divided into two classifications: grit chamber (plain sedimentation) and sedimentation tanks (clarifiers). The criteria for these classifications are the size, quantity, and specific gravity of the suspended solids to be separated. This section discusses the design of a sedimentation process based on both theory and the experience of the author, in addition to field operational data.

It must be emphasized that the efficiency of the sedimentation process is greatly influenced by the level and adequacy of raw water conditioning upstream from the sedimentation process. The key to effective clarification is proper coagulation and flocculation of suspended matter within the raw water.

There are three main configurations for sedimentation tanks: horizontal rectangular basins, upflow sedimentation tanks, and upflow reactor clarifiers with sludge blanket. A rectangular tank with a horizontal flow is generally the favored tank configuration due to its hydraulic stability (if properly designed) and its tolerance of shock loadings. This type of tank also has a predictable performance and is capable of withstanding a flow rate that is twice the ordinary design rate recommended by the regulatory agencies, without significant deterioration in the quality of the settled water. Furthermore, the horizontal rectangular basins are simple to operate and adapt easily to the installation of high-rate settler modules.

Considerations There are many important considerations that directly affect the design of the sedimentation process: overall treatment process, nature of the suspended matter within the raw water, settling velocity of the suspended particles to be removed, local climatic conditions, raw water characteristics, geological characteristics of the plant site, variations in the plant flow rate, occurrence of flow short-circuiting within the tank, type and overall configuration of the sedimentation tank, design of the tank inlet and outlet, type and selection of high-rate settling modules, method of sludge removal, and cost and shape of the tank. These subjects, along with other miscellaneous items, are discussed in detail.

Overall Treatment Process When designing an overall treatment process, the source of the water must be carefully evaluated. If the source is a river with episodes of flash floods, the treatment process train should include a grit chamber so that sand and silt will settle out near the intake. A grit chamber is a plain sedimentation process that removes discrete particles larger than 15 μm in diameter. If the filter bed is designed as an ordinary rapid sand bed, the degree of clarification is important because the turbidity of the filtered water is directly proportional to the settled water turbidity (Figure 3.2.5-1). Thus, a conservatively designed sedimentation process should be used to obtain a settled water turbidity of less than 2 NTU.

As discussed in Section 3.2.4, flocs that are both small in size and physically strong, obtained by adding polymer as a filtration aid, are needed for multimedia or coarse, deep filter beds. With these characteristics in mind, the criteria used in the design of the sedimentation basin must be different from those used in the design of classic rapid sand filters, so that the overall process efficiency is maximized.

Data obtained from many existing plants demonstrate that dual media filters remain very efficient despite filter influent turbidities ranging from 5 to 7 NTU (Figure 3.2.5-2). The direct filtration process also supports this characteristic. The sedimentation process can therefore tolerate much higher hydraulic loading rates when using multimedia or coarse, deep media filter beds. In general, when dual media filters are used, the hydraulic loading rate of the sedimentation process can be twice or even greater than the rate used in a rapid sand filter system.

Nature of Suspended Matter Raw water contains two basic types of suspended matter: particles that are discrete in nature (nonflocculable) and colloidal suspensions. Examples of nonflocculable particles are sand and silt.

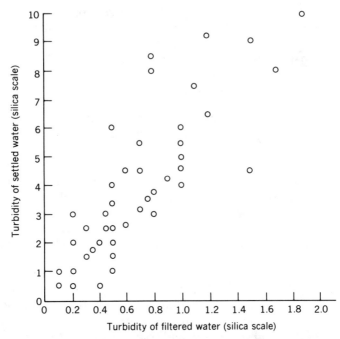

Figure 3.2.5-1   Relationship between turbidity of settled and filtered water on conventional rapid sand filter. *Note*: (1) Filter rate ranged from 4.2 to 6.7 m/h (1.7–2.7 gpm/ft²). (2) Alum flocculation and sedimentation before filtration. No filter aid was applied. (From S. Kawamura, *J. AWWA*, 67(10):535, 1975.)

Colloidal suspensions include clay, microorganisms, and substances that cause color. Colloidal suspensions must be flocculated with coagulant so that their particles will be removed during sedimentation.

Discrete particles may be removed by a cyclone separator, plain sedimentation, or through the use of a grit chamber. However, colloidal suspensions require a different type of settling basin (clarifier), one that has a much more conservative hydraulic loading rate. This is explained in the section covering design criteria.

Settling Velocity of the Particles   The sedimentation process is based on gravitational settling of the particles. For this reason, the design engineer should know the settling velocities of each particle that is to be removed in the anticipated water temperature range.

The settling velocity of each particle in still water may be measured under controlled conditions within the laboratory. Bench scale flocculation tests (jar tests) and the measurement of floc settling velocities are therefore strongly recommended. As previously mentioned, an appropriate safety factor should be applied since floc settling velocities can vary widely depending on several

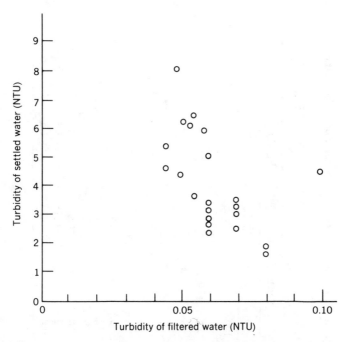

Figure 3.2.5-2 Relationship between turbidity of settled and filtered water on typical dual media filter. *Note*: (1) Filter rate ranged from 10 to 24 m/h (4.1–9.4 gpm/ft²). (2) Polyelectrolyte (as filter aid) dosage ranged from 7 to 23 ppb. (From S. Kawamura, *J. AWWA*, 67(10):535, 1975.)

parameters: type of coagulant employed, mixing conditions during the flocculation process, and the nature of the colloidal matter contained in the raw water. If these tests cannot be performed, for whatever reason, the numbers listed in Table 3.2.5-1 may be used.

According to Hazen (1904), the efficiency of an ideal horizontal flow sedimentation tank is a function of the settling velocity (v_0) of the particle to be removed, the surface area of the tank (A), and the rate of flow (Q) through the basin. This relation can be expressed as

$$v_0 = Q/A$$

Q/A is commonly known as the surface loading or overflow rate and is expressed in m/h or gpm/ft². Once the settling rate of the suspended matter (being removed) is determined and the rate of flow through the tank is known, a physical basin size (area of the tank) can be calculated. A minimum safety factor of 1.5 should be included in the calculation of the area.

According to Hazen, the efficiency of the tank is independent of its depth and detention time. However, a shallow depth theoretically favors settling of the particles. Detention time also affects tank efficiency since flocculant par-

TABLE 3.2.5-1 Settling Velocities of Particles (at 10°C) and Compatible Hydraulic Surface Loading Rates of Sedimentation Tanks

| Type of Particle | Specific Gravity | Particle Size | | Settling Rate | | | Compatible Tank Surface | |
| | | Mesh | mm | mm/s | fpm | | Loading | Approximate |
						m/h	gpm/ft^2
Sand	2.65	18	1.0	100	19.7	360	144
Sand	2.65	20	0.85	73	14.3	263	105
Sand	2.65	30	0.6	62	12.2	223	89
Sand	2.65	40	0.4	42	8.2	151	60
Silt	2.65	70	0.2	21	4.1	76	30
Silt	2.65	100	0.15	15	3.0	54	22
Silt	2.65	140	0.10	8	1.6	29	12
Silt and Clay	2.65	200	0.03	6	1.2	22	9
Silt and clay	2.65	230	0.06	3.8	0.75	14	5.6
Silt and clay	2.65	400	0.04	2.1	0.41	7.5	3
Clay	2.65	—	0.02	0.62	0.12	2.3	0.9
Clay	2.65	—	0.01	0.154	0.03	0.54	0.2
Alum floc	1.001	—	1–4	0.2–0.9	0.04–0.18	0.71–3.3	0.3–1.3
Lime floc	1.002	—	1–3	0.4–1.2	0.08–0.23	1.5–4.3	0.6–1.7

ticles become larger and heavier with time due to the mixing effect in the tank and therefore settle much faster. Consequently, the detention time of the basin is significant with regard to flocculant particles.

Local Climatic Conditions　There are several local climatic conditions that must be considered when designing a sedimentation process: daily and seasonal temperature fluctuations, water and air temperature range, degree of rainfall, and intensity and direction of the winds. If the daily temperatures fluctuate widely, within a day or a week, the basin will most likely experience flow short-circuiting due to a density current. Moreover, regions with severe winter conditions should have the tank either covered or housed in a heated building.

Water treatment plants located in regions of warm or hot weather will have problems with algal bloom and have a heavy growth of vegetation in the watershed. If the region is both warm and sunny, the treatment plant will always have problems with algal growth on the basin walls and launders. This can be prevented by either covering the basin or providing sufficient chlorine residual to the influent water. However, the addition of chlorine should be exercised with extreme caution because of the likelihood that THM formation will exceed the tolerable limits.

Treatment plants located in regions of high rainfall tend to have large amounts of suspended matter in the raw water. The design should therefore include a continuous sludge removal system that will remove the accumulated sludge before it reduces the effective tank volume and before it becomes septic. In cases where the raw water contains a high concentration of grit and silt, a presedimentation basin must be included in the design.

In regions other than the tropics, a majority of the local vegetation shed their leaves into the water source, as well as the basin, during the winter, not only causing organic color to be imparted to the water but also clogging filter bed. Floc formed from highly colored water is very fragile and difficult to settle. Additionally, the pH of this type of flocculated water may become 6 or less. Therefore, both submerged and surface metals and concrete surfaces will corrode unless special design alterations are made.

Regions that continually experience strong winds have design problems of their own. Strong winds can greatly disturb the water flow within the clarifiers and may induce flow short-circuiting, depending on the direction of the wind and the orientation of the basins. Furthermore, if trees are located near the settling basins, the winds may also blow leaves and branches into them.

Raw Water Characteristics　One of the most dramatic effects of raw water characteristics with regard to both horizontal flow and upflow clarifiers is the sudden change in water temperature. A change of 1°F (0.5°C) can cause a severe density flow (Figure 3.2.5-3). If the influent water is colder, it will travel along the bottom and rise upward toward the end of the tank. If the water is warmer, flow short-circuiting will occur at the surface of the tank.

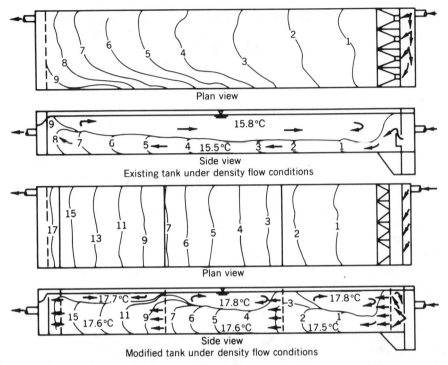

Figure 3.2.5-3 Original and modified tanks under density flow conditions. Observations are based on a hydraulic scale model study. *Note*: (1) Computed detention time is 49 min or 4 h and 15 min for the actual tank. (2) Observed flow-through time is 9 min for the original tank model and 17 min for the modified model. (From S. Kawamura, *J. AWWA*, 73(7):372, 1981.)

This condition is particularly serious when a treatment plant has more than one source and these sources are often switched or if the blending ratio of two or three sources is frequently changed.

Turbidity spikes also cause density currents that are identical to those created by cold influent water. Highly turbid water produces a large volume of floc after flocculation. Since these flocs have a higher specific gravity than the water in the tank, the influent dives to the bottom, creating a distinct density flow. A raw water turbidity of over 30 NTU usually produces this type of density flow.

Attention should also be given to the aggressiveness of the raw water to submerged parts (both concrete and metal). The design engineer should either choose an appropriate type of cement and provide protection by applying a special coating or use noncorrosive material such as Type 18-8 stainless steel. Submerged steel parts must be properly painted and preferably have cathodic protection. Although stainless steel Type 18-8 is significantly more expensive, its use generally will eliminate the need for costly operation and maintenance procedures—such as painting and cathodic protection.

Soil Characteristics Consideration must also be given to the soil characteristics of the plant site. Of special interest are the bearing capacity of the native soil, the presence of geological faults, the level of the underground water table, and the salt content of the soil. Inadequate geotechnical studies have much too often resulted in the subsidence of the clarifiers and the development of large cracks in the basin walls and floor slabs.

A high salt content within the soil quickly corrodes metals, including reinforcing steel bars embedded in concrete. This problem may be minimized by providing a concrete cover that is greater than 2 in. A soil resistivity below 1500 uΩ/cm is generally considered to be highly corrosive and the use of steel shell tanks should be avoided unless special measures for corrosion protection are implemented. Under these conditions, special cement (Type II or V) and a low shrinkage concrete mixture must be specified for the concrete.

Variation in Plant Flow Rate Sudden changes in plant flow rate, also known as hydraulic shock loads, may affect the flow condition and disturb the settling efficiency of the tank. This is especially true with sharp increases in flow rate during short periods of time. Clarifiers that are most susceptible to hydraulic shock loading are the upflow clarifier and the sludge blanket reactor clarifier. The rectangular horizontal flow clarifiers generally tolerate hydraulic shock loadings much better than the previously mentioned types.

Flow Short-Circuiting Three basic types of flow short-circuiting may occur within the clarifiers. The first is caused by improper design. The second, although not serious, is a chronic flow short-circuiting that becomes evident when a significant amount of floc is carried over to the filters. The third type is the creation of a density flow within an ordinary sedimentation tank. This density flow is caused by a difference in density between the influent water and the water within the tank; the influent water may be colder or warmer, contain a higher concentration of suspended solids, or create a salinity pulse, that is, an influx of NaCl due to the intrusion of seawater.

The first type of flow short-circuiting is typified by a tank design that has both a poor inlet design and a short distance between the inlet and outlet. Other factors that exacerbate this problem are a large tank width in conjunction with a short tank length and a deep water depth. A tank designed in this manner will always have a high Reynolds number and very low flow stability.

The second type of flow short-circuiting is observed in many sedimentation tanks whose plans are based on the prevailing tank design criteria set by the regulatory agencies. The ordinary plant visitor may not notice the flow short-circuiting but plant operators observe that (during the day) the influent water tends to dive down at the inlet of the tank and rise at the effluent, carrying much floc with it. The operators call this phenomenon "noonday turnover." In actuality, the most serious flow short-circuiting tends to occur mainly from midnight to dawn. In this case, the influent water short-circuits at the surface due to its slightly higher water temperature. As long as the top of the tank is exposed to the weather, both the first and second types of flow short-circuiting cannot be controlled effectively by the placement of diffuser walls at both the inlet and outlet of the tank.

The third type of flow short-circuiting is a severe case of the second. This density flow problem is commonly caused by switching from one source of water supply to another, suddenly changing the mixing ratio of two and sometimes three different sources, abruptly shifting the reservoir intake elevation, and having a high raw water turbidity (over 50 NTU). This particular type of flow short-circuiting may also be induced by strong winds blowing along the longitudinal axis of the tank or by a salinity pulse. Nothing can stop this type of density flow once it has started.

This third type of flow short-circuiting can be minimized by installing intermediate diffuser walls perpendicular to the flow direction, in the middle or at two-thirds of the tank length. The minimization is achieved by increasing the detention time and mixing effect in front of each intermediate diffuser wall (see Figure 3.2.5-3). The magnitude of the density current may be evaluated by Harleman's formula:

$$v = \left(8g \, \frac{\Delta\rho}{\rho} \, \frac{h \, s}{f(1 + \alpha)} \right)^{0.5}$$

where v = velocity of the density flow (fps),
  g = acceleration of gravity (32.2 ft/s^2),
  $\Delta\rho$ = density difference between the two liquids (lb/ft^3),
  ρ = density of the influent (lb/ft^3),
  h = depth of the density current flow (ft),
  s = slope of the channel bottom,
  f = Darcy–Weisbach friction factor,
  α = correction factor for kinetic energy ranging from 0 to 1 (0.43 for turbulent flow).

A similar equation is recommended by the Japan Society of Civil Engineers (1963) for computing the velocity of the density flow:

$$v = \left(2g \, \frac{\Delta\rho}{\rho} \, \frac{\Delta h \alpha}{k} \right)^{0.5}$$

where v = velocity of density flow (m/s),
  $\Delta\rho$ = difference in density (g/cm^3),
  ρ = density of the influent (approximately 1.0 g/cm^3),
  g = acceleration of gravity (9.81 m/s^2),
  Δh = depth of the density flow (m),
  α = correction factor for kinetic energy (approximately 0.5),
  k = dilution factor (normally 1.5–2).

According to field measurements conducted by the author of this book, the difference in water temperature (near the surface of the tank and the influent water) is usually 0.2–0.5°C during the day and the flow velocity of the density flow is 2.6–6 fpm (0.8–1.8 m/min) despite a designed flow velocity of 1.3 fpm (0.4 m/min).

Type of Sedimentation Tank Upflow clarifiers and reactor clarifiers perform very well if both the raw water characteristics and the hydraulic loading rates are constant. The reactor clarifier is best suited for lime or lime–soda ash softening of the water because of the seeding effect of the unit. Consequently, lime softening of underground water is the perfect application for the reactor clarifier. However, these two types of sedimentation tank are very susceptible to hydraulic and solid shock loading. Wind and uneven exposure to sun can also disturb the effectiveness of the clarification process.

The upflow clarifier and the reactor clarifier are usually compact and predesigned by the equipment manufacturer. The overall cost may therefore be cheaper than using custom designed rectangular tanks. Furthermore, the sludge can easily be removed from these tanks due to the continuously fluidized sludge blanket. On the other hand, a significant amount of metal is used on the inside of these tanks and the problems of corrosion and maintenance must be addressed. It is necessary to emphasize that upflow and reactor clarifiers are not capable of functioning at rates well over the designed rates because floc removal is based on the balance between particle settling velocity and vertical flow velocity.

Unlike the upflow and reactor clarifiers, the horizontal rectangular basin can tolerate both hydraulic and solid shock loadings. In fact, rectangular basins can handle a flow rate that is 50–100% over the designed rate without significantly degrading the quality of the settled water. The flow condition of the basin, under higher flow rates, is more stable because of an increase in the Froude number. Most large municipal water treatment plants use horizontal rectangular clarifiers primarily because of their flexible performance, predictable settling efficiency, and minimum maintenance cost. A comparison between the upflow and rectangular clarifier is presented in Table 3.2.5-2.

Overall Configuration The preferred configuration of the sedimentation basins is a battery of rectangular tanks sharing common side walls. The upstream portion of each basin contains a flocculation process and each rectangular tank is connected by a common influent and effluent channel; both channels span all the basins. This configuration offers several advantages: a reduction in construction cost, minimum site requirement due to the lack of space between the basins, easy distribution of the plant influent water to each basin, minimum floc breakage, minimum floc settlement between the flocculation and sedimentation basins, simple refilling of the basins with clean settled water from the effluent end via a valve, easy single-point application of water

TABLE 3.2.5-2 A Selection Guide for Some Basic Types of Clarifier

Type of Clarifier	Some Design Criteria	Advantages and Disadvantages	Proper Application
Rectangular basin (horizontal flow)	Surface loading: 0.34–1 gpm/ft² (0.83–2.5 m/h) Water depth: 3–5 m Detention time: 1.5–3 h Width/length: >1/5 Weir loading: <15 gpm/ft (11 m/m·h)	1. More tolerance to shock loads 2. Predictable performance under most conditions 3. Easy operation and low maintenance costs 4. Easy adaptation to high-rate settler modules a. Subject to density flow creation in the basin b. Requires careful design of the inlet and outlet structures c. Usually requires separate flocculation facilities	Most municipal and industrial water works Particularly suited to larger capacity plants
Upflow type (radial-upflow type)	Circular or square in shape Surface loading: 0.5–0.75 gpm/ft² (1.3–1.9 m/h) Water depth: 3–5 m Settling time: 1–3 h Weir loading: 10 gpm/ft (7 m/m·h)	1. Economical compact geometry 2. Easy sludge removal 3. High clarification efficiency a. Problems of flow short-circuiting b. Less tolerance to shock loads c. A need for more careful operation d. Limitation on the practical size of the unit e. May require separate flocculation facilities	Small to mid-sized municipal and industrial treatment plants Best suited where the rate of flow and raw water quality are constant

| Reactor clarifiers | Flocculation time: approx 20 min
Settling time: 1–2 h
Surface loading: 0.8–1.2 gpm/ft²
$\quad$ (2–3 m/h)
Weir loading: 10–20 gpm/ft
$\quad$ (7.3–15 m/m·h)
Upflow velocity: <50 mm/min | 1. Incorporates flocculation and clarification in one unit
2. Good flocculation and clarification efficiency due to a seeding effect
3. Some ability to take shock loads
a. Requires greater operator skill
b. Less reliability than conventional due to a dependency on one mixing motor
c. Subject to upsets due to thermal effects | Water softening
A plant that treats a steady quality and quantity of raw water |
| Sludge blanket clarifiers | Flocculation time: approx 20 min
Settling time: 1–2 h
Surface loading: 0.8–1.2 gpm/ft²
$\quad$ (2–3 m/h)
Weir loading: 10–20 gpm/ft
$\quad$ (7.3 to 15 m/m·h)
Upflow velocity: <10 mm/min
Slurry circulation rate: up to 3–5 times the raw water inflow rate | 1. Good softening and turbidity removal
2. Compact and economical design
3. Tolerates limited changes in raw water quality and flow rate
a. Very sensitive to shock loads
b. Sensitive to temperature change
c. Several days required to build up the necessary sludge blanket
d. Plant operation depends on a single mixing flocculation unit motor
e. Higher maintenance costs and a need for greater operator skill | Water softening
Flocculation/sedimentation treatment of raw water with a constant quality and rate of flow
Plant treating a raw water with a low content of solids |

Notes: (1) The reactor clarifiers and the sludge blanket type clarifiers are often considered to be in the same category.
(2) Surface loading: m/h = (m³/m²·d) ÷ 24; therefore, 60 m³/m²·d = 1 gpm/ft².
(3) Weir loading: m/h·m = (m³/d·m) ÷ 24; therefore, 175 m³/d·m = 10 fpm/ft.

treatment chemicals to the combined settled water (prior to the filters), and minimum yard piping.

Another configuration that is often adopted is a circular array of reactor clarifier basins arranged in a grid point fashion or that follow the topography of the plant site. The only disadvantage with this type of configuration is that each tank requires individual inlet and outlet yard pipings and adequate spacing must therefore be provided between the basins. The reactor clarifier basins have a flocculation zone in the center and a sedimentation zone in the outer portion; they are proprietary units designed by the equipment manufacturers. The reactor clarifier basin can be either circular or square in shape.

The configuration of the basin is strongly influenced by the shape, topography (slope), and geographic condition of the plant site. Thus, the design engineer may not have much of a choice in determining the basin configuration.

Inlet and Outlet of the Basin   Regardless of the shape and the type of clarifier, hydraulic control of the basin influent is one of the most important design factors. As expounded in many technical articles, any flow imbalance at the basin inlet will lead to flow short-circuiting, jetting, turbulences, and an overall lack of hydraulic stability within the settling zone. Although various baffling methods have been tested for distributing the water to the settling basin at the basin inlet, the most simple and effective method to date is the perforated baffle. The most effective perforated baffle, for a specific basin shape, can only be determined through hydraulic scale model studies. However, certain general rules may be used in designing the baffle walls for most basins. The author conducted an extensive hydraulic scale model study in the late 1950s and determined the general design criteria for the perforated baffle. Based on this study the major design requirements are as follows:

1. The ports should be uniformly distributed to the baffle wall, which in turn covers the entire cross section of the basin.
2. A maximum number of ports should be provided so that the length of the jets is minimized and the dead zone between ports is reduced.
3. The headloss through the ports should be approximately 0.12–0.35 in. (0.3–0.9 mm) to equalize the flow distribution across the entire cross section of the tank inlet with minimum floc breakage.
4. The headloss through the ports should be less than 0.4 in. (1 cm) to prevent floc breakage.
5. The size of the ports should be uniform in diameter, 3–8 in. (0.075–0.2 m), to avoid clogging by algae and other debris.
6. The ports should be spaced approximately 10–20 in. (0.25–0.5 m) on center to provide structural strength for the diffuser wall.
7. Port configuration should be arranged in such a manner that the parallel jets will direct the flow toward the basin outlet.

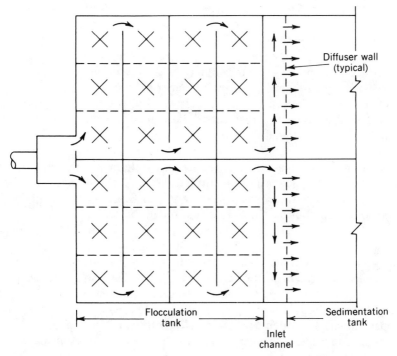

Figure 3.2.5-4   Inlet channel between flocculation and sedimentation tanks.

8. If the basins are fed from a common channel or if there is an inlet feed channel zone between the flocculation and sedimentation basins, the inlet diffuser wall should be located approximately 6.5–8 ft (2–2.5 m) downstream of the inlet. The inlet zone significantly increases floc size by allowing the proper gentle mixing effect to occur (Figure 3.2.5-4).

The water exiting the sedimentation basin should be uniformly collected across an area that is perpendicular to the proper flow direction. Beginning in the late 1950s a number of long launders (skimming troughs) were installed in the sedimentation tanks in an attempt to achieve uniform collection of the settled water. This practice is quite varied for the upflow clarifiers but generally not for the rectangular settling tanks. The installation of long launders in horizontal flow rectangular tanks is virtually ineffective when either a bottom-flowing or surface-running density current exists in the tank.

As previously discussed, most sedimentation basins exhibit a bottom-flowing density current; the basin end wall deflects this current, which then rises to the surface, carrying floc, and flows into the end portion of each launder. This phenomenon has been observed in both operational basins and in hydraulic scale model studies. In terms of design, this phenomenon suggests that transverse launders placed away from the end wall of the basin might be more effective than finger (long) launders. A few operational plants have supported this proposal. Yet, long launders have three major advantages when used in conjunction with rectangular basins: the water level of the tank remains virtually constant, regardless of changes in the plant flow rate; wave action created by winds is minimized; and high-rate settler modules, such as tube settlers, are easily hung from the launder system.

V-notched weir plates are generally attached to the launders. The weirs create a free discharge flow that enters the launder system, along with any floating matter. Consequently, there is a tendency to break the fragile alum floc. Unlike the weirs, submerged orifices do not allow floating scum to flow into the launder and they also tolerate less accurate installation. Therefore, in some cases, submerged orifices are drilled into the side walls of the launders. The floc breakage characteristics of submerged orifices may be slightly better than those of weirs but the flow velocity through the orifice must range from 1.6 to 2.5 fps (0.5–0.75 m/s) in order to provide a controlled headloss of approximately 1.5 in. (38 mm). Some floc breakage is therefore unavoidable.

The previously held idea that the installation of perforated baffles at the tank outlet is equivalent to the installation of inlet baffles (to collect settled water uniformly) still persists. On the surface this concept appears to be reasonable and sound. However, observations of hydraulic scale model studies and operational plants demonstrate that the effluent baffle is unexpectedly ineffective. The primary problem is that the effluent baffle is only capable of providing an insignificant correction factor for a density current. It is therefore common to see numerous flocs downstream of the effluent baffle, although the water upstream may be very clear. For this reason, the use of the effluent baffle at the end part of the basin is not recommended. In fact, the effluent baffle is now seldom employed.

High-Rate Settling Modules   The term high-rate settler refers to any small sized inclined tubes or tilted parallel plates which permit effective gravitational settling of suspended particles within the modules. The detention time in the settling modules ranges from 5 to 20 min, depending on the type of module and the surface loading rate. Surface loading is conveniently calculated from the area of the basin that is covered by the settling module and is generally 2–3.5 gpm/ft^2 (5–8.8 m/h). The settling efficiency of modules with such high hydraulic loading values is equivalent to a conventional sedimentation basin with a designed detention time of 3–4 h and a surface loading rate of 0.5 gpm/ft^2 (1.3 m/h).

The concept of an ideal settling tank (and the advantages of shallow tanks) was first advocated by Hazen (1904) in the beginning of the 20th century. Camp (1953) later created the "double-try" settling tank while applying this concept. Around 1955 the first high-rate settling module was commercialized in Japan under the name "Uno Separator"; the module consisted of sloping parallel plates spaced 25–50 mm apart. Similar modules were later marketed in Sweden under the name "Lamella Separator" and a tube settler module was also marketed by Micro Floc, Inc. (1969) and other firms (United States).

Shortly after the release of the first high-rate settler, a modified parallel plate module was released. This version had a series of deflector baffles attached to one side of the plates, which were designed to promote flocculation within the module (Figure 3.2.5-5), thereby suppose to permit greater efficiency in clarification. However, data from side-by-side pilot studies suggest that the effectiveness of this module appears to be only marginally better than modules without baffles.

Although the high-rate settlers are designed to deliver a certain level of performance, they do not always function according to their specifications. The primary reasons for this abberant behavior are (1) poor flocculation, (2) uneven flow distribution to the inlet of the settler module, (3) heavy algal growth in the modules, and (4) heavy scale ($CaCO_3$) deposits on the module. The heavy scaling clogs the flow paths and, more importantly, eventually leads to corruption of the system.

Depending on the type of parallel-plate settling module, water enters either horizontally and parallel to the plates or upward along the 60° tilted plates. In the tube settlers the direction of flow is always upward at 60° to the horizontal plane, although some earlier prototypes have their plates at a 15° angle. Under normal circumstances, the settled floc slides down the inclined plates or tubes to the tank bottom for subsequent mechanical sludge removal. The design of the high-rate settler modules must address the following issues: settling velocity and characteristics of the suspended matter, flow velocity within the settler module, surface loading, Reynolds number and Froude number for the flow within the module, selection of the appropriate sludge collection unit (to be installed beneath the settler module), spacing of the launder when installed above the settling module, and the supporting system.

Experience has shown that data from small-scale pilot plants (area of 1 m or less) cannot be used as meaningful design criteria because pilot studies often give very optimistic loading rates; the studies are usually conducted under carefully controlled and ideal conditions. The actual design criteria should be more conservative because of the potential for nonuniform flow conditions at the inlet to the settler modules as the result of density flow, improper tank design, or unfavorable coagulant dosage or flocculation efficiency. Depending on the surface area of the tank that is covered by the settler modules, the maximum surface loading value should be limited to 2–2.5 gpm/ft² (5–6.3 m/h) for cold regions and 3–3.5 gpm/ft² (7.5–8.8 m/h) for warm to hot regions.

The design of the inlet (to the settler module) is very important because it is essential to have a uniform flow into the entire module. In the case of a rectangular tank, the first quarter of the basin length is generally devoid of settler modules. This allows heavy floc to settle naturally and also provides an inlet buffer zone that improves the inlet flow condition to the settlers. In tube settler modules, an average flow velocity of approximately 0.5–0.65 fpm (0.15–0.2 m/m) is normally used in settling most alum floc. Moreover, an approaching flow velocity of approximately 2 fpm (0.6 m/mm) should be used in the tank upstream of the module installation.

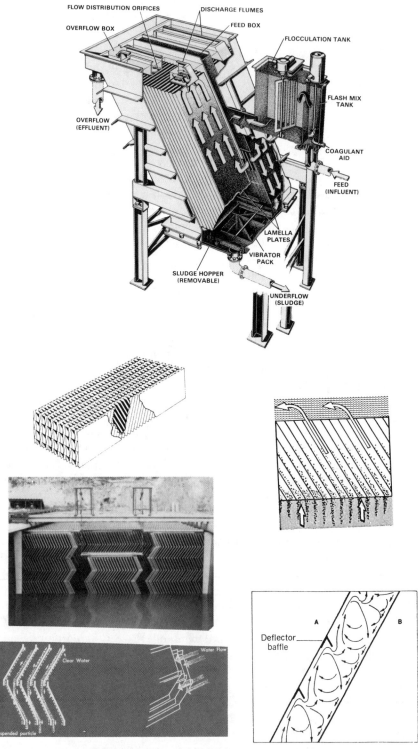

Figure 3.2.5-5 Several types of high-rate settler module.

The detention time of the settler module varies as a function of the type of module. Tube settlers generally have a detention time of 3.5–5 min, while tilted plate settlers provide a longer period of time; some units even require 15–20 min. The Reynolds and Froude number are not considered to be part of the design criteria, yet they act as good guides in designing the high-rate settlers. The Reynolds number for the settlers should be kept below 200, preferably around 50, and the Froude number should be greater than 10^{-5}.

When selecting a continuous sludge removal system, the number of available alternatives is rather limited. The most applicable types of sludge collector, for the sedimentation process, are the chain-and-flight (fiberglass reinforced plastic) and the cable-operated underwater bogies with squeegees. It should be noted that these types of sludge collecting system generally require a minimum water depth (in the tank) of 12 ft (3.7 m) to accommodate their installation.

Use of the skimming trough system is required because it uniformly collects clarified water from the area of the tank that is covered by the settler modules. However, the settler modules may be hung from the launders, provided that the launders are designed to tolerate this strain. The launders are usually spaced 10–13 ft (3–4 m) on center and hydraulic loading is usually less than 17 gpm/ft (12.5 m³/m · h). Figure 3.2.5-5 depicts a few high-rate settlers.

Sludge Removal Method   The settling characteristics of alum or ferric floc are typified by three basic stages: the hindered settling, transition, and compacting stages (Figure 3.2.5-6). During the hindered settling stage the floc slides down the smooth surface of the plastic plates (or tubes) which are inclined at a minimum of 60° from the horizontal plane. In the transition and compaction stages the floc no longer slides down as easily and the angle of repose (of the sludge) is generally 90°. In most cases only a few hours are necessary for the settled floc to reach the compaction stage.

A design flaw that is commonly made by inexperienced engineers is to provide a series of shallow angled sludge hoppers on the bottom of the basin. In this scheme, each hopper is fitted with a pipe containing a shut-off valve as the means of sludge removal. According to this design the valve is opened two to three times a day and the sludge is washed out with the water. However, even when the side walls of the hoppers are built 45° to 60° from the horizontal, this particular design does not work. The reason for the design failure is that a water passage, the same size as the pipe diameter, is created in the settled sludge when the valve is opened and the majority of the sludge therefore remains settled and undisturbed. In order to salvage this design, the sludge must be kept in a fluid condition at all times. Consequently, the bottom of the sedimentation tank should be designed as a flat surface with or without a slight slope toward the location of the sludge hopper, depending on the method of mechanical sludge removal.

In the United States the three continuous sludge removal systems that have been proved and widely used in recent years are the center-pivoted rotating

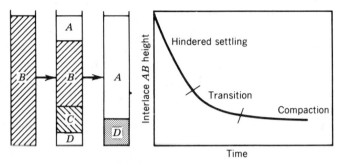

Figure 3.2.5-6 Settling characteristics of alum floc. A, supernatant zone; B, discrete settling zone; C, hindered settling zone; D, compression zone.

rakes, the chain-and-flight collector, and the traveling bridge. The center-pivoted rakes have a good history of service and low maintenance cost. Yet, there are four major drawbacks to this particular system: (1) a large amount of metal is used for the raking assembly; therefore it has corrosion and associated maintenance problems; (2) the center-pivot portion of the tanks must be very deep (over 20 ft) with a conically shaped bottom, which significantly increases tank construction cost; (3) even if the expandable arms are used to cover the corners of the square area, there is inadequate sludge removal in the corners; and (4) the capital cost for the system can potentially be high if the rectangular tank has a length to width ratio of 3:1 or larger. Furthermore, plant operators of treatment plants that only have the front half of the tank covered with the rotating rakes are not satisfied with its performance. The chain-and-flight collector has a maximum length of approximately 230 ft (70 m) per unit. The traveling bridge collector can be installed to any length of tank. It must be emphasized that a tank equipped with high-rate settler modules requires a compatible continuous sludge removal unit due to the problem of rapid sludge accumulation. This high rate of sludge accumulation may completely obstruct modules such as tube settlers unless the sludge is continuously removed. Figure 3.2.5-7 depicts the sludge removal units that are available from reputable equipment manufacturers.

Cost   When the geological and civil works are normal for a plant site, an ordinary rectangular sedimentation basin will cost from \$70 to \$80 per square foot of the basin surface area (mid-1980 cost on the West Coast). As a rule of thumb, the cost of the flocculation/sedimentation basin process is approximately 20% of the total plant construction cost. The reactor clarifiers are rather compact units and the manufacturers claim that they are cost effective, rendering a 10–20% savings in cost. Studies of water treatment plant bids show that the actual costs of these two types of clarifier are generally the same. Yet, the reactor clarifiers are preengineered by the manufacturers, up to 150 ft (45 m) in diameter, and the design cost of these units is therefore virtually eliminated—excluding the foundations and pipings.

High-rate settlers can be effective if they are both properly designed and installed. There is a vast difference in cost between the tube settler modules ($50/ft^2) and some parallel-plate settlers ($300/ft^2). Thus, selection should be based on a sound engineering evaluation.

The selection of the sludge removal units is likewise affected by cost. If the length of the rectangular sedimentation tank exceeds 230 ft (70 m), the traveling bridge is more cost effective: two chain-and-flight units (in series) are required to service the same distance. A traveling bridge unit can easily span 100 ft (30 m) and the capital cost for the equipment is significantly reduced by having one bridge span two or three tanks. This type of arrangement can be achieved because the width of each tank is usually less than 40 ft (12 m). Conversely, the chain-and-flight unit has a maximum width of approximately 20 ft (6 m). The tank width should therefore be the aliquot of 16–20 ft (5–6 m) in order to economize on equipment cost. Generally, the chain-and-flight unit is more economical than the traveling bridge only if the length of the tank is less than 230 ft (70 m); even if an additional cross collector is placed at one end of the tank, there is still a savings of approximately 30–50%.

Intermediate Diffuser Baffle Walls   A rectangular tank with one or two intermediate diffuser walls was first introduced in Japan by the author in 1952 and a number of functional large basins were designed soon after. The intermediate diffuser walls help to minimize the formation of density currents because the gentle vertical mixing effect upstream of the diffuser wall reduced the temperature difference between the influent water and the water in the tank. This gentle mixing also significantly promotes the growth of floc size and weight. Consequently, the floc settles very well, both in the first compartment and upstream of each diffuser wall. Furthermore, stable flow patterns occur downstream of each diffuser wall and the wind effect on the flow in the tank is reduced. It must be emphasized that overuse of intermediate diffuser walls creates difficulties during the installation of the continuous sludge removal equipment.

The aforementioned type of basin was installed in the cities of Tokyo, Kobe, and Kanagawa (Japan) under bogie type sludge collectors. The collectors were placed between the diffuser walls (across the flow direction), in contrast to the longitudinal direction of most ordinary tanks.

Okuno et al. (1980) (Tokyo Metropolitan Public Works) compared the operational efficiency of four types of full-scale wastewater sedimentation tank: a central feed and peripheral collecting circular tank, a peripheral circular tank, a rectangular tank with a tray (double-decked), and a rectangular tank with two intermediate diffuser walls. The results of that study showed that tanks with two intermediate diffuser walls were far superior and that the double-deck tanks did not perform satisfactorily (Figure 3.2.5-8), primarily due to the unstable flow characteristics of the tank.

Figure 3.2.5-7 Mechanical sludge collection systems. (a) Chain-and-flight type. (b) Traveling bridge with suction headers. (c) Cable-operated underwater bogie sludge collector.

Figure 3.2.5-7 (Continued)

Shape of the Tank When first designing a rectangular tank it appears that there is much freedom in selecting the shape of the basin. However, experience has shown that basins that are both wide and deep tend to exhibit flow instability and a distinct density flow pattern. On the other hand, a basin that is narrow, shallow, and long will have flow stability and a minimal amount of flow short-circuiting. Rectangular tanks with a 180° turn at the midlength and crescent shaped tanks are both ineffective because of flow short-circuiting: these designs can only be salvaged if a very long detention time is provided. The actual basin design must also take the following items into consideration: the installation of sludge removal equipment, the potential installation of high-rate settler modules, the presence of turbulence caused by winds, the scouring of bottom sediments, and the cost of the tank.

The following guidelines can help to determine the physical shape of the sedimentation tanks: (1) the water depth should be 10–16 ft (3–5 m); (2) the ratio between the length (L) and the width (W) should be $L/W = 6:1$, with a minimum of 4:1; (3) the ratio between width (W) and the water depth (H) should be $W/H = 3:1$, with a maximum of 6:1.

From an economical as well as a practical viewpoint, the preferred height of the tank freeboard is approximately 2 ft (0.6 m). The freeboard acts as a wind barrier and helps to prevent wind-induced waves from splashing on the walkway.

The flow characteristics of the sedimentation basin can be estimated by the Reynolds (Re) and the Froude numbers (Fr):

$$\text{Re} = \frac{vR}{\nu} < 2000$$

$$\text{Fr} = \frac{v^2}{gR} > 10^{-5}$$

where v = displacement flow velocity (m/s),
R = hydraulic radius, or $R = A/P$ (m),
A = area of flow passage (m^2),
P = wetted perimeter (m),
ν = kinematic viscosity (m^2/s),
g = gravity constant (9.81 m/s^2).

It is rather surprising to discover that an ordinary basin has a Reynolds number that is over 15,000 and a Froude number that is less than 10^{-6}. Consequently, the flow characteristics of an ordinary basin are far more inferior than most engineers would expect. The flow characteristics of an ordinary basin can be significantly improved if longitudinal baffles are installed parallel to the side walls of the basin. This arrangement divides the basin into a series of narrow channels, thereby decreasing the flow passage area (A). The proper installation of longitudinal baffles in this manner will reduce the Reynolds number to 50% of that of ordinary basins and will increase the Froude number to the desired range of 10^{-5}.

The minimum width between the longitudinal baffles should be 10 ft (3 m) to allow for the installation of sludge removal equipment, for easy maintenance, and for economic considerations. The baffles may be composed of thin concrete walls or even wooden planks.

One of the least desired shapes is a basin that has a 180° turn at its midlength. A significant number of operational basins have been designed in this manner, primarily for space-saving reasons. These tanks have either a horizontal or vertical turn at the midlength and always exhibit poor efficiency. This inefficiency is due to the presence of turbulence near the area where the water must make the turn and the existence of large "dead spaces" within the basin.

Upflow and Reactor Clarifiers The upflow and reactor clarifiers are proprietary units that have their basic size and blueprints preestablished by the equipment manufacturers (based on the flow rates). These units can be grouped into three categories: simple upflow clarifiers, reactor clarifiers, and sludge blanket reactor clarifiers.

The upflow and reactor clarifiers offer several advantages over the rectangular basins: (1) compact design and economical use, since no design effort is required by the engineer; (2) simpler sludge removal; and (3) efficient

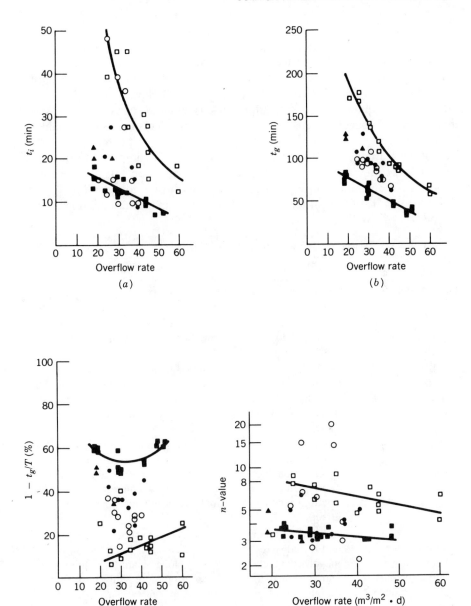

Figure 3.2.5-8 Hydraulic characteristics of five types of clarifier: ○ CFPC, central feed peripheral collect; ● PFPC, peripheral feed peripheral collect; ▲ RDD-1, rectangular double deck-1; ■ RDD-2, rectangular double deck-2; □ RTMD, rectangular tank with multidiffuser wall. (a) Short circuit. (b) Mean retention time. (c) Dead space ratio. (d) Fair's n (stabilization index). (Adapted from Okuno et al., "Analysis of Existing Clarifiers and Design Considerations for Better Performance," Tokyo Metropolitan Government Report, 1980.)

clarification due to the seeding effect. This seeding effect is achieved through the recirculation of the floc and the adsorption of the floc by the sludge blanket, as long as the flow rate and the raw water quality are fairly constant; (4) better clarification efficiency due to improved flocculation and adsorption effect by the sludge blanket; and (5) capable of delaying degradation of effluent water quality, in case of improper coagulant dosage or even a short interruption of coagulant feed, due to the buffering effect of the sludge blanket zone.

However, there are also several disadvantages: (1) compared to the horizontal tanks they require more stringent operational control; (2) there is a rapid loss of efficiency during hydraulic and solid overloading and hydraulic shock loading; (3) frequent temperature fluctuations induce distinct flow short-

TABLE 3.2.5-3   Selection Guide of Proprietary Clarifiers

Type of Unit	Advantages	Disadvantages
Detritus tank	Potential low capital cost	Flow short-circuiting, potential clogging of the grit removal pipe, corrosion of the raking system
Center feed and peripheral collection	Low capital cost, easy sludge removal	Flow short-circuiting, limitations on the practical size of the unit, less tolerant of shock loading
Peripheral feed and peripheral collection	Good performance for raw water with a high suspension of solids	Potential for flow short-circuiting, performance relies on the manufacturer's design, limitations on the practical size of the unit
Reactor clarifier with high recirculation and mechanical sludge plow	Compact and economical, tolerant of shock loading, good clarification due to a seeding effect	Corrosion of underwater metal parts, less reliable due to its dependency on one drive unit, requires extensive yard piping to connect each unit, practical size limitations
Reactor clarifier with sludge blanket zone and mechanical sludge collector	Good lime softening and turbidity removal, compact and economical design, several manufacturers to choose from	Very sensitive to shock loading, requires 2–4 days to build up the necessary sludge blanket, requires greater operator skills, corrosion of underwater metal parts
Reactor clarifier with hydraulic flocculation and sludge removal	Good performance under the proper conditions, no maintenance of mechanical equipment	Sludge will choke the unit if heavy solids are present in the raw water, no operator control over the flocculation process, process functions will terminate if the water dousing system fails, produced by only one manufacturer (patented)

circuiting; (4) if the unit is not equipped with a positive raking mechanism, it is difficult to remove the sludge during periods of high turbidity; and (5) more time is required to produce the sludge blanket whenever the sludge blanket clarifier is drained. Moreover, a significant amount of metal is used on the interior of these tanks and the problem of corrosion and maintenance must be addressed.

These types of clarifier are usually employed in industrial and municipal applications, such as lime softening of groundwater and clarification of turbid water. It must be emphasized that the flow rates and the raw water quality in these situations are nearly constant at all times. Table 3.2.5-3 is a selection guide for these units. Refer to Figure 3.2.5-9 for illustrations of each type of unit.

Type and Selection Guide   This section presents a brief summary of the typical sedimentation tanks used in the water treatment process and a few guidelines for making the proper selection.

Available Alternatives

Horizontal flow	Long rectangular tanks
Horizontal flow	Center feed tanks (circular or square)
Horizontal flow	Peripheral feed circular tanks
Upflow clarifiers	Proprietary units
Reactor clarifiers	Proprietary units, recirculation units, sludge blanket units
High-rate settler modules	Proprietary items, tube settlers, parallel-plate modules

Selection Criteria   The selection of the sedimentation tanks should be based on the following criteria:

1. The type of suspended matter that is to be removed: for example, grit, silt (plain sedimentation), chemically flocculated suspended matter, or biological floc.
2. The overall process train and the role of sedimentation: for example, pretreatment for rapid sand filters, pretreatment for multimedia high-rate filters, clarification of lime softened floc prior to filtration, or removal of sand and silt upstream of the intake pumps.
3. The topography and ground conditions of the plant site.
4. The potential for hydraulic shock loading and the degree of fluctuation in the quality of the influent water.
5. The requirements for the quality of the settled water with respect to subsequent processes.

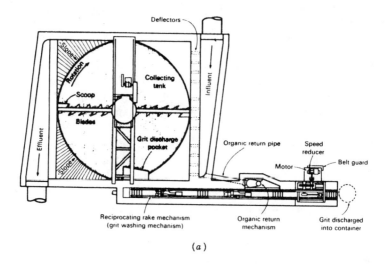

(a)

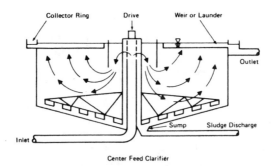

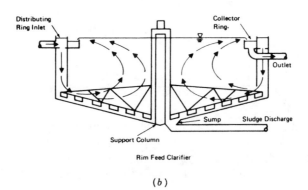

(b)

Figure 3.2.5-9 Various types of clarifier. (a) Detritus tank with grit washer (Dorr–Oliver). (b) Concept of center feed and peripheral feed clarifiers. (c) Reactor clarifier with high recirculation (Infilco). (d) Sludge blanket type of reactor clarifier (Graver Water Company). (e) Reactor clarifier with hydraulic flocculation and sludge removal (Degremont).

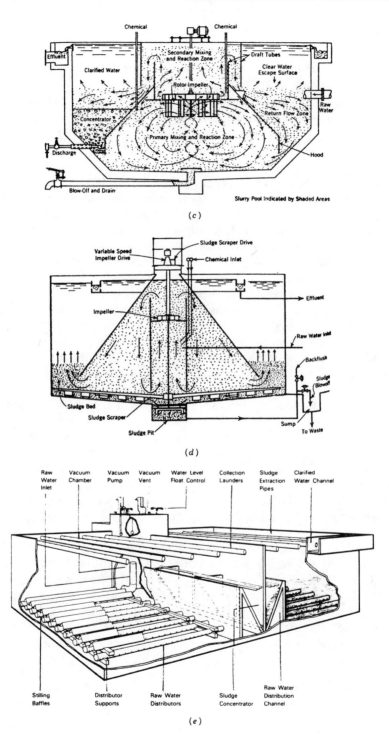

(c)

(d)

(e)

Figure 3.2.5-9   (Continued)

6. The nature and amount of sludge that will be produced.
7. The local climatic and geological conditions.
8. The utilization of site dimensions and considerations with respect to future plant expansion.
9. The capital required for the sedimentation tanks and the operation and maintenance costs.
10. The time period that is allowed for the design and building of the treatment plant.

Order of Preference   The order of preference, in selecting the sedimentation process, is established by both the performance and the operation and maintenance costs of each type. Based on past experience, the order of preference among the various types of plain sedimentation tank is (1) a long rectangular tank, (2) a rectangular tank containing high-rate settler modules, and (3) proprietary units. Similarly, the order of preference among the clarifiers is (1) a long rectangular tank, (2) a rectangular tank containing high-rate settler modules, and (3) a reactor clarifier.

A rectangular tank with a horizontal flow is generally favored because of its hydraulic stability (if properly designed) and its tolerance of shock loadings. This type of tank also has a predictable performance and is capable of withstanding a flow rate that is twice the designed rate without significant deterioration in the quality of the settled water. Additionally, this type of tank is easily operated, has low maintenance, easily adapts to the installation of high-rate settler modules, and uses space efficiently, especially with the common wall design.

The use of reactor clarifiers should be considered and evaluated if: (1) the plant flow rate and the raw water quality are relatively constant; (2) the raw water turbidity is always between 5 and 300 NTU; (3) using lime softening or other processes that benefit from the seeding effect; (4) the units are installed in a housing; (5) there are time constraints on both the design and construction, as well as budget constraints; and (6) the owner prefers their use.

Three types of clarifier are not recommended for removing alum floc because of their hydraulic instability: horizontal flow with center feed, peripheral feed, and the simple upflow clarifiers. The use of rectangular tanks with trays (including the double-deck type) is generally not recommended unless the available site area is very limited since they do not meet the expected level of settling efficiency, have high construction costs, and present difficulties when installing mechanical sludge removal equipment on each tray (unless the clearance is over 6 ft).

Discussion of Alternatives

Horizontal Flow and Long Rectangular Basins   As previously discussed, under the section "Overall Configuration," the most cost effective and space-saving configurations are rectangular basins arranged longitudinally and side-by-side. The important design requirement is a uniform distribution of feed water at the entrance of the basin. Effluent control by means of a series of long skimming troughs and a baffle wall is not as effective as some textbooks and design guides published by regulatory agencies suggest. These subjects are described under the section entitled "Inlet and Outlet of the Basin."

Two major pieces of basin equipment commonly used are the sludge collection system and high-rate settler modules (see previous discussion). Proper installation of the tube settler modules requires a minimum tank depth of 12 ft, depending on the type of sludge collection system under the modules. A typical installation profile is illustrated in Figure 3.2.5-10.

Sludge Collection Systems   The manufacturers of rectangular and circular basins offer several different types of continuous mechanical sludge collectors. The main types are (1) chain-and-flight, (2) traveling bridge with sludge scraping squeegees and a cross collector at one end of the tank, (3) traveling bridge with sludge suction headers and pump or siphon mechanism, (4) float supported sludge suction headers pulled by wires, and (5) underwater bogies with squeegee (pulled by wires).

The selection of the proper sludge collection system depends on the local weather conditions; the nature and quantity of suspended solids in the raw water, the type and amount of coagulant used, the shape of the tank, the installation (or lack) of high-rate settlers, aesthetics, and cost.

In regions of cold weather the formation of ice on the surface of the water prohibits the use of moving parts. Therefore, the traveling bridge and float supported sludge headers should not be used in these regions, unless the tanks are either covered or housed in a heated building. There are even a few case histories where the traveling bridge has been derailed by gale force winds.

Other important factors that should be considered are the nature of the suspended solids and the quantity of solids expected to settle in the tank. For example, (1) water containing large amounts of freshwater clams will have

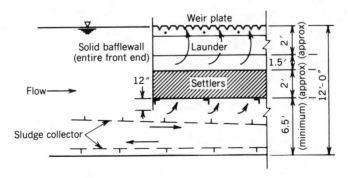

Figure 3.2.5-10   Required tank depth for tube settler installation.

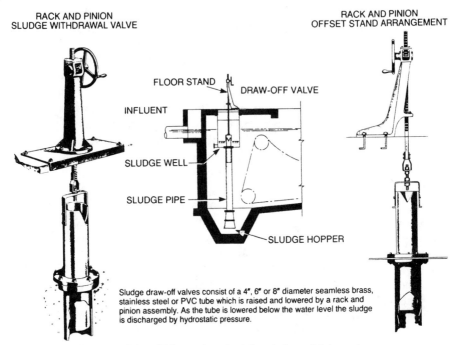

RACK AND PINION
SLUDGE WITHDRAWAL VALVE

RACK AND PINION
OFFSET STAND ARRANGEMENT

FLOOR STAND

DRAW-OFF VALVE

INFLUENT

SLUDGE WELL

SLUDGE PIPE

SLUDGE HOPPER

Sludge draw-off valves consist of a 4″, 6″ or 8″ diameter seamless brass, stainless steel or PVC tube which is raised and lowered by a rack and pinion assembly. As the tube is lowered below the water level the sludge is discharged by hydrostatic pressure.

Figure 3.2.5-11 Telescopic valves for sludge withdrawal.

shells settle in the grit chamber and settling tanks; (2) suction type sludge collector systems require the orifices of their headers to be periodically cleaned because the size of the orifices is usually less than $\frac{3}{4}$ in. (20 mm) and therefore the orifices become frequently clogged; (3) lime softening of hard water tends to produce large amounts of heavy sludge, which may become troublesome for suction type units; and (4) the chain-and-flight units require frequent maintenance whenever large amounts of abrasive grit are handled on a continuous basis (frequent maintenance may also be indicative of improper installation and/or use of inferior parts).

The shape of the tank also limits the type of sludge collection unit that may be installed. A unit with a center-pivoted rotating rake is best suited for circular or square basins. Since most sedimentation tanks are long and rectangular in shape, the traveling bridge or chain-and-flight units are commonly employed. The choice between the two types is determined by cost, aesthetics, and the presence of settler modules in the tank.

A chain-and-flight unit can service a maximum tank length of approximately 200 ft (60 m) if it is run on a continuous basis, to remove a heavy accumulation of sludge. However, one manufacturer claims that it can provide a unit that can cover 300 ft (90 m), but it must be operated in an intermittent fashion, such as in the removal of a light accumulation of sludge. If the basin length is longer than these limits, two sets of collectors are required. Basins containing high-rate settlers will have space limitations and therefore require either the chain-and-flight or underwater bogies to be installed.

Unlike the chain-and-flight modules, the traveling bridge collectors can service any length of tank. However, it is more cost effective if the basin length exceeds 260–300 ft (80–90 m). Yet, residents living near these types of plant may complain that the traveling bridge collectors are unsightly. The bridge is often travel on rails with the cog drive mechanism but it can be on solid rubber tire wheels without rails.

A suction system with a traveling bridge is generally capable of withdrawing more concentrated sludge than the scraper. This type of unit also brings the sludge to a channel that is above the water level, thereby allowing the plant operator to visually inspect the consistency of the sludge and smell any extraordinary odor. Telescoping valves achieve the same effect when used in conjunction with a sludge scraper unit (Figure 3.2.5-11). If the sludge is sucked up by means of a pump or siphon, the discharge rate of the sludge should be designed to be approximately 10 gpm per linear foot (120 L/m·min) of the tank width.

In summary, a traveling bridge with suction headers is the best choice for large tanks that do not have high-rate settlers but have normal levels of alum or ferric sludge deposits. In regions with moderate to warm weather, plant sites far from residential areas are prime candidates for the traveling bridge type of suction collector. Chain-and-flight units may be used under any conditions. However, their application to very long and wide tanks is not cost effective because the maximum length of one chain-and-flight unit is limited to 20 ft (6 m). Additionally, water containing large amounts of abrasive grit would necessitate a high maintenance rate. If cost control is a major issue, float supported suction headers and underwater bogie units may be considered as alternatives.

The traveling speed of the continuous sludge removal unit is usually 1 fpm (0.3 m/min). This speed ensures that sludge resuspension does not occur. Units that can lift the sludge collection header onto a return path may travel at speeds that are 5–10 times higher than on the sludge collection path.

For developing countries or small plants in isolated areas, the bottom of the sedimentation tank should have a slope of at least 1:300 toward the tank inlet, to ensure easy manual sludge removal through the use of water jets. If mechanical sludge scraper equipment is used, the slope may range from 1:300 to 1:600. When a sludge suction mechanism is adopted, the entire bottom of the tank may be level. Operators of plants using this type of system have never made adverse comments regarding the design of the basin.

The underflow rate (associated with sludge removal) of the horizontal flow and long rectangular basins is usually 0.1–0.2% of the plant flow rate. The concentration of solids in the sludge is 0.5–2%, depending on the type of mechanical unit employed and operational adjustments.

Design Criteria This section presents the basic design criteria for three types of clarifier: grit chamber, rectangular sedimentation tank, and sedimentation tank with high-rate settlers.

Grit Chamber (Removal of sand and silt)

Type	Rectangular tank with horizontal flow
Minimum size of grit to be removed	0.1 mm
Minimum number of tanks	Two*
Water depth	10–16 ft (3–5 m)
Mean flow velocity	10–15 fpm (3–4.5 m/min)
Detention time	6–15 min
Surface loading	4–10 gpm/ft² (0.4–1.0 m/h)
Length/width ratio (*L/W*)	1:4 to 1:8
Water depth/length ratio	Minimum of 1:8

* One tank with a bypass channel may be used if the anticipated amount of grit is small.

Rectangular Sedimentation Tank (Removal of flocculated suspended matter)

Minimum number of tanks	Two
Water depth	10–15 ft (3–4.5 m)
Mean flow velocity	1–3.5 fpm (0.3–1.7 m/min)
Detention time	1.5–4 h
Surface loading	0.5–1.0 gpm/ft² (1.25–2.5 m/h)
Length/width ratio (*L/W*)	Minimum of 1:4
Water depth/length ratio	Minimum of 1:15
Launder weir loading	12–18 gpm/ft (9–13 m³/m·h) (could possibly be 20 times higher)
Sludge collector speed	
For the collection path	1–3 fpm (0.3–0.9 m/min)
For the return path	5–10 fpm (1.5–3 m/min)

Sedimentation Tank with High-Rate Settler (Tube settler)

Minimum number of tanks	Two
Water depth of the tank	12–15 ft (3.6–4.5 m)
Surface loading of the tank area covered by the settler	1.5–3 gpm/ft² (3.8–7.5 m/h)
Mean velocity in the tube settler module	Maximum of 0.5 fpm (0.15 m/min)
Detention time in the settler module	Minimum of 4 min
Launder weir loading	5–20 gpm/ft (3.8–15 m³/m·h)

Example Design Calculations

Example 1   Estimated Density Flow Velocity

Given

Water depth of the sedimentation tank	4 m (13.1 ft)
Water temperature of the tank	20°C
Water temperature of the influent	19°C

Determine   The estimated velocity of the density flow in the sedimentation tank.

Solution   Use the density flow formula recommended by the Japan Society of Civil Engineering (1963):

$$v = \left(2g \frac{\Delta\rho}{\rho} \frac{\Delta h\alpha}{k} \right)^{0.5}$$

Assume that

$$\alpha = 0.5$$

$$k = 2$$

$$\Delta h = 2 \text{ m}$$

$$\Delta\rho = 0.99823 - 0.99823 = 0.0002 \text{ g/cm}^3$$

Therefore,

$$v = \left(2 \times 9.81 \times \frac{0.0002}{0.99823} \times \frac{2 \times 0.5}{2} \right)^{0.5} = 0.044 \text{ m/s}$$

or $v = 2.66$ m/min

Example 2   Grit Chamber Design

Given

Plant flow rate	100 mgd (4.4 m³/s)
Minimum size of the sand to be removed	0.1 mm
Water temperature	50°F (10°C)

Determine

(i) The number and shape of the tanks.
(ii) The size and water depth of each tank.
(iii) The configuration of the inlet valves and diffuser wall.

Solution

(i) Provide two rectangular tanks for the most cost effective design.
(ii) The size of each tank is calculated as follows. The settling velocity of 0.1 mm of sand is

$$v_0 = 1.57 \text{ fpm} \quad \text{or} \quad 8 \text{ mm/s} \quad \text{(Re. Table 3.2.5-1)}$$

Make the water depth of each tank 12 ft (3.7 m) and the width of each tank 40 ft (12 m). The mean flow velocity is

$$Q/a = 4650 \div (12 \times 40) = 9.7 \text{ fpm}$$

$$L = K(h/v_0)v, \quad \text{where } K = 1.5$$

$$L = (1.5) \left(\frac{10}{1.57}\right) \times 9.7 = 92.7 \text{ ft}$$

Say 100 ft (30 m) for the tank length.

Detention time: $\quad V/Q = (12 \times 40 \times 100) \div 4650$

$$= (\text{ft}^3) \div (\text{cfm}) = 10.3 \text{ min}$$

Surface loading: $\quad Q/A = 34{,}750 \div (40 \times 100)$

$$= (\text{gpm}) \div (\text{ft}^2)$$

$$= 8.7 \text{ gpm/ft}^2 < 10 \text{ gpm/ft}^2 \quad \text{O.K.}$$

Check the calculations via the Reynolds (Re) and Froude (Fr) numbers.

$$\text{Reynold's number:} \quad Re = \frac{vR}{v}$$

where $R = A/P = \dfrac{12 \times 40}{40 + (2 \times 12)} = 7.5 \text{ ft}$

$$v = 1.41 \times 10^{-5} \text{ ft}^2/\text{s}$$
$$v = 9.7 \div 60 = 0.162 \text{ fps}$$

Therefore,

$$Re = \frac{0.162 \times 7.5}{1.41 \times 10^{-5}} = 86{,}200 \quad \text{OK for grit chamber}$$

$$\text{Froude number:} \quad Fr = \frac{v^2}{gR}$$

$$= \frac{0.162^2}{32.2 \times 7.5}$$

$$= 1.09 \times 10^{-4}$$

(iii) For the inlet valves and diffuser walls, use two 36 in. rectangular butterfly valves at each tank inlet to facilitate easy removal of grit in the inlet channel. Avoid using sluice gates because they tend to "freeze up" with infrequent operation. The velocity through the valve is

$$v = Q/A = (50 \times 1.55) \div (2 \times 3^2) = 4.3 \text{ fps}$$

$$\text{Headloss through the valve} = \frac{1}{2\,g} \left(\frac{38.75}{0.7 \times 9} \right)^2$$

$$= 0.59 \text{ ft}$$

Provide a straight line diffuser wall that is parallel to the end wall of the inlet with a space of 6 ft. A total of 420 orifices (6 in. diameter) are required for the diffuser wall in order to allow a flow velocity of 2 fps through each orifice at a flow rate of 100 mgd. The orifices should be spaced 15 in. apart (center to center) in vertical rows and 21 in. (center to center) for the horizontal direction. The ratio of the orifice area to the tank cross section is approximately 10%. The headloss through the orifice is approximately 1.5 in. at 100 mgd. Figure 3.2.5-12 illustrates the configuration of the grit chamber.

Example 3 Rectangular Sedimentation Tank Without a Tube Settler

Given

Plant flow rate	2 m³/s (46 mgd) maximum daily rate
Suspended solids to be removed	Alum floc
Measured floc settling velocity	30 mm/min at 10°C
Expected settled water turbidity	2 NTU or less for rapid sand filters
Minimum ambient temperature	−3°C
Potential future plant expansion	Up to 4 m³/s
Location of plant site	Far from residential areas

Determine

(i) The number of tanks.
(ii) The size of each tank.

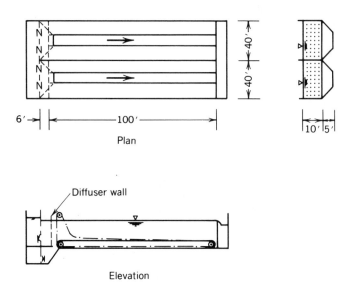

Figure 3.2.5-12 Sketch of the grit chamber.

(iii) The configuration of the tank inlet and diffuser wall.
(iv) The type of sludge removal system.
(v) The configuration of the effluent structure.

Solution

(i) Either two or three tanks may be selected. Although three tanks provide more operational flexibility (one tank may be "off-line" due to repairs), two tanks should be chosen for this example because (1) future plant expansion will provide the additional tanks; (2) the daily maximum flow rate is 2 m³/s and it is safe to assume that the daily winter flow rate will be approximately 1.0 to 0.7 m³/s (more suited to two tanks); and (3) the two tank design is more cost effective.

(ii) Considering that the standard maximum width of the chain-and-flight sludge collector is 6 m (20 ft), an aliquot of 6 m should be evaluated for the width of the tank: 12 m, 18 m 24 m.

In the case of the traveling bridge sludge collector, a tank width in the range of 12–30 m (40–100 ft) is considered to be cost effective. The tank should be 18 m (60 ft) wide to maintain a good tank length to width ratio.

Based on floc settling velocity measurements (bench scale studies conducted during the winter seasons), the hydraulic surface loading rate of the tank can be set at 1.8 m/h (0.72 gpm/ft²) with a floc settling velocity of 30 mm/min. Although one might think that this value is too conservative for the winter season (a flow that is less than 1 m³/s), the plant may use only one tank to treat the plant flow, thereby allowing the tank loading rate to remain normal. Consequently, 1.9 m/h (0.75 gpm/ft²) is used as the design criterion of the tank. This number includes a comfortable safety margin for all seasons.

The required total tank surface area is

$$(2 \times 3600) \div 1.9 = (m^3/h) \div (m/h) = 3790 \ m^2$$

Since each tank has a width of 18 m, the required length would be

$$3790 \div (2 \times 18) = (m^2) \div (m) = 105 \ m \quad (343 \ ft)$$

The water depth of each tank should be 4 m (13 ft) to facilitate the possible installation of tube settler modules in future.

Check the values:

$$\text{Detention time at } Q_{max} = \frac{(18 \times 105 \times 4) \times 2}{2 \times 60}$$

$$= \frac{m^3}{m^3/min}$$

$$= 126 \ min \ (2 \ h \ and \ 6 \ min)$$

$$\text{Detention time at } Q_{ave} = 126 \ min \times 1.5^*$$

$$= 189 \ min \ (3 \ h \ and \ 9 \ min)$$

* The maximum daily flow rate is usually 1.5 times the average daily flow rate.

$$\text{Tank length to width ratio} = 105 \ m \ to \ 18 \ m$$

$$= 5.8{:}1 > 4{:}1 \quad \text{O.K.}$$

$$\text{Tank length to water depth ratio} = 105 \ m \ to \ 4 \ m$$

$$= 26.3{:}1 > 15{:}1 \quad \text{O.K.}$$

$$\text{Mean flow velocity:} \quad Q/A = (2 \times 60) \div 2 \times (18 \times 4)$$

$$= (m^3/min) \div (m^2)$$

$$= 0.83 \ m/min < 1.1 \ m/min \quad \text{O.K.}$$

$$\text{Reynolds number:} \quad Re = \frac{vR}{v}$$

where $v = 0.83 \ m/min = 1.39 \times 10^{-2} \ m/s$

$$R = \frac{4 \times 18}{(2 \times 4) + 18} = 2.77 \ m$$

$$v = 1.31 \times 10^{-6} \ m^2/s \ at \ 10°C$$

Therefore,

$$Re = \frac{1.39 \times 10^{-2} \times 2.77}{1.31 \times 10^{-6}}$$

$$= 29,620 \quad \text{No}$$

Re is too high and the flow in the tank is highly turbulent. The preferred Reynolds number is less than 18,000.

$$\text{Froude number:} \quad Fr = \frac{v^2}{gR}$$

where $v = 1.6 \times 10^{-2}$ m/s
$R = 2.77$ m
$g = 9.81$ kg/s²

Therefore,

$$Fr = \frac{(1.39 \times 10^{-2})^2}{9.81 \times 2.77}$$

$$= 7.1 \times 10^{-6} < 10^{-5} \quad \text{No}$$

The number is too small and the flow is therefore unstable.

How much will the flow characteristics in the tank improve if longitudinal baffles (partition walls) are installed in each tank?

Trial 1: One Longitudinal Wall in the Middle

$$R = \frac{(4 \times 9) \times 2}{(9 + 2 \times 4) \times 2} = \frac{m^2}{m} = 2.21 \text{ m}$$

Since v is practically unchanged,

$$Re = \frac{1.39 \times 10^{-2} \times 2.12}{1.31 \times 10^{-6}} = 22,670$$

This Reynolds number is better than before but still too high.

$$Fr = \frac{(1.39 \times 10^{-2})^2}{9.81 \times 2.12} = 9.2 \times 10^{-6} < 10^{-5}$$

The Froude number is still a little too low.

Trial 2: Two Longitudinal Walls 6 m Apart (on Center) Forming Three Channels

R = 1.71 m	v is practically unchanged
Re = 18,280	Tolerable
Fr = $1.2 \times 10^{-5} > 10^{-5}$	O.K.

Trial 3: Six Longitudinal Walls 3 m Apart (on Center) Forming Seven Channels

R = 1.1 m	Flow velocity (v) is a little faster but practically unchanged
Re = 11,850	Fair
Fr = 1.8×10^{-5}	Good

As demonstrated, a tank with a long and narrow channel is hydraulically superior. It is therefore understandable why a tube settler or a parallel-plate settler performs well. Thus, from a practical perspective, the choice would be to use two longitudinal baffles to create three channels that are 6 m (20 ft) in width. The revised tank length to width ratio (L/W) is 105 m to 6 m or 17.5:1.

In summary, the features of each tank are as follows:

Average water depth	4 m (13 ft)
Tank width	18 m (60 ft) with two longitudinal baffles
Tank length	105 m (343 ft)
Detention time	2 h and 6 min
Surface loading	1.9 m/h (0.75 gpm/ft²)
Mean flow velocity	0.83 m/min (2.7 fpm)

(iii) When the tank is fed by a common inlet channel, two basic types of settling tank inlet may be used: flow splitting weir and submerged ports with valves. If the flocculation tank is the head end of the sedimentation tank, a baffle or diffuser wall may be used to separate the two tanks.

In the case where the tanks are fed directly from a feed channel, the design of the channel should be large enough to avoid producing significant headloss from one end of the channel to the other. This condition is particularly important for the flow splitting weir.

The weir inlet has problems not only with leakage every time a tank is isolated with stop logs but also with sludge accumulation in the feed channel. From a practical aspect, the best decision is to place submerged ports at the bottom elevation of the inlet channel by means of butterfly valves. The inlet ports should provide approximately 0.10 m (4 in.) of controlled headloss to ensure even flow distribution to each tank. The design of the inlet channel is delineated in Section 3.2.3 under "Example Design Calculations" (inlet channel).

Provide three submerged holes with butterfly valves as the tank inlet. Each hole should be 0.6 m in diameter (24 in.) so that the flow velocity through each orifice is

$$v = \frac{Q}{\Sigma a} = \frac{2}{6 \times (0.785 \times 0.6^2)}$$

$$= 0.708 \text{ m/s} \quad (2.3 \text{ fps})$$

$$\Delta h = \frac{1}{2g} \left(\frac{q}{ca} \right)^2$$

$$= \frac{1}{19.62} \left(\frac{2 \div 6}{0.7 \times 0.283} \right)^2$$

$$= 0.086 \text{ m} \quad (3.4 \text{ in.}) \quad \text{O.K.}$$

The diffuser wall should provide approximately 0.23 m/s (0.75 pfs) of flow velocity through each port at the maximum daily flow rate.

$$\text{Required total port area} = \frac{Q}{v} = \frac{2 \text{ m}^3/\text{s}}{0.23 \text{ m/s}}$$

$$= 8.7 \text{ m}^2$$

The total number of 0.12 m (diameter) ports that are required is (8.7 m²)/[0.785 × (0.12 m)²] = 770.

Provide 10 rows of ports in the horizontal direction with a spacing of 0.4 m and 39 rows in the vertical direction, spaced 0.45 m apart (13 rows in each 6 m segment). In doing this, the actual total number of ports is 390 for each tank or a total of 780 (total area of 8.82 m²). See Figure 3.2.5-13 for the design.

Check the values:

Flow velocity in each port at Q_{max}:   2 m³/s ÷ 8.82 m = 0.227 m/s

Headloss through diffuser wall: $\dfrac{1}{2g} \left(\dfrac{2}{0.7 \times 8.82} \right)^2$

$$= 0.0053 \text{ m} \quad (0.2 \text{ in.})$$

(iv) The traveling bridge with pump suction is the most preferred type of sludge removal system because: (1) the tank length is over 100 m and therefore the traveling bridge is the most cost effective; (2) one unit can span two 18 m (width) tanks and this is very cost effective; (3)

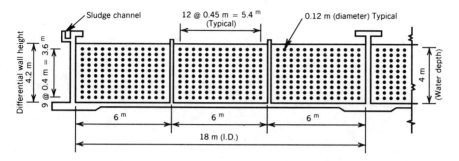

Figure 3.2.5-13 Inlet diffuser wall (elevation).

heavy ice formation (which prohibits movement of the bridges) is not an issue since the lowest anticipated temperature is only $-3°C$; (4) the tank bottom does not have to be sloped with this system, thus simplifying the design and minimizing construction cost; and (5) the aesthetics of the bridge (the industrial equipment look) is not an important issue because the plant is far from residential areas.

(v) Configuration of the effluent structure. Some regulatory agencies require an extensive launder system to be included in the plant design. An acceptable hydraulic weir loading rate is approximately 10.8 m³/ m·h (15 gpm/ft) for the launders. The required weir length (based on the criterion) is

$$(2 \times 3600) \div 10.8 \quad \text{or} \quad 666 \text{ m for two tanks}$$

A few engineers also believe that the installation of a baffle wall, one similar to the influent diffuser wall, minimizes the density flow. However, experience indicates that neither the long launders nor the baffle wall (located at the effluent end) significantly improves the settling efficiency of long rectangular settling tanks. Consequently, the end walls of the tanks may be built as an overflow of the effluent wall.

It is also necessary to provide three 0.2 m (diameter) holes equipped with butterfly valves in the end wall near the bottom elevation of the tank. These valves are used in backfilling the tank after it is cleaned and backflashing during the cleaning process. Backfilling of the tank is especially important for flocculation–sedimentation combination tanks that have vertical shaft flocculators; the flocculators will not function until the water level in the tank reaches half the depth of the tank. If turbid water is introduced from the inlet side during backfilling of the tank, the unflocculated water will flow to the end of the tank until the water level reaches half the depth of the tank. The unpretreated water will therefore flow directly into the filters after the tank is filled.

Example 4   Rectangular Tank with Tube Settlers

Given   The same criteria apply as in Example 3 with the addition of the following:

Dual media filters   After the sedimentation process
Minimum water temperature $-15°C$
Location of plant site   Near residential areas and the area
                                                              of the site is small.

Determine

(i) The number of tanks.
(ii) The required tank area that will be covered by the settler modules.
(iii) The size of each tank.
(iv) The configuration of the tank inlet and diffuser wall.
(v) The configuration of the launders and sludge collection system.

Solution

(i) The number of tanks is the same as in Example 3—two tanks.
(ii) The required tank area that is to be covered by the tube settler module may be calculated. The following relation can be established from Figure 3.2.5-14:

$$S_0 = \frac{Q}{A} \frac{W}{h \cos\alpha + w \cos^2\alpha}$$

Since tube settler modules have $\alpha = 60°$,

$$S_0 = \frac{Q}{A} \frac{W}{0.5 h + 0.25 W}$$

where A = surface area of the settling tank that is covered by the
          settler module (m²),
      Q = flow rate of the tank (m³/s).

A standard tube settler module has the following dimensions: $h = 0.55$ m (1 ft 9 in.) and $W = 0.05$ m (2 in.). Therefore,

$$S_0 = \frac{Q}{A} \frac{0.05}{(0.05 \times 0.55) + (0.25 \times 0.05)} = \frac{Q}{A} (0.174)$$

The floc settling velocity, as measured in the laboratory, is 0.0003 m/s. However, actual tank conditions are not as well controlled and

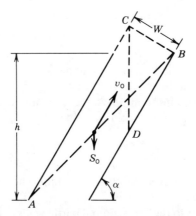

Figure 3.2.5-14   Element of tube settler.

therefore a value of $S_0 = 0.00025$ m/s is used, providing an approximate safety factor of 83%:

$$S_0 = 0.00025 = \frac{0.174}{A}$$

$$A = 696 \text{ m}^2 \quad \text{or} \quad 7490 \text{ ft}^2 \text{ for each tank}$$

Since the width of the tank is assumed to be 18 m, the length of the tank that is to be covered by the tube settler is 696 m² ÷ 18 m = 38.7 m. Make it 38 m.

The actual tank area for the tube settler is therefore 18 m × 38 m = 684 m² per tank.

Check the values: the surface loading for the tank area that is covered by the settler module is

$$\frac{Q}{A} = \frac{1 \times (60)^2}{684} = 5.3 \text{ m/h} \quad (2.1 \text{ gpm/ft}^2)$$

Flow velocity in the settler module is

$$v_0 = \frac{Q}{A \sin \alpha} = \frac{1 \times 60}{684 \times 0.866} = 0.1 \text{ m/min}$$

The hydraulic radius of the tube settler module is

$$R = \frac{A}{P} = \frac{(0.05)^2}{4 \times 0.05} = 0.0125 \text{ m}$$

$$\text{Re} = \frac{(1.69 \times 10^{-3}) \times (1.25 \times 10^{-2})}{1.519 \times 10^{-6}} = 14 < 50 \quad \text{Very good}$$

$$\text{Fr} = \frac{(1.69 \times 10^{-3})^2}{9.81 \times (1.25 \times 10^{-2})} = 2.3 \times 10^{-5} > 10^{-5} \quad \text{Good}$$

Since a standard sized module has tubes that are 0.61 m (2 ft) long, the detention time is 0.61 m ÷ 0.1 m/min or 6.1 min.

(iii) Two identical tanks that are 18 m wide will be used as discussed in Example 3. It is common practice to have 75% of the tank area covered by the settler modules and the remaining 25% left as open space, to settle heavy floc and to improve the flow inlet condition to the settler. The total tank length should therefore be

$$L = 38 \text{ m} \div 0.75 = 50.7 \text{ m}$$

Say 50 m. The water depth in the tank should be 4 m (13 ft) as illustrated in Figure 3.2.5-10.

With respect to the longitudinal baffle walls, the flow characteristics of the tank are no longer critical because the flow characteristics in the tube settler are significant. Thus, only one longitudinal baffle wall is required at the center line of the tank.

In Summary, the features of each tank are as follows:

Tank width   18 m
Tank length   50 m
Water depth   4 m
One longitudinal baffle wall $(L/W) = 5.5:1$
Tube settler modules are installed over 75% of the tank area (38 m from the end wall)

(iv) The tank inlet and diffuser wall are the same as in Example 3.

(v) Since an extensive skimming launder system is required with the tube settler modules, provide six launder troughs for each basin. Each launder must cover the entire length of the settler module; thus, the length of each launder should be 38 m.

The flow rate in each launder trough is $1 \div 6 = 0.167$ m³/s. For a trough with a rectangular section and a level invert,

$$Q = 1.38 \, bh_0^{1.5}$$

where b = width (m),
      h = water depth at the far end of the trough.

Select 0.5 m as the width for the launder trough:

$$h_0^{1.5} = \frac{0.167}{1.38 \times 0.5} = 0.242$$

$$h_0 = 0.39 \text{ m}$$

Make the interior height of the trough 0.5 m (0.11 m freeboard). Each trough should be supported by pedestal columns spaced 3.8 m (12.5 ft) on center.

The chain-and-flight sludge collection system must be employed in this design because of the use of the tube settler modules, the formation of thick ice during severe winter conditions, and for aesthetic reasons. Since one tank is partitioned into two channels (9 m wide), two sets of chain-and-flight units are installed in each channel.

The end wall of each tank should have two 0.25 m (diameter) holes with butterfly valves for backfilling of the basin. Ideally, the entire basin should be housed to protect the system from extremely cold temperatures. An illustration of the sedimentation tank is provided in Figure 3.2.5-15.

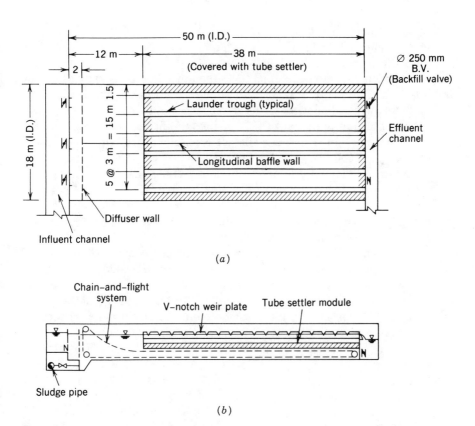

Figure 3.2.5-15 Sedimentation tank with tube settler modules. (a) Plan. (b) Elevation.

Operation and Maintenance   The operational staff of water treatment plants should be aware of the following items:

1. Floc settling condition in the tank.
2. Abnormal phenomena.
3. Equal hydraulic loading to each tank.
4. Schedule for sludge removal.
5. Special design features of the tank.

If the suspended solids are well flocculated, a majority of the floc should settle in the first half of the tank. Consequently, visual observation of the floc settling condition (from the walkway) is important and the coagulant dosage or flocculator mixing intensity should be adjusted accordingly; inadequate flocculation will allow small sized floc to leave the flocculation tanks. However, turbulence in the basin increases floc size in the area from the middle to the end of the tank. As a result, there is a heavy carryover of floc to the filters.

For many plants it is common for the water in the middle of the tank to be clear yet have a lot of floc rise at the end of the tank. This type of pattern indicates the presence of a distinct density flow. Plant operators have only three possible choices to counteract this situation. The first is to block off the last 10–20 ft (3–6 m) of V-notches or submerged orifices, on the launders. This will only work if launders are used to collect both settled and clear water from the midsection of the tank. This simple modification often significantly improves the situation. The second choice is to feed 0.2–0.5 mg/L of anionic polymer to the second stage of the flocculation process. The polymer acts as a flocculation aid that helps to produce readily settleable floc. The last choice pertains to plants that have more than one source of water: minimize the switching of water sources or minimize the changes in the blending ratio of the two sources. By doing so, the flow short-circuiting through the basin (the product of a density current) can be reduced.

It is very important that plant operators detect and evaluate any abnormal phenomena occurring within the tank. This includes the sudden appearance of floating sludge (sludge bulking), scum, midge fly larvae, and algae growth on the tank walls and baffles. Sludge bulking occurs when gas accumulates in the settled sludge as a result of anaerobic decomposition by bacteria or supersaturation of the raw water (by air) so that it floats rather than sinks. Sludge bulking is usually controlled by prechlorination: 2–3 mg/L of chlorine. Midge fly larvae bloom occurs during certain seasons if the sludge is not frequently removed. Algae will grow on the tank walls, the launders, and the baffles during warm months, especially in nutrient-rich water. Both the algae and larvae problems can be controlled effectively by prechlorination (1–2 mg/L). However, recent limitation on the use of THM restricts the use of prechlorination for many plants. A viable alternative is to provide occasional shock treatments with 2–3 mg/L of chlorine.

The accumulation of heavy scum can be attributed to many causes: changes in the raw water quality, changes in the coagulant (from alum to ferric chloride), and supersaturation of the raw water by air or gas. Supersaturation is the most difficult problem to solve. The only practical solution is to prevent the entrainment of air in the raw water.

Another phenomenon that should be noted on a regular basis is corrosion. Concrete surfaces and submerged metal parts will severely corrode with repeated exposure to a combination of aggressive water, chlorine, and alum. In this situation, a proper coating must be applied to these surfaces or, in the case of the metals, cathodic protection should be implemented. A severe pitting type of corrosion may occasionally be observed in submerged stainless steel parts (Type 316). This occurs only under anoxic conditions and can be minimized by specifying the use of Type 304 stainless steel or by making the conditions aerobic.

The design of the tank is based on equal hydraulic loading to all the tanks. Yet, this may not occur due to erroneous design or improper adjustment of the weir and valve openings. Plant operators should therefore check the flow rate through each tank by visual observation or by measuring the water height over the effluent weirs. An uneven flow rate to each tank can be roughly corrected by adjusting either the flow splitting weirs at the tank inlet or the inlet valve openings.

Certain factors can create undesirable flow patterns. For example, uneven leveling of each launder trough or effluent end weir wall can cause a skewing flow to be created. Uneven tank subsidence may tilt the tank and thereby create horizontal flow short-circuiting. Improper design of the inlet channel and the tank inlet will also cause a great imbalance in the flow rate among several tanks that are in parallel. In certain plants a difference of as much as 50% has been observed, even though the elevations of the launder troughs at each end of the tank were virtually the same.

Perhaps one of the most important operational issues, for the sedimentation tank, is the optimization of the sludge withdrawal process. In general, tanks that lack high-rate settler modules will have 60–80% of the sludge deposited in the front half of the tank, provided that the flocculation process is functioning well. Sludge sweeping should therefore be performed more frequently for the front half of the tank; a traveling bridge can be programmed to sweep the front half two or three times before running the entire length (in order to sweep the last half). In contrast, other types of sludge scraping mechanism, such as the chain-and-flight unit, cannot perform partial sweeps.

The proper sludge sweep cycle and the duration may be determined from several trial runs, by noting the concentration of the solids and the texture and smell of the sludge (for freshness). The most appropriate types of unit for this test are the sludge suction sweep units and the telescoping valves (for sludge discharge) with chain-and-flight collectors. These units allow the plant operator to determine the proper sludge sweeping cycle by allowing easy observation and testing of the sludge consistency.

TABLE 3.2.5-4   A Guide for Traveling Bridge Operation

(a)   A Half-length (front half) sweeping (47 metres)
(Table shows sweeping cycle and corresponding sweeping speed)

		Sweeping Speed (m/min)				
		0.75	1.0	1.5	2.25	3.0
Total sweeping Time (1) (minutes)		62	47	31.3	21	15.7
Volume of Sludge Removed (2) (1) x 4.8 m^3/min		297.6	225.6	150.2	101	75.4
Required cycle of 1/2 length sweeping per day.	1% solid sludge (1310 m^3/d)	4.4	5.8	8.7	-	-
	1.5% solid sludge (873 m^3/d)	2.9	3.8	5.8	8.6	-
	2% solid sludge (655 m^3/d)	2.2	2.9	4.3	6.5	8.7
	2.5% solid sludge (524 m^3/d)	1.8	2.3	3.5	5.2	7.0

▓▓▓   Recommended Range

(b)   A Full length sweeping followed by 1/2 length sweeps

Total pump capacity = 4.8 m^3/min (Total of two bridges)
(Table shows an appropriate sweeping speeds under certain conditions)

		Sludge Conentration (% solids)			
		1%	1.5%	2%	2.5%
Total volume of sludge to be removed (m^3)	Case 1*	218.3	145.5	109.2	87.3
	Case 2$^+$	327.5	218.3	163.8	131
Req'd pumping time to remove the sludge by this path (min.)	Case 1*	45	30.3	22.8	18.2
	Case 2$^+$	68.2	45.5	34.1	27.3
Req'd sweeping speed (m/min) for each case (approx)	Case 1*	2.1	3.1	4.1^x	4.8^x
	Case 2$^+$	1.4	2.1	2.8	3.4^x

Note: *Case 1: Five consecutive 1/2-length sweepings before the
                    full length sweeping.
        +Case 2: Three consecutive 1/2-length sweepings before the
                    full length sweeping
        xBeyond the capability of travelling bridge

Other methods of testing include the installation of magnetic or ultrasonic sludge density and flow meters in the sludge discharge line or the installation of a piece of transparent tubing in the sludge line to allow visual observation. However, the second method (transparent pipe) is simply not practical because the pipe becomes coated with sludge in a short period of time.

Table 3.2.5-4 is an example of a plant operator's operational guide for the traveling bridge fitted with sludge suction pumps. In areas where the water table is high, the tank cannot be drained at will, unless the tank is designed to resist uplift. For this reason, some plant designs include an underdrainage system for each tank. Each underdrain line has an observation point where the operator can visually inspect the flow of water, to see if the water is running or standing. This type of system allows the operators to check any tank for leaks and allows them to note unusually high underground water tables.

Lastly, the operator should be familiar with certain design features of the tank. For example, the diffuser wall should be designed to have a restricted water passage (facilitated by uniformly distributed ports) to produce uniform flow distribution. In a few cases, the plant operators had a passageway cut in the diffuser wall (3 by 6 ft) to allow easy access for the maintenance crews. In another case, the bottom 1–2 ft of the baffle wall were completely removed in an attempt to make the task of sludge removal and tank drainage more convenient. All these modifications caused the diffuser wall to become nonfunctional.

Since the majority of plant operators are not fully aware of the actual flow characteristics of conventional sedimentation basins, Appendix 9 (Tracer Test) has been added to this book. As shown in the figures in Appendix 9, the actual flow-through time (t) is only about 30% of the computed mean detention time (T); the performance of most sedimentation tanks is represented by these graphs. Although plant operators cannot improve this situation, they should be aware of this fact and should try to improve the settling efficiency of the tank by improving flocculation.

BIBLIOGRAPHY

ASCE, AWWA, CSSE, *Water Treatment Plant Design*, AWWA, New York, 1969 and 1990.

Bernardo, L. D., et al., "Use of Perforated Baffle at the Inlet of Rectangular Settling Basins," *J. AWWA*, 72(9):528(1980).

Camp, T. R., "Sedimentation and Design of Settling Tanks," *ASCE Trans.*, 111:895–958(1946).

Camp, T. R., "Studies of Sedimentation Basin Design," *Sewage & Industrial Wastes*, 25:1–12(1953).

Culp, K. Y., et al., "Tube Clarification Process, Operating Experiences," *ASCE Proc.*, 1179(January 1955).

Edzwald, J. K., et al., "Removal of Humic Substances and Algae by Dissolved Air Flotation," EPA/600/2-89-032, USEPA (1989).

Hazen, A. M., "On Sedimentation," *ASCE Trans.*, 63:45 (1904).

Hudson, H. E. Jr., *Water Clarification Processes—Practical Design and Evaluation*, Van Nostrand Reinhold, New York, 1981, pp. 123–138.

Huisman, L., *Sedimentation and Floatation*, Delft University of Technology, Delft, The Netherlands, 1973.

Ingersol, A. C., et al., "Fundamental Concepts of Rectangular Settling Tanks," *ASCE Proc.*, 1179(January 1955).

Kawamura, S., "Hydraulic Scale Model Simulation of the Sedimentation Process," *J. AWWA*, 73:372–379(July 1981).

Kawamura, S., et al., "Evaluation of Launders in Rectangular Sedimentation Tanks," *J. WPCF*, 58(12):1124(December 1986).

Monk, R. D. G., et al., "Designing Water Treatment Facilities," *J. AWWA*, 79(2):45(1987).

Montgomery, J. M., Consulting Engineers, *Water Treatment Principles and Design*, Wiley, New York, 1985, pp. 135–151, 523–532.

Okuno, N., et al., "Analysis of Existing Clarifiers and Design Considerations for Better Performance," Tokyo Metropolitan Government Report, 1980.

O'Melia, C. R., "Particles, Pretreatment, and Performance in Water Filtration," *J. Environ. Eng. ASCE*, 111, No. 6 (December 1985).

Walker, J. Donald, "Sedimentation," in *Water Treatment Plant Design for the Practicing Engineer*, R. L. Sanks, ed., Ann Arbor Science Publishers, Ann Arbor, MI, 1979, pp. 149–182.

Weber, W. J., *Physicochemical Process for Water Quality Control*, Wiley-Interscience, New York, 1972, pp. 111–138.

Yao, K. M., "Design of High Rate Settlers," *ASCE Sanit. Eng. Div.*, 99:621–637 (October 1973).

Yee, L. Y., et al., "Inlet Design for Rectangular Settling Tank by Physical Modeling," *J. WPCF*, 57(12):1168(1985).

Zabel, T., "The Advantages of Dissolved Air Floatation for Water Treatment," *J. AWWA*, 77(5):46(May 1985).

3.2.6 Granular Medium Filtration

Purpose In a water treatment process train, the fundamental system that removes particulate matter is filtration. The most common filtration process employs a granular medium of a certain size and depth. The pretreated water passes through the filter bed where a majority of the particulates are removed in the top portion, as well as throughout the entire depth of the bed.

The primary objective of the design engineer should be to deliver a filter design that provides a steady production of high-quality water, with minimal capital and minimal operation and maintenance costs. When designing filters, most engineers rely on either their own experience, the experience of others, or textbook design criteria. However, engineers should expand certain design variables, such as the size and thickness of the medium, filtration rate, filter underdrains, filter washing conditions, and available headloss for filtration. These variables must be developed based on the raw water quality, type of filter bed, and desired filtered water quality.

It should be noted that there is a basic common denominator among the coagulation, flocculation, and filtration processes. Under certain raw water conditions, adequate treatment of the raw water can be carried out in the filter alone, and the need for ordinary flocculation and sedimentation processes may possibly be eliminated.

This section focuses on the common granular media filters. Other types of filter are discussed in Sections 3.2.7 and 7.8.

Considerations   Several factors must be taken into consideration when designing a proper granular medium filtration process. They are local conditions, design guidelines set by regulatory agencies such as the State Department of Health, site topography, plant size, raw water quality, type of pretreatment process, new and proven types of filter, provisions for future modification or addition of filters, type of filter wash system, control of the filtration rate, type of filter bed, chemical application points, and other miscellaneous items. These subjects are discussed in detail.

Local Conditions   Perhaps the single most important issue in filter design is the local condition. Factors such as the availability of qualified technicians and raw materials, the local climate, and the soil and geotechnical characteristics of the plant site, all affect the design of the granular medium filtration process.

For example, a basic type of filter should be chosen for a treatment plant located on a small resort island or in a town of a developing country. Under these conditions, the best type of filter is one that has a simple control system. It is tragic to see that many design engineers fail to take this into consideration by designing treatment plants with expensive filters and sophisticated control systems for regions lacking both qualified plant operators and support personnel. Not only are these plants producing substandard quality water, they are rapidly deteriorating due to the lack of qualified maintenance personnel. The filter bed design should also be determined by the availability of filter materials at a reasonable cost.

Climate is the second issue that must be studied prior to filter design. In regions of cold weather, the filters must be either covered or housed in a building to prevent the water from freezing. Although the control cabinets are supposed to be hermetically sealed, plants located in regions experiencing severe dust storms should have a housing over the filter control system. The filter bed should also be covered to prevent debris and sand from entering the filter, thereby preventing both operational and maintenance problems.

Lastly, the soil and geotechnical characteristics of the plant site should be analyzed. Of primary interest are the level of the underground water table and the condition of the soil because the filters are relatively deep and heavy structures. There are a few documented cases where the filters became tilted or subsided as a direct result of a high groundwater table and a poor foundation.

Large cities that are surrounded by a wealth of technology and qualified technicians are prime candidates for treatment plants equipped with both a first-class filter system and a sophisticated backup system. Some of the filters used in the treatment plants of these large metropolitan areas operate at a rate that is 10–13 gpm/ft² (25–32 m/h) above the average filtration rate in the United States, which is 5–6 gpm/ft² (13–15 m/h).

Guidelines Set by the Regulatory Agencies   Guidelines for designing treatment plants can be found on both the state and county levels. They are especially stringent with respect to filter design. A few of the requirements are worth noting: maximum filtration rate, continuous monitoring of the filter effluent turbidity, rewash (filter to waste) requirements, and the types of filter that can be employed. For example, the State of California discourages the use of pressure filters in public water treatment plants because the condition of the filter bed cannot readily be observed. Therefore, the bed could potentially be in poor condition without the operator's knowledge. California also discourages the use of declining rate filtration control systems because of the likelihood for high turbidity breakthrough at the beginning of the filter cycle because of a filtration rate that is twice or more the nominal rate at the beginning of the process.

A few states also dictate the physical features of the filters such as media size, depth of the filter bed, and the spacing and height of the wash trough. The "Ten States Standards" and the rules set by the State of Utah are examples of this type of design control. On the national level, the EPA is primarily concerned with the removal of harmful microbes, such as *Giardia lamblia*, *Cryptosporidium*, and *Legionella*, and viruses from the water supply. This can be achieved through coagulation followed by filtration and disinfection.

Site Topography   The topography of the plant site must be carefully evaluated prior to design since it determines the location, type, and arrangement of each unit in the filtration process. For example, the headloss across a water treatment plant typically ranges from 15 to 25 ft. Therefore, if the plant site is flat, the filter structure and clearwell must be located deep within the ground, unless a pump is placed in the middle of the process train. If a gravity flow is to be maintained across the plant, the only other alternative would be to elevate the pretreatment unit processes above the ground.

Topography also dictates the type of filter backwash facility that may be used. If a suitable hill exists on the plant site, the wash tank can be built on it. However, a direct pump wash or self-backwash system should be seriously evaluated if the plant site is flat. This minimizes problems with freezing and avoids the unsightly appearance of the elevated wash tank. Additionally, the energy effectiveness of the self-backwash filters is gained. The ideal plant site has a constant slope of 2–3%, which allows the filter wash tank, filters, and wash-waste holding tank to be easily situated and economically built since there is no need for excessive excavation. The slope also helps to satisfy the required water level of each process. A steeply sloped topography not only severely restricts the arrangement of the unit processes but greatly increases the probability of land slide.

Plant Size   The size of the water treatment plant is instrumental in determining both the physical and performance requirements of the filters. Plant size usually dictates the total number of filters, the size of each filter, the degree of instrumentalization of the filter control system, and the method by which the filtration rate is controlled. Very large plants, those processing over 200 mgd (9 m³/s), can easily have their total number of filters exceed 30. However, the total number of filters may be limited if either the size of each filter is increased (up to a practical limit of 2000 ft², i.e., 100 m²) or if a high filtration rate such as 8 gpm/ft² (20 m/h) is adopted or if both are implemented.

The capacity of the treatment plant also aids the design engineer in selecting the most appropriate type of filter. For example, plants operating on a small scale (less than 3 mgd or 30 L/s) can generally use proprietary filters supplied by the equipment manufacturers, also known as "package filters." These types of filter are generally cost effective, easy to operate and maintain, and perform well for plants of this size as long as the package filters are properly selected, that is, based on a rigorous engineering evaluation.

Raw Water Quality and the Type of Pretreatment   The raw water characteristic should also be scrutinized when designing the filter bed since these characteristics determine the type of pretreatment, which in turn determines the type of filter bed. For instance, if the underground water contains high levels of iron or manganese, the best method for removing these compounds is by oxidation with chlorine or potassium permanganate, followed by pressure filters: the safe yield from a well is generally less than 3 mgd (0.13 m³/s). This method is the primary choice for well water application because it preserves the pressure of the pumped water and also allows for backwashing of the filters using the effluent from other pressure filters on the line; additional backwash pumps are therefore not required. Iron and manganese removal may also be achieved through direct filtration with preoxidation by chlorine, ozone, or permanganate. Green sand has been used as a filter medium in this type of situation, although sand may be used as an alternative.

To optimize the performance of the filtration process, the filter bed design should complement the pretreatment process. For instance, under conditions where direct filtration or in-line filtration is applicable, the filter bed should be designed with a larger storage capacity for suspended solids than conventional beds. Therefore, a reverse graded or a coarse and deep bed should be selected.

The presence or the lack of a high concentration of suspended solids in the raw water requires special consideration when selecting the type of filter bed. Filtration of algae laden water is very difficult even after flocculation and sedimentation due to the rapid clogging of the filters by algae. A coarse, deep bed or a reverse graded filter bed should be evaluated under these circumstances. Conversely, a common monosand bed (often referred to as a rapid sand bed) may be considered when the filter influent contains a minimal amount of suspended solids and if the quality of the filter influent is very good: the result of a conservatively designed pretreatment process. Since the bed depth of these types of filter is relatively shallow (less than 24 in.), surface filtration, not in-depth filtration, plays a major role in the monosand beds. A two-stage filtration should also be evaluated if the raw water quality is good (i.e., has low turbidity), thereby allowing both the flocculation and sedimentation processes to be eliminated.

Due to the limited amount of water for domestic use in large urban regions, tertiary treatment of sewage as a water re-use has become an important issue. The filter for biologically treated sewage behaves somewhat different from that used in the water supplied field, although the physical characteristics are the same. The graphs in Appendix 15 are provided to illustrate the typical performance of tertiary filters.

New and Proven Types of Filter Like any other type of equipment, new forms of filtration systems, filter media, and filter control systems become available on the market at any given time. Yet, the design engineer must be extremely careful with regard to using the new products. Although these new products are appealing, there is a likelihood that problems will emerge after a year or two of continuous operation. For example, one company marketed small sized aluminum chips as a filter medium. According to the manufacturer, chlorine (added to the filter influent) slowly dissolved the aluminum and allowed coagulation and flocculation to occur within the filter bed, eliminating the need for coagulant. The concept was very appealing and the filters performed well during the first 3–6 months of operation. However, after 9–12 months the filter bed became a large chunk of fused aluminum chips. Similar unforeseen problems have also occurred with new types of filter underdrain system.

Despite convincing laboratory test results, engineers should remember that real filters are subjected to many different conditions, which cannot be duplicated in either a laboratory or a piloting facility. Since most critical problems occur within a year or two of operation, it is prudent to wait until these problems are isolated and resolved. Design engineers should not hesitate to research new products, but they are strongly advised to select items that are proved to be reliable, with low maintenance and ease of repair.

Allowing for Future Filter Modifications or Additions The design engineer should anticipate an increasing demand on the plant output by including a provision for future modification of the filters or addition of more filters. The process should therefore be designed so that future construction will have minimal impact on plant operations.

In many cases, water treatment plants increase their operating capacity by adding basins and filters in stages. If the original treatment unit processes are not designed as a module, the new filters must be added onto the original filter bank. In situations such as this, the filter influent channel, effluent channel, and backwash system must be designed to accommodate the new flow rate, but based on the original plant specifications.

Construction during plant expansion should have minimal impact on the existing facilities. For this reason, the module expansion scheme must have a carefully conceived master plan for the yard pipings. The plan should maintain equal distribution of water flow to each module and should not have any physical interference between the major pipe lines.

The need to provide for future filter modifications has become more of a reality due to the EPA's proposal to make future water treatment regulations more stringent. It is also probable that the old monosand bed will be replaced with dual media reverse graded beds, such as anthracite and sand, to increase the output of the filters.

Presently, the most important consideration is the possibility that granular activated carbon (GAC) beds will be required to adsorb objectionable organic compounds—an EPA rule that is anticipated to be implemented in the near future. One method would be to replace the existing filter medium with GAC medium. However, the GAC bed requires an empty bed contact time (EBCT) of 15–30 min, thus necessitating a bed depth of 6–8 ft, as long as the filtration rate is at a level rate of 2–4 gpm/ft^2 (5–10 m/h). Under no circumstances can such a depth be provided by ordinary filters, unless the filter cells are initially designed with the option to convert to GAC adsorption filters.

The deep bed filters also demand special design features, such as air and water backwash (also referred to as air-scour wash); the air-scour backwash system requires a specific type of filter underdrain system. Moreover, recent prototype studies indicate that deep bed filters should be designed with both an air-scour wash and surface wash system whenever polymers are used as part of the coagulant to prevent mudball formation.

Filter Washing System   There are four basic schemes in filter washing: elevated wash tank, direct pump, self-backwash (Greenleaf type), and continuous backwash (Hardinge or Dyna Sand filters). The first two systems are the traditional and proven filter washing schemes. The self-backwash filter may be either a proprietary item or custom designed by the engineer. The continuous backwash filters are proprietary items. Filter selection is based on plant capacity, site topography, local conditions, and other factors such as energy efficiency, the wash-waste handling situation, and owner preference.

The elevated wash tank and direct pump wash systems have ample flexibility in adjusting the backwash rate. The elevated tank scheme is particularly suited for a plant site with high ground on which to locate the tank. The advantages of the elevated tank scheme are its effective use of energy, a small capacity pump that fills the tank within an hour or two, and emergency reserves of pressurized water in case of power outage. However, the elevated tank system requires a large tank to satisfy the two filter wash requirements: a large and long backwash pipe and a minimum elevation of 33 ft (10 m) above the filter wash troughs. Conversely, the direct pump wash system eliminates the need for a large elevated wash tank and control of the wash rate is relatively easy. However, both the size of the pump and the required horsepower become quite large for large sized filters. The direct pump system is therefore best suited for small filters or air-scouring wash systems since these must change the backwash rate two or three times during a washing cycle.

With respect to the washing system, the basic alternatives are backwash alone, a combination of surface wash and backwash, ordinary air-scour wash, and a simultaneous air and water wash. Backwash alone cannot adequately maintain clean filter bed conditions unless coagulation and flocculation are achieved through ozonation (alone) or with a minute amount of alum (without polymers). The filtration of recycled water at large aquariums is a prime example of this. Rapid sand filters, ordinary dual media beds, and tertiary filters are all effectively cleaned by backwash supplemented by surface wash. In contrast, coarse deep bed and tertiary filters are more suited to ordinary air-scour wash or simultaneous air and water wash. The latter requires special buffer plates around each wash trough, unlike the regular air-scour wash, which requires neither the wash troughs nor the baffles.

Some authors claim that the air-scour wash is a far superior system than surface wash. But this claim loses credibility after reviewing the performance of several operational plants: the data show that a surface wash system that is properly designed and operated can match or even exceed the performance of the air-scour system. For example, a review of four plants in California revealed that tertiary filters fitted with dual-arm surface agitators were devoid of "mud balls" and had clean bed conditions despite 15 years of operation. In contrast, filters in both the United States and Europe which employed air-scour wash systems occasionally had a significant amount of mud balls whenever polymers were used as coagulant or filtration aid.

Design engineers should not be misled by the impressive boiling action at the surface of the filter bed during air-scour wash because this violent boiling action actually occurs in only the top 6–8 in. of the filter bed. Therefore, mud balls that have descended below this boiling zone remain unbroken. It is worthwhile to mention that several water treatment plants in the United States, originally designed with ordinary air-scour filters, have added the surface wash system in response to the problem of mud ball formation.

Filtration Rate Control   There are basically two modes of filtration: constant-rate filtration and declining-rate filtration. In spite of claims to the contrary, both the constant-rate and declining-rate filters are capable of producing filtrates with less than 0.1 NTU turbidity. However, the colloidal particles must be properly conditioned in the pretreatment process and the filter bed must be both properly designed and maintained.

Only four basic types of rate control system have supplied consistent and reliable performance: (1) constant-rate filtration with a flow meter and a flow modulation valve; (2) constant level filtration with equal-flow splitting inlet weirs, a water level sensor, and a flow modulator valve; (3) declining-rate filtration with a submerged inlet that may or may not be fitted with a weir to control the effluent level; and (4) constant-rate filtration with equal-flow splitting inlet weirs and a weir to control the common effluent level. These four basic types of system are represented in Figure 3.2.6-1.

Each of these control systems has both good and bad features. The last two systems are simple and require very low maintenance. However, only a plant with more than six filters can fully maximize these positive features. The first two systems have been the standard flow control measures for the past 30 years and are still quite popular among design engineers. Therefore, a large percentage of operational filters are equipped with these types of system, despite their high capital cost of approximately $20,000 per set (average).

Of the four rate control systems, the first has the highest maintenance cost and consequently its popularity has been steadily diminishing. Another negative aspect is that the filtration mode changes from the designed constant-rate filtration to declining-rate filtration when the control system fails. Many old constant-rate filters have been operating in the declining-rate mode because their flow control systems are beyond repair. This is particularly true for treatment plants in developing countries.

The declining-rate filtration system is very simple to design and build and generally produces good quality water. However, in order to meet the present high standards for filtered water quality, this mode of filtration requires vigilant plant operators because the water level in the filter cells tends to fluctuate widely and creates the potential for partial drainage, overflow, and potential initial turbidity breakthrough at beginning the filtration cycles, unless the design has been made very carefully. It is precisely because of these problems that some regulatory agencies, such as in the State of California, discourage the use of declining-rate filtration in new filter designs.

One operational problem associated with the declining-rate filters is that there are no clear parameters for filter washing: the headloss for all the filters on-line is nearly equal; there is no flow metering device to indicate the filtration rate of each filter; and the turbidity of the filtered water does not increase with time. Thus, operators usually wash the filters based on the length of the filter run.

It is an undeniable fact that very few new filters in industrialized countries are designed as declining-rate filtration systems, because there are no positive means of controlling the filtering conditions. Yet, the declining rate system

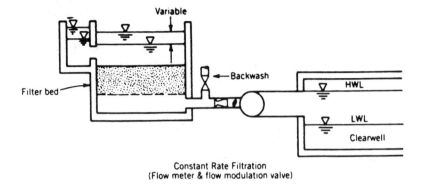

Constant Rate Filtration
(Flow meter & flow modulation valve)

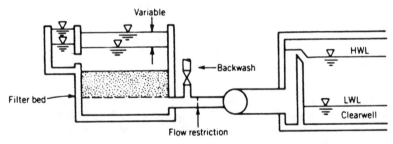

Constant Level Filtration
(Influent control, level sensor & modulating valve)

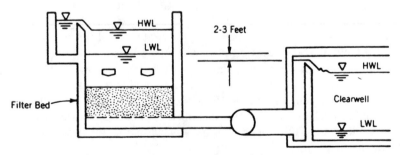

Declining Rate Filtration
(No influent control, no modulating valve, an orifice plate)

Rising Level, Self-Backwash Filters
(Influent control, no modulating valve, no backwash piping)

Figure 3.2.6-1 Basic filter control systems. (Adapted from *Water Treatment—Principle & Design* by J. M. Montgomery, Consulting Engineers, Wiley, New York, 1985.)

still satisfies the needs of many filtration plants especially in developing countries because of its simplicity and low maintenance requirements and the less stringent governmental requirements.

The fourth type of system, constant rate with equal loading to each filter with a rising water level, is capable of backwashing a filter by using the effluent from the rest of the filters on the line. These are commonly referred to as self-backwash filters. A few equipment manufacturers supply these types of filter with four filters per module. The major drawbacks of the self-backwash filters are the limited flexibility in adjusting the backwash rate and the deeper filter cells; they are at least 4 ft deeper than the normal filter design. However, there are positive features to this type of system: the lack of a flow modulation valve; the simple opening and closing of the inlet and waste-wash valves to initiate filter washing; the ability to function without the system of backwash pumps, tanks, and wash lines; and the easy determination of headloss through the filter (by measuring the water level in a filter cell). Since the water depth increases with the duration of the filter run, the problem of air-binding is minimized. The self-backwash filter is therefore particularly suited for situations where the raw water is saturated with air that is entrained in the water. Moreover, this type should always be evaluated because of its limited use of mechanical equipment, including backwash pumps and valves. Presently, there are three large plants, with capacities exceeding 400 mgd (19.5 m/s), using this type of filter: two in California and another in Manila, Philippines.

Type of Filter Bed

The selection of a filter bed is first contingent on the availability of bed material. Other factors such as the primary purpose of filtration, the type of filtration process, and the filtration rate of the system also influence the decision.

The first step in selecting a filter bed is to study the local conditions. In the United States, almost any type of common material is available at a reasonable cost: sand, crushed anthracite coal, granular activated carbon (GAC), and even garnet sand. This is not the case for most developing countries and a few industrialized nations. For instance, a large part of Africa and Southeast Asia use sand and gravel as the primary filter materials because other types are simply too expensive.

The second step in selecting an appropriate filter medium is to delineate the primary purpose of the filtration process. Sand and the combination of sand and anthracite coal are the two basic and proven alternatives with respect to ordinary water purification. Alternatively, the more expensive garnet grains may be used in conjunction with sand or anthracite coal. If the main purpose is the filtration of sewage, as in the tertiary filtration of sewage, monosand or mono-anthracite coal is the most appropriate filter bed medium. A dual media filter bed may also be considered because of its proven performance, but the disadvantage is that it tends to have problems with sludge accumulation in the sand–coal interface. However, this problem may be eliminated by employing air-scour backwash or dual-arm surface agitators to wash the sludge out of the interface. However, the air-scour wash system requires the filter bed to be restratified with a high backwash rate because the sand and the coal are intermixed during the wash, and its advantages of having smaller backwash pipe lines, valves, and waste-wash tank are consequently lost. Conversely, a monomedium bed does not require restratification (i.e., does not require a wash rate step-up control system), and the advantages of the air-scour wash can be fully utilized.

The filtration process also affects the selection of the filter bed because of the special requirements of each type of process. The direct and in-line filtration processes must have filter beds with a large floc holding capacity. A reverse graded filter bed, such as the dual media or coarse deep bed, satisfies this requirement. In two-stage filtration the filter bed of the first stage acts as a roughing filter and also carries out the flocculation process. Data obtained from pilot filter tests and actual installations using the two-stage filtration process indicate that the first-stage filter bed must be composed of pea sized gravel (3–5 mm), but the second-stage filter bed may be designed in the same fashion as an ordinary filter.

The fourth step in the selection of the filter bed is based on the filtration rate. For slow sand filtration the most appropriate choice is a fine grained monosand bed since the filtration rate for the slow sand filters is less than 0.2 gpm/ft^2 (0.5 m/h). Rapid sand filtration, with filtration rates ranging from 2 to 3 gpm/ft^2 (5 to 7.5 m/h), usually have medium sized sand (0.5 mm E.S.). Filters displaying filtration rates of 5–10 gpm/ft^2 (12.5–25 m/h), commonly known as high-rate filters, always consist of a reverse graded filter bed or a deep, large grain, monomedium bed. The details of these filters are discussed elsewhere in this chapter.

Chemical Application Points   Disinfectant, filter aids, adsorbant (for taste and odor control), and alkali chemicals (for pH control) are the most common chemicals added to the filter influent to aid the process of filtration. Three important design considerations must be addressed with regard to this subject: minimize the number of chemical application points, recognize that certain chemicals have the potential to "break through" into the filtered water, and ensure that the chemicals are adequately distributed in the water. Lastly, the design engineer should be cognizant that the locations of the application points determine the type of chemicals that may be applied.

The first item is to minimize the number of chemical application points. High-rate filtration often employs a polymer as a filtration aid. Polymer and chlorine are fed to the filter influent to promote filter efficiency and keep the filter bed clean. If the plant design provides only one water passage between the pretreatment process units and the filter bank, then these two chemicals may be fed to the water at this single location via a mixing device. In many treatment plants there are several passages between the settled water channel and the filters. This requires the chemicals to be distributed to these several application points and necessitates the use of additional chemical feeders and flow splitting equipment.

The second important issue is the potential breakthrough of the added chemicals into the filtered water; this phenomenon degrades the quality of the filtered water. Some treatment plants routinely add alkali chemicals such as lime or sodium hydroxide to the filter influent to raise the pH to 8 or above, following flocculation with alum and sedimentation. This practice always produces a high aluminum concentration in the finished water due to the pH and the basic characteristic of the aluminum hydroxide (see Figure 3.2.3-1). Another common practice is to add powdered activated carbon (PAC) to the filter influent as a means for taste and odor control. However, a significant amount of carbon breaks through the filter and appears in the finished water. Particle counters may be used to monitor the filter effluent, but turbidimeters are virtually incapable of detecting carbon leakage.

The third issue, adequate mixing of the chemicals at the application point, can be resolved by including either an appropriate type of chemical diffuser pipe in conjunction with a baffled channel or an in-line static mixer (for the pipe).

The last issue, with respect to the application of chemicals to the process water, is the selection of the feeding points. The filter influent, the filter effluent, and the backwash line are the three potential points. Chemicals that may be fed to the influent are chlorine, polymers, and to some extent PAC. Chlorine, alkali chemicals, corrosion inhibitors such as sodium hexametaphosphate, the fluoride compounds (for fluoridation) should ideally be applied after filtration because of their function and effectiveness. The two chemicals that may be fed to the backwash water to sterilize and condition the filter beds are chlorine and polymers. It should be noted that the recent EPA standards may force many treatment plants with filters lacking a filter-to-waste function to feed polymers to the backwash line to minimize the initial turbidity breakthrough of the filters.

Miscellaneous Items   Design engineers must also consider several additional issues when designing a proper granular medium filtration process: the use of wash troughs; the amount of allowable headloss for filtration; and the type of filter underdrain, type of filter, and waste wash water handling facility.

In the United States the wash troughs are an essential part of the filter design due to the predominant use of a high backwash rate. Filter backwashing, with or without surface jets as the auxiliary scouring system, necessitates the use of wash troughs because the prompt removal of the dirty wash waste is essential in minimizing both the washing time and the formation of mud balls in the filter bed. Conversely, an ordinary air-scour wash system uses a low backwash rate and the washing time is therefore twice as long as high-rate backwashing. The most air-scour system in Europe does not have wash troughs since the filters are effectively cleaned, as long as the width of the filter is less than 12 ft (4 m) and no polymer application to filter influent;

this type of system has been operating in Europe since the early 20th century. The choice between the two types of washing system is contingent on the type of filter, especially the size and depth of the filter bed, and the designer's preference.

When designing the filtration process, engineers should realize that the various filters provide different amounts of allowable headloss. Gravity filters generally have an available headloss of about 8–10 ft (2.5–3 m). Other types of filters, such as proprietary, valveless, or monovalve packaged filters, do not provide more than 6 ft (1.8 m) of available headloss. Automatic backwash filters (Hardinge type) typically provide a maximum of 1.0 ft (0.30 m) of available headloss.

Filter underdrain systems primarily differ in their filter washing design and the types of filter required by each system. Most municipalities in the United States use filters that are large in size and operate with a wide variety of underdrain systems, ranging from clay tile block, precast concrete laterals, false bottoms with strainer, to porous plates. In order to realize the full potential of the treatment plant processes, design engineers should select an underdrain system that is the most compatible with the size of the filter, the physical characteristics of the filter media, and the type of filter washing system that is implemented. If air-scour is chosen as the washing system, the most common and accepted type of underdrain is a false bottom fitted with long leg strainers. However, there are several disadvantages in using this type of system: the slits of the strainer heads may become clogged by filter media or debris left in the pipes and the false bottom during construction. There have also been a few instances where the false bottom floors became loose and lifted up after repeated filter washings. To avoid clogging the strainer heads, it is essential to thoroughly clean the underdrains and pipes after construction. Moreover, a slit size of a minimum of $\frac{5}{36}$ in. (3.5 mm) or an orifice size that is larger than $\frac{5}{18}$ in. (7 mm) in diameter should be selected whenever possible.

The four essential factors in selecting the filter underdrain system are uniform flow distribution of the backwash, durability, reliability, and cost effectiveness. The flow distribution of the backwash water in the filter cell is kept uniform by (1) making the orifice or slits of the filter underdrain system small enough to introduce a "controlled" headloss and (2) decreasing the flow velocity in the pressurized conduit upstream of the underdrain system so that the hydraulic grade and the energy lines of flow entering the underdrain system are fairly uniform. The false bottom underdrain system usually provides the best performance and has minimal maintenance costs.

The overall capacity of a water treatment plant determines the type of filter that may be included in the filtration process. For instance, automatic backwash filters, self-backwash filters, and pressure filters are frequently used for small plants—those processing less than 15 mgd (0.65 m^3/s). These types of filter are proprietary items marketed by a number of manufacturers. They offer advantages such as low capital cost, minimal headloss across the filters (except in pressure filters), good filtered water quality, and simple operation and maintenance. Medium to large plants—those processing over 10 mgd—generally use gravity downflow filters with dual media filters constructed of reinforced concrete structures.

Upflow filters, where the water passes upward through the filter bed, have also been used on a limited basis. One obvious advantage of the upflow filters is that they can achieve coarse to fine filtration with a single medium such as sand, and thereby eliminate the need for more costly multimedia beds. However, one serious disadvantage of the upflow filter is that the filter bed expands as it becomes clogged and allows previously removed solids to escape into the effluent. An extra deep bed (minimum of 6 ft) containing coarse sand (1.5–2 mm) or a system of restraining grids, placed at the surface of the filter bed, may be used to overcome this problem if normal sized medium is used. This particular type of design is seldom considered for municipal treatment plant applications because of the extreme care that must be exercised during operation. Therefore, its use is restricted by the regulatory agencies.

If a treatment plant undergoes expansion, both the size and the elevation of each new filter must match the existing filters so that the old backwash facility may still be used. This not only reduces the cost of expansion but also maintains the continuity in operating procedures. For ordinary filters, the size of the waste washwater handling facility is directly proportional to the plant capacity and the size of each filter cell. The automatic backwash filters sequentially clean a small portion of the filter bed and thereby constitute a continuous filtration process; this type of filtration process produces a small amount of wash-waste on a continuous basis. For this reason, the waste can easily be handled by a very small facility. In contrast, the large filters used at large plants require a sizable handling facility because these filters produce a significant amount of instantaneous waste flow during a washing period. Regardless of the type of filter, 2–3% of the plant flow rate is waste flow and the designer must therefore evaluate alternative types of filters and the size of each filter, whenever the plant site does not have enough area to accommodate a large waste handling facility.

Type and Selection Guide   This section briefly summarizes the common filter units and provides guidelines for making the proper selection.

Available Alternatives   There are three basic types of granular medium filters: (1) slow sand filters, (2) rapid sand filters, and (3) high-rate filters with either a coarse deep monomedium or reverse graded multimedia bed. Both the rapid sand and high-rate filters require chemical coagulation and, in most cases, clarification prior to filtration so as to optimize filter performance. Coagulation and clarification pretreatment remove excessive amounts of suspended solids, color, and some mineral components. The pretreatment also removes a large portion of microorganisms such as algae and *Giardia*.

A commonly encountered design flaw is that the type and degree of pretreatment does not match the filter requirements. For instance, if the source is a large lake with consistently good water quality, a slow sand filtration or a direct in-line filtration system, as well as a two-stage filtration system with high-rate filters may be adopted. However, if the quality of the raw water is generally good with occasional high turbidity spikes, algal bloom, or taste and odor problems, a complete treatment process must be considered; the treatment process train should include a clarification process and a short (60–120 min) settling time. In this type of situation the sedimentation tanks may be either bypassed or used as a large channel between the flocculation and filtration process, but only during periods of good raw water quality, by using the direct filtration mode and by employing small amounts of coagulant. From these examples, it is evident that design engineers must review both the raw water characteristics and the seasonal variations in the water quality and should seriously consider running a bench scale or pilot study to evaluate the treatability of the raw water. Only then can a proper pretreatment and filter scheme be established.

The filters can be classified into four descriptive categories based on the filtration rate, the type of filter washing that is required, the type of filtration rate control, and the type of filter bed that is selected. As previously discussed, the slow sand, rapid sand, and high-rate filters are the three commonly recognized classifications. Yet, from the viewpoint of the design engineer, a more detailed division is required.

The first group is the gravity filters: slow sand, rapid sand, and high-rate filters containing a multimedia or a coarse, deep medium bed. The second division is the pressure filters, which encompass proprietary items such as the horizontal single cell, horizontal multicell, vertical single cell, and vertical biflow filters. The third group is the automatic backwash filters: these are exclusively composed of proprietary items. Lastly, the custom designed filters include standard rapid sand filters, multimedia high-rate filters, coarse deep bed high-rate filters, self-backwash gravity filters, constant-rate gravity filters, declining-rate gravity filters, GAC adsorption filters, and ion exchange resin filters.

Selection Guide   Before designing a filtration system, engineers must establish the following items: type of filter, size and number of filters, filtration rate and terminal headloss, control of the filter flow rate, characteristics of the filter bed, type of filter wash system and wash trough, filter underdrains, auxiliary scouring, and filter appurtenance systems. The specification of these design elements should be based on the following conditions:

1. Local conditions, such as weather, the level of available technical service, and capability of operators, and the topography of the plant site.
2. Regulatory constraints.
3. Plant capacity: does the plant process more than or less than 10 mgd ($0.5 \, m^3/s$)?

4. The quality of the water source: turbidity, color, temperature, amount of dissolved air, algae count, taste and odor, and fluctuations in the water quality.
5. The final quality of the process water, that is, the drinking water quality standards.
6. The type of overall treatment process train: conventional complete treatment, direct filtration, iron and manganese removal, organic compound removal, or demineralization.
7. The condition of the plant site and the effective use of space.
8. The amount of available hydraulic headloss across the filters.
9. The initial capital cost and the cost of operation and maintenance.
10. Owner preference.
11. The potential for future plant expansion and modification.

TYPE OF FILTER   Small to medium sized plants processing less than 15 mgd (0.65 m^3/s) may use proprietary type filters; these should be evaluated along with the custom designed filters. The proprietary units offer several advantages: lower capital cost, easy operation and maintenance, a filtered water quality ranging from acceptable to good, and shorter plant design and construction periods due to the predesigned feature of these units. The main disadvantage is that they generally have a limited life span of 20–30 years. There is also a possibility that future use of certain proprietary units may be restricted because they may not be able to meet the more stringent guidelines of the regulatory agencies. Other disadvantages include their questionable aesthetics and limited flexibility with respect to filter design and operation. Automatic backwash filters, self-backwash filters with a module design, and steel shelled gravity or pressure filters are all proprietary types of filter.

The most common types of filter that are included in the design of large treatment plants, those processing a minimum of 15 mgd, are the gravity-downflow and the dual media (sand and coal) filters. The dual media filter with a reinforced concrete cell structure is the standard design that is used by engineers because it has a life cycle that spans 50 years or more.

SIZE AND NUMBER OF EACH FILTER   The size of each filter and the total number of filters are interrelated subjects. The maximum filter size is dictated by the difficulties encountered in providing uniform flow distribution of the backwash water over the entire filter bed, by the economically feasible size of the filter backwash tank and pumps, and the cost of the waste-wash handling facilities. In general, the practical maximum size of an individual filter bed is approximately 1500 ft^2 (150 m^2). A minimum of four filters should be provided for medium to large sized plants, those processing over 15 mgd. In contrast, very small plants processing a maximum of 2–3 mgd (0.09–0.13 m^3/s) may use as few as two filters if there are financial constraints. The fewer the total number of filter, the more cost effective it is to construct. However, this means that the larger plants require very large sized individual filters. Refer to the design criteria section of this chapter for the equation used to determine the required number of filters in relation to the plant capacity. If a self-backwash type of filter is selected, the minimum number of filters per module must be four, regardless of the plant size, and preferably six. Otherwise, the system may not be capable of producing the required volume of backwash water (to wash one filter) during periods of low plant flow. However, these numbers may be disregarded if either an auxiliary backwash pump or tank is provided to supplement the inadequate production of wash water.

It is important to remember that if a plant is designed with a small number of filters the filtration rate in the remaining filters substantially increases whenever one or two of the filters are placed off-line due to washing or repair. The design engineer must therefore select an adequate number of filters and the proper size for each filter, so that the filtration rate of the remaining units will not be excessive when at least one is out of service. It is recommended that no more than 33%, and preferably a maximum of 15–20%, of hydraulic surcharge be allowed to flow to the remaining filters whenever one unit is off-line. This practice minimizes any turbidity breakthrough that may occur during shock loading.

Conditions: Raw water: Turbidity = 1.9–2.1 TU, Alkalinity = 120 mg/L, pH = 8.1, W.T. = 13°C
Treatment process: Direct filtration (pilot filter with/dual media bed)
Coagulants: Alum = 2.5 mg/L, CatFloc–T = 0.35 mg/L

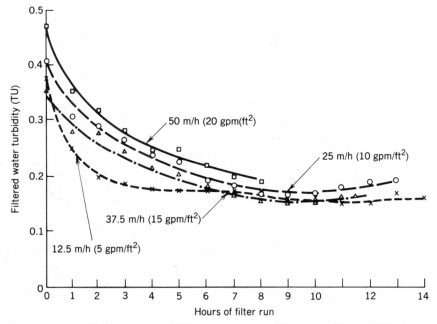

Figure 3.2.6-2 Filtered water turbidity versus hours of filter run. (From MWD Southern California—Internal Report, 1978.)

Conditions: Raw water: Turbidity = 1.9–2.1 TU, Alkalinity = 120 mg/L, pH = 8.1, W.T. = 13°C
Treatment process: Direct filtration (pilot filter with dual media bed)
Coagulants: Alum = 2.5 mg/L, CatFloc–T = 0.35 mg/L

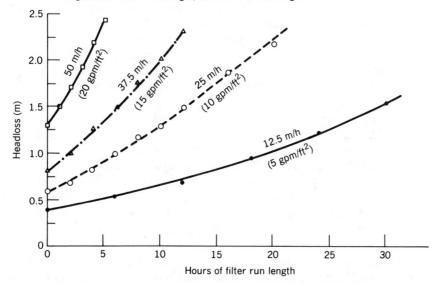

Figure 3.2.6-3 Headloss versus filter run at four different filtration rates. (From MWD Southern California—Internal Report, 1978.)

FILTRATION RATE AND TERMINAL HEADLOSS   Contrary to popular belief, the filtration rate of the high-rate filters does not significantly affect the quality of the filtered water for rates up to 10 gpm/ft² (25 m/h). This is true only if the suspended colloids (in the raw water) are properly coagulated by a combination of coagulant and a small amount of polymer and if the excess suspended solids are removed during pretreatment. However, the rate at which the headloss is developed during higher filtration rates is quite dramatic; so much that filtration rates over 10 gpm/ft² are considered to be impractical for normal dual media filters. Figures 3.2.6-2 and 3.2.6-3 illustrate the performance characteristics of the high-rate filters. When processing physically weak floc such as alum floc, the quality of the filter effluent tends to degrade as the filtration rate rises above 4 gpm/ft² (10 m/h). This phenomenon is more frequent during periods of cold weather and is more apparent for the rapid sand filters. Consequently, a designed filtration rate of 6 gpm/ft² (15 m/h) is generally recommended for ordinary high-rate filters and 3 gpm/ft² (7.5 m/h) for rapid sand filters. A filtration rate of 10 gpm/ft² (25 m/h) is reasonable for coarse, deep bed, high-rate filters if a small amount of polymer is applied as a filter aid. However, the design engineer should discuss the chosen filtration rate with the local regulatory agency prior to design.

Ordinary rapid sand filter beds generally have a filtration rate of 2 gpm/ft² (5 m/h), which translates to a Reynolds number of approximately 1.0; this number is well within the limits of a laminar type of flow through a granular bed. The typical high-rate filter with a standard dual media filter bed has a filtration rate of 6 gpm/ft^2 (15 m/h) and a Reynolds number of approximately 10, which is still within the upper limits of a laminar flow characteristic. Both the coarse, deep bed filter and the first-stage roughing filters of a two-stage filtration process have filtration rates of 15 gpm/ft^2 (37.5 m/h). The Reynolds numbers of these two types of unit are 30–40; these figures border on the laminar flow regime (far end) and the turbulent type of flow (beginning). This transitional type of flow may explain why high-rate filter beds have remarkable flocculation characteristics, in addition to the ability to retain a majority of the flocculated suspended matter when a small amount of coagulant or flocculant (polymer) is added to the filter influent. These characteristics are generally attributed to semilaminar types of flow.

The available headloss for filtration is generally designed to be 8–10 ft (2.5–3 m) for gravity filters, but only 1 ft (0.3 m) for automatic backwash filters; the latter are predesigned by the equipment manufacturers. Pressure filters usually provide a terminal headloss of over 30 ft (9.3 m), but their application is not practical for large treatment plants because of the large number of required units.

The filtration rate and terminal headloss for a particular type of filter and filter medium design should be selected after thoroughly analyzing the economic trade-offs: the total required area of the filter bed, the available hydraulic loss across the filter process, the anticipated terminal headloss prior to turbidity breakthrough in the filter bed, and the length of the filter run. All these factors must fall within the limits dictated by the quality of the effluent water. It is important to mention that the effluent turbidity of gravity filters often begins to increase when the net headloss is over 6 ft. Consequently, the filters must be backwashed, despite an available headloss of 8 ft or more, if the turbidity of the filtrate is to be maintained at less than 0.25 NTU. Regardless of the conditions, filter washing should be initiated automatically or manually by any of the following: turbidity breakthrough, the run time of the filter, or terminal headloss.

CONTROL OF THE FILTER FLOW RATE   The two basic schemes for controlling the filter flow rate are constant-rate and declining-rate filtration. In recent years, the most popular scheme is the constant-rate filtration system since it provides better operational control over the filters, has proven performance, and is overwhelmingly preferred by plant operators.

Declining-rate filtration has some merits but lacks positive filtration control measures. Thus, the owners of modern filtration plants seldom choose this type of system.

A typical scheme for the constant-flow type of control, developed in recent years, is equal loading self-backwash filtration (see Figure 3.2.6-1). This design scheme should be evaluated for any situation because of its simplicity, efficient use of energy, uncomplicated instrumentation, simple filter control system, and simple operation and maintenance. This system also eliminates the need for backwash headers, valves, pumps, and an elevated washwater storage tank.

DEPTH, SIZE, AND COMPOSITION OF THE FILTER BED   Filter efficiency is determined by factors such as certain physical characteristics of the filter bed, the conditions of the bed, and the effectiveness of coagulation pretreatment. The design engineer only has control over the physical parameters of the filter bed: media size and shape, bed porosity, total surface area of the filter media grains, and bed depth to media grain size ratio. Once the size of the media and the depth of the bed are selected, both the bed porosity and the total surface area of the media grains becomes fixed. Thus, selection of the appropriate size and depth of media are of crucial importance to the design engineer.

There are three basic methods in selecting the proper depth and size of the media: on the basis of existing data obtained from filters treating the same or similar type of raw water, on the basis of data obtained from pilot plant studies, and on the accepted ratio between the bed depth and media size. The first option is often used because it is a foolproof method. However, it is both crude and unscientific. The second approach is tempting but requires a minimum of 6 months to obtain reliable data based on the different raw water qualities exhibited during both warm and cold weather. Consequently, a significant sum of money (at least $100,000) must be budgeted to complete the pilot study. Moreover, a pilot study may not always be feasible for various reasons: a limited schedule for completing the project, unavailability of fresh raw water because of the remoteness of the project site, and incomplete water sources due to factors such as the drilling of wells or the construction of a new reservoir (dam).

The third method is based on the relationship between the established size of the media grain and the filter bed. This relationship has been studied since the mid-20th century by two or three researchers using pilot filter columns in conjunction with actual filter performances. This method is considered to be the proven technique in selecting the appropriate depth and size of the media and a detailed discussion is presented in the section covering design criteria.

The relationship between the size and the depth of the filter media implies that the available surface area of the filter bed media grains is a significant factor in selecting the type of filter bed. Ordinary high-rate filters have a porosity ratio of 0.45, a sphericity of 0.8, and a bed depth to effective media size ratio of 1000. Thus, the total surface area of the filter media grains per unit area is approximately 3000 ft^2 (3000 m^2) in area per square foot (square meter of bed area). When a dual or multimedia filter bed is specified, the size of the medium grains and the specific gravity of each layer become extremely important with respect filter backwashing and the potential for media loss. If an inappropriate combination of media is specified, only a fraction of the filter bed will be adequately cleaned. The remaining part of the bed will be either compacted and dirty or washed away at a particular backwash rate. In order to ensure that the two filter media particles with size d_1 and d_2 and densities ρ_1 and ρ_2 have the same terminal velocity, the following relationship should exist:

$$\frac{d_1}{d_2} = \left(\frac{\rho_2 - \rho}{\rho_1 - \rho}\right)^{0.667}$$

where ρ is the density of water.

Figure 3.2.6-4 will aid engineers in selecting the proper combination of common filter materials at a particular backwash rate. Figure 3.2.6-5 compares the performances of five different types of filter bed.

FILTER WASH SYSTEMS AND WASH TROUGHS The two basic types of filter backwash system are (1) fluidized bed backwash with or without surface wash and (2) air-scour with partial fluidized backwash. The first type has been used

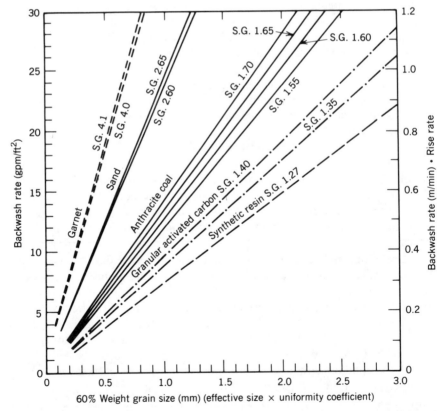

Figure 3.2.6-4 Appropriate filter backwash rate at water temperature of 20°C (68°F).

extensively in the United States, South America, and Japan. With this system the backwash troughs are arranged 6–8 ft apart (1.8–2.4 m) and 2.5–4 ft (0.8–1.2 m) above the filter bed. The second type is primarily practiced in Europe and typically does not provide wash troughs above the filter bed. Rather, a single overflow side wall is provided for the discharge of the back-wash waste into a channel and the top of the side wall is usually about 20 in. (0.5 m) above the filter bed.

The system of concurrent water and air wash (air-scour) has very limited application. This method applies air and water during the entire wash period; the backwash rate is high enough to fluidize the bed; and special baffled plates are provided on either side of each trough to minimize the loss of media during air-scour (Figures 3.2.6-6). The air-scour wash system is particularly suited for washing tertiary filters of sewage treatment plants, which filter strong biological floc.

If a fluidized bed backwash system is selected, the spacing and height of the troughs must be determined carefully to ensure that the dislodged floc is efficiently washed away and so that the process of backwash and surface wash does not flush filter media over the sides of the troughs.

According to the "Ten States Standards," (1968) the distance from the top of the trough to the surface of a rapid sand filter bed should be equivalent to the rate of backwash expressed in inches per minute. The rules also specify that the troughs should be spaced in such a manner that the floc does not travel more than 3 ft (horizontally) before reaching a trough. These criteria are sufficient when designing rapid sand filter beds but are inadequate when designing high-rate filters with multimedia filter beds. If the troughs are designed according to these standards, there will be an excessive loss of the anthracite coal layers. Therefore, for dual media filters, the top of the troughs should generally be located 3.5–4 ft (1.1–1.2 m) above the surface of the filter bed and should be spaced 6–10 ft (1.8–3 m) apart.

There are basically two types of wash trough: those with a shallow and wide cross section and those whose cross section is both narrow and deep with a U-shaped or slightly V-shaped bottom. Wash troughs with a wide cross section tend to have higher upflow velocities whenever the backwash flow exceeds the elevation of the trough bottom; thus, the suspended solids are quickly removed. Troughs that are both narrow and deep may be constructed with thinner walls due to a higher moment of inertia and greater structural integrity. However, the bottom of the wash troughs should not be flat (for both types) because froth and sludge tend to accumulate beneath the trough. This accumulated mud then falls back onto the filter bed and forms mud balls.

Engineers must design the wash troughs so that they will accommodate the maximum expected wash rate for both backwash and surface wash, in addition to a 4–6 in. (10–25 cm) free-fall into the main collection gullet at the lower end. The bottom of the trough (interior) may be either horizontal or sloping. A horizontal bottom is much simpler to construct and install. The

Filter Media Specifications

Name	Material	Effective Size	Depth (in.)	Uniformity Coefficient	L/D
A—Small dual media	Anthracite	1.0	20	1.45	1016
	Sand	0.5	10	1.3	
B—Intermediate dual media	Anthracite	1.48	30	1.5	1023
	Sand	0.75	15	1.2	
C—Large dual media	Anthracite	2.0	40	1.5	1016
	Sand	1.0	20	1.25	
D—Mixed media	Anthracite	1.0	18	1.45	1306
	Sand	0.42	9	1.5	
	Garnet	0.25	3	1.25	
E—Monomedium	Anthracite	1.0	40	1.4	1016

Note: L/D corresponds to the ratio of the depth of the medium (*L*) to the effective size of the medium (*D*).

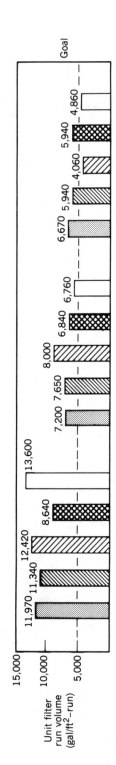

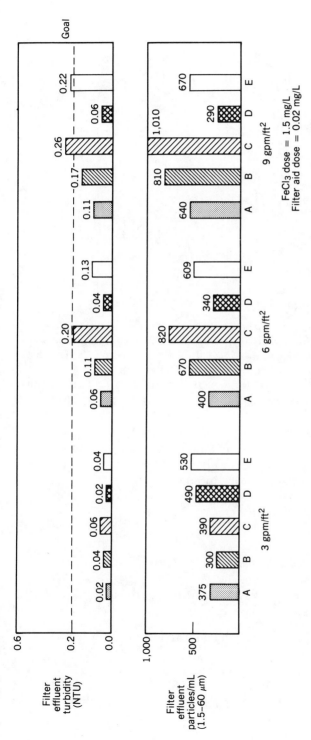

Figure 3.2.6-5 Results of filtration rate and medium type comparison tests. (Adapted from Elekutria WTP pilot study report by J. M. Montgomery, Consulting Engineers.)

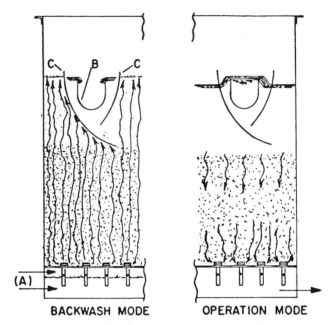

BACKWASH MODE   OPERATION MODE

Figure 3.2.6-6 Special baffles for concurrent air and water wash troughs. Schematic diagram shows the operation of the MULTIWASH Filter Process. During backwash (left) water and air are simultaneously passed through the media (A). The dirty backwash water is collected by the backwash trough (B) after air is separated from the water stream by the exclusive MULTIWASH baffles (C). After backwash the filter is placed back into normal service (right). (Courtesy of General Filter Company.)

width of the filter cell should be less than 20 ft (6 m) so that "off-the-shelf" types, such as plastic troughs reinforced with fiberglass, may be used. The required height for a trough with level inverts and a rectangular cross section may be computed by using the following formula:

$$h_0 = \left(\frac{Q}{2.49\ B}\right)^{0.667}$$

where Q = total rate of discharge per trough (cfs),
$\ \ \ \ \ \ \ B$ = interior width of the trough (ft),
$\ \ \ \ \ \ \ h_0$ = depth (ft) of the process water at the upstream portion of the trough.

A freeboard of at least 3 in. (7.5 cm) should be added to h_0. If metric units are used, the above formula becomes

$$h_0 = \left(\frac{Q}{1.38\ B}\right)^{0.667}$$

where Q is in m³/s and h_0 and B are in meters.

Unlike the fluidized bed backwash system, the European standard air-scour wash does not use a series of wash troughs but has a single V-shaped trough along the side wall opposite the wash-waste overflow wall, which provides the surface sweeping function (Figure 3.2.6-7). If the wash waste is to be effectively removed without the use of wash troughs, there is a limitation to the width of each filter cell: 16–20 ft (5–6 m) is considered to be the maximum width. The design engineer should also be aware that filter media migration has frequently been reported in cases where the filters lack the regular wash troughs; in this case, media migration is caused by the momentum of the filter influent.

FILTER UNDERDRAINS The selection of the appropriate type of filter underdrain must be based on whether or not air-scour wash is chosen as the filter washing system. The underdrain system used in air-scour wash is specifically designed for this purpose. It is therefore prudent to use one of the proprietary types, one that has a long history of success, instead of a custom designed underdrain; this will ensure that the requirements of the system are met without failure. Figures 3.2.6-8 and 3.2.6-9 depict a few commercially available types of underdrain: false bottoms with long leg strainers and fiberglass reinforced plastic blocks with channels (interior).

The strainer system is the predominant type of air-scour wash filter underdrain because of its long history of effectiveness and dependability. It is extremely important to select strainers that are physically strong. Furthermore, they preferably have a mechanism for adjusting the stem height, even if this feature requires the expenditure of more money. Strainers that are physically weak or brittle tend to break during construction or filter backwashing, resulting in costly and lengthy repairs. Even when the bottom slab of the underdrain is not level, due to imprecise construction or uneven settlement of the filter structure, certain types of strainer allow the height of the strainer legs to be adjusted by approximately $\frac{3}{4}$ in. (20 mm). Another important feature of the strainer type of underdrain is that it has a false floor on which the strainer nozzles are mounted.

Engineers are recommended to design the air-scour underdrain as a monolithic reinforced concrete floor since this will reduce the potential for un-

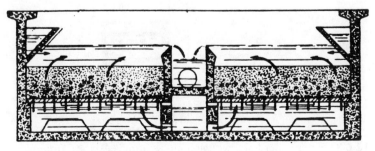

Figure 3.2.6-7 Air-scour wash filter. (From Degremont, *Water Treatment Handbook*, Halsted Press, New York, 1979.)

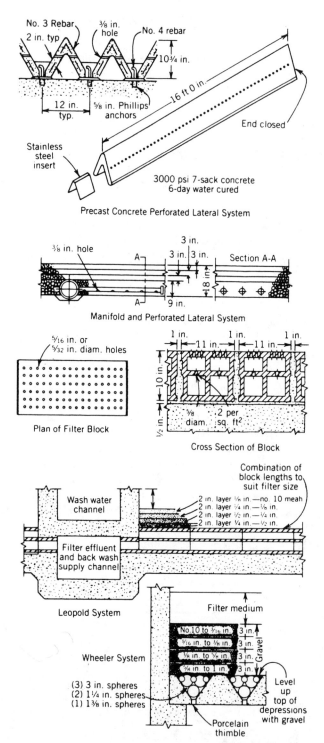

Figure 3.2.6-8 Filter underdrains with non-air-scouring. (Adapted from *Water Treatment—Principle & Design* by J. M. Montgomery Consulting Engineers, Wiley, New York, 1985.)

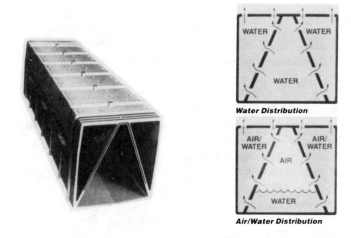

Leopold's Universal Underdrain Block
(courtesy of F.B.Leopold Co. - Bulletin ASU-100)

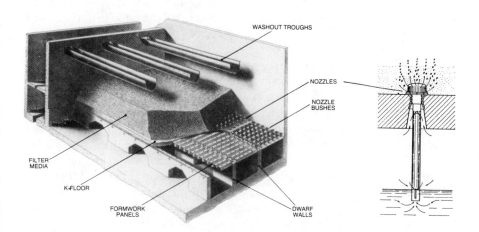

PCI's K-Floor with Nozzles
(coutesy of Paterson Candy International)

Figure 3.2.6-9 Filter underdrains for air-scouring. (a) Leopold's underdrain block (b) PCI's K-Floor with nozzles

derdrain upset. As previously described, a significant number of filters have had problems with large amounts of air emanating from the structural joints during air-scouring and their floors have consequently become damaged. This was especially true in cases where precast concrete blocks were secured with anchor bolts and between the blocks was cement grouted.

The plenum should have a minimum height of 2 ft (0.6 m) to facilitate inspection of the nozzle legs and the condition of the false bottom. This height also allows the plenum to be constructed easily.

Lastly, the design engineer must give special consideration to the size of the strainer head slits. If these slits are less than 0.1 in. (0.25 mm) wide, they will easily become clogged by fine media grains and other debris that may be left in the backwash line or plenum during construction. The engineer is therefore advised to select the largest possible width. The appropriate slit size is half the effective size of the filter medium covering the strainer heads. Thus, if the effective size of the sand is 1 mm, the size of the slits should be 0.5 mm. Manufacturers often recommend covering the strainers with 6 in. (150 mm) of pea sized gravel because this practice minimizes clogging of the strainers and distributes the backwash flow more efficiently, but only for certain types of strainer nozzle.

The plastic block type of underdrain of the air-scour wash system has been used successfully since 1980. This system requires a specifically graded gravel bed, with an approximate depth of 12–15 in. (30–38 cm), to be located directly above the underdrain because the size of the orifice is too large to retain the filter media. However, plant operators must strictly follow the manufacturer's instructions for both the wash rates and the sequence of the air and water. Otherwise, the gravel bed will be upset and the expensive process of reconstructing the filter gravel bed will be required. Recently, the manufacturer recommends a porous plate in lieu of the specifically graded gravel bed directly above the plastic underdrain blocks.

There are a number of alternative types of underdrain system available for the regular backwash system: precast concrete laterals, pipe laterals, and Leopold's block system (Figure 3.2.6-8) are proven types that are used extensively. Unlike the air-scour wash, these types of underdrain are not limited to the proprietary types, provided that they are properly designed. Moreover, any reputable contractor can fabricate most of these underdrains on the job site. The design of the filter underdrain system is based exclusively on the hydraulics of the filter backwash. The details are discussed elsewhere.

AUXILIARY SCOURING OF THE FILTER WASH   Auxiliary scouring is absolutely essential during filter backwash whenever coagulants are used in the pretreatment process, especially if polymer is used as a filter aid. Backwash alone cannot maintain the filter beds in a reasonably clean condition for longer than several months, regardless of adjustments in the backwash rate and control of the bed expansion rate.

Regular surface wash systems are capable of adequately cleaning ordinary rapid sand and dual media filters. The advantages of this type of system are its simplicity, effectiveness, and easy operation and maintenance. Unlike air-scour wash, it does not require a delicate auxiliary scour and backwash sequence, nor does it require the backwash rates to be adjusted during a wash cycle. The air-scour wash system should be employed for any deep bed filters with coarse media and may be the preferred system for ordinary filter beds processing biologically pretreated sewage.

The two basic types of surface wash system are the fixed grid and rotating arms. Both types have been used since the mid-20th century and have been proved effective when properly designed, fabricated, installed, and operated. The fixed grid type of surface wash may be designed by the engineer and requires much less maintenance because of the lack of moving parts. The rotating arm types of system should be neither designed by engineers nor fabricated in the machine shop because of their sensitive design and specifications.

The three types of rotating agitator are surface agitators, subsurface agitators, and dual-arm agitators. The first is primarily used in conjunction with the rapid sand filters. The second and third types are specifically designed for dual or multimedia filter beds. In the United States, only one manufacturer has had over 20 years of successful operation for the three rotating arm systems. The mechanical rake system has been tried but its bulky mechanism, the limited freedom in design, and its frequent maintenance requirements have made this system unpopular.

Figures 3.2.6-10 and 3.2.6-11 illustrate the effectiveness of surface wash.

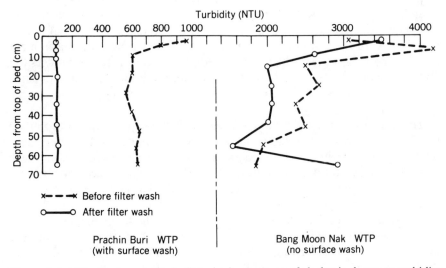

Figure 3.2.6-10 Sludge profile in filter beds. Amount of sludge is shown as turbidity (NTU).

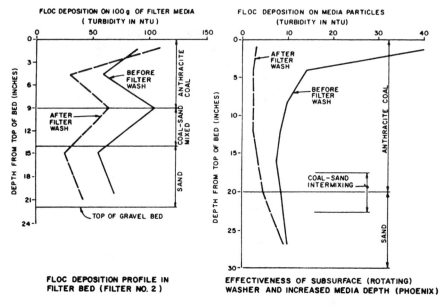

Figure 3.2.6-11   Effectiveness of surface wash system.

APPURTENANT SYSTEMS   The major filter appurtenant systems are the filter backwash tank, pumps, auxiliary scouring systems, wash flow control system, and wash-waste handling systems. Since the last item is a subject in itself, it is discussed separately in Section 4.2.

The filter backwash tank must have a capacity that is large enough to hold a sufficient amount of water for two filter washes. The tank must also provide an adequate depth of water above the tank exit so that a vortex will not be formed toward the exit pipe. Formation of a vortex leads to the entrainment of air in the wash water, which in turn reduces the flow rate and hinders the functioning of the flow rate controller. Therefore, it is always a good practice to provide an antivortex assembly in the wash tank at the entrance of the wash main.

The lowest water level in the backwash tank should generally be about 40 ft (12 m) above the top of the wash-water troughs located in the filter bank. Under normal conditions, the fixed grid type of surface wash can obtain water from the backwash tank. In contrast, the rotating arm surface wash system requires a minimum of 80 psi (5.5 kg/cm^2 or 540 kPa) and therefore usually requires a booster pump system.

The source of the filter backwash may be one of the following: an elevated wash tank, the pressurized main water distribution line, the water from the direct pump of the clearwell, or the effluent from the rest of the filters on the line (as in the self-backwash filters). The choice among these alternatives depends heavily on factors such as site topography, cost considerations, and aesthetics.

The elevated tank is the most common source of backwash. Depending on the topography of the site, the tank is either on stilts or situated on a hill and may also be installed over the control building. The tank may also be filled by a small capacity pump between backwashes.

An alternative to the elevated tank system is the direct pump backwash. In general, this system is rather expensive to operate due to its high electric consumption. In fact, some plants cannot operate the filter backwash pumps during periods of peak power demand. Therefore, the direct pump wash system must be accompanied by a standby power generator. It is also advisable to place a pressure relief standpipe in the pump discharge line to avoid filter underdrain failure, which may result from excessive pressure. The power demand of the direct pump wash system may be reduced by designing the filter to backwash only half a filter cell at a time. This allows the plant to use a wash pump that is only half the usual size.

The air-scour backwash filters are often designed with a pump backwash system for two reasons: the backwash rate is half that of an ordinary fluidized backwash system and the wash rate is a two-step system. Two-step backwash may be achieved by using one pump for the first wash and a second pump during the second step, rather than an elaborate flow rate control system. Air-scour wash is attained by employing air blowers. The most common type of blower is the positive displacement blower ("Roots" type), despite its high level of operational noise.

The backwash flow control system is necessary for setting the proper washing rate, regardless of the type of auxiliary scour employed during the backwash. The two most common systems are the mechanical flow rate controller and the master flow control valve, which is located in the backwash main. The mechanical flow rate control system is a combination of a flow meter and a flow-modulating butterfly valve. The second system is generally a plug or ball valve, which is opened in a slow and controlled manner to purge air out of the filter bed at the beginning of a wash and to slowly increase the wash rate to a maximum (preset) rate; the preset rate is controlled by a stopper located at the valve operator. Use of the master flow control valve is recommended for most cases. However, the mechanical flow rate controller should be employed if the water level in the tank varies over 10 ft (3 m) during a filter wash or if the backwash pipeline is both large in size and short in length. Otherwise, there will be an insufficiently controlled headloss and a large variation in the backwash rate (during the washing cycle) will result. Note that the backwash valve should be opened slowly to avoid disturbing the gravel bed (filter) and the underdrain system.

If a self-backwash filter design is adopted, the filter washing system may be dramatically simplified since it does not require a pump system, a backwash flow control system, a backwash line, or valves.

Design Criteria This section covers the basic design criteria for granular medium filters: the number of filters, the size of each type of filter, the characteristics of the filter bed, the filtration rate, the headloss across the filters, the type of filter backwashing system, the specifications for the gravel support bed, the basic hydraulics, and a basic filter design scheme.

Number of Filters The minimum number of filters for very small plants processing less than 2 mgd (90 L/s) is two. If the capacity of the plant exceeds 2 mgd, the minimum number of filters becomes four. Note that the required number of filters is calculated from the equation

$$N = 1.2Q^{0.5}$$

where N = total number of filters,
$\quad\quad Q$ = maximum plant flow rate in mgd.

Size of Each Filter The dimensions of four types of filter are discussed below.

ORDINARY GRAVITY FILTERS

Width of a filter cell	10–20 ft (16 ft average)
	(3–6 m)
Length to width ratio	2:1 to 4:1 (3:1 average)
Area of a filter cell	250–1000 ft^2 (600 ft^2 average)
	(25–100 m^2)
Depth of the filter	12–20 ft (17 ft average)
	(3.2–6 m)

Note: The filters for medium to large plants, those processing a minimum of 20 mgd (0.9 m^3/s), are generally composed of two cells per filter and have a central gullet down the center.

SELF-BACKWASH FILTERS (GRAVITY FILTER)

Width of a filter cell	10–20 ft (16 ft average)
Length to width ratio	2:1 to 4:1
Area of a filter cell	250–800 ft^2 (25–80 m^2)
Depth of the filter	18–25 ft (22 ft average)
	(5.5–7.5 m)

Note: The minimum number of filters is four, but six to eight are preferred. Each filter normally consists of only one cell due to limitations in the backwash water supply.

AUTOMATIC BACKWASH FILTERS (HARDINGE TYPE)

Width of a filter	Standard 16 ft (5 m)
Length of a filter	Up to approximately 120 ft (37 m)

Depth of a filter cell	7–11 ft (2.1–3.3 m)
Width of each cell	8–25 in. (0.2–0.64 m)
Depth of filter a bed	11–48 in (16 in. standard) (0.25–1.2 m)
Traveling speed of the washing carriage	21 ipm (0.53 m/min) standard and indexing available for deeper than 16 in. bed depth.

PRESSURE FILTERS

Diameter of a filter cell	4–20 ft (10 ft average) (1.2–6 m)
Length of a filter cell	8–50 ft (20 ft average) (2.4–15 m)
Depth of a filter bed	2–3 ft (0.6–0.9 m)

Note: Both horizontal and vertical cells are available. However, the vertical cells are more common for larger sized units because they are cleaned to a greater extent during filter washing. The multicell (four-cell) horizontal units have the following advantages: a self-backwash system, space-saving features, and cost effectiveness.

Filter Bed: Types of Medium and Depth   Silica sand and anthracite coal are the most commonly used types of filter medium. The American Water Works Association has set the quality standards for both these materials; refer to AWWA Standards B100. Garnet, ilmenite, pumice, and synthetic materials may also be used as filter media but to a lesser extent due to their limited availability and relative expense. The depth of the filter bed is a function of medium size and generally follows the relationship shown below:

$l/d_e \geq 1000$ for ordinary monosand and dual media beds

$l/d_e \geq 1250$ for trimedia (coal, sand, and garnet) and coarse monomedium beds (1.5 mm $< d_e >$ 1.0 mm)

$l/d_e \geq 1250–1500$ for very coarse monomedium beds (2 mm $< d_e >$ 1.5 mm)

where l = depth of the filter bed in mm,
d_e = effective size of the filter media.

Notes

1. If the turbidity of the filtered water must be less than 0.1 NTU, without the use of polymer as a filtration aid, it is recommended that the l/d_e

ratio be increased by 15%. Table 3.2.6-1 illustrates the general design criteria for common filter beds.

2. Design engineers should anticipate stringent drinking water standards for future plant operation and are therefore advised to include a provision for replacing the regular filter beds with GAC medium. Due to the longer contact time required by the GAC bed, an EBCT of a minimum of 15 min, the depth of the regular filter bed must be made deeper to accommodate the GAC medium. A depth of 8 ft (2.4 m) is needed to provide 15 min of contact time at a loading rate of 4 gpm/ft^2 (10 m/h).

Filtration Rate As previously mentioned, regulatory agencies restrict the maximum filtration rate for a filter design. However, the agencies will consider a higher filtration rate if a well managed long-term pilot study can prove that the higher rate satisfies all the requirements. The proven and accepted filtration rates for common filter beds are listed in Table 3.2.6-1.

Engineers should realize that an acceptable filtration rate is a function of factors such as the size of the medium, the degree of pretreatment, and whether polymer is used as a filtration aid. Design engineers who must comply with very conservative design criteria should also include a provision for the possible increase in filtration rate. Modern dual media filter beds are capable of accommodating rates that are 6–8 gpm/ft^2 (15–20 m/h), whereas the coarse, deep beds may accommodate filtration rates of 8–13 gpm/ft^2 (20–30 m/h). Hence the size of the influent valve, effluent valves, pipes, and channels must be properly selected when implementing the higher flow rate.

Headloss Across the Filter

TOTAL HEADLOSS ACROSS EACH FILTER

Ordinary gravity filters	9–15 ft (12 ft average) (2.7–4.5 m)
Pressure filters	50–100 ft (75 ft average) (15–30 m)
Automatic backwash filters	2–3 ft (0.6–0.9 m)

NET HEADLOSS AVAILABLE FOR FILTRATION

Ordinary gravity filters	6–12 ft (9 ft average) (1.8–3.6 m)
Pressure filters	25–50 ft (7.5–15 m)
Automatic backwash filters	0.5–1.0 ft (0.15–0.3 m)

Filter Washing

BACKWASH WITH SURFACE WASH   The appropriate backwash rate should be determined by the specific gravity of the media, the size of the media grains, and the water temperature. Figure 3.2.6-4 may be used in determining the backwash rate.

BACKWASH RATE

Ordinary rapid sand bed	15–18 gpm/ft^2 (0.6–0.74 m/min)
Ordinary dual media bed (trimedia included)	18–22 gpm/ft^2 (0.74–0.9 m/min)
Ordinary GAC bed	12–16 gpm/ft^2 (0.5–0.65 m/min)

SURFACE WASH RATE

Fixed nozzle type—flow rate	3–4 gpm/ft^2 (0.12–0.16 m/min)
—pressure	8–12 psi (55–83 kPa)
Rotating arm type—flow rate (single arm)	0.5–0.7 gpm/ft^2 (0.02–0.03 m/min)
—pressure	70–100 psi (480–690 kPa)
Rotating arm type—flow rate (dual arms)	1.3–1.5 gpm/ft^2 (0.05–0.06 m/min)
—pressure	80–100 psi (500–690 kPa)

Note: The pressures shown are at the water jet discharge points.

AIR-SCOUR BACKWASH

BACKWASH RATE

Air-scour stage for ordinary beds	2–4 gpm/ft^2 (0.08–0.16 m/min)
Air-scour stage for coarse deep beds	8–10 gpm/ft^2 (0.3–0.4 m/min)
Rinse stage for ordinary beds	10–12 gpm/ft^2 (0.4–0.5 m/min)
Rinse stage for coarse deep beds	16–24 gpm/ft^2 (0.65–1.0 m/min)

TABLE 3.2.6-1 Types of Medium and Their Application

Filter Medium	Type of Filter	Medium Design Criteria	Advantages/Disadvantages
Fine sand	Slow sand filter 0.05–0.17 gpm/ft² (0.13–0.42 m/h) filtration rate	Effective size: 0.25–0.35 U.C.: 2–3 Depth: 3.3–4 ft (1.0–1.2 m) S.G.^{a} ≥ 2.63	1. Simple design and construction 2. Good effluent quality without pretreatment a. Requires a large filter bed area b. Applicable only for good quality raw water c. Requires frequent scraping off of surface layer (every 20–30 days)
Medium sand	Rapid sand filters 2–3 gpm/ft² (5–7.5 m/h) filtration rate	Effective size: 0.45–0.65 U.C.: 1.4–1.7 Depth: 2–2.5 ft (0.6–0.75 m) S.G. ≥ 2.63	1. A proven and widely accepted filtration process 2. A wide application range if pretreatment is provided a. Rather short filter runs due to surface filtration b. Always a need for coagulation pretreatment and an auxiliary washing system
Coarse sand	High-rate filters 4–12 gpm/ft² (10–30 m/h) filtration rate Direct filtration	Effective size: 0.8–2.0 U.C.: 1.4–2.0 Depth: 2.6–7 ft (0.8–2 m) S.G. ≥ 2.63	1. An effective high-rate filtration process with very long filter runs 2. A wide application range with polymer pretreatment a. Auxiliary wash system is limited to air-scour type b. Requires deep filter cells and a special underdrain

Multimedia coal–sand dual or coal–sand–garnet trimedia	High-rate filters 4–10 gpm/ft² (10–25 m/h) filtration rate Direct or in-line filtration	*Sand* Effective size: 0.45–0.65 U.C.: 1.4–1.7 Depth: 1 ft (0.3 m) *Anthracite coal* Effective size: 0.9–1.4 U.C.: 1.4–1.7 Depth: 1.5 ft (0.45 m) S.G. ≥ 1.5 to 1.6 *Garnet* Effective size: 0.25–0.3 U.C.: 1.2–1.5 Depth: 0.25 ft (0.0075 m) S.G. ≥ 4.0–4.1	1. An effective high-rate filtration process with long filter runs 2. A proven and widely accepted filtration process a. Either surface wash or air-scour wash is required as an auxiliary washing system and a polymer is required as a filter aid b. Proper selection of each medium is important c. Requires a high backwash rate for restratification
Granular activated carbon (GAC)	Removal of organic contaminants 3–6 gpm/ft² (7.5–15 m/h) filtration rate Contact time: 15–30 min	Effective size: 0.5–1.0 U.C.: 1.5–2.5 Depth: 6–12 ft (1.8–3.6 m) S.G. ≥ 1.35–1.37	1. A proven and accepted process for specific removal of organic contaminants (i.e., taste and odors, THMs, and pesticides) 2. Can also operate effectively as a conventional filter a. Must be regenerated or replaced when adsorption capacity is depleted b. High initial and maintenance costs
Proprietary type media	Variety of types, including green sand and synthetic media	Depends on the purpose	1. Design and efficiency guaranteed by the manufacturer a. Limited number of suppliers b. Mostly patented items

[a]S. G. = specific gravity.

AIR-SCOUR RATE

For ordinary filter beds	2.5–3 cfm/ft^2
	(0.75–0.9 m/min)
For coarse deep beds	3–4 cfm/ft^2
	(0.9–1.2 m/min)

Underdrain System   Selection of the underdrain system varies with the type of filter system that is chosen. Normal backwash and self-backwash filters may use the same type of underdrain system but with one important modification, the latter has a larger orifice size. Air-scour backwash filters are most compatible with the polyethylene blocks or the strainer type of underdrain system. Although there are many other types of filter underdrain system available, those listed here are considered to be the proven and most effective types.

NORMAL BACKWASH FILTERS: WITH OR WITHOUT SURFACE WASHING

Type	Headloss at Ordinary Backwash Rates	Orifice Size (Diameter)	Remarks
Precast concrete laterals (tepees)	3–5 ft (1–1.5 m)	$\frac{5}{16}$–$\frac{3}{8}$ in. (8–10 mm)	• 12 in. lateral spacing • 3 in. orifice spacing on either side of the lateral • Maximum lateral length of 16 ft
Dual-parallel lateral blocks (tile blocks)	2–6 ft (0.6–1.8 m)	$\frac{5}{32}$–$\frac{1}{4}$ in. (4–6 mm)	• 12 in. lateral spacing • 48 or 18 orifices per ft^2 area • Maximum lateral length of 50 ft
Pipe laterals	3–5 ft (0.9–1.5 m)	$\frac{1}{4}$–$\frac{3}{8}$ in. (6–10 mm)	• 12 in. lateral spacing • Orifices are spaced 3–4 in. apart and 45° down-angle from the horizontal on both sides of the lateral • Maximum lateral length of 20 ft
Strainer	4–7 ft (1–2 m)	0.25–0.75 mm	• Plenum or lateral bottom • Strainers spaced 6–10 in. apart • A space less than 10 in. for the lateral

SELF-BACKWASH TYPES OF FILTERS   The only difference in design criteria for these types of underdrain, as compared to normal backwash filters, is the required headloss across each filter. This in turn affects the orifice size of the underdrain system. The required headloss, at the designed backwash rate, should be 0.5–1 ft (0.15–0.3 m) while maintaining no more than a ±5% of water flow uniformity across a filter bed area. In the case of precast concrete laterals, this requirement is satisfied by using orifices that are $\frac{1}{2}$ in. (12 mm) in diameter and spaced 3 in. (center to center) on each side of the laterals. All other types of underdrain must be designed specifically to meet the required headloss.

AIR-SCOUR BACKWASH FILTERS   The backwash filters listed below are the proven and most effective types available.

Type	Headloss During Washing	Orifice Size	Remarks
Dual-lateral blocks (poly-ethylene blocks)	25 and 40 in. under si-multaneous air and water back-wash at 3 cfm/ft² con-stant air rate and 15 and 20 gpa/ft² back-wash rate	$\frac{1}{4}$ in. (6 mm)	• 12 in. lateral spacing • 23 dispersion orifices per ft² of bed area • Maximum jointed lat-eral length is 50 ft • Gravel layers required on top of the blocks
Strainer	15–24 in. under 20 gpm/ft² wash rate	0.25–0.77 mm (slit size)	• The preferred height of the plenum bottom is 2 ft • Durable and proven types of strainer should be selected • A gravel bed is usually not required above the strainers

Notes

1. The porous plate type of underdrain system has been used successfully in conjunction with the automatic backwash filter, a proprietary unit. However, this type of underdrain will eventually become clogged if moderate to hard water (over 100 mg/L of $CaCO_3$) is continuously filtered. Periodic cleaning of the underdrain with acid, 5% solution of HCl, is therefore mandatory.

2. The lateral type of underdrain should provide air release holes or slits at the top, the end, and midpoint of each lateral. Moreover, each end of the lateral should be anchored in order to prevent the laterals from moving during air-scouring.

3. The most important design consideration for the underdrain system is the hydraulics during backwash. The flow velocity at the entrance to each lateral should not exceed 3.5 fps (1 m/s) and the velocity at the entrance of the manifold is preferably less than 5 fps (1.5 m/s) so that a nearly uniform flow distribution is maintained in the filter bed.

4. A plenum beneath the filter underdrain system ensures better flow distribution. Yet, the underdrain slab should be monolithic in nature to minimize the problems of the underdrain system.

GRAVEL SUPPORT BED   For most underdrain systems the following values are commonly used:

Layer Number	Passing Screen Size	Retaining Screen Size	Depth of Layer	Note
1	$\frac{1}{2}$ in. (40 mm)	$\frac{3}{4}$ in. (20 mm)	4–6 in. (100–150 mm)	Bottom layer
2	$\frac{3}{4}$ in. (20 mm)	$\frac{1}{2}$ in. (12 mm)	3 in. (75 mm)	
3	$\frac{1}{2}$ in. (12 mm)	$\frac{1}{4}$ in. (6 mm)	3 in. (75 mm)	
4	$\frac{1}{4}$ in. (6 mm)	No. 6 sieve (3 mm)	3 in. (75 mm)	
5	No. 6 sieve (3 mm)	No. 12 sieve (1.7 mm)	3 in. (75 mm)	Top layer

Total depth: 16–18 in. (400–450 mm)

For the dual-parallel lateral block type of underdrain (Leopold, Inc.), the following values are the most appropriate:

Layer Number	Passing Screen Size	Retaining Screen Size	Depth of Layer	Note
1	$\frac{3}{4}$ in.	$\frac{1}{2}$ in.	3 in.	Bottom layer
2	$\frac{1}{2}$ in.	$\frac{1}{4}$ in.	3 in.	
3	$\frac{1}{4}$ in.	No. 6 sieve	3 in.	
4	No. 6 sieve	No. 12 sieve	3 in.	Top layer

Total depth: 12 in. (300 mm)

For the dual-lateral block type of underdrain with air-scour wash (Leopold, Inc.) the following values are recommended:

Layer Number	Passing Screen Size	Retaining Screen Size	Depth of Layer	Note
1	$\frac{3}{4}$ in.	$\frac{1}{2}$ in.	2 in.	Bottom layer
2	$\frac{1}{4}$ in.	$\frac{1}{8}$ in.	2 in.	
3	$\frac{1}{8}$ in.	No. 10 sieve (2 mm)	2 in.	
4	$\frac{1}{4}$ in.	$\frac{1}{8}$ in.	2 in.	
5	$\frac{1}{2}$ in.	$\frac{1}{4}$ in.	2 in.	
6	$\frac{3}{4}$ in.	$\frac{1}{2}$ in.	2 in.	Top layer

Total depth: 12 in. (300 mm)

Notes

1. Some manufacturers and textbooks recommend using a gravel bed that is shallower than 16 in. for ordinary underdrains, including the "Wheeler Bottom." However, numerous case histories have proved that shallow gravel beds tend to become disrupted.

2. Larger sized gravel, such as the bottom two layers, must be at least 3–4 in. (7.5–10 cm) in depth to compensate for any inaccuracies that may occur during construction.

3. The size and depth of the first layer should be adjusted according to the type of underdrain that is selected. Precast "tepees" and pipe laterals (6 in. or 15 cm in diameter) must have a depth of 6 in. for the first layer to ensure that the laterals are covered.

4. Most strainer nozzle types of underdrains completely eliminate the need for a gravel bed if the nozzles are spaced closely together. If the slit size is greater than $\frac{2}{100}$ in. (0.5 mm), it is recommended that the nozzle domes be covered by a layer of pea sized gravel to ensure good distribution of the backwash flow.

5. If air in the underdrain system and main backwash line is not purged prior to backwash, the composition of the gravel bed may be disrupted during backwashing by the explosive movement of large volumes of compressed air.

6. A reverse graded gravel bed, of the type illustrated for the dual-lateral block type of underdrain with air-scour wash, was first developed in Chicago by John Baylis. It effectively prevents gravel bed upset by using controlled amounts of air flow with backwash. However, if this type of gravel bed is used in conjunction with ordinary backwash, the fine gravel (middle) layer is likely to accumulate sludge.

7. The quality of the gravel must meet the standards set by the AWWA (refer to AWWA Standards B100). The use of limestone must be avoided especially when the filter bed is acidified in preparation for cleaning.

Basic Hydraulics

FLOW VELOCITY   The maximum flow velocities (approximate) sustained by the filter piping or channels and valves are as follows:

	Ordinary Filters (fps)	Self-Backwash Filters (fps)
Influent channel	2	2
Inlet valve	3	5
Forebay channel	0.5	0.5
Effluent valve	5	2
Effluent channel	5	2
Backwash main	10	3
Backwash valve	8	5
Surface wash line	8	8
Wash-waste main	8	8
Wash-waste valve	8	8
Filter-to-waste valve* (at the designed filtration rate)	17	17
Inlet to the filter underdrain lateral	4.5	4.5

Note: The flow velocity in the influent pipe or channel may be 6 fps (2 m/s) if flow rate controllers are located in each filter effluent line or if the filtration system is a declining rate system.

* The size of the rewash pipeline and valve is based on a flow velocity of approximately 5–6 fps (1.5–1.8 m/s) at a filtration rate of 2 gpm/ft^2 (5 m/h).

MISCELLANEOUS ITEMS

1. The water depth above the filter bed must be a minimum of 6 ft (1.8 m).
2. The approximate initial headloss across an ordinary filter, at 60°F (15°C), is as follows:

Filtration Rate (gpm/ft^2)	Headloss (ft)	Type of Filter Bed
2	0.8	Rapid sand
3	1.3	Rapid sand
4	1	Dual media
6	1.3	Dual media
8	1.8	Dual media
10	2.5	Dual media

3. The lowest water level in the backwash tanks should be approximately 35 ft (10.5 m) above the lip of the washwater troughs. The capacity of the backwash tank must be large enough to hold water for two filter backwashings.

4. The elevation of the weir, which controls the filter effluent water level, should be set at a height that produces the appropriate hydraulic gradient at the top of the filter bed at the designed filtration rate of the filters.

5. The maximum water level in the wash-waste holding tank should be properly set so that the backup of water in the forebay of the filter is prevented when the most remote filter is undergoing backwashing. The end of each wash trough must have a "free flow" at all times in order to maintain even flow conditions through each trough.

6. The number of wash troughs is determined by the characteristics of the filter bed and the method of filter washing. Most high-rate filters with dual media beds lose a significant amount of coal whenever surface wash is employed. This can be minimized by shortening the overlap in the duration of surface wash and backwash, and if an adequate distance is provided between the top of the filter bed and the lip of the trough. The height and spacing of the troughs are generally determined by using the following criteria:

   a. The distance between the lip of the trough and the top of the bed is $(0.5L + D) > H_0 < (L + D)$.

   b. The spacing of the troughs, from center to center, is $1.5 H_0 > S < 2.5H_0$, where L is the depth of the filter bed media, D is the depth of the trough, H_0 is the distance between the lip of the trough and the top of the bed, and S is the spacing distance of the troughs from center to center.

   c. A smaller H_0 may be used for coarse, deep filter beds by using the equation $2 \text{ mm} < ES > 1.5 \text{ mm}$. However, a larger H_0 should be used for dual filter beds with small sized media ($ES < 0.85$ mm for coal) due to a larger bed expansion rate.

   d. The criteria shown are not applicable to the filters with none fluidized air scouring wash system especially for coarse and deep medium filter beds.

The wash troughs may be deleted, copying the European design (Figure 3.2.6-7), if air-scour wash is adopted. However, a coarse, deep bed filter with air-scour will require the use of wash troughs; the troughs may be spaced beyond what is allowed by the formula since it applies to regular backwash systems. The wash troughs improve the distribution of the filter influent and consequently media migration does not occur. This is possible because filtration rates above 8 or 10 gpm/ft^2 (20 or 25 m/h) induce a high inflow rate, thus changing the hydraulic characteristics of the influent.

Basic Filter Design Scheme   This section provides a few general tips and briefly describes some basic filter designs.

GENERAL   The possibility of contaminating the treated water, due to leakages, cross connections, or operator error, is minimized by providing a clear physical separation between the unfiltered water line and the filtered water line. Engineers are therefore advised not to locate these pipelines or conduits in the same filter pipe gallery. The photo picture shows an example of unfavorable filter pipe gallery design.

When designing the filtration system, the engineer must consider the accessibility of the various mechanical items. The pipe gallery should have a clear and wide passage so that operators can maneuver vehicles, which may be necessary to provide adequate maintenance and repair work for large scale plants. All major valves and valve operators should be situated in such a manner that the plant operator can visually observe and easily service them.

Additionally, an access hole should be provided in the filter underdrain to facilitate inspection.

An air release valve should be furnished at the end of the washwater line and an air relief/intake valve should also be supplied for the filtered water line and the main wash-waste lines. If air is allowed to accumulate in these pipe lines, the flow may be reduced by an air blockage and the explosive escape of the air could damage the filters.

The top of the forebay walls should be higher than the highest water level in the filter cell. This design feature minimizes filter media migration, which is initiated by the inertia of the influent flow. However, most European air-scour wash filters do not follow this criterion and the majority of these filters therefore have problems of varying degrees with filter media migration.

Engineers should try to simplify the flow rate control system by avoiding high maintenance items, such as mechanical flow rate controllers, unless it is absolutely necessary.

Problems associated with air-binding are diminished when a minimum water depth of 6 ft (2 m) is provided above the filter bed. This depth also induces a higher available headloss for filtration. The problem of air-binding may also be prevented by installing a control weir for the effluent level of the filter; this scheme will provide a positive back pressure in the entire bed, at all times. Additionally, this control weir prevents the disturbance of the gravel bed by entrained air during backwash by continually flooding the filter bed. A more detailed discussion of air-binding is presented later in this chapter.

Based on the above considerations, a typical filter arrangement will be similar to that illustrated in Figure 3.2.6-12.

BASIC FILTER STRUCTURES When the total number of filters exceeds six, they are generally arranged on both sides of the filter gallery, primarily for cost saving reasons. Figure 3.2.6-13 depicts five basic design schemes. Alternatives 1 and 2 pertain to the self-backwash filters and differ only in the design of the filtered water collection main. Alternative 3 is a conventional type of filter with an equal loading and constant level control system. Each filter in this scheme has a flow modulating valve in its effluent line but lacks a flow metering device.

Alternatives 4 and 5 relate to air-scour wash filters. Although the filter-to-waste line is not shown in alternative 5, it can easily be provided at the intersection of the backwash and filter effluent valves. One important consideration is the elevation of the air supply headers. The headers should be located at least 2 ft (0.6 m) above the highest water level in the filter cells, thereby reducing the chance of water back siphoning into the air line.

Example Design Calculations Three examples are discussed in this section. They concern the selection of the most appropriate filtration process and filter design and present the best method for determining the appropriate combination of filter media for multimedia beds.

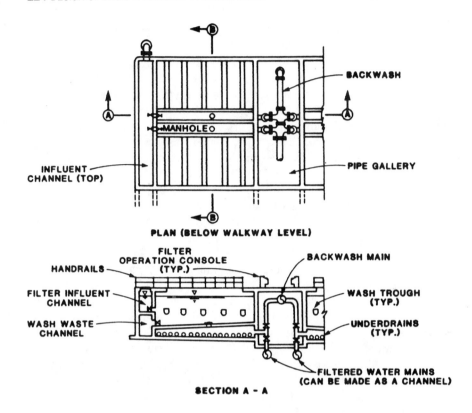

PLAN (BELOW WALKWAY LEVEL)

SECTION A - A

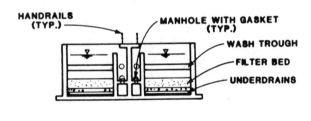

SECTION B - B

Figure 3.2.6-12 Basic filter design scheme.

Example 1   Filtration Process

Select the most appropriate type of filtration process and overall process train based on the source of the water and the following local conditions:

Case 1. A treatment plant in the Midwest has an ultimate capacity of 90 mgd (4 m³/s). The source is a large river that exhibits high turbidity spikes, up to 500 NTU, during the summer months. The source

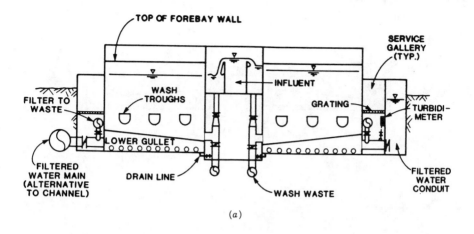

(a)

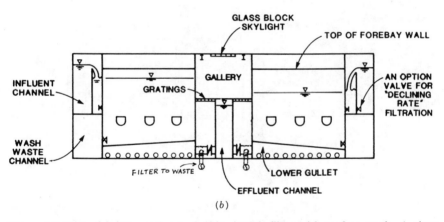

(b)

Figure 3.2.6-13 (a) Alternative 1: Self-backwash filter with surface wash. A clear separation of filtered and unfiltered pipings; the structure is approximately 25 ft deep. (b) Alternative 2: Self-backwash filter with surface wash. A central filtered water channel arrangement; the structure is approximately 25 ft deep. (c) Alternative 3: Conventional type filter with surface wash. Approximately 17 ft for filter cell and 25 ft height for pipe gallery. (d) Alternatives 4 and 5: Air-scour wash filters. Approximately 17 ft for filter cell and pipe gallery, except Alternative 4, which has 25 ft for height of gallery.

    also has an average coliform count of 780 MPN, an average TOC of 3.5 mg/L, a seasonal algae bloom, and an associated taste and odor problem that is primarily a function of diatoms.

Case 2. A 150 mgd (6.5 m³/s) plant located in the Northwest obtains water from a large lake that consistently has good water quality and an average turbidity of 1.3 NTU. However, turnover of the lake water occurs twice a year. During these periods the turbidity level is

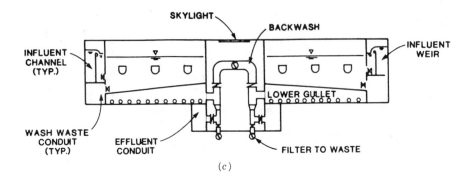

(c)

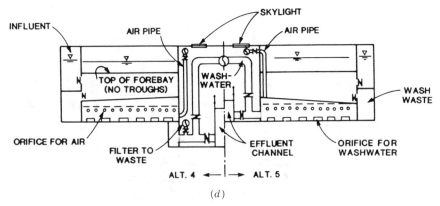

(d)

Figure 3.2.6-13 (Continued)

expected to rise as high as 10 NTU and the iron and manganese levels may become as high as 1.2 mg/L and 0.2 mg/L, respectively. Additionally, the algae count can be expected to be as high as 1000 ASU/mL.

Case 3. A small town in a developing Asian country requires a 3 mgd (0.13 m³/s) plant. The raw water may be obtained from a local lake. The quality of this water is good except during the monsoon season. Although accurate records are not available, the turbidity is expected to be approximately 50 NTU during the monsoon season.

Case 4. A town proposes to treat 5 mgd (0.22 m³/s) of water obtained from three wells. The well water contains an average of 2.2 mg/L iron and 0.4 mg/L manganese and has a hardness of 550 mg/L. Other elements, including turbidity, are always under the MCL of the drinking water standards. The plant is located in Alaska.

Case 5. A large camp site, located in a national park in the Central Mountain region, requires a maximum of 3 mgd (0.13 m³/s) drinking water during the tourist season. The water source is a local impoundment fed by streams. The water quality is reasonably good but has occasional turbidity spikes of up to 50 NTU during thunderstorms. The water contains moderate coliform and algae counts and may contain *Giardia*. The plant is shut down from November to April.

Solution

Case 1. A coarse, deep monomedium filter bed with air-scour wash is the most appropriate choice and a provision for future conversion of these filter beds to GAC beds should seriously be considered. The bed should have 6–8 ft (2–2.4 m) of sand or anthracite coal monomedium, having an effective size of 1.5–2 mm, depending on the depth of the bed. The monomedium should have a uniformity coefficient of approximately 1.5 and a filtration rate of 6–8 gpm/ft² (15–20 m/h); these figures are reasonable choices. A filter bed that is 6 ft (1.8 m) deep provides an EBCT of 15 min at a filtration rate of 3 gpm/ft² (7.5 m/h). However, if the rate is 4 gpm/ft² (10 m/h), the depth of the bed must be 8 ft (2.4 m) in order to maintain the same contact time.

 The overall process train should be preozonation, coagulation, flocculation, sedimentation, filtration, and chloramination. The provision for converting the filter bed to GAC adsorption should include the addition of filter cells to compensate for the lower allowable filtration rate of the GAC filters. Since the GAC filter effluent contains very high counts of bacteria, additional post-ozonation may be implemented prior to postchloramination. A provision for feeding hydrogen peroxide during preozonation and the addition of a surface wash system may also be included in the design to enhance efficiency of ozonation and to eliminate the problem of mud ball formation within the deep filter bed.

Case 2. In this particular case the recommended system is a standard dual media filter bed with surface wash as the auxiliary scour during backwash. The self-backwash type of filter is selected because of its simplicity, energy efficiency, and proven performance. These characteristics minimize the operation and maintenance efforts. The treatment process train includes prechlorination, coagulation, and roughing filters prior to the dual media filters and chlorination process; this type of process train is called two-stage filtration. A bypass channel should be provided around the roughing filters because the plant will be operated as an in-line filtration process during most of the year. However, the roughing filter must be employed during seasons with high algae counts and periods of high turbidity.

A provision for future addition of the preozonation and chloramination processes, as well as granular GAC capping of the filter beds, should be included to control taste and odor problems. Direct filtration is not the appropriate choice for this situation since this process has no defense against very short filter runs due to algae blooms.

The roughing filter of the two-stage filtration performs as a flocculation process and also removes 50–75% of suspended matter. The process of two-stage filtration is discussed in detail in Section 3.2.7.

Case 3. Considering the local conditions and the raw water quality, a slow sand filtration with a presedimentation reservoir should be selected as the treatment scheme. This system is simple to operate and maintain and does not require costly maintenance and repairs. The presedimentation reservoir should provide a storage period of 1–2 days because this will facilitate the removal of large suspended particles during periods of turbidity spikes (in the raw water).

Case 4. The appropriate choice for this situation is a standard dual media bed or a medium size monosand bed, with surface wash scour during filter wash. The first choice for the overall treatment process is lime softening (lime flocculation and clarification) by means of a sludge blanket type of reactor clarifier, followed by acid pH control or carbon dioxide (for recarbonation) then clarifier, all prior to the filters. Lime softening reduces the water hardness to a range of 80 mg/L and effectively removes iron and manganese at the levels dictated in the problem.

The second choice for the overall treatment process begins with the oxidation of iron and manganese with either chlorine or potassium permanganate, or both. The second step is pretreatment to the dual media bed filtration, followed by side stream softening via an ion exchange process. Lastly, the softened and filtered water are blended to produce a water with medium hardness. Flocculation and sedimentation are preceded by the oxidation of the reduced metals.

The pH of the pretreatment process water should be kept in the range of 8–8.5 by using ferric iron coagulant, cationic polymer, or both. Water softening may be achieved through a cation exchange resin process or a membrane technique, with proper pH adjustment and the addition of a scale inhibitor, due to the rather small output of the treatment plant. All processes must be housed since the ambient temperatures are extremely cold during the winter months.

Case 5. If enough land is available, a slow sand filtration process should be considered because of its simplicity, reliability, and low operation and maintenance costs. However, it is very unlikely that the area required by the slow sand filtration will be available in the mountainous location. Due to the size of the plant, the seasonal operation, and the limited maintenance services, a proven type of semiautomatic plant is recommended.

The filter bed should be a simple monosand bed (rapid sand bed) because of the low initial cost and simple operation. The primary choice for the process is two-stage filtration. The second is a package-type of flocculation/sedimentation pretreatment with a self-backwash type of pressure filter. The last choice is a roughing filter followed by proprietary automatic backwash filters. The last process requires skilled and knowledgeable operators. The selection of the treatment process is strongly influenced by the requirements set by the local regulatory agency.

Example 2   Filter Design

Given   A 50 mgd (2.2 m³/s) filtration plant with an ultimate capacity of 100 mgd (4.4 m³/s) is located in a quiet and good residential area in the Northeast. The plant site has an average 2% slope (one way). The treatment process train is a conventional complete process and pumps lake water that has a reasonably good quality throughout the year.

Determine

  (i) The total number of filters.
 (ii) The type of filter bed, the filtration rate, and the filter washing scheme.
(iii) The size of each filter.
 (iv) The filter arrangement.
  (v) The size and type of the major filter valves.
 (vi) The filter influent channel, pipings, and central gallery.
(vii) The type of filter media, gravel bed, and underdrain.
(viii) The backwash rate, the type of wash troughs, and auxiliary scour wash.
 (ix) The plan and sections of the filter.
  (x) The elevation of the filter effluent control weir and the configuration of the effluent channel.
 (xi) The device used to measure the backwash flow rate.
(xii) The filter piping scheme that allows for the filter-to-waste feature and continuous sampling of the filter effluent.

Solution

(i) The total number of filters in the final stage of this plant is $N = 1.2$ $(100)^{0.5} = 12$. Therefore, six of these filters are constructed in the initial stage.

(ii) Due to the relatively good quality of the raw water, a standard dual-media filter bed is preferred. This type of filter bed has a history of reliable performance when used in conjunction with surface wash and does not employ noisy air blowers for the air-scour process, thereby preserving the peace of the neighborhood.

The designed filtration rate, with all the filters on line, is 6 gpm/ ft² (15 m/h). Under the given conditions, this rate is widely accepted by most regulatory agencies. Note that the size of the major pipes and channels should be 50% larger than the designed rate to accommodate a potentially higher loading rate to each filter during future operation.

Filter washing is achieved through normal backwash: a self-backwash filter design with surface wash as the auxiliary scour during backwash. The self-backwash scheme is selected because it resolves important factors, such as aesthetic and noise considerations that are dictated by the neighborhood, the possibility of the water freezing should an elevated tank be selected, and the slope of the site. The average 2% slope reduces the need for deep excavation since the filters must be located at a lower elevation.

The most energy efficient wash tank system is the elevated wash tank. However, it should only be considered if the plant site has a suitable hill on which it can be situated. If the wash tank is an elevated tank on legs or a high cylindrical tank, this results in a negative visual impact to the neighborhood.

(iii) The total required filter bed area at 100 mgd is 11,583 ft² (1080 m²) or (100 × 695) gpm ÷ 6 gpm/ft². Since the total number of filters is 12, each filter bed area is 965 ft² (90 m²). Each filter will have two filter cells and a central gullet; this configuration ensures favorable hydraulic characteristics during backwash. The filter area of each filter cell is 965 ÷ 2 = 482.5 ft² (45 m²). Therefore, make each filter cell 15 ft (4.5 m) wide and 32 ft (10 m) long, so that the area of each cell is 480 ft² (45 m²).

We check the actual filtration rate by calculating:

69,500 gpm ÷ (480 × 2 × 12) ft²

$$= 6.03 \text{ gpm/ft}^2 = 15 \text{ m/h} \quad \text{O.K.}$$

(iv) The filters are arranged as illustrated in Alternative 1 of Figure 3.2.6-13. A double row of filters is constructed along a central pipe gallery where the piping for the filter influent and wash waste are located. The filtered water channels are situated opposite from the central gallery so that there is no possibility of contaminating the filtered water by cross connection. The filtered water may be collected into a channel instead of a large, expensive pipeline and fittings because of the high flow rate of this system and to allow the valves to be accessible for easy maintenance and repair.

(v) The sizes and types of the major filter valves are listed below:

$$\text{Influent flow rate} \quad 6.03\,\text{gpm/ft} \times 960\,\text{ft} = 5{,}790\,\text{gpm} = 13.2\,\text{cfs}$$

$$\text{Backwash flow rate} \quad 22\,\text{gpm/ft} \times 960\,\text{ft} = 21{,}120\,\text{gpm} = 48.2\,\text{cfs}$$

$$\text{Surface wash flow rate} \quad 3\,\text{gpm/ft} \times 960\,\text{ft} = 2{,}880\,\text{gpm} = 6.6\,\text{cfs.}$$

Based on these data, the following selections are made:

Valve	Size and Type	Velocity Through the Valve	
Filter influent	12 in. butterfly valve	$v = 4.2$ fps	O.K.
Filter effluent	36 in. butterfly valve	$v = 1.9$ fps	O.K.
Backwash (maximum)	Filter effluent valve is also the backwash valve for the self-backwash filters	$v = 6.8$ fps	O.K.
Surface wash	12 in. butterfly valve	$v = 8.4$ fps	O.K.
Filter-to-waste	12 in. butterfly valve	$v = 16.8$ fps	O.K.
Wash-waste drain	36 in. butterfly valve	$v = 7.7$ fps for back and surface wash	O.K.

Note: The sluice gate and the gate valve are bulky and expensive items that require frequent maintenance and repair, in contrast to the butterfly valves specified by the AWWA standards.

(vi) In order to establish an approximately equal water level in each inlet channel, the flow velocity at the beginning of the influent channel must not exceed 1 ft/s (0.3 m/s). Since each filter is equipped with identically sized rectangular weir inlets at the same elevation, this ensures that all filters on the line receive almost the same flow rate (i.e., equal loading).

The influent channel is located directly above the pipe gallery and should be approximately 22 ft (6.6 m) in width in order to allow for a low flow velocity. The gallery therefore has ample space for inspection and maintenance of the valves and other equipment. For this type of filter, the total depth of the filter structure is approximately 25 ft (7.5 m), which allows a sufficient water depth to be provided above the bed.

As illustrated in Figure 3.2.6-14, the channel width, the height of the gallery, and the location of the inlet and wash-waste pipelines are all dictated by the dimensions of the pipe fittings and the valves, in addition to the thickness of the concrete walls.

Constructability, the ease with which the valves can be removed during major repairs, the wall pipes for leakage water stops, and the

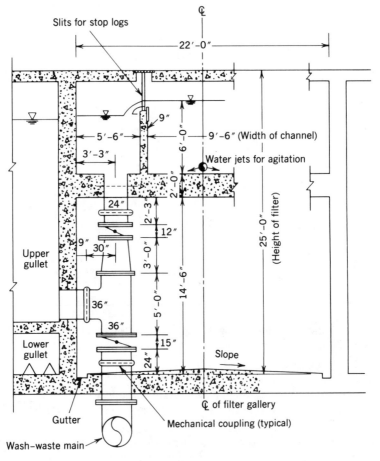

Figure 3.2.6-14 Filter piping, influent, and pipe gallery design (only the left half, from the center line, is shown).

slope of the floor toward the drainage gutters are all important design considerations. Unfortunately, many inexperienced design engineers disregard topics such as providing a minimum space requirement of 9 in. (22.5 cm) between the wall and the adjacent pipe flanges, for easy installation of the nuts and bolts, for supplying mechanical couplings to facilitate minor adjustments in pipe alignment and easy removal of the valves, and, lastly, for adequate drainage of water from the gallery floor.

The influent channel has a cross-sectional area that is 6 × 9.5 = 57 ft² (5.3 m²). Since the settled water is divided among the six first-stage filters and the six final-stage filters, the 100 mgd flow is split into two directions. The maximum flow velocity in the influent channel is therefore

$$(50 \times 1.55) \text{ cfs} \div 57 \text{ ft}^2 = 1.36 \text{ fps} \quad \text{or} \quad 0.4 \text{ m/s} \quad \text{O.K.}$$

(vii) The specifications for the filter media area as follows:

	Anthracite Coal (Top Layer)	Filter Sand (Bottom Layer)	Remarks
Effective size (mm)	1.0 ± 0.05	0.55 ± 0.02	For purchase specifications
Uniformity coefficient	≤1.5	≤1.5	For purchase specifications
Specific gravity	1.65 ± 0.05	≥2.62	After 24 h of soaking
Depth (in.)	20	10	After scraping off the top layer of fine particles

Notes.

1. When each medium is first placed in the filters, an extra 10% should be added to the depth to allow for the removal of the fine particles (scraping) after backwashing, as specified below. The process of scraping is vital for the proper functioning of the filter bed. If the fine particles are not removed from the surface of the bed, it will result in very short filter runs.

2. Following the initial placement of the sand, the filter should be backwashed twice at a rate of 22 gpm/ft² (55 m/h). After the first backwash, the top $\frac{1}{2}$ in. of medium should be removed. Repeat the backwash process and scrape an additional $\frac{1}{2}$ in. off the top. Thus, the final depth of the sand layer will be 10 in. (25 cm).

3. Place the coal layer on top of the sand layer and backwash the filter bed three times at 22 gpm/ft² (55 m/h), scraping off the top layer after

each wash. Remove the top 1 in. after the first wash, $\frac{1}{2}$ in. after the second wash, and $\frac{1}{4}$ in. after the third wash.

4. The AWWA filter material specifications (AWWA standards B100) should be used for the other necessary properties of the media. The gravel bed is composed of five layers of graded gravel and the total depth is 18 in. (45 cm). See item B(a) for the details.

For this example the filter underdrain system of choice is the precast concrete laterals with a triangular cross section (Figure 3.2.6-8). Dual lateral blocks or pipe laterals may be employed if the owner prefers their use. Regardless of the type of underdrain, the headloss should not exceed 12 in. (0.3 m) at the designed backwash rate. In addition, a uniform flow distribution must be maintained over the entire filter bed; the difference should be less than 5%. The higher the headloss, the deeper the filter structure. Precast concrete laterals that are 14.5 ft (4.4 m) long are placed at 12 in. (0.3 m) intervals and both sides of the sloped walls should have orifices that are $\frac{1}{2}$ in. in diameter and 3 in. on center.

Each filter cell is 32 ft (9.6 m) in length and there are 64 laterals in each filter.

The maximum backwash flow rate for each lateral is

$$21,120 \text{ gpm} \div 64 = 330 \text{ gpm} \quad \text{or} \quad 0.753 \text{ cts}$$

The cross-sectional area of the lateral is

$$(0.615 \times 0.6) \div 2 = 0.185 \text{ ft}^2$$

The flow velocity at the inlet of the lateral is

$$0.753 \div 0.185 = 4.1 \text{ pfs} < 4.5 \text{ fps} \quad \text{O.K.}$$

The total number of orifices ($\frac{1}{2}$ in. diameter) per lateral is

$$2 \times 58 = 116 \quad \text{or} \quad 7424 \text{ orifices per filter}$$

The flow rate from each orifice is

$$0.753 \div 116 = 0.0065 \text{ cfs}$$

The cross-sectional area of each orifice is

$$(0.5/12)^2 \times 0.785 = 0.00136 \text{ ft}^2$$

The backwash flow velocity through each orifice is

$$0.0065 \div 0.00136 = 4.77 \text{ fps}$$

The headloss through each orifice hole is

$$\Delta h = K \frac{v^2}{2g} = 2.4 \frac{4.77^2}{64.4} = 0.85 \text{ ft} < 1 \text{ ft} \quad \text{O.K.}$$

It is imperative that each lateral be anchored to the floor even if the weight of the lateral exceeds the normal uplift force that is generated during an average backwash. The orifices should be located in the upper portion of the inner triangular section and directed horizontally so that air will not be trapped in the laterals and so that the water jets are not directed toward the surface of the filter bed. Both ends of each orifice should be reamed to provide rounded edges, minimizing the headloss. Computer analysis of this type of underdrain reveals an approximate change of 5% in the flow distribution.

(viii) The designed backwash rate is the maximum wash rate and should only be used during the summer months when the water temperature is high. Based on Figure 3.2.6-4 and the specified filter media, the maximum backwash rate is 22 gpm/ft² (55 m/h). The maximum backwash flow rate is 21,120 gpm or 30.6 mgd (1.35 m³/s). However, the average backwash rate required to backwash one filter is approximately 19 gpm/ft² (48 m/h) or 26.2 mgd. This means that the plant flow rate should be at least 31 mgd, if the maximum backwash rate is used. Due to the design of the self-backwash filter, 27 mgd must be filtered to produce a backwash flow rate of 19 gpm/ft². Whenever the plant is operating at less than the minimum plant flow rate, required to backwash one filter, the required rate may be met simply by increasing the plant flow rate to the rate required to wash one filter.

Each filter is provided with four wash troughs composed of fiberglass reinforced plastic. The lightweight and durable troughs facilitate easy installation and simple adjustment of their level during construction and provide trouble-free maintenance.

The maximum flow rate for each trough is

$$(21,120 + 2880) \text{ gpm} \div 8 = 3000 \text{ gpm}$$

By using the trough sizing diagram (Figure 3.2.6-15) a trough with a width of 21 in. and an overall height of 24 in., with a half-circle shaped bottom, is selected.

The elevation of the trough should be high enough so that the media grains that are kicked up by the surface wash jets are not easily washed out during the simultaneous surface and backwash period.

The height (H_0) of the troughs above the filter bed is

$$(0.5 \times 2.5 + 2) > H_0 < (2.5 + 2)$$

$$3.25 \text{ ft} > H_0 < 4.5 \text{ ft}$$

Make it 4 ft

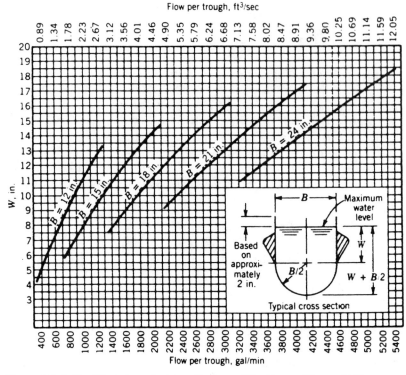

Figure 3.2.6-15 Wash trough sizing diagram. (Courtesy of Leopold Company.)

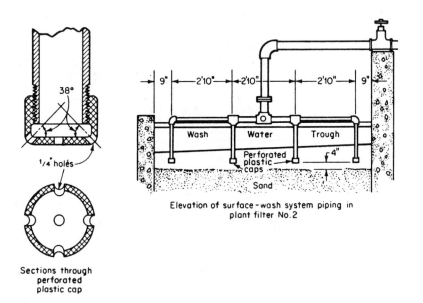

Figure 3.2.6-16a An example of the Baylis type surface wash system.

The spacing (S) of the troughs is

$$(1.5 \times 4) > S < (2.5 \times 4)$$

$$6 \text{ ft} > S < 10 \text{ ft}$$

Use a spacing of 8 ft on center

Auxiliary scouring will be provided by a surface wash system. The first choice is either a fixed nozzle or rotating arm type. The second choice is either a surface wash or a subsurface wash unit. Despite its effectiveness, the Baylis type of auxiliary scouring (Figure 3.2.6-16a), consisting of a series of vertical pipes with nozzle caps at each end, will not be adopted because an intense jet of water will directly hit the gravel bed and upset the stratification, whenever the caps are lost; several case histories exist for this problem.

In this application a fixed nozzle type of auxiliary scouring is used simply because of the fewer maintenance requirements. Three inch diameter schedule 80 PVC or stainless steel lateral pipes should be horizontally arranged at 4 ft intervals and 3 in. above the surface of the bed. The orifices must be $\frac{1}{4}$ in. (6 mm) in diameter at approximately 8 in. intervals. The jets should be directed 25° below the horizontal plane so that the coal and sand interface will be scoured.

The subsurface wash systems have a slightly better performance when used in conjunction with dual media beds. Yet, the experience over the last 20 years has shown that the difference is marginal. In fact, the subsurface wash system potentially has problems with the nozzles becoming clogged whenever the rubber caps are lost. The flow rate for each filter during surface wash is 2880 gpm or 6.6 cfs. Since there are 8 laterals in each cell and 16 laterals in each filter, the flow rate to each lateral is

$$2880 \div 16 = 180 \text{ gpm} \quad \text{or} \quad 0.41 \text{ cfs}$$

Therefore, the flow velocity in the 3 in. lateral is 8.2 fps, which is a reasonable figure.

Each lateral in the surface wash system should have $\frac{1}{4}$ in. orifices drilled in a staggered fashion at 8 in. intervals on either surface of its 14.5 ft length. The entire width of the filter cell, 15 ft, may not be covered by the orifices due to the maneuvering space that is required to install fittings, such as elbows and endcaps, and possible repair of the piping system.

The total number of orifices per lateral is

$$[(14.5 \times 12) \div 8] \times 2 = 44$$

The flow rate through each orifice is

$$0.41 \div 44 = 0.00932 \text{ cfs}$$

The area of each orifice is

$$(0.25 \div 12)^2 \times 0.785 = 0.00034 \text{ ft}^2$$

while the flow velocity at each orifice is

$$0.00932 \div 0.00034 = 27.4 \text{ fps}$$

In order to satisfy the design criteria, the jet velocity for surface scouring should be between 20 and 30 fps (6–9 m/s).

The loss of hydraulic head through the orifice is

$$\Delta h = 2.4 \left(\frac{27.4^2}{64.4} \right) = 28 \text{ ft} \quad \text{or} \quad 12 \text{ psi}$$

The source water may be either settled or filtered water. The latter is more common and must be supplied by a pump. The total dynamic head of the pump is a function of the size and length of the supply pipe and the elevation of the water source. The pump generally has a minimum total dynamic head of 65 ft (20 m).

(ix) The plan, sections, and details of the designed surface wash system are shown in Figure 3.2.6-16b.

(x) The elevation of the effluent level control weir is an important factor in setting the filter backwash rate control weir. The hydraulics of the filter backwash is evaluated on a wash rate of 22 gpm/ft² (48.2 cfs).

The 36 in. butterfly valve (BV) entrance loss, BV loss, and exit loss are given by

$$h_1 = (0.5 + 0.25 + 1.0) \frac{v^2}{2g} = 1.75 \frac{(6.8)^2}{2g} = 1.25 \text{ ft}$$

Headloss in the lower gullet is

$$v = 48.2 \div 5^2 = 1.93 \text{ fps maximum} \qquad \text{(negligible loss)}$$

Entrance to the underdrain lateral is

$$h_2 = 0.5 \frac{v^2}{2g} = 0.5 \frac{(4.1)^2}{2g} = 0.13 \text{ ft}$$

Headloss in the underdrain lateral is negligible.

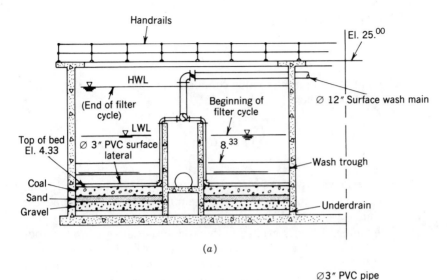

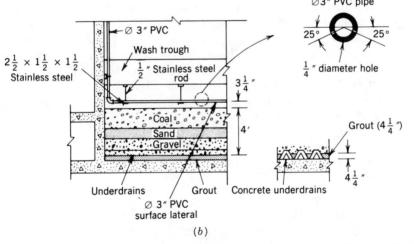

Figure 3.2.6-16b Filter structure design. (a) Cross section of the filter. (b) Details of the surface wash system.

Headloss at 0.5 in. orifices of the underdrain is given by

$$h_3 = 2.4 \frac{(4.77)^2}{2g} = 0.85 \text{ ft}$$

Headloss through the gravel bed is

$$h_4 = 0.38 \text{ ft}$$

obtained from a pilot study conducted at 16°C.

Headloss through the filter bed is

$$h_5 = 1.42 \text{ ft}$$

obtained from a pilot study conducted at 16°C.
Total headloss is

$$\sum_{i=i}^{i=5} i = 4.03 \text{ ft}, \quad \text{say } 4.0 \text{ ft}$$

Since a headloss of approximately 4 ft (1.2 m) will be provided during
one backwashing at a rate of 22 gpm/ft^2 (55 m/h), the effluent level
control weir should be set so that the water level in the effluent
channel is 4 ft (1.2 m) higher than the top of the washwater troughs.
In order to adjust the overflow elevation to a 4 ft span, two weirs,
10 ft long (3 m), and two down-opening slide gates are installed in
the wall of the filter effluent channel. As illustrated in Figure 3.2.6-
16c, Section A–A, the top of the notched concrete wall should be
set at an elevation of 10.25, that is, 1 ft 11 in. above the top of the
wash troughs, which are located in the filters. The operator may
manually raise the down-opening slide gates from an elevation of
10.25 to 13.25 whenever the backwash rate must be adjusted. Under
these circumstances the gate acts as a weir. The required hydraulic
head for a plant flow rate of 50 mgd, using two weirs that are 10 ft
(3 m) in length, is

$$H^{1.5} = \frac{(50 \times 1.55) \div 2}{3.33 \times 10} = 1.16$$

Thus, H is 1.1 ft.

The top edge of the slide gates should roughly be set at an elevation
of 11.23 (say 11.25) if the plant is to produce a wash rate of 22 gpm/
ft^2 (55 m/h) when the plant flow rate is 50 mgd.

The effluent channel is 6 ft (1.8 m) wide with an average water
depth of 12 ft (3.6 m). At a plant flow rate of 50 mgd, the maximum
flow velocity in the channel is

$$(50 \times 1.55) \text{ cfs} \div (6 \times 12) \text{ ft} = 1.08 \text{ fps} < 2 \text{ fps}$$

Note: As described earlier, a new filter bank, identical to this filter
bank, will be added at a future date. The settled water will flow into
the filters between the two filter banks. In the future, this second
filter bank will have an identical filter effluent level control system
and the filter effluent from the two banks will be combined. The flow
rate will be measured by a Venturi type meter.

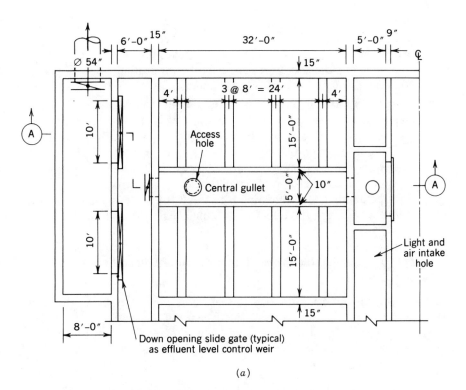

(a)

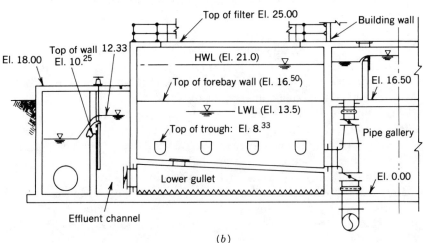

(b)

Figure 3.2.6-16c Filter structure design. (a) Plan at elevation 17.00. (b) Section A–A.

(xi) The rate of backwash flow is measured by a Venturi type meter which is placed in the 54-inch filter efflueut main. When one filter is undergoing backwash, the meter will indicate a decrease in flow rate. This decreased flow rate is the actual wash rate for the filter. The backwash rate may be adjusted by either raising or lowering the down-opening slide gates of the filter effluent channel.

(xii) Alternative 1 of Figure 3.2.6-13 should be adopted if filter-to-waste is practiced at the beginning of each filter cycle or if there is continuous sampling of filtered water from each filter. The design in Figure 3.2.6-16c should be modified to provide a service gallery. The volume of the waste water can be quite large; however, the quality of this water is quite good. Therefore, it is better to recycle it back to the filter influent, rather than directing it to the wash-waste holding tank and thereby reducing the effective holding capacity of the tank.

Example 3   Selecting the Proper Combination of Filter Media for Multimedia Beds

Given   A dual media filter bed is to be designed. However, the local regulatory agency has limited the effective size of the sand to 0.45 mm and has restricted the uniformity coefficient to less than 1.7. The available filter medium, anthracite coal, has a specific gravity of 1.63, determined by the AWWA's standard test method.

Determine   The effective size and uniformity coefficient for the coal media.

Solution   First, the design engineer should select the lower uniformity coefficient of sand; this figure may be obtained from the supplier. Once this is obtained, the optimum filter performance can be anticipated and the sand and coal mixing zone can be reduced.

Any supplier of filter media should be able to provide sand and coal with a uniformity coefficient of 1.5 without increasing the unit cost. However, a uniformity coefficient of less than 1.3 should not be specified simply because either the suppliers will not bid on the project or the unit cost may be unjustifiably high with few compensating benefits.

Engineers should set the sand specifications at an effective size of 0.45 mm, a uniformity coefficient of 1.5, and a specific gravity of 2.63. The matching coal size may be determined by either computation or by using the nomograph:

$$\frac{d_1}{d_2} = \left(\frac{\rho_2 - \rho}{\rho_1 - \rho}\right)^{0.667}$$

where d_1 = 0.45 mm,
$\quad d_2$ = the size that must be determined,
$\quad \rho_1$ = 2.65, the density of particles of size d_1,
$\quad \rho_2$ = 1.63, the density of particles of size d_2,
$\quad \rho$ = 1.0, the density of the fluid.

$$\frac{0.45}{d_2} = \left(\frac{1.63 - 1}{2.63 - 1}\right)^{0.667}$$

or d_2 = 0.85 mm, the effective size of coal

Alternatively, Figure 3.2.6-4 may be used. First, locate the size of 0.675 on the horizontal scale. Then, move vertically to a S.G. of 2.63, which may be established between 2.60 and 2.65 for sand. Move horizontally to the right to a S.G. of 1.63, which is assumed to be between 1.6 and 1.65 of anthracite coal. Finally, move down vertically to the horizontal scale and find 1.27. The effective size is therefore 1.27 ÷ 1.5 = 0.85 mm.

Operation and Maintenance   This section discusses the various aspects of evaluating regular filter performance, optimizing the pretreatment process, performing jar test procedures, and evaluating and adjusting the filter washing process. Other important operational parameters and common filter problems are also covered.

Regular Filter Performance   In order to optimize the filtration process, plant operators should routinely evaluate the filter performance. The three indicators used in evaluating plant performance are the filtered water turbidity, the length of the filter run, and the ratio of the volume of backwash water used to the volume of the filtered water. Under normal conditions, the filtered water turbidity should always be less than 0.5 NTU, with 0.2 NTU as the target turbidity level.

The second item, filter run length, is a function of size of filter medium and the influent water quality, which in turn is a function of the solids content, water temperature, filtration rate, and the condition of the filter bed. A treatment plant that is both well designed and operated should have a filter run length longer than 24 h whenever the filtration rate is less than 6 gpm/ft^2 (15 m/h). If a reverse graded filter bed is selected, the filters should be run for an average of 2 days at a rate of 4–6 gpm/ft^2 (10–15 m/h), as long as the treatment process is not a direct or in-line type of filtration.

Shorter filter run lengths can be attributed to eight major reasons: the accumulation of fine media particles on the surface of the filter bed, an effective size of filter media which is too small for the filtration rate, too much floc and suspended matter (in the filter influent), a filter bed that is dirty and full of mud balls, an overabundance of filter clogging algae in the water source, air binding, a higher than optimum dosage of polymer (as either a flocculant aid or filter aid), and recycling substandard clarified filter wash-waste to the filter influent. With respect to direct and in-line filtration, extremely short filter runs can easily be induced by unusually high levels of suspended matter in the raw water and/or overfeeding of coagulant (i.e., over 6 mg/L of alum).

One of the best indicators of filter performance is the ratio of the water used for filter washing to the amount of water that is filtered prior to filter washing. Under normal conditions, this ratio is less than 3%; a ratio that is less than 2% is considered to be very good. If the ratio exceeds 5%, the filter performance is judged poor. Engineers should realize that filter performance is much better during months with warm ambient temperatures due to faster floc formation and quicker floc settling in the settling tanks. A well maintained filter may have a ratio of 2.5–3% during the winter, but only 1.5–2% in the summer.

Another indicator of filter performance is the volume of water that is filtered per unit area of filter bed during a filter run. This is also referred to as the unit filter run volume (UFRV). The UFRV is the product of the filtration rate (gpm/ft^2) and the filter run length (minutes) and is expressed as gallons per square foot of filter bed area. In general, a UFRV that is less than 5000 gal is unacceptable because of the extremely short filter run length. A UFRV that is over 10,000 is indicative of normal filter performance.

Pretreatment Optimization   Plant operators have two methods by which they can optimize filter performance. The first involves the coagulant, while the second concerns the mixing energy level and mixing time. In both cases, jar tests are used to evaluate the various factors, especially for the conventional treatment process.

In order to optimize the coagulant, the operator must evaluate both the type of coagulant and the dosage. Alum and polymerized aluminum chloride generally do not produce floc that is physically strong enough to endure the high-rate filters. Therefore, small amounts (15 to 30 µg/L) of anionic or nonionic polymer are usually added to strengthen the floc and to condition the filter beds. Ferric iron salts, such as ferric chloride or ferric sulfate, should always be evaluated as a primary coagulant whenever the raw water has a pH that is over 7.5 (high) and an alkalinity that is over 50 mg/L. The smaller dose of ferric coagulant produces floc that is both physically stronger and heavier than those produced with alum. Moreover, ferric floc is virtually insoluble in water over a wide range of pH: a range of 5–11 for ferric floc, compared to 5.5–7.2 for alum floc.

Whenever the required dosage of metallic coagulant is low (less than 6 mg/L), the solubility of the floc becomes an important factor in the filtered water quality. Many treatment plants using a low dosage of alum in treating water with a high pH exhibit high concentrations of residual aluminum in the filtered water. Treatment plants that practice direct filtration or in-line filtration should therefore give special consideration to this issue. Plant operators should remember that a high dosage of alum during periods of high turbidity also creates a high concentration of residual alum because of a suppressed pH (a pH that is less than 5.5). Consequently, proper pH control by alkali chemicals is required during the flocculation process.

Lastly, evaluation of cationic polymers, approved by the EPA for the purpose of water treatment, is essential in selecting the type of coagulant to be employed. The positive aspects of the cationic polymers are that they are not affected by the pH of the process water; they produce a tougher floc; and the sludge is easier to handle than those produced by alum. Since the effective dosage is small, rarely exceeding 2 mg/L, the volume of sludge that must be handled and disposed of will be reduced.

In many cases, a combination of alum and cationic polymer is used because of better clarification and cost effectiveness. Most cationic polymers cost approximately eight times more per unit weight of the product than alum. Consequently, the polymer dosage is usually limited to 0.5 mg/L and is supplemented by 2–4 mg/L of alum. This combination provides both good performance and cost effectiveness when applied to a direct filtration process. As indicated earlier, final selection of the appropriate coagulant is a function of the effectiveness of the chemical(s) as a coagulant, the cost of treating the water, the problems associated with disposal of the sludge, and the ease with which the coagulant is handled by plant operators.

The second important factor in optimizing the coagulant is the determination of the optimum dosage. Generally, the optimum coagulant dosage for a water treatment plant is the dosage that results in a filter influent turbidity that is less than 1 NTU or the dosage that produces the lowest supernatant turbidity after jar testing. Due to cost considerations, plant operators usually select the minimum dosage required to produce the second lowest supernatant turbidity (based on jar test results). This practice is justifiable for rapid sand filtration because a straight relationship exists between the filtered water turbidity and the settled water turbidity. However, modern filter beds, such as the reverse graded beds and coarse, deep beds, require a very different way of thinking. The modern high-rate filters require polymer to be applied to the filter influent to maximize their efficiency; the second flocculation (incidental flocculation) actually occurs in the filter inlet channels, pipes, and filter bed (primarily in the filter bed). The polymer may be used with or without metal salt coagulants, such as alum, and the dosage that produces a supernatant turbidity (after jar testing) of less than 1 NTU is generally not the best for the overall filter performance. Data obtained from operational high-rate filters indicate that the optimum coagulant dosage, that which produces properly coagulated water (does not produce large floc), is only one-half to one-third the optimum dosage required by rapid sand filtration. The proper coagulant dosage for high-rate filters generally yields a supernatant turbidity of 2–4 NTU by the jar test. Figure 3.2.6-17 summarizes the relation between proper coagulant dosage and types of filter (treatment process).

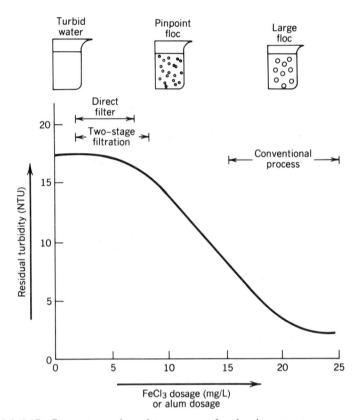

Figure 3.2.6-17 Proper coagulant dosage range for the three treatment processes.

The most difficult situation in selecting the optimum coagulant dosage is encountered with the in-line and direct filtration processes. The dosage at which pinpoint floc is produced is already an overdose that will induce a short filter run. Several equipment manufacturers and researchers have suggested that, for these types of filtration process, plant operators use pilot filter column tests to determine the proper coagulant dosage. However, experience has shown that very few operators of plants equipped with the pilot filter columns use them regularly enough to warrant the cost. Plant operators respond that the pilot columns are simply not practical enough.

Another option is to use the small glass tube pilot filter. This type of test filter is composed of a half dozen 1 in. (diameter) glass tubes. The glass tubes are filled with small sized sand to a depth of 1 ft (0.3 m). The filterability of the supernatant, obtained from jar tests, is determined by passing it through the pilot filter. Yet, plant operators rarely employ this apparatus on a regular basis because it is both unreliable and time consuming. The small glass pilot filter yields unreliable data because the sand in the glass columns is not ripened, as in actual filter beds. An alternative to the sand has been employed by the author and others. In this case, filter paper (Whatman No. 1) is used as the filter medium. This option is much quicker and the results correlate reasonably well with actual filter bed performance. It must be emphasized that the ordinary jar test procedure is a good control tool for the plant process but only for conventional processes. It does not work well for direct or in-line filtration, unless the flocculated water is first filtered through filter paper, such as Whatman No. 1 or a similar type of paper with a pore size of approximately 1–2 µm.

The second method in optimizing filter performance involves the mixing energy level and the mixing time ($G \times t$). Although these parameters are supplied to the operator by the design engineer, there is still some room for adjustment.

Recommended Jar Test Procedures   Although some plant operators believe that jar testing does not yield reliable results, it is a useful test for controlling coagulant dosage. Factors such as the mixing conditions, alum solution strength, the method of chemical feeding, and the flash mixing system all affect the correlation between the jar test results and the actual plant performance.

If the mixing conditions ($G \times t$) of the jar tests are not similar to the actual process, the results may be quite different. Most modern treatment plants use a 50% solution of liquid alum. However, a 1% solution of liquid alum is often used in the jar tests, potentially distorting the correlation between the jar test and the actual plant process. Experience has shown that a 0.1% solution of alum is better suited for the jar tests as long as it is freshly made (Kawamura, 1973). The feeding condition of the alum (i.e., continuous feed or intermittent feed), the actual type of flash mixing system, and the presence of flow short-circuiting in the actual tank also greatly affect the correlation between the jar test results and the plant performance.

Refer to Appendix 7 for a recommended jar test procedure.

Evaluation and Adjustment of Filter Washing Procedure   In order to optimize the filtration process, the condition of the filter bed must routinely be evaluated. Operators may maintain the proper bed conditions by adjusting the filter washing procedure. There are three basic methods by which to evaluate the effectiveness of the filter washing procedure: (1) visually inspecting the filter bed before and after filter washing, (2) measuring the turbidity of the backwash waste at 1 min intervals after initiating backwash, and (3) core sampling of the filter bed both before and after filter washing.

A filter bed that is properly maintained is clean (i.e., practically devoid of mud balls) but is also in ripened condition; the media grains of a ripe filter bed are coated with a proper amount of floc or polymer. During visual inspection of the filter bed, plant operators must look for mud balls, cracks, and any other unusual conditions. The bed is in poor condition with mud balls and mud accumulation if the operator finds lumps and cracks on the surface of the filter bed prior to filter washing and lumps even after washing. The most obvious indicators of a dirty filter bed are the existence of worms and debris on its surface. If a filter cell exhibits any mounding, concaved areas, or obvious shifts in the media, there is either a scouring action by filter influent or a maldistribution of the backwash water and boiling action in the filter bed. The operator should immediately investigate the causes of these unfavorable phenomena.

The second method, measuring the turbidity of the backwash waste, is used to evaluate the effectiveness of the filter washing procedure in removing floc from the filter bed. A valuable tool in assessing the washing efficiency and the optimum duration of the filter washing procedure is the turbidity profile of the backwash waste plotted against wash time. A low profile with a low peak curve is indicative of ineffective washing. A high profile curve with a high peak curve is characteristic of effective washing. Figure 3.2.6-18 illustrates these two types of curve.

Many plants wash their filters for an excessively long period of time, until the operator can clearly see the surface of the bed. This method of washing is actually detrimental to the filtration process because the filter often exhibits a distinct initial turbidity breakthrough, for a period of 15–30 min or longer, until the bed regains its ripened condition. Studies based on both pilot scale and actual filters indicate that this initial turbidity breakthrough may be minimized or even eliminated by terminating filter washing when the turbidity of the wash waste ranges from 10 to 15 NTU; this usually occurs after 5–6 min of regular washing (not air-scour washing). Adverse effects, such as a shorter filter run length and the accumulation of mud within the filter bed, will not occur if the filter wash rate is properly selected and a surface wash is implemented.

The energy input generated by the backwash process, and by either surface wash or air-scour, may be estimated by using the following equations:

Energy input due to the backwashing process:

G ranges from 300 to 400 s^{-1}

$$G = \frac{\rho \, g v_b \, \Delta h}{\mu L}$$

Energy input due to the water jets of surface wash:

G ranges from 1000 to 1300 s^{-1}

$$G = \frac{\rho g v_s \, \Delta p}{\mu \alpha L}$$

Energy input generated by air scouring:

G ranges from 400 to 500 s^{-1}

$$G = 9 \left(\frac{Q_a \log(H + 34) \div 34}{\mu V} \right)^{0.5}$$

where ρ = fluid density,
   g = acceleration of gravity (32.2 ft/s^2)
   v_b = backwash rate (fps),
   Δh = headloss across the bed by backwash (ft),
   μ = absolute viscosity of water (lb-s/ft^2),
   L = depth of the fluidized (expanded) bed (ft),
   v_s = surface wash rate (fps),
   Δp = hydraulic head applied to the bed by the water jets (ft),
   α = 0.25 for regular surface wash agitators and 0.5 for dual-arm agitators,
   Q_a = air flow rate (cfm),
   H = water depth (ft),
   V = volume of the filter bed (ft^3).

When backwash is applied to a filter bed, the headloss begins to increase with an increase in the wash rate. Yet, when a specific backwash rate is reached and the wash rate continues to increase, the hydraulic headloss will begin to decrease due to fluidization of the filter bed. This characteristic is illustrated in Figure 3.2.6-19. Data obtained from both pilot and actual filter studies show that the optimum backwash rate is approximately twice the wash rate at which maximum headloss is attained. For instance, in Figure 3.2.6-19 the proper wash rates for sand II and sand I are 20 and 15 gpm/ft^2, respectively. Beyond these optimum rates there are no appreciable benefits, primarily because the filter bed is highly fluidized; the media grains are no longer in contact with each other and the scrubbing action of the filter bed is therefore eliminated. This same logic can be applied during the rinse period after air-scour wash. However, operational experiences indicate that the backwash rate at which the maximum headloss is produced is the most effective concurrent washwater rate because the majority of the sludge, attached to the media grains, is stripped during air-scour and transported into the water. The recommended filter washing procedures, with auxiliary scouring, are illustrated in Figure 3.2.6-20. The mud deposition profile of this figure, both before and after filter washing, clearly demonstrates the effectiveness of the washing procedure. Based on the above discussion, plant operators should be able to evaluate their present washing practices and make the necessary adjustments.

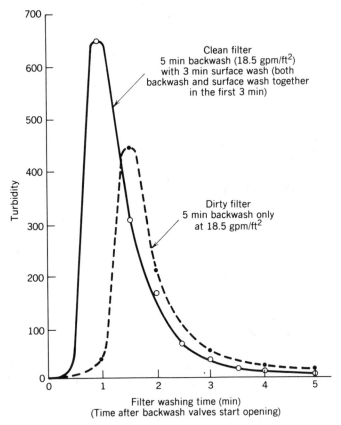

Figure 3.2.6-18   Backwashing waste characteristics (conventional sand filters). *Note:* The water level in the filter cell was lowered approximately 6 inches above the top of the filter bed prior to filter washing.

The third method used to evaluate the effectiveness of the washing conditions is core sampling. Core sampling allows plant operators to establish a mud profile for the entire depth of the filter bed. Core samples should be obtained from selected filter beds at regular intervals according to a rigorous schedule. This procedure is very important because it allows the operator to check the actual depth of the medium, to obtain a medium size distribution profile across the entire depth of the filter bed, to evaluate filter washing efficiency, to check the filter bed conditions, and to evaluate the movement of the filter gravel. These data can only be obtained by taking core samples from multiple locations within the filter cell. Moreover, it is crucial to obtain samples at various depths from the same location. If one core sample is taken across the entire depth of the filter, from one location, it may not be representative of the filter bed since it does not show the accumulation of fine medium at the surface of the bed.

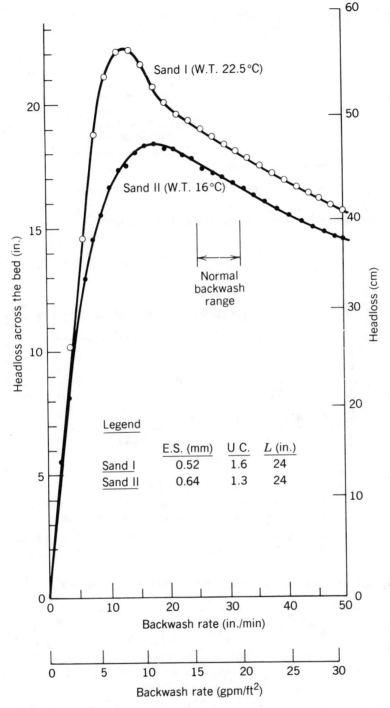

Figure 3.2.6-19 Relation between backwash rate and headloss. (Adapted from P. A. Norman, *Backwash Instability in Rapid Sand Filters*, M.S. thesis, University of New South Wales, Australia, 1970.)

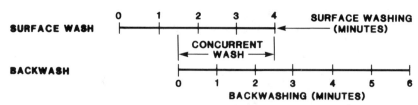

NOTES: 1. Draw down water level to approx. 6 inches above the bed before surface wash starts.

2. Longer concurrent wash will result in higher media loss.

A. RECOMMENDED SURFACE AND BACKWASH SEQUENCE

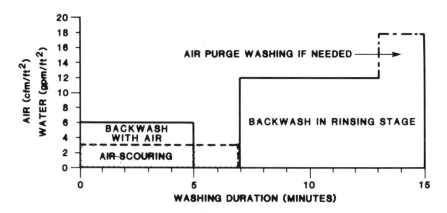

NOTES: 1. Draw down water level to approx. 2 inches above the bed.

2. The first slow backwashing must be stopped before the water level gets up to two inches below the top of wash waste weir.

3. The backwash in rinsing stage would require higher than 12 gpm/ft^2 to purge air trapped in the bed for last 2 to 3 minutes.

B. NORMAL AIR-SCOURING WASH FOR COMMON SIZE MEDIA BED

Figure 3.2.6-20 Recommended sequence of auxiliary washing.

For ordinary filter beds, the most simple and practical sampling tool is a thin walled pipe that is 5 ft (1.5 m) long and approximately 1.5 in. (3.8 cm) in diameter. The inner wall of the tube must have an appropriate amount of friction; a sample of the thin layer of the filter bed will not remain in the tube if the inner surface is too smooth. Therefore, PVC piping is not suitable for this type of sampling. The most appropriate type is the thin walled galvanized conduit pipe used in electrical wiring. A minimum volume of 15 in.^{3}, 250 mL per sample, is required for proper evaluation of the filter bed.

The recommended sampling techniques and a simple quantitative analysis of the mud that accumulates on the surface of the filter bed are described in Appendix 8.

Mud Deposition Profile   Both the effectiveness of the filter backwash process and the condition of the filter bed may clearly be evaluated from the mud profile graphs. For example, a filter without a surface wash system cannot adequately wash out floc that is retained during a filter cycle, thus allowing the sludge to accumulate slowly. Conversely, a filter with a surface wash system is effectively cleaned throughout the entire depth of the bed, as in-dicated in Figure 3.2.6-10. The mud profiles of dual media beds from two different treatment plants are illustrated in Figure 3.2.6-11. The graph on the left shows a slight increase in mud accumulation within the coal–sand inter-face, despite the presence of a surface wash system. The filter in the right graph has a subsurface wash system and the interface zone virtually has no increase in mud accumulation.

A filter bed that is properly conditioned is also in a ripened stage; each medium grain is coated with a thin film of coagulant hydroxide or polymer. A bed that is too clean usually exhibits a distinct turbidity breakthrough at the beginning of each filter cycle, lasting anywhere from 30 to 60 min. Based on the author's experience, a turbidity of 30–60 NTU, determined after filter washing by the sludge retention analysis test method, is indicative of a clean, as well as ripened, bed. A filter bed that is slightly dirty, less than ideal but no need to concern about, will have a turbidity that is between 60 and 120 NTU. However, a filter bed exhibiting a turbidity over 120 NTU is dirty, and both the filter washing system and filter washing conditions must be reeval-uated. A turbidity that is over 300 NTU is indicative of a mud ball problem.

The mud deposition analysis procedure, described in past articles written by the author (1975), advises using 50 g samples of filter media instead of 50 mL of sample, primarily because of time constraints. This procedure is fine with respect to monomedium filter beds; however, it is more accurate to use 50 mL of sample when analyzing dual or multimedia beds because the volume of each medium is quite different for the same weight of sample. This drastic difference in volume is due to the difference in specific gravity for each medium.

When mud balls are found in the filter bed, measure the percent volume of mud balls in the bed. If the percentage is less than 0.1%, the filter bed is clean. If the mud balls are 0.1–0.5% by volume, the filter bed is in good condition; 0.5–1.0% is indicative of a fairly clean bed; while 1–5% indicates the filter bed is in bad condition. If the volume of mud balls is over 5%, the bed must be replaced with new media.

Restoration of the Filter Bed   Filter bed medium is commonly lost after repeated filter washings. The rapid sand filters generally do not lose a significant amount of sand; a normal loss of 1–2% of the total bed depth (original) occurs annually. Yet, the loss of anthracite coal is much more significant; an annual loss of 5–7% is not unusual for ordinary dual or trimedia beds.

Excessive media loss is often caused by the following reasons: air-binding in the filter bed; an unnecessarily long and overlapping surface wash and backwash (over 3 min); an excessive backwash rate, especially during months of cold water temperatures; a mismatch in the size of the coal and sand; a wash trough elevation that is too low; improper sequential control of both the air-scour and rinsing backwash processes; and leakage of the filter media through the underdrain system. If the media loss is excessive, the plant operator should first locate and correct the cause. The original depth of the filter bed should be restored by adding the appropriate media to the surface of the filter bed and backwashing the filters a minimum of two runs; this will fluidize the bed and ensure a proper gradation profile.

Many engineers and plant operators ask what the tolerable amount of media loss should be. Losing up to 20% of the original depth or a loss of 6 in. (15 cm) in depth, whichever is the first, is generally a tolerable loss if total bed depth is approximately 30 in. Filters with surface wash systems begin losing the advantages of surface wash when the distance between the top of the bed and the elevation of the jet nozzles exceeds 6 in. Filter performance also begins to decrease when media loss is greater than 20% of the original depth.

The purchase of replacement sand for rapid sand filters is quite simple; specify the same size medium grain as the original but with a lower uniformity coefficient (1.5) to avoid the short filter run lengths characteristic of fine medium grains. However, dual and multimedia beds must have the replacement media properly matched in physical characteristics so that all the layers will fluidize to the same degree during backwash. The specific gravity and size of both anthracite coal and garnet grains are very important considerations. Based on past experience, the specific gravity of sand is always near 2.63 but this is not the case for commercially available anthracite coal and garnet grains. Over the past 20 years the specific gravity of anthracite coal, when used as filtering material, has ranged from 1.50 to 1.72. Therefore, the size of the available anthracite coal must be adjusted to match that of the sand. Figure 3.2.6-4 will help determine the complementary media sizes. First, locate the representative sand size ($0.55 \times 1.5 = 0.83$ mm). Then, move vertically up to the specific gravity of the sand (2.63). Next, move horizontally to the right, to a specific gravity of 1.65 (assuming that this is the available type of anthracite), and vertically down to 1.53, the representative particle size of the coal. Assuming that the uniformity coefficient is specified as 1.5, we find that the effective size of the coal is

$$1.53 \div 1.5 = 1.02 \text{ mm}, \quad \text{say 1 mm}$$

The following formula may also be used to compute the matched size of two different types of filter bed medium:

$$\frac{0.55}{d_2} = \left(\frac{1.65 - 1}{2.63 - 1}\right)^{0.667}$$

or $d_2 = 1.015$, say 1.0 mm

If other kinds of filter material, such as garnet or GAC, are to be used in conjunction with sand, the appropriate media sizes may be found in the same manner. As demonstrated in Example 1 of the Design Problems, it is imperative that the media specifications do not allow for a wide range of media sizes and specific gravities.

Gravel Bed Upset in the Filter Major disturbances in the gravel bed are usually caused by errors either in the design phase, in the construction phase, or during operation. Plant operators have no control over the first two. Improper design is the responsibility of the design engineer. The obvious design flaws are improper specifications for the gradation and thickness of each gravel layer. Based on case histories, disturbances in the gravel bed and leakage of filter media through the supporting gravel bed are generally caused by improper construction of the gravel bed, including the omission of one or even two layers. Gravel bed upsets are also caused by improper operation of the backwash process.

The most common operational error is to quickly initiate full-scale backwashing for a filter containing no water. This type of error usually occurs when the filter is being filled with water for the first time, by an individual who does not have adequate knowledge of the filter, by the contractor, or by an inexperienced operator. The composition of the gravel bed may also be disturbed by backfilling a filter that has been drained, at full backwash rate without slowly operating the backwash valve. In either case, the combination of the quickly escaping compressed air (in the filter underdrain system), the short-circuiting of the water, and the escape of air can cause major disturbances in the arrangement of the gravel bed. It may even damage or break the underdrain system. Once the gravel bed is disturbed, it is irreversible, until reconstructed. It is therefore essential to slowly backfill the filters until the water level covers the surface of the filter bed. The initial backfilling rate should not exceed 5 gal/ft^2. The backwash valves must also be opened very slowly to avoid lifting the filter media bed off the gravel layer. Otherwise, an open space will be formed between the media bed and the gravel bed. If this should occur, the filter bed will break at several points and a boiling action will occur and the upper portion of the supporting gravel bed will be disturbed (Figure 3.2.6-21). If the plant operator notices a distinct boiling action during filter backwashing, the filter must be drained, and the condition of both the gravel bed and the inside of the underdrain system must be inspected. The filter bed should be dug up by using a 4 ft by 4 ft bottomless box as a shoring frame. Failure of the gravel bed is also clearly evidenced when filter media accumulation is found both inside and downstream of the underdrain system.

During the mid-20th century, John Baylis proposed the use of a reverse graded gravel bed. Although this type of bed is effective in preventing the movement of gravel, it has problems with heavy mud accumulation in the fine gravel layers of the gravel bed (middle layer) for ordinary backwashing filters.

Air-Binding   Air-binding is a phenomenon in which large amounts of air bubbles accumulate in the filter bed. Air dissolves in the water near or even over the saturation point. The pressure in the upper portion of the filter bed

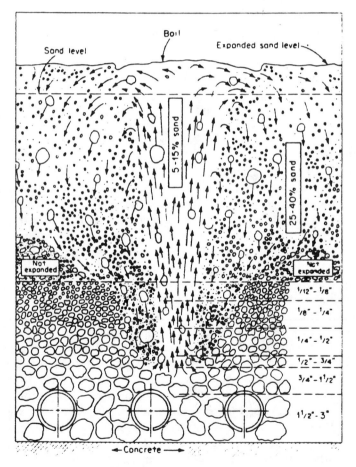

Figure 3.2.6-21   Filter gravel bed upset by operational error. (From AWWA, *Water Quality and Treatment*, 3rd ed., McGraw-Hill, New York, 1971.)

may be reduced to below atmospheric pressure due to the creation of negative pressure during the latter stages of filtration. When this occurs, the air in the water no longer remains dissolved but accumulates within the filter bed, rapidly increasing the headloss throughout the bed. Figure 3.2.6-22 illustrates the development of negative pressure within the filter bed. If a high water level is maintained in the filter, air-binding will be delayed. Therefore, a shallow filter with a water depth that is 4–5 ft (1.2–1.5 m) above the filter bed and a filter that lacks an effluent water control weir tend to promote air-binding. This type of filter is shown in Figure 3.2.6-1 and represents a constant-rate filtration scheme.

When air-binding occurs in the filter bed, gaseous air violently escapes when the filter effluent valve is closed. This phenomenon causes media loss. Therefore, the water in the filter should be lowered to a level below the top of the wash troughs and the filter waste valve should not be opened until the majority of the air has escaped from the filter bed. Moreover, filter washing should be delayed until most of air has escaped.

The degree of air-binding may be reduced or even eliminated if filter washing is frequently initiated, whenever the headloss reaches 4–5 ft. This practice prevents the creation of negative pressure in any part of the filter bed. In many cases, air-binding tends to occur during certain seasons such as spring, when there is a high degree of dissolved air. Air-binding may also be a function of the degree of aeration due to hydraulic power plant activity upstream of the plant or the degree of cascading at the intake.

Air-binding will not occur if the filters are properly designed. For example, the declining-rate filters and the rising level self-backwash filters shown in Figure 3.2.6-1 can eliminate air-binding due to a high water depth (above the bed) and/or because the effluent water level control weir of the clearwell is situated high enough so that negative pressure is not created during a normal filtering cycle.

Filter Performance Surveillance   The four basic tools used in monitoring the filter performance are the turbidimeter, pH meter, particle sizing and counting equipment, and filter headloss profile monitor. Each type is discussed in detail.

Turbidimeters are used not only to monitor the filter effluent turbidity of individual filters but also to obtain samples from the middle (depth) of the filter bed and the filter wash waste. Monitoring the turbidity of the water at the coal–sand interface, located approximately half-way down the filter bed, is recommended because the turbidity of this water can give the operator advance warning of a turbidity breakthrough across the filter bed. The filter turbidity of the wash waste should be measured because it is a good parameter for adjusting the duration of filter washing.

Nephelometric measurement of turbidity is required because the standard turbidity measurement in the United States is expressed in NTU. The nephelometric technique measures the degree by which the incident light beam

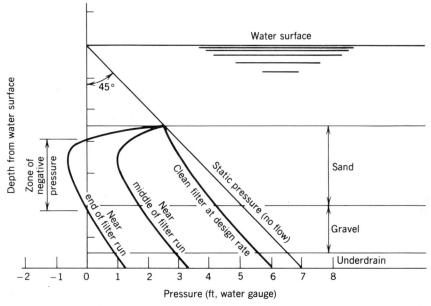

(a) Pressure versus depth in a gravity filter at various times

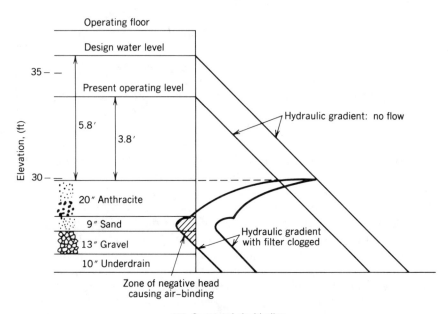

(b) Causes of air-binding

Figure 3.2.6-22 Air-binding in the filter bed. (From H. E. Hudson, *Water Clarification Process*, Van Nostrand Reinhold, New York, 1981 bottom Figure only.)

is scattered by the suspended particles; the light beam is at a right angle to the particles. Several companies manufacture turbidimeters but the most popular model is produced by the Hach Company (United States). Extreme care must be taken not to introduce any air bubbles to the turbidimeter because this will cause inaccurate readings. Each individual filter effluent line should be monitored by a reliable turbidimeter that is also capable of measuring very low levels of turbidity; the new EPA drinking water standards demand a turbidity of less than 0.5 NTU 95% of the time. If this primary standard is violated, a harsh penalty will be enforced by the EPA.

The pH meter is indirectly related to filter performance because it detects the potential for corrosion and gives early warning of other problems in filter performance. If the pH of the filtered water is above 8, a significant amount of dissolved aluminum will break through the filter bed. The same situation will occur if the pH is below 5. A low pH also gives rise to corrosion problems and special attention should be directed to the problem of lead corrosion. A high filtered water pH will also reduce the effectiveness of chlorine disinfection. It is standard practice to measure the pH of the combined filtered water for the reasons given above.

A unique tool in evaluating filter performance, in conjunction with the turbidimeter, is the particle sizing and counting equipment. This analyzer is especially valuable in evaluating the filtered water characteristics that cannot be detected by the turbidimeter. For example, the turbidity readings may be the same for the filtered water of plant A and plant B. Yet, the particle size and the number of particles in each of the waters may be quite different. Since the design of the turbidimeter is based on nephelometry—that is, it measures the degree to which the incident light is scattered by the suspended particles—the size, shape, and color of the suspended particles greatly affect the reading. The turbidimeter therefore is incapable of measuring the true content of dark suspended particles such as PAC. However, a good correlation generally exists between the turbidity and the particle count as illustrated in Figure 3.2.6-23. Based on the Hiac Criterion Model PC-320 or equivalent units, particle counts of less than 100 per milliliter are generally considered excellent. Particle counts of $100-500$ mL^{-1} are good, but counts over 500 are unfavorable.

The newest particle sizing and counting equipment is sold by Met One (Model 203) and a similar unit by another are able to measure 0.5 to 15 μm range. This is important because not only most bacteria, Giardia and Crystosprodium are in this range but also most filters with granular beds are considered to be somewhat ineffective to remove the particulates of this range. The unit is rather expensive, over $20,000 per unit. This type of unit is actually best suited for study and research and is too expensive to be used in each filter as a surveillance tool for daily filter performance. However, it is advantageous to have one unit in the plant process control laboratory or on one of the selected filters.

The filter headloss profile monitor supplies the plant operator with a head-

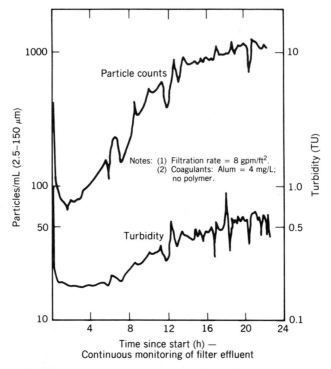

Figure 3.2.6-23 Filter performance evaluation by turbidity and particle counts with continuous monitoring of filter effluent (Adapted from MWD Southern California—Internal Report, 1978.)

loss profile across the filter bed. The monitor should be arranged so that there are several pressure taps located at various depths of the filter bed. These pressure taps should be connected to glass tubes, creating a piezometer board. If "high-tech" equipment is included in the plant design, the headloss profile may be shown on a cathode ray tube (CRT) in the control room, thus allowing for continuous observation and giving early warning of air-binding conditions. Although this unit is a good performance monitoring tool, it is not necessary for all the filters to be supplied with one. One select filter should be designed to receive this unit if the plant owner expresses interest in its use.

Common Filter Behavior and Control Measures   In contrast to all other processes in an ordinary water treatment process train, the filtration process is generally a batch process. The filtering cycle usually lasts from 1 to 3 days, followed by filter washing. The washing cycle lasts a total of 15–30 min, including the drawdown of water level prior to washing and refilling with water. After the washing cycle is completed, the filter cycle is repeated. The special performance behavior of the filter may be attributed to the batch operation process. As illustrated in Figure 3.2.6-23, the filtered water often exhibits a sharp turbidity breakthrough at the beginning of the filter cycle. It is usually very short, 10–20 min. After this period the turbidity of the filtered water drops below 0.5 NTU and eventually settles in the vicinity of 0.15 NTU; this level is both the practical and the upper turbidity limit defined by most regulatory agencies. However, the turbidity of the filtered water gradually increases with time and exceeds 0.5 NTU after 24 h. The turbidity breakthrough at both the beginning and late stages of a filtering cycle is characteristic of most constant-rate filters employing alum coagulation. Design engineers and plant operators must be cognizant of this particular filter behavior and should analyze the methods used to control the turbidity breakthrough.

The initial turbidity breakthrough may be controlled by any of four methods: wasting the filtered water, slowly opening the filter effluent valve, adjusting the filter washing practice, or adding polymer to the filter backwash water. The first two methods are related to the design of the filter system. The last two are operational manipulations, except proper design of the polymer feed system to the backwash water. The first method isolates the filtered water from the clearwell for a period of 10–15 min until the turbidity of the water falls below 0.5 NTU. The second slowly opens the filter effluent valve, taking 10–20 min to arrive from a fully closed valve position to the appropriate open position—a position that permits the designed filtration rate to occur. The third method, adjusting the filter washing practice, terminates the filter wash cycle when the turbidity of the wash waste is approximately 10–15 NTU, thus maintaining the ripened condition of the filter bed. The last method applies approximately 0.2 mg/L of nonionic polymer during the last 3 min of backwash. This last method is more of a "trial-and-error" type of technique and is not always successful because the polymers are very selective with respect to the characteristics of the suspended matter.

Turbidity breakthrough in the late stages of the filter cycle may be controlled by strengthening the floc and by increasing the adsorption capability of the filter bed. This may be achieved by feeding cationic polymer as a coagulant, with or without alum, or by adding minute amounts of nonionic polymer to the filter influent (as a filtration aid); alum must be the sole coagulant. Plant operators must be careful to limit the dosage of the nonionic polymer (filtration aid) to 0.015–0.025 mg/L, since overfeeding will result in extremely short filter run lengths. According to Figure 3.2.6-24, the late-stage turbidity breakthrough of the previous filter is completely eliminated when the coagulant is switched from alum to cationic polymer. The data shown in Figures 3.2.6-23 and 3.2.6-24 are the performance results of actual dual media filters.

Operational Procedures Following Plant Shutdown   If the treatment plant is shut down for more than 3 days during any season (for any reason) or if the plant flow rate is only 10–20% of the designed rate for a period of 1 month or more, the majority of the filters may be drained or all the filter

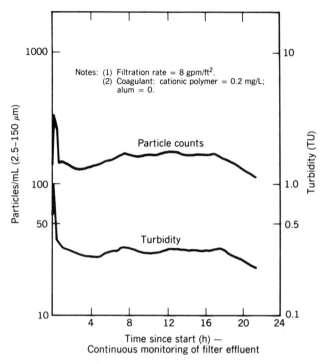

Figure 3.2.6-24 No turbidity breakthrough except at beginning. There is continuous monitoring of filter effluent. (Adapted from MWD Southern California—Internal Report, 1978.)

effluent valves may be closed, allowing the water to stand over the filter bed. However, if the plant is only shut down for 2 or 3 weeks, the filters should not be drained because it will take a considerable period of time to regain the proper (mature) filter bed conditions if the bed has been allowed to dry up. When a plant is treating an extremely low flow rate for long periods of time, a few operators choose to use only one or two filters and the rest are left on standby; these standby filters contain standing water. However, it is better to use all the filters, allowing water to move through all the filters, since this will prevent an explosive growth of microorganisms in the filter bed and will also prevent anaerobic conditions from occurring in the filter bed. Anaerobic conditions often lead to objectionable taste and odor problems. In any case, if the filters are left standing for over 3 days without being drained (during warm months), they should be backwashed prior to reintroduction into the filtration cycle. If the filters cannot be backwashed, due to an insufficient amount of water, it is essential to completely drain the standing water and to backwash the filters for approximately 3 min or more after backfilling with backwash water. Since a rate of 2 gpm/ft² (5 m/h) is equivalent to 3.2 in./min (8 cm/min), it would take approximately 30 min to completely drain a filter, assuming that the water depth in the filter is 8 ft (2.4 m) and the average rate of the drain is 17 ft/h (5 m/h).

If the filter underdrain system or the gravel beds are reconstructed or disturbed by workers, the filter-to-waste line should also be employed. For example, if a portion of the filter underdrain is repaired, a filtered water turbidity of almost 30 NTU or more will result from the disturbance of the gravel bed, which is usually quite filthy after several years of service. The design of the filter system should therefore include a filter-to-waste line, even though it may not be used for normal daily filter operations.

Liquefaction of Filter Bed Due to Strong Earthquakes An interesting and important behavior of granular medium filters is the potential for high turbidity breakthrough during strong earthquakes (magnitudes above 5) due to scrubbing and liquefaction phenomena. Plant operators should be aware of this fact and whenever these phenomena are observed, they should (1) stop plant operations, if possible, or (2) waste the filter effluent through the filter to the waste line in order to reduce the plant flow rate. The importance of these emergency procedures is illustrated by the following case history: Two modern water treatment plants in southern California, located near the epicenter of a 5.5 magnitude earthquake, recorded an effluent turbidity as high as 3.0 NTU immediately after the quake.

Before the system is put back into operation, each filter should be lightly backwashed for about 3 min.

BIBLIOGRAPHY

AWWA, *Water Quality and Treatment*, 3rd ed., McGraw-Hill, New York, 1971.

Baylis, J. R., "Nature and Effect of Filter Backwashing," *J. AWWA*, 51:1433 (November 1959).

Cleasby, J. L., "Declining-Rate Filtration," *J. AWWA*, 73(9):484 (September 1981).

Cleasby, J. L., et al., "Effectiveness of Backwashing of Wastewater Filters," *J. Environ. Eng. Div. ASCE*, 104(EE4):749 (August 1978).

Degremont, *Water Treatment Handbook* 5th ed., Halsted Press, New York, 1979.

Fujita, K., "Hydraulics of Filtration Theory," *J. Jpn. WWA*, 445:2 (August 1973).

Hudson, H. E., *Water Clarification Processes*, Van Nostrand Reinhold, New York, 1981.

Ives, K. J., *The Scientific Basis of Filtration*, Noordhoff-Leyden, Groningen, 1975.

Kawamura, S., "A Study on Filter Washing Efficiency," in *Proceedings of the 6th Annual Japan WWA Conference*, May 1955.

Kawamura, S., "A Pilot Study on Filter Bed and Underdrain System," in *Proceedings of the 14th Annual Japan WWA Conference*, May 1963.

Kawamura, S., "Coagulation Considerations," *J. AWWA*, 65(6):417 (June 1973).

Kawamura, S., "Design and Operation of High-Rate Filters," *J. AWWA*, 67(10):535 (October 1975).

Kawamura, S., "Two-Stage Filtration," *J. AWWA*, 77(12):42 (December 1985).

Monk, R. D. G., "Design Options for Water Filtration," *J. AWWA*, 79(9):93 (September 1987).

Monk, R. D. G., "Improved Methods of Designing Filter Boxes," *J. AWWA*, 76(8):54 (August 1984).

Norman, P. A., Backwash Instability in Rapid Sand Filters, A thesis for M.S. Univ. of New South Wales, Australia (1970).

Weber, W. J., *Physicochemical Processes*, Wiley-Interscience, New York, 1972.

3.2.7 Specific Types of Filter

This section discusses the slow sand filters, proprietary filters (in general), the two-stage filtration process, and a miscellaneous type of proprietary filter—the automatic backwash filter.

Slow Sand Filters Slow sand filters have been used in water treatment since the early 19th century. They have been proved to be effective under various conditions as long as the filters are properly designed and applied to appropriate situations. The system is simple, cost effective, reliable, and easy to operate. Thus, remote areas and small to mid-sized water supply systems in developing countries should seriously consider their use.

The filtration rate of slow sand filters is 50–100 times slower than ordinary rapid sand and high-rate filters. Consequently, a considerably larger area (filter bed) is necessary to produce the same amount of water.

A unique feature of the slow sand filters, as described in much of the literature, is the presence of a thin layer of medium on the surface of the filter bed, known as the "schmutzdecke"—loosely translated as "dirty mat." This special layer contains a large variety of microorganisms and enables these filters to remove bacteria by a factor of 10^3–10^4 and *E. coli* by a factor of 10^2–10^3. The schmutzdecke also removes organic matter and reduces the turbidity of the raw water.

The slow sand filters offer several advantages: (1) they are simple to build and operate and are cost effective; (2) the filtration system is reliable and produces good quality drinking water; (3) the system does not require highly trained operators and has minimal power requirements; and (4) the filters can tolerate hydraulic and solid shock loadings provided that they are not excessive. Yet, the disadvantages tend to restrict their use: (1) they require a large amount of land; (2) the filters are easily clogged by excessive amounts of algae; and (3) intermittent operation of the filters may degrade the quality of the filter effluent by promoting anaerobic conditions within the filter bed.

The slow sand filter system is a batch process and the filters must therefore be periodically cleaned. Since the majority of the suspended solids (in the raw water) are removed in the top 0.8–1.2 in. (2–3 cm) of the bed, the filter must be cleaned by scraping a thin layer of sand off its surface. Backwashing is not employed, since this will hydraulically restratify the bed. The filter bed may require scraping every 2–6 months, depending on the turbidity and the concentration of algae in the raw water.

The filter cleaning operation is usually performed on a manual basis and completed in 1–2 days. Since manual cleaning of the filters does not require highly trained individuals, this particular feature is best suited for regions with an abundance of inexpensive labor. Industrialized nations should use mechanized systems, such as tractors with special skimming buckets or traveling bridges fitted with scraper mechanisms, to make the cleaning operation more economical.

After the bed is cleaned, a period of up to 1 week is required to ripen the filter bed (build the schmutzdecke) so that an acceptable quality of water is produced. Treatment plants that use the slow sand filters and process raw water with seasonally high turbidity spikes and algal blooms must use pretreatment with chemical flocculation and settling, in conjunction with prechlorination.

Acceptable Raw Water Quality for Slow Sand Filtration   The acceptable raw water quality for slow sand filters is generally based on the same criteria used for the direct filtration process of high-rate filters. The criteria are listed in Table 2.3.5-1 in Chapter 2. Slow sand filters can tolerate raw water with turbidities greater than 15 NTU as long as the turbidity spikes are less than 50 NTU and last no more than 2–3 days. However, if the spikes exceed these criteria or if frequent high algal counts are to be expected, an appropriate pretreatment process train must be implemented.

General Design Criteria   The general design criteria for the slow sand filter system are listed below.

Required filter bed area	0.5–0.15 m^2 per capita per day or 0.55–1.6 ft^2 per capita per day
Number of filters	$n = 0.25\,Q^{0.5}$, where n is the number of filters and is greater than or equal to 2, and Q is the plant flow rate (in m^3/h)
Area of each filter bed	Less than 3000 m^2 with an average of 100–200 m^2, or less than 32,500 ft^2 and an average of 1100–2,150 ft^2
Filtration rate	0.1–0.2 m/h (0.04–0.08 gpm/ft^2)
Filter bed depth	1–1.5 m (3–5 ft)
Support gravel depth	0.3–0.45 m (four layers)
Water depth above the bed	1–1.5 m
Height of the freeboard	≥0.2 m
Height of the filter box	2.5–4 m, average of 3.2 m (8.2–13 ft, average of 10.5 ft)

Filter sand
    Effective size   0.15–0.35 mm
    Uniformity coefficient   Less than 3, preferably near 2
Filter gravel
    Top layer   0.4–0.6 mm with a depth of 10 cm
    Second layer   1.5–2.0 mm with a depth of 10 cm
    Third layer   5–8 mm with a depth of 10 cm
    Bottom layer   15–25 mm with a depth of 10 cm
Underdrains   The slow sand filtration system generally has a main manifold and many laterals. The laterals may be porous drain tiles or glazed pipes laid with open joints, but PVC pipes are more often used.
                                      4 m (13 ft) maximum lateral spacing.
Filter effluent level control weir   The effluent weir should be set at least 0.2 m above the top of the initial height of the filter bed

Figure 3.2.7-1 is a longitudinal section of a slow sand filter with a filter effluent level control weir and major valves. The filter effluent level control weir may be a down-opening slide gate, for more flexible operation, or a telescoping tube type valve.

Operation and Maintenance of Slow Sand Filters   This section discusses the start-up, filtration, filter cleaning, and resanding processes.

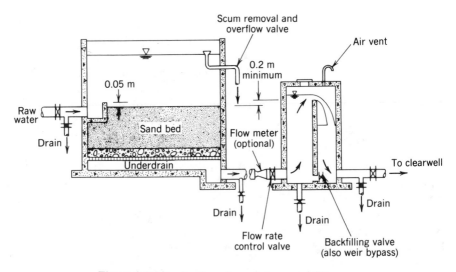

Figure 3.2.7-1   Section view of slow sand filter.

START-UP   When construction of the filter is complete, the sand in the bed is both clean and dry. In this condition, the filter is incapable of water purification because it must first undergo a maturation process to build up the schmutzdecke. The maturation process has two distinct phases: backfilling and gradual initiation of filtration.

The first step, slow backfilling, drives the air out of the sand bed. When the raw water reaches a level of approximately 1 m above the surface of the bed, backfilling is terminated and the filter inlet valve is opened, introducing raw water from the inlet instead of the outlet. The backfilling process is extremely important because it prevents the filter bed from being scoured or disturbed by the cascading effect of the raw water onto the filter bed. When the water reaches normal operational levels, the second step is initiated.

The second step in the maturation process is gradual initiation of filtration. This procedure is achieved by partially opening the filter effluent valve; the suggested filtration rate is 10–30% of the designed filtration rate. During this step the filtered water must be continuously removed via a drain valve until the quality of the water meets the drinking water quality standards. This step usually takes several weeks in regions of hot weather and considerably longer in colder regions due to the suboptimal levels of bacterial nutrients and low temperature.

The maturation or ripening period may be shortened by seeding the new filter sand with schmutzdecke from another filter. Filter bed maturation may be detected as a slight, yet distinct, increase in the headloss across the filter bed. Once the bed becomes ripe, the effluent valve should be opened to produce the designed flow rate and the quality of the effluent water should be rechecked. When the effluent water meets the drinking water quality standards, the drain valve is closed and all the filtered water is directed to the water supply system. The quality of the effluent water must always be checked whenever the filtration process is interrupted for more than 1 or 2 weeks.

FILTRATION PERIOD   The main task of the operator is to control the flow rate through each filter by first adjusting the opening of the filter effluent valve, and by then making the necessary adjustments at the filter influent valve. At the beginning of each filtration cycle, the filter effluent valve should be at a partially closed position since there is very little headloss across the filter bed. The headloss increases with the accumulation of suspended matter on the surface of the bed—with the creation of the schmutzdecke. It is therefore necessary to provide some type of flow indicator, near the filter effluent valve, which allows the operator to gauge the flow rate and adjust the valve opening so that the proper flow rate can be met. The simplest method is to install a gauge staff upstream of the filter effluent level control weir. It may be a gauge glass or a float in a floatwell, but it must be located outside the box housing the control weir since the operator must be able to easily check the reading while making the flow adjustments. Alternatively, a flow meter may be located immediately upstream of the filter effluent valve. This is an ideal configuration and all filters should implement this scheme to provide more accurate flow control, with a few exceptions. The exceptions, due to cost consideration, are small plants and developing countries.

The filter influent valve should also be adjusted periodically to keep the influent and effluent flow rates in balance. If the operator fails to do so, an exorbitant amount of water will flow through the overflow pipe or the water above the filter bed will be very shallow.

Plant operators must maintain the filters in good working order by routinely monitoring the filter system. For instance, algal blooms in the raw water should be controlled at the source by adding 0.15 mg/L of copper sulfate or through other means, before the filters become clogged. Operators should routinely clean the filters and their surroundings, in addition to removing any floating scum, leaves, and debris. The operator should also routinely analyze samples of both the raw water and the filtered water at prescheduled intervals.

FILTER CLEANING   As previously mentioned, the slow sand filters may require cleaning every 2–6 months, depending on the algal concentration and the turbidity of the raw water. The cleaning process is initiated at the end of the last filter run and is composed of three phases: draining the filter bed, removing the sand, and reripening the bed.

The end of a filter run is characterized by a rapid increase in headloss across the bed that is simultaneous with a reduction in the filter effluent flow rate, despite a fully opened effluent valve. There will also be a slight degradation in the quality of the filter effluent water. The filter influent valve is closed at the end of a filter run and the effluent valves are opened to drain the water level; the water level above the bed will drop to the height of the outlet weir overnight. The filter effluent valve should then be closed and both drain valves, just downstream of the filter influent valve and upstream of the filter effluent valve, should be opened to speed the drainage process. The water level must be 6–8 in. (15–20 cm) below the surface of the bed before the cleaning operation can be initiated. As soon as the top of the bed is sufficiently dry, the second step in the cleaning operation may be started. The scraping process should not be delayed because the guano of scavenging birds contaminates the filter bed to a greater depth.

The scraping operation removes the top 0.8–1.2 in. (2–3 cm) of the bed by means of manual labor or mechanical equipment. In either case, the schmutzdecke and the surface sand adhering to it should be quickly stripped and carefully stacked into ridges. If the bed is manually cleaned, the waste material should be removed with square blade shovels and the debris transported by wheelbarrow or conveyer belt. If the filter beds are large and the cost of personnel is high, the scraping may be performed by a tractor fitted with a specifically designed scraping blade. Both mechanical equipment and wheelbarrows must be run on protective planks so that the top portion of the filter bed is not damaged and the microorganisms (in the bed) are not killed.

Upon completion of the scraping operation, the surface of the filter bed should be leveled.

If the schmutzdecke is largely composed of filamentous algae, the cleaning operation will be easy. However, if the predominant species are nonfilamentous organisms, such as diatoms, the cleaning process will be more difficult. Workers must take precautions to avoid being exposed to harmful microorganisms contained in the schmutzdecke. Likewise, the workers must be careful not to contaminate the filter bed with bodily excretions or spit.

The third step in the cleaning process is reripening of the filter bed. This process is the same as described in the start-up process, although the maturation period is greatly accelerated.

RESANDING   After approximately 20–30 scrapings, the depth of the filter bed is reduced to 20–28 in. (0.5–0.7 m). This is the minimum depth that should remain above the supporting gravel bed. When the bed is reduced to this level, the filter should be drained and new sand must be added to restore the original medium depth.

The resanding operation does not simply add new sand on top of the original bed. First, 12–20 in. (0.3–0.5 m) of the old bed is moved to one side of the filter. Next, new sand is added on top of the old sand; the depth of this old sand layer should be approximately 0.2 m. Finally, the new sand layer is covered by the old sand, which was initially set aside. This practice retains most of the valuable microorganisms, thus shortening the reripening period.

In regions where sand is difficult to obtain or is expensive, it is common to recycle the sand that was removed during the cleaning process. This contaminated layer is washed by water jets or by a hydraulic educator system and stored for future resanding operations. The scrapings must be washed immediately after removal or it will become putrid and produce an objectionable odor. It is quite difficult to remove the coating of debris if the scrapings are allowed to dry.

Sludge removed from the scrapings contains numerous bacteria and organic substances; thus, it is very odoriferous and attracts many flies. It is therefore prudent to schedule the recycling and resanding operations during the winter season.

Proprietary Type Filters   Many water treatment equipment manufacturers supply various types of filter, mainly for industrial purposes, but a few market filters specifically designed for water treatment plants. In the United States, the proprietary filters are primarily used in small sized plants (those processing less than 15 mgd). However, this is not necessarily true for some overseas countries.

The two basic types of proprietary unit used in water treatment are the gravity filter and pressure filter. Gravity filters may be composed of a variety of materials but pressure filters are cylindrical pressure vessels that are exclusively fabricated from steel plates.

The use of pressure filters may be considered if the high water pressure of the plant inlet is to be preserved, in order to avoid having to repump to obtain the water pressure necessary to distribute the filtered water. Since pressure filters use an equivalent of approximately 15 psi (10.5 m) of line pressure, provided that the inlet line pressure is over 35 psi (25 m) of water head, the filter effluent line still has enough high pressure to backwash the filters without the use of wash pumps. Additionally, there is enough line pressure remaining in the system to transmit filtered water to a remote distribution reservoir. If gravity filters are used under these same conditions, the filter effluent line virtually has no line pressure left. Engineers should therefore evaluate the possible use of pressure filters when designing small to mid-sized plants under the aforementioned conditions, despite the fact that state regulatory agencies generally discourage the use of pressure filters because of the difficulty in inspecting the filter bed conditions.

In the water supply business, one of the best applications for pressure filters is the filtration of pumped deep-well water containing iron and manganese. This type of filter will remove these compounds through oxidation by chlorine or potassium permanganate and the well pumps usually supply a high level of extra pressure to the pump discharge line.

The proprietary filters that are generally used in public water treatment plants are the open-top gravity filters, since these allow easy observation and maintenance of the filter bed. Yet, each manufacturer markets a somewhat unique filter: self-backwash, continuous automatic washing, air-scour wash, bumping bed, coarse media bed for oil removal, ultra high-rate, and so on. The size and details of the proprietary filter units are generally preengineered by the manufacturer, based on a wide range of flow rates; these units are not cataloged based on filter size and appurtenances, but on flow rates. Since the manufacturer has predesigned the blueprints, the completion time of the project is much shorter than when using custom designed filters. Additionally, many of these filters are self-contained units that require rather simple power and pipe connections to make them functional. Another advantage of these units is that the manufacturers always give the customers a limited guarantee on both the performance of the units and the equipment. However, the customer usually cannot request any significant design modifications without forfeiting the guarantee. This can be a potential problem when the job site is in a remote area; repair and replacement of certain parts may be difficult if they are patented items. Moreover, many proprietary filters are fabricated with metal plates and corrosion protection is therefore a serious problem. Although proprietary filters may be custom designed, this negates the advantage of using these units: the savings in time and money.

The decision to use proprietary filters versus custom designed units is the responsibility of either the engineer or the owner. If the engineer is authorized to make the decision, he/she must conduct a thorough evaluation of the units that are on the market. Special attention should be given to the number of years the unit has been in operation, feedback from plant operators using these types of unit, quality control, follow-up servicing, cost of each type of unit, and the financial stability of the manufacturer. The engineer may choose to conduct prequalification of certain manufacturers prior to bidding in order to avoid purchasing an inferior unit, although the unit may meet the specifications. Figure 3.2.7-2 depicts the proprietary type filters.

Two-Stage Filtration   The two-stage filtration process has historically been applied to the slow sand filtration process to improve the overall efficiency of filtration. Not until the late 1970s did this process gain attention as a substitute for the high-rate filters used in both direct filtration and in-line filtration; these two types of filter are susceptible to turbidity spikes and algal blooms in the raw water. The two-stage slow filtration process has also been successfully used in Europe in the treatment of poor quality raw water, as well as raw water with seasonal degradations in quality.

The two-stage filtration process for high-rate filtration is composed of two distinct filtration steps: the first-stage filters and the final-stage filters. These stages consist of a bank of coarse bed roughing filters and a bank of ordinary high-rate filters, respectively. The ability of the coarse media bed to act as a flocculation system has been recognized and demonstrated in a number of studies since the early 1970s. The function of the roughing filter is flocculation: to produce floc with the proper characteristics for the subsequent high-rate filters, in addition to removing approximately 50–80% of the suspended matter. The filters therefore depend on coagulant dosage and the nature of the raw water.

The major advantage of the two-stage process are the substantial savings in coagulant—only 4–6 mg/L of alum is required—and the small amount of sludge production. In contrast to the direct and in-line filtration processes, the two-stage process is also capable of withstanding turbidity spikes of short duration and algal blooms. A few equipment manufacturers have recognized these advantages and have commercialized this system with proprietary designs and patents. The one disadvantage of these filters is that wash waste is produced by two sets of filters.

Refer to Table 2.3.6-1 for a general idea of the applicable raw water quality for the two-stage process. Figure 3.2.6-17 contains the proper range of coagulant dosage for the two-stage filtration process in relation to direct filtration and a conventional complete treatment process. Figure 3.2.7-3 illustrates an example of the two-stage filtration system.

General Design Criteria

ROUGHING FILTER (FIRST-STAGE FILTERS)

Filtration rate	12–15 gpm/ft^2 30–37.5 m/h)
Filter medium	Pea gravel, crushed anthracite, or plastic chips
Size of medium (d)	3–6 mm, effective size

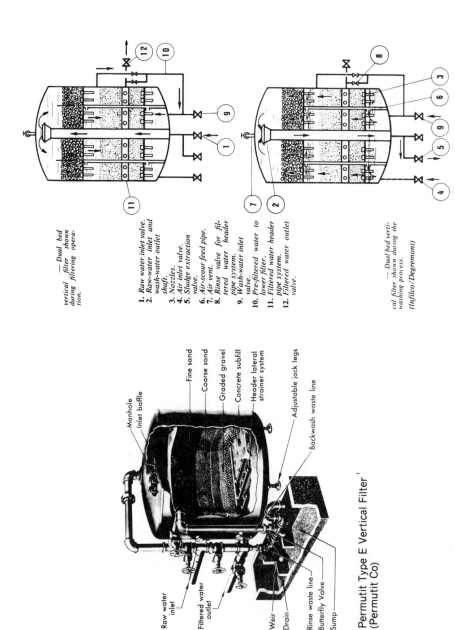

— Dual bed vertical filter shown during filtering operation.

1. Raw water inlet valve.
2. Raw-water inlet and wash-water outlet shaft.
3. Nozzles.
4. Air inlet valve.
5. Sludge extraction valve.
6. Air-scour feed pipe.
7. Air vent.
8. Rinse valve for filtered water header pipe system.
9. Wash-water inlet valve.
10. Pre-filtered water to lower filter.
11. Filtered water header pipe system.
12. Filtered water outlet valve.

— Dual bed vertical filter shown during the washing process.

(Infilco/Degremont)

Manhole
Inlet baffle
Fine sand
Coarse sand
Graded gravel
Concrete subfill
Header lateral strainer system
Adjustable jack legs
Backwash waste line

Raw water inlet
Filtered water outlet
Weir
Drain
Rinse waste line
Butterfly Valve
Sump

Permutit Type E Vertical Filter
(Permutit Co)

Figure 3.2.7-2 Proprietary pressure filters. (From Bulletins of Permutit Company and Infilco-Degremont International.)

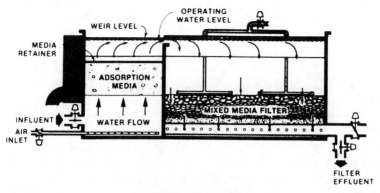

FILTRATION MODE

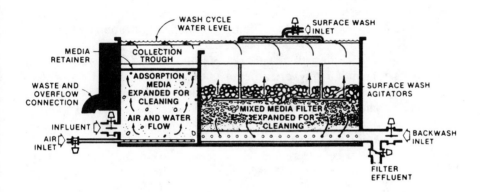

BACKWASH CYCLE

Figure 3.2.7-3 Automatic backwash filter (ABF) produced by Environmental Elements Corporation.

Uniformity coefficient	≤ 1.5
Bed depth (l)	$500 < l/d > 250$
Filter washing	Air-scour backwash
Backwash rate	15–20 gpm/ft^2
	(37.5–50 m/h)
Air-scour rate	3–5 cfm/ft^2
	(55–90 m/h)
Maximum available headloss for filtration	1 m (3.3 ft)
Underdrains	Air-scour type
Washwater troughs	Use of the troughs is preferred
Direction of flow	Downflow is normal; however, upflow and horizontal flow have been applied

FINAL FILTER (SECOND-STAGE FILTERS) The design criteria for the final filters are exactly the same as those of the high-rate filters. Refer to Section 3.2.6.

Miscellaneous Types of Filter Diatomite, biflow, and automatic backwash filters such as the Hardinge filter, ABF (manufactured by Infilco-Degremont and other companies), and Dyna Sand (by Parkson Corp.) and other firms are the other types of filter that are used with limited applications in the processing of the public drinking water supply. Of these units, the ABF is the only type that has considerable acceptance for industrial applications, as well as potable water plants for small public water supplies when the use of package plants is appropriate.

When the patent on the ABF design expired in the 1960s, a total of four manufacturers began marketing very similar units. The competition helped to generate a much wider acceptance of this type of filter, especially for use in tertiary wastewater filters. Yet, certain states discourage the use of ABF for potable water because of its declining-rate mode of filtration and the lack of a provision for filter to waste mechanism.

The original ABF design has a depth of only 11 in. (0.28 m) of 0.6 mm (effective size) sand. However, a depth of 16–24 in. (0.4–0.6 m) of dual media is now also available. The ABF filter bed is supported by a porous plate underdrain and is divided into numerous filter cells that are approximately 9 in. wide. Each cell is individually washed by a traveling bridge that is equipped with a backwash pump and a wash-waste pump, thus allowing the rest of the filter cells to remain in service. If filter bed depth exceeds 16 in., the traveling bridge stops at each filter cell (indexing) to complete filter backwash. The standard width of one automatic backwash filter is 16 ft (5 m) and the recommended maximum length of each filter is 150 ft (45 m).

The basic principle on which the automatic backwash filters operate is a surface filtration that facilitates shallow penetration of floc into the filter bed. The filter therefore operates at a low headloss of 6–12 in. (0.15–0.3 m) with a filter rate of 2–3 gpm/ft^2 (5–7.5 m/h). Filter backwashing is initiated by either a headloss or a timer. A typical wash cycle is every 2–6 h and the surface of the bed never becomes compacted because of this frequent wash cycle. Consequently, surface wash is not required and air-binding does not occur.

The advantages of this filter are that it does not need a backwashing facility, in the regular sense, and the wash waste is very easy to handle due to the low flow rate: only 200 gpm (0.23 m^3/h) for a period of 30–90 min. One drawback, with respect to potable water production, is the need for a housing over the filter bank. The housing is necessary because of the potential for contamination, since the filtered water flows in an open top channel.

The maintenance requirements of the ABF are very simple: periodic inspection and lubrication of the drive mechanism and pumps and occasional touch-up painting. Yet, if the process water has a medium to high hardness, the porous plate underdrains will become clogged by calcium carbonate deposits over a period of 15–20 years, depending on the conditions. Experience

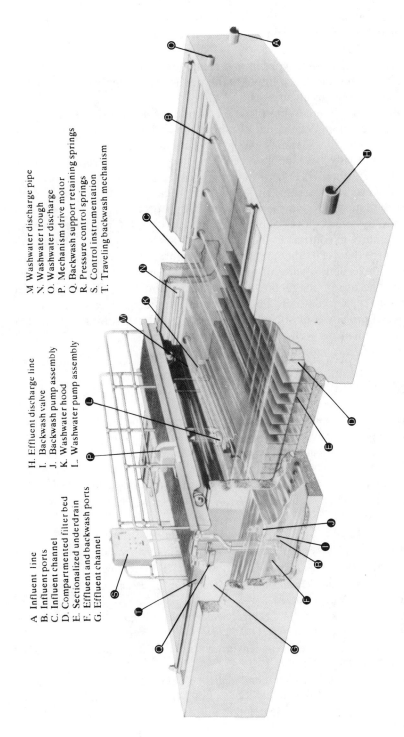

A Influent line
B Influent ports
C Influent channel
D Compartmented filter bed
E Sectionalized underdrain
F Effluent and backwash ports
G Effluent channel

H Effluent discharge line
I Backwash valve
J Backwash pump assembly
K Washwater hood
L Washwater pump assembly

M Washwater discharge pipe
N Washwater trough
O Washwater discharge
P Mechanism drive motor
Q Backwash support retaining springs
R Pressure control springs
S Control instrumentation
T Traveling backwash mechanism

Figure 3.2.7-4 A commercialized two-stage filtration system produced by Neptune Microfloc, Inc.)

275

has shown that an acidification treatment—application of 5–10% hydrochloric acid, restores the capacity of the filters. Figure 3.2.7-4 presents a schematic view of an ABF.

Discussion of the other types of miscellaneous filter will not be attempted since the application of these filters to the public water supply is very limited due to their serious disadvantages.

One of the newly emerging filtration techniques for potable water treatment is the membrane technique: ultrafiltration and nanofiltration. The pore size of these membranes is small enough to separate solids from the raw water but may potentially be capable of filtering out bacteria, viruses, *Giardia*, humic substances, and THMs virtually without the use of coagulant or chemicals. This new water treatment system has been evaluated by pilot scale plants in the late 1980s and the results are rather promising.

This new system requires only approximately 10 m (15 psi) of equivalent driving pressure, in contrast to the reverse osmosis technique, which requires a driving force that is 20 times greater. Furthermore, the membrane technique produces an excellent quality of water with small amounts of chlorine, with or without the use of coagulant; however, approximately 20% of the water is wasted as rejection water. Should this technique become practical, the conventional process could be greatly altered, possibly resulting in a smaller space requirement, as well as fewer operating personnel, due to the installation of a high-level automatic control system.

BIBLIOGRAPHY

Burns, D. F., et al., "Particulate Removal on Coated Filter Media," *J. AWWA*, 62:121 (February 1970).

Cleasby, J. L., "Slow Sand and Direct In-Line Filtration of a Surface Water," *J. AWWA*, 76:44 (December 1986).

Degremont, *Water Treatment Handbook*, 5th ed., Halstead Press, New York, 1979.

Dillingham, J. L., et al., "Optimum Design and Operation of Diatomite Filtration Plants," *J. AWWA*, 58:657 (June 1966).

Graham, N. J. D., ed., *Slow Sand Filtration*, Ellis Horwood, New York, 1988.

Huisman, L., et al., *Slow Sand Filtration*, World Health Organization, Geneva, 1974.

Kawamura, S., "Two-Stage Filtration," *J. AWWA*, 77:42 (December 1985).

Sank, L. R., ed., *Water Treatment Plant Design for the Practicing Engineer*, Ann Arbor Science, Ann Arbor, MI, 1978.

Schulz, C. R. and Okun, D. A., *Surface Water Treatment for Communities in Developing Countries*, Wiley, New York, 1984.

Vaillant, C. J., "Simple Water Treatment Processes," in *IWSA Southern Asia Regional Conference*, Bombay, October 1981.

Bersillon, J. L., et al., "Ultrafiltration Applied to Drinking Water Treatment: Case of Small System," *AWWA Annual Conf.*, Los Angeles, CA (June 1989).

Jacangelo, J. G., et al., "Assessing Hollow-Fiber Ultrafiltration for Particulate Removal," *J. AWWA*, 81(11):68 (Nov. 1989).

3.2.8 Disinfection Processes

Purpose The purpose of disinfection process is to kill the growing form of pathogenic microorganisms, not necessarily the resistant spore forms, through the use of chemicals or ozone. Pathogens are defined as any type of microorganism capable of producing disease. The process of disinfection should not be confused with sterilization, which refers to the total destruction or total removal of all microorganisms from the treated medium, including the spore forms.

Strictly speaking, heat treatment, ultraviolet irradiation, and gamma or x-ray irradiation should be included in the disinfection process. However, these processes are generally practiced by various industries and are not used in treating the public water supply, primarily because they do not have the residual disinfecting power required to combat postcontamination within the distribution systems. Thus, these types of disinfecting process are not discussed in this book.

Considerations The major considerations in selecting a disinfection process are the presence of surrogate organisms in the drinking water supply, the feasibility of using alternative disinfectants, the disinfectant residual–contact time relationship ($C \times t$); the formation of disinfectant by-products and their magnitude, the quality of the process water, the safety problems associated with the disinfectants, and the cost of each disinfection alternative. Each of these considerations is discussed in turn.

Surrogate Organisms Coliform organisms, particularly *Escherichia coli* (*E. coli*), are continuously present in the human intestine in large numbers. Billions of these organisms are excreted by an average human or animal per day. Thus, their presence in the process water is indicative of fecal pollution. The coliform group of bacteria is defined as all anaerobic and facultative anaerobic, gram-negative, non-spore-forming bacilli that produce acid and gas through the fermentation of lactose.

The most common method of assessing the safety of the drinking water supply is the coliform test. However, microbiologists generally agree that coliforms may be more rapidly inactivated at much lower disinfectant dosages than some enteric viruses and protozoan cysts. Consequently, the EPA has proposed that the standard plate count be instituted as a routine test to screen for bacteria regrowth. The presence of a large population of organisms, as indicated by the plate count (heterotrophic bacteria), would certainly be indicative of an ineffective disinfection process. The heterotrophic plate count (HPC) for the water in distribution system should be less than 10 colonis per mL. The EPA is also emphasizing the inactivation of protozoan cysts such as *Entamoeba histolytica* and *Giardia lamblia*, as well as the *Cryptosporidium* and *Legionella* viruses, and enteric viruses.

Alternative Disinfectants   Chlorine, chloramines, chlorine dioxide, and ozone are the practical alternative disinfectants. The selection process is based on the effectiveness of the alternative as a disinfectant, the formation of harmful by-products, the residual disinfecting capabilities, and cost effectiveness.

CHLORINE   Chlorination of potable water has been practiced in the United States since 1903; therefore, much knowledge is available concerning the application technology and its effectiveness. Despite growing concerns regarding THM, a suspected carcinogen that forms during the chlorination process, chlorination will undoubtedly continue to be the primary water disinfectant for years to come, especially in developing countries. The basis for this reasoning is that all potable water should provide disinfectant residual in the distribution system and chlorine can easily satisfy this requirement. Additionally, chlorine has been proved to be effective as a disinfectant at reasonable cost for almost a century.

Chlorine is generally applied to the process water as an aqueous chlorine solution: chlorine gas is hydrolyzed to form hypochlorous acid (HOCl) and hypochlorite ion (OCl^-) in ammonia free water. Hypochlorous acid in water is commonly known as free residual chlorine. The relative concentration of these two species is a function of the pH of the water and, to some degree, the temperature (Figure 3.2.8-1). The source of the chlorine is usually liquified chlorine gas. However, small plants commonly use calcium hypochlorite, sodium hypochlorite, or chlorinated lime as sources of chlorine.

The free residual chlorine rapidly and indiscriminantly reacts with many substances in the water, including microorganisms. Thus, effective disinfection and minimum THM formation may be achieved if chlorine is added to clarified water such as filtered water.

Many factors affect the disinfection efficiency: oxidizable substances in the process water, particulate concentration, pH, temperature, contact time, and the level of the residual chlorine. The ideal disinfecting condition has a low pH (6–7), a relatively high water temperature (20–25°C), a longer contact time (over 30 min), and higher levels of residual chlorine (over 0.5 mg/L).

According to Morris (1975), hypochlorous acid is 100 times more effective than hypochlorite ion as a germicide for enteric bacteria and amoebic cysts. Yet, hypochlorous acid is less effective in combatting viruses, as well as amoebic cysts and bacterial spores, when compared with other types of disinfectant such as ozone or chlorine dioxide. Although hypochlorite ion is much less effective than hypochlorous acid when compared with monochloramine, its germicidal power is greater by a factor of 2.

Chlorination is a proven, reliable, and mature technology. However, THM formation and public safety are major concerns for both the engineer and the general public. Studies have shown that THM formation is substantially reduced when the application point of chlorine is changed from the head of the plant to postfiltration. The potential leakage of chlorine gas has been minimized with improvements in the chlorine feeders and chlorine storage. Im-

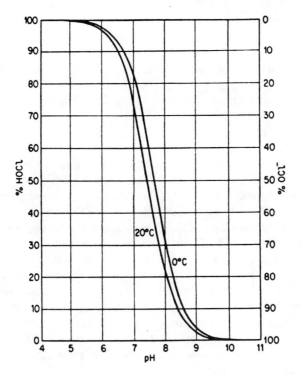

Figure 3.2.8-1 The effect of pH on the distribution of hypochlorous acid and hypochlorite ion in water.

provements in both technology and operating practices will allow for a more reliable and safe disinfecting process.

CHLORAMINES   Chloramination of drinking water has been used successfully in many water utilities and has proved to be an effective disinfecting process. However, chloramines are less effective as bactericides and viricides when compared to chlorination. Moreover, neither the enteric viruses nor protozoan cysts are effectively inactivated in short periods of time (less than 30 min) at low dosages (1–2 mg/L). Yet, chloramination does have the advantage of producing very small amounts of THM—levels that are insignificant to human health.

During the late 1970s, the EPA originally proposed that chloramination should not be used as the primary disinfectant due to its suboptimal performance as a bactericidal and viricidal agent, but this proposal was later modified, leaving implementation to the discretion of each state. Thus, the chloramination process may be used as a predisinfectant and/or a postdisinfectant prior to distribution.

One of the most sensible uses of chloramination is in combination with other disinfectants, particularly in the case of postdisinfection since the chloramine residuals last much longer than free chlorine and do not form significant amounts of THM. Chloramination is normally practiced at a ratio of approximately 1 part of ammonia to 3 parts of chlorine (using a mg/L base) to ensure monochloramine formation. Thus, if a postchlorinated water has 0.9 mg/L of free residual chlorine, 0.3 mg/l of ammonia to be added. However, if a 0.5 mg/L of ammonia is added prior to postchlorination, a minimum of 1.5 mg/L of chlorine must be added.

CHLORINE DIOXIDE   Chlorine dioxide has been used as a primary disinfectant by several large systems in Europe. However, it has had limited application as a water disinfectant in the United States; it has been used primarily as a means of taste and odor control.

Chlorine dioxide in water is an effective bactericide, equal to or slightly better than hypochlorous acid. For example, compared to free chlorine, chlorine dioxide is more effective as a bactericide and viricide at a higher pH range (8.5–9) than at a pH of 7. One definite advantage of chlorine dioxide is that it requires the pH of the filtered water to be above 8, thereby minimizing lead corrosion and the corrosion of other metals, in accordance with the recent EPA guideline. Also in this category is the disinfection of softened water by the excess lime softening process. Another advantage of chlorine dioxide is its ability to maintain a residual in the distribution system for an extended period of time.

The disadvantages of chlorine dioxide disinfection are its higher operational costs, relative to chlorine, and the possible toxicity of chlorite (ClO_2) and chlorate (ClO_3) ions; these form during the production of chloride dioxide. Additionally, excess chlorine added to the water will form THMs and undesirable by-products whenever chlorine is used in excessive amounts to ensure a complete reaction with the sodium chlorite. This practice was commonly employed in the United States to produce chlorine dioxide.

Since the EPA recommends that the combined residual concentration of chlorine dioxide and chlorite be 0.5 mg/L or less, whenever chlorine dioxide is used as a disinfectant, optimization of the generator performance is extremely important. Chlorine dioxide is generally produced by the reaction between chlorine gas and sodium chlorite solution. Preengineered and packaged chlorine dioxide generating equipment is offered by a few equipment manufacturers.

OZONE   Among the commonly available oxidants used in water treatment, ozone is the most powerful oxidant and disinfectant. Ozone has been used as a disinfectant worldwide in over 1100 plants, including very small installations, primarily in Europe. In practice, ozone is often used in conjunction with a secondary disinfectant such as chlorine or chloramine, since it is not capable of providing a residual disinfectant to combat the proliferation of microorganisms within the distribution system. Although ozone is used as the sole disinfectant in a few European countries and other parts of the world, it is not the recommended method of disinfection in the United States precisely because of the lack of protection against postcontamination.

The effectiveness of ozone disinfection is a function of the pH and temperature of the process water and the method of ozone application. The required residual–contact time relation ($C \times t$) of ozone disinfection is approximately 1.5% of free chlorine residual. Yet, the type of mixing, the degree of mixing, and the rate of mass transfer greatly influence the effectiveness of ozonation because ozone has a low solubility in water.

The advantages of using ozone as a disinfectant are its strong disinfecting power, a significant reduction in the formation of THM precursors prior to final chlorination, a limited number of by-products (aldehydes are a by-product that may have adverse effects to human health), and the low dosages (1–2 mg/L) required to complete disinfection. However, ozone also has the following disadvantages: the process does not provide residual protection after ozonation due to the rapid decomposition of ozone; the ozone must be produced on-site because it cannot be stored; and the ozonation system has a high capital cost ($2000/lb of ozone production per day for most medium size plants).

If cost is not a factor, the effectiveness of ozone disinfection may be enhanced by the use of a catalyst, such as ultraviolet light or hydrogen peroxide. However, the use of these catalysts may not be justifiable in all cases of ozone disinfection because of their marginal effectiveness.

Disinfectant Residual–Contact Time Relationship   Two factors are of primary importance in disinfection: the concentration of the disinfectant residual and the contact time. The relationship may be expressed as

$$C \times t = \text{constant}$$

where C = disinfectant residual in mg/L,
t = contact time in minutes.

The important issue is that a low concentration of disinfectant, with a long contact time, accomplishes the same goal as using a high residual concentration with a short contact time. However, other factors such as the pH and temperature of the process water also affect the efficiency of the disinfection process.

The EPA issued a draft guideline in 1987 for the required $C \times t$ under various conditions. Table 3.2.8-1 lists the recommended disinfectant residual–contact time relationship for disinfecting filtered water to inactivate *Giardia lamblia*. However, a lack of maximum disinfectant dosage limitations makes this grid line somewhat impractical.

TABLE 3.2.8-1 $C \times t$ **Values for Achieving 90% Inactivation of** *Giardia Lamblia*[a]

	pH	Temperature					
		0.5°C	5°C	10°C	15°C	20°C	25°C
Free chlorine[b]	6	49	35	26	18	13	9
	7	70	50	37	25	19	12
	8	101	72	54	36	27	18
	9	146	104	78	52	39	26
Ozone	6–9	0.97	0.63	0.48	0.32	0.24	0.16
Chlorine dioxide	6–9	20	13	10	5	5	3.3
Chloramines (preformed)	6–9	1295	737	675	505	366	260

[a]Adopted from *Guidance Manual* (March 31, 1989).
[b]*Ct* values will vary depending on concentration of free chlorine. Indicated *Ct* values are for 1.0 mg/L free chlorine. (For other free chlorine concentrations, see *Guidance Manual*.)

Disinfection By-products All disinfectants react with any element that is oxidizable in water. Thus, the water supply industry is now faced with the problem of controlling disinfection by-products. THMs and other halogenated organics are produced by chlorination. Aldehydes are formed by ozonation; other compounds such as haloacids, haloacetonitriles, haloketones, and bromate may also be formed depending on the type of elements present in the process water. Although the magnitude of the ozonation by-products is generally below the level of health concerns, it may only be a matter of time before science discovers which ozonation by-products significantly affect human health.

The EPA has currently set the maximum allowable level of THM as 0.1 mg/L. However, this level will most likely be further reduced in the near future.

Quality of the Water Being Processed If the water being processed contains turbidity, color, ferrous iron, manganese, and organic substances, all disinfectants described in this section will react with them. Consequently, the effectiveness of disinfection becomes very low and a significant amount of disinfection by-products will most likely be formed. Turbidity and the presence of other particulate matter in the process water are very significant with respect to effective disinfection because the microorganisms living within these materials may not be killed. Thus, water entering the disinfection process must have both a high quality and a low turbidity. The water should therefore be filtered prior to final disinfection so that the disinfecting agents can work effectively without producing significant amounts of by-products.

Safety of Operation All disinfectants are harmful to human health and animal life. The design and handling of the disinfection system should therefore strictly adhere to the local, state, and federal (OSHA) safety rules and codes, in addition to those listed in the chemical handling manuals provided by the chemical industries.

If a chlorination system is located in a residential area, a neutralization system (for chlorine gas) should seriously be considered to combat chlorine gas leakage. The decision to integrate a neutralization system into the overall plant design should be weighed for the sake of public safety, even though the governmental codes do not specify their use. Certain municipalities also prohibit the transport of chlorine gas to specific areas of the city because of safety considerations. Under these conditions, the choice of disinfectant is limited to hypochlorites of calcium or sodium. However, chlorine gas and hypochlorites may be generated electrolytically from brine or salt solutions using specially designed cells and an ordinary power source.

The locations of the ozone manufacturing facilities and the ozone contact basins are also dictated by safety considerations. These are generally isolated from the operations building because of the potential for fire and health hazards. No matter the type of disinfectant, plant operators should plan and frequently practice safety maneuver drills.

Cost   Cost is a prime consideration in selecting the type of disinfection process. The approximate construction costs for the chlorination, chloramination, chlorine dioxide, and ozone disinfecting systems, as established by the EPA and slightly modified, are listed in Table 3.2.8-2. The cost does not include the construction of supporting systems, nor does it include the construction of the contact tank.

As Table 3.2.8-2 clearly indicates, construction of the ozonation system is by far the most expensive of the four alternatives. Refer to Table 3.2.8-3 for the total cost of disinfection, including the operation and maintenance of the feed system for each of the four alternatives. These costs were developed by the EPA.

Design Criteria   The design criteria for the chlorination, chloramination, and chlorine dioxide systems are discussed below. The fourth alternative disinfection system, ozonation, is discussed in Section 4.8 of this book. Therefore, no attempt is made to describe it here.

TABLE 3.2.8-2   Cost of Constructing Disinfection Systems

Plant Size (mgd)	Chlorine (2 mg/L)	Chloramine (2 mg/L)	Ozone (1 mg/L)	Chlorine Dioxide (1 mg/L)
1	$15,000	$20,000	$55,000	$25,000
10	$32,000	$40,000	$150,000	$50,000
100	$130,000	$145,000	$800,000	$170,000
150	$150,000	$170,000	$1,200,000	$200,000

[a]Data from EPA Publication No. 600/2-79-162b, August 1979. The costs are based on ENR Index 2850.

Chlorination System

Dosage	1–5 mg/L (2.5 mg/L average)
Number of chlorine feeders	Minimum of two; one standby is required. Standard sized feeders are 10, 20, 30, 50, 75, 100, 150, 250, 500, 1000, 1500, 2000, 3000, 4000, 5000, 6000, and 8000 lb/day and in the 20:1 feed range.
Evaporators	These are normally required if the daily dosage exceeds 1200 lb/day (550 kg/day). A safe yield of chlorine gas from one 1 ton cylinder is 400 ppd (180 kg/day) without an evaporator under normal temperature.
Gas cylinder weighing scale	Minimum of one, as required by the regulatory agencies
Chlorine expansion tank	Required for liquid chlorine line
Chlorine gas strainer	Required
Chlorine gas pressure regulation valve	Required
Chlorine gas pressure gauges	Required
Chlorine eductors	Required
Water supply to the eductor	Dual supply of pressurized water or a standby booster pump is required. Additionally, a device to prevent backflow should be provided.
Chlorine solution rotometers	Optional
Chlorine gas leak detectors	Required
Chlorine residual analyzers	Required
Chlorine containers	One ton cylinders are preferred whenever the plant uses over 200 ppd (100 kg/day) of chlorine. A 1 ton cylinder weighs 1.9 tons, including the weight of the cylinder. A 20 ton stationary tank or a 17 ton tank trailer system is used for large plants.
Housing	Chlorine storage and feeder rooms shall not be located underground. The cylinders and feeders should be located in separate rooms. The inventory should include enough chlorine cylinders for a minimum of 15 days of operation.
Ventilation	Each room shall have an air ventilation system that provides one complete air change per minute. The air outlets should be located near the floor and ex-

TABLE 3.2.8-3 Costs of Disinfection

Plant Size (mgd)	Costs of Disinfection (¢/1000 gal)				
	Chlorine (2 mg/L)	Chloramines (2 mg/L)	Ozone (1 mg/L)		Chlorine Dioxide (1 mg/L)
			Air Feed	Oxygen Feed	
1	3.8	7	6	7.8	8.5
10	1.7	1.9	2.2	2.3	2.2
100	0.48	0.9	1.4	1.36	1.7
150	0.45	0.8	1.2	1.15	1.5

*a*Data taken from EPA Publication No. 600/2-81-56, 1981. The costs do not include the contact tanks.

	hausted in a safe direction. The air inlet should be through gravity type louvers located near the ceiling.
Inspection window	The door and the wall of both the chlorine storage and feeder rooms should have clear glass windows to permit viewing without having to enter the rooms.
Access door	All access doors shall be outside the building. If access is required from within the main building, the door must have an airtight seal.
Heating	The chlorination building shall have a heating system that will maintain an ambient temperature of 15°C (60°F). This feature is vital if a system does not have a chlorine evaporator.
Safety equipment	Eye wash, shower, and gas masks of the canister type must be provided in compliance with OSHA regulations.
Chlorine neutralization system	Optional (Ceilcote Co., Inc., of Ohio, Ametec, Inc., of Ontario, Canada, and EST Corp. of Quakertown, PA, manufacture the scrubber). see Appendix 13
Chlorine cylinders	All cylinders shall be anchored by chains for safety reasons. This measure must be implemented in plants located in earthquake zones.
Application points	Chlorine application points should be provided at the head of the plant, the filter influent, and the filtered water, even though chlorine may not be applied on a continuous basis.

Figure 3.2.8-2 is a schematic of a chlorine feed system.

The measured chlorine gas released from the feed system is generally dissolved in water by means of a hydraulic eductor, then transported to the application point as chlorine solution (approximately 3000 mg/L). However, in cases where the distance between the feeder and feed point is over 150 ft, the eductor is placed at the injection point. Under these circumstances, the chlorine gas is pulled through the PVC feed pipe under a slightly negative pressure for reasons of safety (in case the pipe is cracked). This scheme becomes problematic in regions experiencing cold winter weather because the cold temperature around the feed pipe causes the warm gas to condense, forming a very strong chlorine solution. Consequently, the gas flow is interrupted. Moreover, the potential for a severe chlorine leak is created due to the dissolving of the PVC joint solvent. For these reasons, design engineers must specify the feed pipes to be well insulated and the feed line is encased by an additional pipe.

Chloramination System   In the chloramination system, ammonia is added prior to or after chlorination. The source of ammonia may be anhydrous ammonia, aqua ammonia, or ammonium salts such as ammonium sulfate. However, the use of ammonium salts is limited to small installations because of their hygroscopic nature and high cost; the cost does not outweigh the fact that the ammonium salts are safe to handle. Generally, aqua ammonia is safer to handle than anhydrous ammonia and is also generally more cost effective due to its wide availability. In regions where the ambient temperature exceeds 90°F (32°C), it is recommended that 25% ammonia solution be stored in a tank located outside to avoid problems associated with ammonia vaporization.

The main design criteria for an aqua ammonia feed system are listed below.

Dosage	0.2–1.5 mg/L (30% of chlorine residual)
Number of ammonia feeders	Minimum of two; one standby is required.
Type of feeder	Solution feeder (metering pump) for 25 or 30% strength aqua ammonia.
Storage tank(s)	A minimum of 15 days storage: a minimum shipment of 4000 gal if in bulk shipment. The tank(s) shall be composed of steel or fiberglass. The tanks shall be painted white and have a sun shield in order to prevent the contents from heating up. The tank will be slightly pressured (25 psi) to prevent the escape of ammonia vapor.
Safety features	Eye wash, shower, and a dike around the storage tank for containment purposes.

Figure 3.2.8-3 illustrates the aqua ammonia feed system.

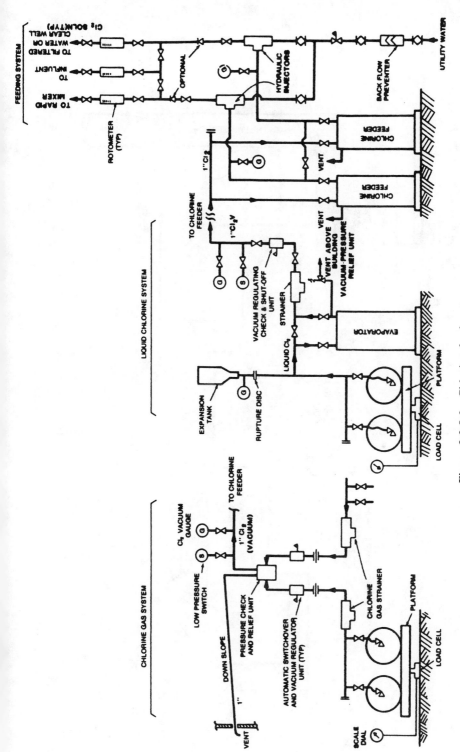

Figure 3.2.8-2 Chlorine feed system.

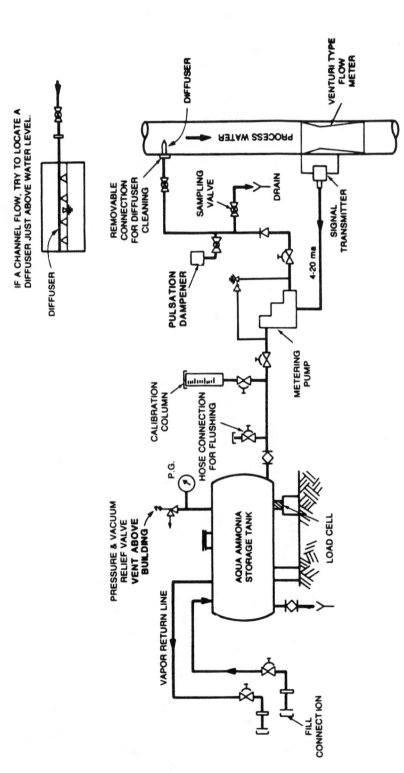

Figure 3.2.8-3 Aqua ammonia feed system.

Chlorine Dioxide   Chlorine dioxide is prepared by reacting sodium chlorite with chlorine gas in a glass column packed with ceramic Rashig rings. A self-contained chlorine dioxide feed system is commercially available. Bottled chlorine dioxide gas is not commercially available due to its explosive nature.

Theoretically, a sodium chlorite solution containing 1.34 kg of sodium chlorite will react with 0.5 kg of chlorine to produce 1 kg of chlorine dioxide. If the reaction is to be completed in a short period of time, say, less than 1 min, the pH of the mixture should range from 2 to 4. Acid feeding is therefore usually required.

Dosage	0.5–1 mg/L (the maximum dosage is limited by the EPA).
Number of feeders	Minimum of two; one standby is required.
Sodium chlorite solution tank	A fiberglass tank that is large enough to produce a 1 day supply of 17% solution at maximum dosage.
Number of chlorite tanks	Minimum of two.
Chlorite feeder	A solution metering pump with a minimum of two feeders; one standby is required.
Safety features	Eye wash, shower, and gas masks of the canister type as dictated by the OSHA regulations. Additionally, some caution must be exercised when handling sodium chlorite because it becomes combustible on contact with organic matter.

Refer to Figure 3.2.8-4 for a schematic representation of the chlorine dioxide feed system.

Disinfectant Contact Tank   The design of the disinfectant contact tank is contingent on the type of primary disinfectant selected. The $C \times t$ values provided by the EPA clearly indicate that the size of the contact tank drastically varies between chloramines, ozone, and chlorine. Nonetheless, the fundamental design rule is to minimize flow short-circuiting within the contact tank.

A recent EPA draft guideline manual suggests that the actual contact time be verified for a contact tank by running tracer tests. When these tests are performed, they establish that most existing clearwells, which serve as contact tanks, have a severe flow short-circuiting problem and that the actual flow-through time is often only 10% of the calculated detention time. The contact tank must therefore be designed to have as near a plug flow condition as possible. One of the best tank designs is identical to a baffled channel type of flocculation tank (refer to Section 3.2.4). Refer Appendix 9 for tracer test.

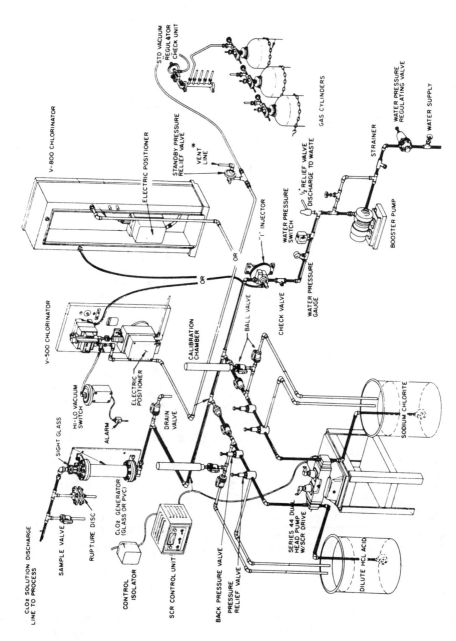

Figure 3.2.8-4 Schematic illustration of chlorine dioxide feed system produced by Wallace & Tiernan Company.

Number of tanks	An absolute minimum of one; two tanks are preferred.
Type of flow path	A serpentine flow path with or without baffles.
Water depth	12–16 ft (3.5–5 m)
End of the tank	An overflow wall shall be provided to maintain a constant water depth.
Detention time	This depends on the type of disinfectant and the $C \times t$ requirement.

Example Design Calculation

Example 1

Given A new 50 mgd plant (daily average flow rate) located in Denver, Colorado, requires a new chlorine disinfection system. Assume that the minimum, average, and maximum dosages are 1.5, 2.5, and 5 mg/L, respectively.

Determine

(i) The considerations for the system design.
(ii) The number and capacity of the chlorine feeders and the appurtenant units.
(iii) The size and number of chlorine containers (storage).

Solution

(i) There are various design considerations for the new chlorine disinfection system: the potential interruption of the chlorine supply due to severe winter weather conditions; the extreme toxicity of chlorine; the need to strictly adhere to the safety codes and OSHA regulations; specifying a large curvature for the access road, in addition to providing an adequate maneuvering space in front of the chlorine storage facility for the oversized trucks that deliver the chlorine cylinders; providing an adequate height (14 ft) below the monorail to facilitate the unloading of chlorine cylinders from the delivery trucks; and providing a heating system in the chlorine storage and feeder rooms.

(ii) The number and capacity of the feeders are as follows:

Maximum plant flow rate	$1.5 \times 50 = 75$ mgd
Maximum dosage	$(5 \times 8.34) \times 75 = 3128$ ppd
Average dosage	$(2.5 \times 8.34) \times 50 = 1043$ ppd
Minimum dosage	$(1.5 \times 8.34) \times (0.25 \times 50) =$ 156 ppd
Ratio of maximum to minimum dosages	$3128{:}156 = 20{:}1$

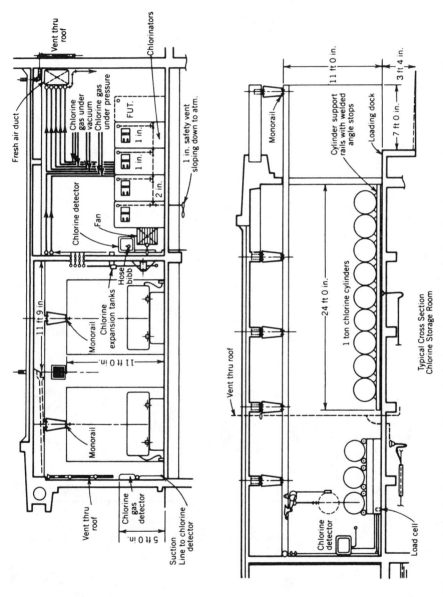

Figure 3.2.8-5 Typical cross sections of chlorine storage room (bottom) and chlorinator room (top). (Courtesy of J. M. Montgomery, Consulting Engineers.)

Note: Commercially available chlorine feeders generally have a 20:1 feed range for manual control and a 10:1 feed range for automatic control.

Based on these figures, select three identical 1500 lb/day feeders. This choice allows for the maximum and minimum dosages and the third feeder serves as an automatic control standby unit.

Additionally, provide two identical 4000 lb/day chlorine evaporators. The maximum chlorine dosage cannot be handled without the use of evaporators.

(iii) The optimum storage of chlorine is based on the maximum dosage of the disinfectant at the average plant flow rate. Since the delivery of chlorine may be interrupted during the winter months by snow storms, a 2 week stockpile must be provided under the above conditions; the plant flow rate during the winter months is less than half the average annual rate. Thus, actual storage (under normal dosage) will be approximately 60 days.

$$[(5 \times 8.34) \times 50] \times 15 = 15.6 \text{ tons}, \quad \text{say 16 tons}$$

The storage area should also accommodate the empty 1 ton cylinders. Since each truck carries a minimum of 8 cylinders, provide 8 additional spaces for the empty cylinders. Consequently, a total of 24 storage racks should be provided in the storage room for the 1 ton cylinders.

Figure 3.2.8-5 is an illustration of typical both a chlorine storage and feeder room.

Operation and Maintenance   The operation of the disinfection process is contingent on the type of disinfectant that is employed, the plant size, regional characteristics, and the raw water characteristics. Since chlorine is the primary disinfectant in water treatment, this discussion is centered on the process of chlorination.

If prechlorination can be practiced without producing significant levels of THM, a residual chlorine concentration of approximately 0.3 mg/L should be maintained in the settled water at all times. This practice will prevent the growth of algae on the basin walls and will prevent the growth of microorganisms in the filter bed. The level of the residual chlorine in the clearwell should be at least 0.5 mg/L so that the chlorine residual at the end of the distribution system will be 0.2–0.3 mg/L at all times. Plant operators must record the level of residual chlorine in the clearwell and distribution system at regular intervals, since a log of these figures must be submitted to the local regulatory agency on a monthly basis.

The chlorine feeders and appurtenant systems must be operated and maintained in strict accordance with the operation manual furnished by the equipment manufacturers and the manual provided by the Chlorine Institute; this includes the handling of chlorine cylinders. If the chlorine feed system should fail, the unchlorinated water must not be allowed to enter the distribution system, because the Safe Drinking Water Regulations demand a minimum residual chlorine of 0.2 mg/L to be present at the end of the distribution system.

Maintenance of the chlorination system requires a thorough understanding of the literature provided by the manufacturer, in addition to the operation and maintenance manual prepared by the design engineers. The daily, weekly, and monthly operation and maintenance procedures are generally described in these manuals. It is imperative that operators do not attempt to perform any unfamiliar maintenance work or work that is beyond their qualifications. Many serious accidents have occurred because someone has ignored this very basic rule.

In the case of chlorine leakage, corrective actions must only be taken by trained operators wearing the appropriate safety equipment. It is essential that all plant operators be trained to repair chlorine leaks. Operators should always work in pairs when searching for and repairing chlorine leaks. All other personnel must be evacuated until the conditions are safe. If the chlorine leak is large, the fire department and police must be notified immediately; all inhabitants in the adjacent area must be warned and evacuated.

Alkali chemicals such as caustic soda and lime may be used to neutralize very small chlorine gas leaks. For example, stoichiometrically 1.13 lb of caustic soda is required to react 1 lb of chlorine. Therefore a 1-ton chlorine leak would require a theoretical 2260 lb of caustic soda. However, the scrubber manufacturers recommend a 3600 gallon of 7% caustic soda. In the case of hydrated lime, 125 lb are required to neutralize a 100 lb cylinder of chlorine and 2500 lb are required for 1 ton of chlorine under ideal mixing conditions.

It is extremely important that treatment plants establish a formal safety program. All personnel should periodically perform "hands-on" training using the safety equipment. The American Water Works Association provides training materials, including videos. The Chemical Transportation Emergency Center (CHEMTREC) provides immediate advice for those at the scene of an emergency and quickly alerts experts for more detailed assistance and appropriate follow-up whenever the situation becomes uncontrollable. CHEMTREC has a toll-free number (1-800-424-9300), which all plant operators should keep in a readily accessible place. An example of chlorine gas scrubbing system is shown on Appendix 10-7 (chem-7).

REFERENCES

Aieta, E. M., et al., "A Review of Chlorine Dioxide in Drinking Water Treatment," *J. AWWA*, 78(6):62 (1986).

AWWA Committee Report, "Disinfection," *J. AWWA*, 74(7):376 (1982).

Barnes, D., "Alternatives to Chlorination for Water Disinfection," *Water J.* (*Australia*), p. 12 (September 1983).

EPA, *Municipal Wastewater Disinfection—Design Manual*, EPA/625/1-86/021 (October 1986).

Glaze, W. H., et al., "The Chemistry of Water Treatment Processes Involving Ozone, Hydrogen Peroxide, and Ultraviolet Radiation," *J. Ozone Sci. Eng.*, 9:335 (1987).

Hoff, J. C., "The Relationship of Turbidity to Disinfection of Potable Water," EPA, 370/9-78-00c, USEPA (1977).

Longley, K. E., et al., "The Role of Mixing in Wastewater Disinfection," ASC Meeting, Anaheim, California, 1978.

Metcalf & Eddy, Inc., *Wastewater Engineering: Treatment Disposal Reuse*, 2nd ed., McGraw-Hill, New York, 1979.

Morris, J. C., *Aspects of the Quantitative Assessment of Germicidal Efficiency Disinfection—Water and Wastewater*, Ann Arbor Scientific Publishers, Ann Arbor, MI, 1975.

Shull, K. F., "Experience with Chloramines as Primary Disinfectants," *J. AWWA*, 73(2):101 (1981).

White, G. C., *Handbook of Chlorination*, Van Nostrand Reinhold, New York, 1972.

3.2.9 Instrumentation and Process Control

Purpose   Due to increased public awareness on the subject of safe drinking water, the regulatory agencies have made the drinking water standards more stringent. Consequently, water treatment plants must have more sophisticated processes and equipment, and plant operators are forced to operate and maintain systems that are vastly more complicated than in the past. As the costs of energy, chemicals, and personnel continue to increase, the managers of water purveyors must be concerned with the quality of the water supply as well as improving the efficiency of production and supply. One solution to these issues is to provide the treatment plants and distribution systems with modern supervisory control and data acquisition systems.

The advantage in computerizing treatment plants is that plant operators are supplied with timely and accurate information on which to base their decisions, thus allowing them to achieve effective process control. Process variables such as the raw water quality, plant flow rate, chemical feed rate, and finished water quality may be monitored by on-line instruments, which relay the information to the operator on a continuous basis. This type of instrumentation may also help to reduce equipment damage because the status of each system is continuously monitored and should the situation go wrong the plant is automatically shut down.

Computerization of the treatment plant frees plant operators from monitoring the various gauges or charts and from continuously updating the operational logs. The system required by water treatment plants is generally simple in nature, unlike the complex systems used in the oil, chemical, and rocket industries. The safety and security of the plant personnel and the plant itself may be improved by installing a closed circuit television monitoring system. The installation of modern instrumentation in a treatment plant per-

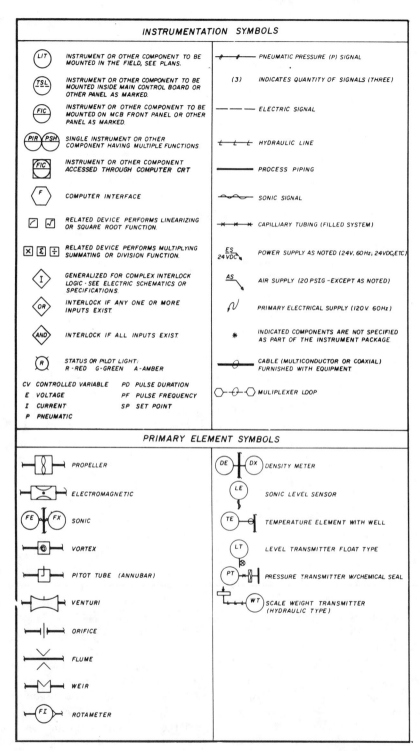

Figure 3.2.9-1 Symbols. (Courtesy of J. M. Montgomery, Consulting Engineers, Inc.)

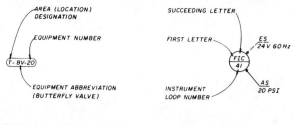

MECHANICAL EQUIPMENT CALLOUT   INSTRUMENT TAG NUMBER

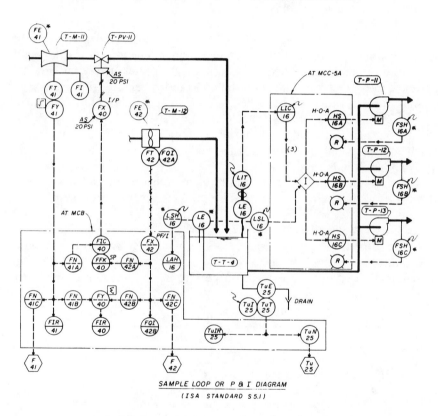

SAMPLE LOOP OR P & I DIAGRAM
(ISA STANDARD S5.1)

SAMPLE INSTRUMENT TAG NUMBERS

TAG NUMBER	FUNCTION
FE - 41	FLOW SENSING ELEMENT (LOOP 41)
FI - 41	FLOW INDICATOR
FIC - 41	FLOW INDICATING CONTROLLER
FIR - 41	FLOW INDICATING RECORDER
FFK - 41	FLOW RATIO STATION
FN - 41	FLOW SIGNAL ISOLATOR
FQI - 41	FLOW INTEGRATING TOTALIZER
FSH - 41	FLOW SWITCH HIGH
FSLL- 41	FLOW SWITCH LOW LOW (EXTREME)
FT - 41	FLOW TRANSMITTER
FX - 41	FLOW SIGNAL CONVERTER
FY - 41	FLOW SIGNAL MATH FUNCTION

Figure 3.2.9-1   (Continued)

mits operators to concentrate on the overall control of the treatment plant and on effective plant operation.

Symbols and Manufacturers The symbols used in designing the instrumentation system are based on those provided by the Instrument Society of America (ISA), the Institute of Electrical and Electronic Engineers (IEEE), and the National Electrical Manufacturers Association (NEMA). Refer to Figure 3.2.9-1 for an example.

The field of instrumentation is a rapidly evolving technology in which new components and equipment are increasingly being geared toward the field of water treatment. Thus, it is very rare that a single manufacturer will be able to furnish the bulk of the necessary equipment. Yet, it is very important to place the overall responsibility of installation and system start-up with a single manufacturer. Firms that have been dominant providers of service in the United States are Fisher and Poter Company, BIF Unit of General Signal, Bristol Babcock, Inc., Westinghouse Company, and Robert Shaw Controls Company. Major manufacturers of instrumentation control equipment such as Foxboro, Leeds, and Northrop may or may not be involved in water treatment plant projects due to the level of competition and the required bidding procedures for services of this nature. However, this is beginning to change.

Considerations The major considerations in designing the instrumentation and control system are the local conditions, the benefits of the system, the selection of the system, the needs of the system, the function of the instru-

TABLE 3.2.9-1   **Benefits of the Instrumentation System**

Purpose	Benefits
Process	Improved process results
	Efficient use of energy
	Efficient use of chemicals
	Automation of some process adjustments
	Greater ability to control complex processes
	Process changes detected in a timely manner
Personnel	Timely and accurate process information
	Safer operation
	Increased security
	Efficient use of labor
	Allows for an overview of plant operations
	Decrease in manual paperwork
	More complete records
Equipment	Increase in running time
	Status is known at all times
	Automatic shutdown to prevent major damage

mentation and control system, the availability of the system, and the understanding and commitment of the operators and management to the system. Each of these issues is discussed in detail.

Local Conditions The instrumentation system is a highly sophisticated item. For this reason, the following issues must carefully be evaluated: the availability of the equipment, the availability of qualified operators, the follow-up services, weather conditions, and the living standard of the region, since the living standard dictates the drinking water quality standards. Due to the high cost of systems acquisition, as well as the cost of repair and replacement parts, the system should only be operated by highly skilled and knowledgeable operation and maintenance personnel. Moreover, the environment of the control room must be strictly controlled to optimize the performance of the computers (microchips); the room must be maintained at a specified temperature and humidity and have a dust-free environment. Extensive instrumentation and control systems are not recommended for developing countries or small plants located in remote areas since they generally cannot meet the criteria listed above.

Benefits of the System Although all plant operators would like to have a modern instrumentation and control system, in each case the benefits of the system must be weighed against the cost factor. For instance, a simple water treatment process such as slow sand filtration cannot justify the installation of high-level instrumentation. Yet, an ozonation system, reverse osmosis process, pumping plant, filter process, and water distribution system all greatly benefit from a highly sophisticated instrumentation and control system. Manual No. OM-6 of *Process Instrumentation and Control System*, by the Water Pollution Federation, lists a number of benefits derived from instrumentation; refer to Table 3.2.9-1.

Selection of the System Although there are various ways to improve plant operations and to achieve the benefits listed in Table 3.2.9-1, modification of an existing, outdated system is usually very difficult, primarily due to outdated sensors, transmitters, and control signalers. In fact, the complete replacement of an old system is often more cost effective than updating certain parts. However, the owner may wish to keep the existing instrumentation system, including the 7 day recording circular chart and the relay logic control system, as a back-up to the new computerized system and/or a training tool for plant operators. In this type of situation, the design engineer should respect the wishes of the owner.

When an engineer is designing a new plant he/she should evaluate the most appropriate degree of instrumentation and control for that particular plant. Although the state-of-the-art system may appeal to both the engineer and owner, these exotic systems may become obsolete after several years due to rapid technological advances, and replacement parts may become unavailable.

The basic rule in selecting a new instrumentation and control system is to anticipate the future needs of the treatment plant, including plant expansion, the potential addition of new treatment processes, and owner preference or corporate policy. If the new system must be compatible with the systems of neighboring districts, engineers should also consider the type of system employed by adjacent water districts and the system used by the city: for example, determine if the city has a main computer that controls the overall distribution of water.

It is important to remember that most water treatment unit processes are slow reacting in nature and have high system lag times. In fact, fast response times create unstable or undesirable operational conditions, such as "hunting" actions. Thus, design engineers should keep this fundamental nature of water treatment process in mind when selecting the instrumentation and process control system. Also, the units and system to be selected should be so called "user friendly" otherwise the design system will never be used effectively.

Needs of the System Design engineers must collaborate with owners to identify the needs of the system. For instance, treatment plants consisting of a simple treatment process only require instrumentation for the plant flow rate, the monitoring of raw and filtered water turbidity, and monitoring of the residual chlorine content of the finished water. Conversely, plants consisting of rather complex processes, such as ozonation, or a large number of filters and pumps require an extensive instrumentation and control system in order to simplify plant operation and to make the plant efficient and cost effective.

The needs of a system may be classified into three categories: essential for plant operation, useful for plant operations, and luxury item. The essential items are plant flow rate control, chemical feed rate control, pump rate control, ozonation system control, filter control, analysis and recording of water quality, control and recording of residual chlorine, and detection of chlorine gas leakage (with an alarm). The useful items are those that reduce the mental and physical activity of the operators or those that can save operational costs. They include computerized data logging, programmable control of the filter backwashing process, closed-loop automatic control of residual chlorine and the pH of the water being processed, and devices that check the coagulant dosages such as streaming current detectors and automatic plant controllers with advanced data handling capability (the latter require extensive custom designed software). All these items are not vital to normal plant operation. Lastly, luxury items are defined as items that make plant operation easier without providing compensatory savings in cost. The design engineer should discuss the needs of the system with both plant operators and the owner prior to making any decisions.

The Task of the Instrumentation and Control System After the preceding items have been considered, the type and purpose of the system must be defined. At a very basic level the system can be used to serve operational, maintenance, safety, and documentation purposes. The operational goals of the instrumentation and control system should be to provide the best control quickly and without human errors. The system can inform the operator of all current activities, where they are happening, and the duration. The system is also capable of making corrective or preventive actions. It can detect and warn the operator of changes in temperature, excessive vibrations, and abnormal levels of power consumption by individual operating equipment and thereby prevent serious damage, costly repairs, and downtime from occurring: These all relate to maintenance work. With respect to the safety aspect of plant operation, a surveillance system is useful in meeting security requirements and the safety of the plant operator, who must operate the equipment alone. Fatal accidents may also be minimized by installing systems that can detect and warn the operator of chlorine gas, ozone, and oxygen leaks. Lastly, documentation is very important for both the operator and plant manager in identifying the control modes and running time, in analyzing the history of repair work, and in taking inventory of the available spare parts.

Availability of System   Three basic types of system are currently available: digital, analog, and control instruments. Digital systems primarily have two positions: on/off, open/closed, high/low, or alarm/normal. This type of signal is used to indicate a change in status or to act as an alarm. The signals may be activated by a position, limit, float, or pressure switch. Analog data systems cover a range of values from liquid or gas flow rates, liquid levels, liquid or gas pressure, to the turbidity or pH of the water. The analog data may be sensed and transmitted in their analog form or may be converted into digital form. The control instruments range from a simple on/off switch to a programmable controller with a computer back-up.

Many factors are involved in selecting the most appropriate type of system for each treatment plant. The characteristics and the need of the treatment process train may determine both the type of control and the complexity of the control system. Factors such as the frequency of information, technical limitations, and local conditions may also limit the type of system designed. Other important considerations are the capital cost of the system and the associated operation and maintenance costs. Lastly, the ease and frequency of calibration, the difficulty in maintenance, and the availability of timely servicing must also be evaluated.

A conventional monitoring and control instrumentation system uses 4–20 milliamp (mA) analog signals (standard) in transmitting measurement and control signals. Voltage or air pressure signals are rarely used today.

Understanding and Commitment   In order for any instrumentation and control system to operate at its fullest potential, the operators must have a complete understanding of the system. The facilities must therefore have well organized training courses that not only familiarize operators with the system but also teach them the steps to be taken when the system fails: after all the system is not infallible. Thus, back-up control and monitoring methods must be an integral part of the plant design and operators must be familiar with these methods.

System maintenance must be conducted by either the operators or maintenance personnel. However, if the system is highly complex, it is in the best interest of the owner to sign a service contract with the equipment supplier or the service agency. Since the operator and maintenance of a sophisticated instrumentation and control system requires an adequate budget just to keep it in service, management must clearly understand the benefits of a properly functioning system. If management is committed to the system, it must also be committed to providing adequate resources for the system: financial resources and the personnel required to design, operate, and maintain it. Without both a clear understanding of the benefits and a thorough commitment to the system, by the operators and management, the instrumentation and control system will not function at its full potential, and many future problems may be anticipated.

Common Measurements   The level, pressure, flow rate, turbidity, pH, and chlorine residual of the process water are major process variables that are measured on a continuous basis in all water treatment plants. Each variable is briefly discussed.

Level measurements are commonly made by float operated transmitters, pressure sensors, capacitance probes, sonic or ultrasonic units, or a bubbler tube system.

The pressure within a system may be measured as absolute pressure, where vacuum is zero pressure; as gauge pressure, where ambient atmospheric pressure is designated as zero; or as differential pressure. The sensors may be U-tube manometers, Bourdon tube gauges, bellow gauges, pneumatic and electric force-balance pressure transmitters, or strain gauge type transmitters.

Flow measurement may be conducted by weirs and flumes, such as the Parshall flume, which is used to measure open channel flow. However, the flow of liquid or gas in pipe lines should be measured by differential flow meters such as the Venturi type flow meter; plate orifice meters, magnetic flow meters, acoustic (ultrasonic) flow meters, vortex shedding flow meters, and propeller or turbine flow meters may also be employed. Yet, the Venturi flow meters are the most commonly used type of flow meter due to their high accuracy ($+0.5-2\%$), low headloss, and wide availability. Engineers should note that selection of the proper size or capacity of a flow meter is vital to obtaining accurate flow measurement. Most flow meters have a turndown (maximum to minimum flow rate) of 5:1 to 10:1. A common mistake of many inexperienced design engineers is to design the meter size based on future maximum flow rates; this practice does not allow for the measurement of low flows at the time of installation.

In the United States, turbidity is measured by nephelometric turbidimeters. The principle of the turbidimeters is to detect and measure the scattered light beam or the portion of a light beam that has not been absorbed by suspended particles, after an incident light has been passed through the process water. Thus, the size, shape, refractive indexes, absorption capacity, and viewing angles of the suspended matter greatly influence the turbidity reading. The nephelometric turbidimeter may be used in situations where the turbidity of the water is approximately 30 NTU. It should not be employed in situations where high turbidities are expected (over 30 NTU) since the meter will go blind at these levels and give false readings.

Turbidity has a fairly linear direct response: zero signal at zero turbidity. True color does not register as turbidity but can cause negative errors. The surface scatter type of turbidimeter may be used for water with high levels of turbidity but must be installed in a vibration-free environment. This type of meter is not appropriate for measuring turbidities lower than 1 NTU because the response is nonlinear. Since any type of turbidimeter will give a false high reading whenever the water sample contains air bubbles, an air bubble trap should be installed in the sample line upstream of the meter.

pH meters are generally composed of three electrodes: a pH sensing electrode, a reference electrode, and an electrode that compensates for the temperature. These electrodes are usually mounted in a chamber and installed in the sample line or submerged in a tank or channel. The most recent models have a signal preamplifier within the electrode assembly to minimize the electrical noise of the high-impedance electrical circuits.

Last, but not least, free and combined chlorine residuals may be measured by residual chlorine analyzers. These units are commercially available from several manufacturers. The principle used in measurement is amperometric titration through oxidation–reduction titration procedures.

Basic Controls   When selecting the type and control mode for a specific application, the engineer must be familiar with the purpose and the hydraulic characteristics of the particular treatment process. There are three basic considerations which engineers should remember when selecting a control system: the nature of the required control, the degree of disturbance that the process can tolerate, and the cost.

There are four types of control method: manual, semiautomatic, automatic, and supervisory. Semiautomatic control requires manual initiation of the automatic function. A typical example would be manual start-up of the automatic backwash sequence. The automatic control method uses sensors, limit switches, times, analytical instruments, controllers, and logic control devices (i.e., programmable controllers) to automatically control the process or equipment. Supervisory control basically refers to a system that is remote controlled: either in-plant or remote from the plant.

The two basic schemes for the control loops are the open loop and the closed loop. Open-loop control does not possess a measurement device within

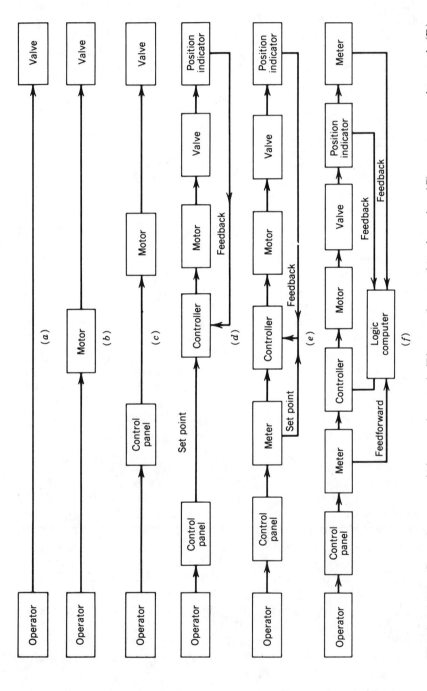

Figure 3.2.9-2 Control systems: (A) manual control; (B) manual control made easier; (C) remote manual control; (D) automatic feedback control; (E) automatic feedforward control; and (F) automatic feedforward–feedback computer control. (Adapted from WPCF Manual No. OM-6, 1978.)

the circuit and the treatment process is not contained in the loop (Figure 3.2.9-2, scheme C). A prime example of this type of control is the flow-paced chlorine feed system because the chlorine feed rate is proportional to the flow rate of the processing water. Chemical feeders commonly use open-loop control because of the ease and effectiveness of the system. In contrast, the closed-loop control circuit includes a device that measures the effect of the process. This information is then relayed to the controller, which compares the result with a set point and performs the necessary adjustments. For example, a closed-loop chlorination system will be produced if a signal is relayed from the residual chlorine analyzer to the controller. Schemes D, E, and F of Figure 3.2.9-2 are examples of closed-loop circuits.

As previously mentioned, water treatment processes usually have a long lag time and are therefore not well suited for closed-loop control systems.

Several types of control mode may be used, depending on the purpose and hydraulic characteristics of the process. The three basic modes are proportional control and its variations, cascade control, and ratio control.

Proportional control is defined as a system in which the controller continuously adjusts a controlling device, such as a valve, to balance the process input with the process demand. An example of proportional control is the control of the filtration rate of filters with respect to the water level of the clearwell. There are also several variations of proportional control: proportional control with reset control or reset control with derivative control.

Cascade control is widely used in the water treatment field, especially in pumping control where the output of one controller adjusts the set point of another controller. In contrast, the ratio control scheme maintains a constant ratio between two variables; the primary variable is normally uncontrolled. The ammonia feed rate of the chloramination process is controlled in proportion to the quantity of the residual chlorine.

Although engineers should select control systems that are simple and cost effective, one type of control system should never be used for chemical feeding: the on/off type of control. The impulse driven control scheme is an on/off type of control which has been widely employed over many years, but it tends to feed chemicals only part of the time and therefore a portion of the water remains untreated. This mode of chemical feeding is detrimental to coagulants such as alum or ferric chloride and to other chemicals such as chlorine and chloramines.

Computer-Based Monitoring and Control   Computer-based monitoring and control equipment uses digital signals. A large number of signals may be transmitted by installing a remote terminal unit (RTU) at each major unit process. The RTU receives analog or digital signals from nearby sensors, then sequences the transmission of these signals so that a single coaxial cable may be used to send and receive a large number of signals. The coaxial cables then transmit these measurement and control signals to a central processing unit (CPU) located in the control center, where they are displayed on a color cathode ray tube (CRT) and logged on a printer or stored on a disk.

Control of the treatment processes is accomplished by coding commands into the video terminal keyboard. The microprocessor-based CPU communicates this information to all the elements of the system: the RTUs and peripherals such as bulk storage devices and printers. The CPU is capable of handling real-time commands; that is, the computer system is capable of handling all multi-user requests, system alarms, process control commands, and much more, almost on an instantaneous basis. The system operator is able to view the system data at the computer terminal.

The use of programmable controllers (PC) and programmable logic controllers (PLC), which facilitate local process control, is another aspect of the computer-based control approach. For example, the filter backwash cycle may be controlled by a PC and the operator will be informed of the progress status of the backwash cycle. Distributed control systems (DCSs) and supervisory control and data acquisition (SCADA) systems have gained popularity in the field of treatment (plant) and water distribution systems and many water purveyors are now adopting their use.

Conventional Versus Computer-Based Systems Conventional monitoring and control instrumentation is a proven and reliable method of operating a water treatment plant. The computer-based monitoring and control system is also a proven technology: however, its use has only become widespread since the late 1970s. Both methods have their advantages and disadvantages. These are listed below.

The conventional monitoring and control system, when compared to a computer-based system, has the following advantages:

1. The technology is familiar, thus the operation personnel do not have to be retrained.
2. Troubleshooting is generally easier for the operators because the system is familiar.
3. Much of the equipment is standardized within the water treatment industry.
4. The overall capital costs are generally lower than for the computer-based system.

However, it also has the following disadvantages:

1. The maintenance requirements are higher due to the use of mechanical relays and timers.
2. Separate hard-wired connections are required for each measurement or control signal.
3. Possible problems with electrical noise may exist if the analog signal wire comes in close contact with a source of high voltage.

4. The incorporation of new instrumentation is more difficult and costly during plant expansion unless the initial plant design and construction are properly planned.

The computer-based system has the following advantages when compared to conventional monitoring and control systems:

1. The system is more reliable and requires less maintenance due to the use of solid-state components which are housed in sealed and noise-proof enclosures.
2. The system is more flexible due to the ability to modify the software to changing needs and the ability to easily incorporate new signals when required.
3. The system is easily expanded to accommodate future process units, in addition to accommodating water distribution data telemetered from locations remote from the water treatment plant.
4. There is a reduction in the number of hard-wired connections; thus, there may be a cost savings in electrical construction due to the fact that the coaxial cables are capable of carrying much more information.
5. Analog (conventional) or digital signals may be incorporated through the use of RTUs.
6. The presentation of operational data is improved through the use of a video terminal capable of color graphics.
7. The operating staff will potentially respond more effectively to impending emergency situations because they will be reacting to trends rather than critical alarms.
8. Reports are automatically generated at the end of each shift, the end of each day, and the end of each month, as required by the regulatory agencies.
9. The staff requirements may potentially be reduced due to the decrease in paperwork and instrumentation maintenance and the ease with which the automatic functions are incorporated.
10. Plant management and planning are improved because the presentation and integration of the plant process data are vastly improved.

However, the computer-based systems also have the following disadvantages:

1. Should the computer system malfunction, the operator could lose both monitoring and control capabilities throughout the plant.
2. The overall capital costs may be somewhat higher than the conventional control system.
3. The operators must be retrained to use the computer system.
4. Key plant functions must usually have conventional controls for back-up operation.

5. Te computer systems are not standardized in the water treatment industry. Thus, the expansion of hardware may be difficult and expensive due to the need to match components. If the new equipment is limited to those offered by the installing manufacturer, the engineer may not be able to obtain competitive bids.

Levels of Computer-Based Monitoring and Control   The size of the computer system required by the plant is not a function of the plant size. Rather, it is determined by the number of analog and digital signals that are to be fed into the computer: the number of signals increases in logarithmic fashion with an increase in the number of tasks.The computer-related tasks are data logging, report generation, alarm, plant graphic displays, analog variable displays, and manual and automatic plant control. There are six computer systems that are different combinations of these tasks:

1. Report generation.
2. Data acquisition and logging/report generation/alarm indication.
3. Data acquisition and logging/report generation/alarm indication/plant graphic display.
4. Advanced display and data handling: data acquisition and logging/report generation/alarm indication/plant graphic displays/analog variable displays.
5. Manual plant control and advanced data handling: data acquisition and logging/report generation/alarm indication/plant graphic displays/analog variable displays/manual plant control.
6. Automatic plant control: data acquisition and logging/report generation/ alarm indication/plant graphic displays/analog variable displays/automatic plant control.

The fifth and sixth computer systems are commonly known as DCSs or SCADA systems.

The first system is a very basic computer system that only generates reports. It is not hard-wired to the plant and the data must therefore be inputted manually. Systems 2 through 6 are built around a mainframe or a powerful personal computer and may be expanded to handle many additional tasks.

The hardware costs of the last five configurations are somewhat dependent on the plant size and layout. Selection of the memory size is determined by the number of analog and digital signals which must be processed and stored by the system. The mode of control and transmission (analog or digital) also influences the size and cost of the computer system. The use of PCs, as opposed to standard relay logic control, increases the cost (PCs are normally considered as part of the system) but reduces the cost of the electrical system because there are fewer relays, circuits, and wiring. If the plant size is large and the unit processes are spread out, the use of RTUs is cost effective since single coaxial cables may be used to transmit both the analog and digital signals.

Other features that should be considered are the standard "vendor" report formats versus custom editable report formats, vendor O&M software versus custom editable O&M software, and the inclusion of a separate engineering or laboratory terminal.

For budgetary purposes, the estimated costs (1989 costs) for the six alternative systems for a 2 m³/s (50 mgd) ordinary conventional process water treatment plant at an ENR index of 5000 are as follows:

System 1	System 2	System 3	System 4	System 5	System 6
$5,000	**$1000,000**	**$125,000**	**$150,000**	**$600,000**	**$800,000– $1,000,000**

If a conventional monitoring and control system is adopted, the estimated cost for systems 1 through 4 would be approximately $220,000. Figure 3.2.9-3 is an example of system 5.

Design of the Instrumentation and Control System   The design of the instrumentation and control system must be conducted by I&C engineers. However, the project engineer should maintain an open line of communication with the I&C engineers and perform a thorough check of the final design. Since the system must be operated and maintained by the plant operators, it is beneficial to elicit input from both the operators and the owner during the design phase.

Design work usually produces the following documents in the sequence listed below. The process flowsheets (PFS) prepared by the project engineer (Figure 3.2.9-4) are the basis for the P&ID diagrams.

- Process and instrumentation diagram (P&ID)
- Process control diagram (PCD)
- Instrumentation and input–output summaries (IIOSs)
- Instrumentation specification sheets (ISSs)
- Logic diagrams
- Panel layout drawings
- Loop interconnection drawings (LIDs)
- Instrument installation details (IIDs)

The P&ID is a functional schematic presentation of the treatment process and the required instruments and controls. Its purpose is to illustrate the function of the treatment plant without referring to the actual hardware; the PCD specifies the monitoring and control loops.

The purpose of the IIOS is to provide continuity between the P&ID and IIOS. These documents specify both the quantity and the characteristics of

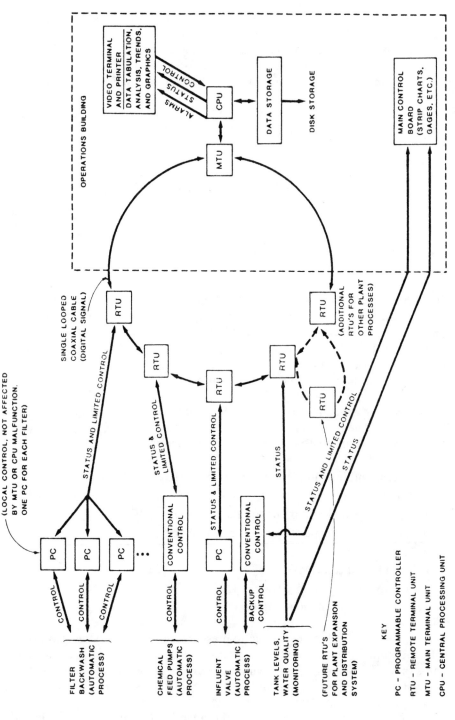

Figure 3.2.9-3 Schematic diagram of monitoring and control system (principal functions).

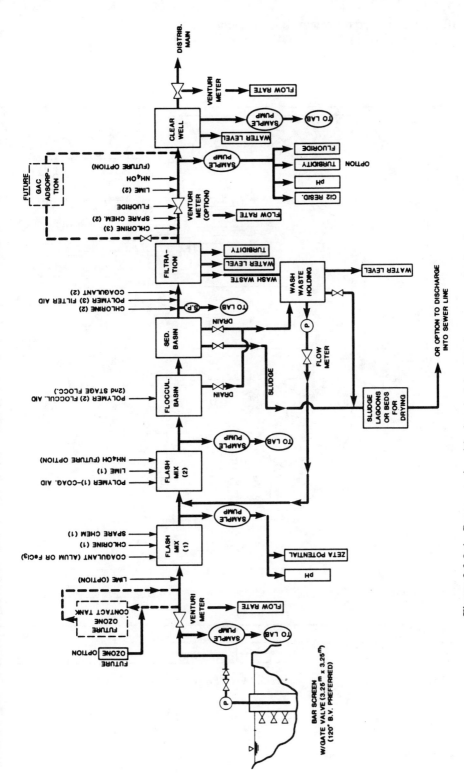

Figure 3.2.9-4 Process train with chemical feed and process control points.

311

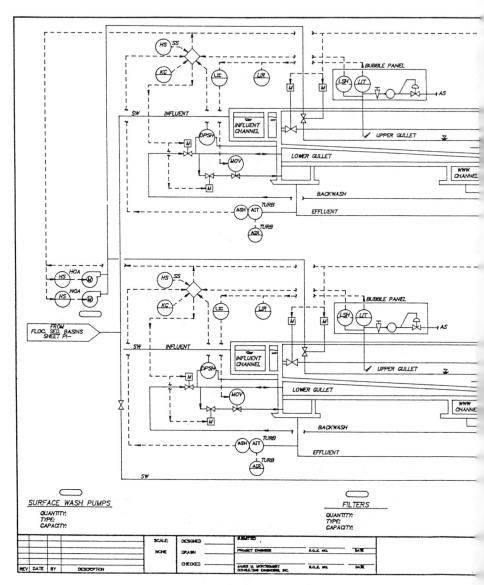

Figure 3.2.9-5 P&ID example.

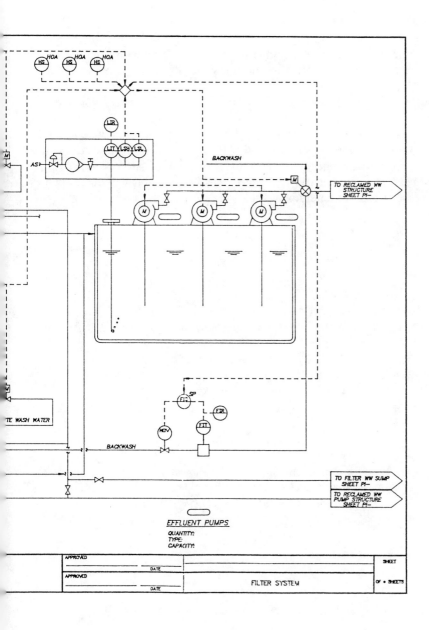

EFFLUENT PUMPS
QUANTITY:
TYPE:
CAPACITY:

FILTER SYSTEM

the analog and digital instrumentation. Similarly, the ISS provides detailed specifications for each task and panel instrument. Logic diagrams for the control panels and programmable logic controls (PLCs) should be obtained by the instrumentation contractor. LIDs are primarily used in projects where the instrumentation and control system must be interfaced with existing control equipment. A typical P&ID for ordinary gravity filters is presented in Figure 3.2.9-5.

Generally, the design engineer should follow the basic design philosophy of KISS—"keep it simple system." There have been several case histories where the plant operator was incapable of handling the situation when there was a sudden and complete failure of the SCADA system during the night or weekend shift. Moreover, plant operators should not be allowed to become overly dependent on automatic control systems because this will make them ill prepared to act in an emergency situation.

BIBLIOGRAPHY

Arndt, D. J., "Making the Right Moves in SCADA Selection," *Water Eng. Management*, p. 25 (October 1985).

AWWA, "Automation and Instrumentation," Manual M-2, 1982.

Babcock, R. H., *Instrumentation and Control: Water and Wastewater Engineering*, Special Issue (1968).

Dobs, D. M., et al., "Quick Payback Comes After PLC Installed," *Water Eng. Management*, p. 25 (October 1988).

Eckman, D. P., *Industrial Instrumentation*, Wiley, New York, 1975.

Heywood, C. H., "SCADA Systems Today and Tomorrow," *Water Eng. Management*, p. 18 (July 1986).

WPCF, "Process Instrumentation and Control Systems," Manual of Practice No. OM-6, 1978.

Liptak, B. G., Editor, *Process Measurement Instrument Engineering Handbook*, Revised Edition, Chilton Book Co., Radnor, Penn. (1982).

Considine, D. M., *Process Instruments and Controls Handbook*, 3rd Edition, McGraw-Hill Book Co., N.Y. (1985).

CHAPTER 4

Subordinate Plant Facilities

4.1 CHEMICAL FEED SYSTEM

Purpose   Most water treatment plants use a variety of chemical engineering processes to produce large quantities of safe drinking water within a short period of time. Therefore, the design of the chemical feed system, including handling, storing, feeding, and mixing, requires special attention.

A properly designed chemical feed system can improve treatment efficiency and thereby reduce operation and maintenance costs. Three basic forms of chemical are used in water treatment: dry, liquid, and gas. It therefore follows that the basic types of chemical feeders are dry, liquid, and gas feeders. Liquid volumetric feeders are preferred over the other two because of their compact size, accuracy, and convenience in handling. Generally, chemical feeders must have a feeding capacity that is broad enough to accommodate the variations in plant flow rate and chemical dosages.

Water Treatment Chemicals   Water treatment chemicals should be of a high-grade, food additive class or those approved by the EPA for the treatment of drinking water. The quality standards for the major water treatment chemicals are established by the American Water Works Association (AWWA). The chemicals may be classified according to their application in water treatment. The chemical names, commercially available form, and approximate unit cost (Bulk purchase: 1989) are listed below. Some important characteristics of these chemicals may be found in Table 4.1-1 and Appendix 13 (Chem. 1-7).

Coagulants and Flocculants

Aluminum sulfate (alum)	Dry or 49% liquid	$250/ton as dry form
Ferric chloride	35–45% Liquid	$200/ton as 42% solution
Ferric sulfate	Dry form	$350/ton
Cationic polymers	Liquid form	$1,500/ton
Anionic polymers	Primarily dry form	$5,000/ton
Nonionic polymers	Primarily dry form	$5,000/ton
Lime (for softening)	Dry form	$70/ton

TABLE 4.1-1 Major Water Treatment Chemicals

Name	Suitable Materials	Available Strengths	Solubility (g/100 mL)	Major Suppliers	Bulk Weight or Dry Chemical in Solution (lb/gal)	Remarks
Aluminum sulfate (alum)	PVC, Rubber, FRP, 316 SS	Dry (17% Al_2O_3) or 49% Liquid alum (LA)	78 @10°C	Allied Chemicals, Stauffer Chemical	60 lb/CF (dry) 5.5 lb of dry alum/gal of 49% solution	Mostly liquid alum, S.G. = 1.33 (LA) corrosive
Ferric chloride	PVC, Hypalon, Hast.C, FRP	35–45% Solution	Infinite	Dow Chemical, Pennwalt Corporation, DuPont	4.6 lb dry F.C./gal of 42% solution	S.G. = 1.46 (42% solution) very corrosive
Ferric sulfate	PVC, Rubber, FRP, 316 SS	Dry (68–78%) or 50% liquid	20.5 @ 10°C	Stauffer Chemical, Tennessee Chemical	70 lb/CF (dry) 6.3 lb dry F.S. in 1 gal 50% solution	Hygroscopic, S.G. = 1.5 (50% solution) corrosive
Cationic polymer	FRP, PVC, 316 SS	Mostly liquid	Infinite	Dow Chemical, Nalco Chemical, American Cyanamid	8.5–9.5 lb/gal	Eye and skin irritant, may be corrosive
Anionic polymer	FRP, PVC, 316 SS	Mostly dry	Infinite	Dow Chemical, Nalco Chemical, American Cyanamid	45–55 lb/CF	Eye and skin irritant
Nonionic polymer	FRP, PVC, 316 SS	Mostly dry	Infinite	Dow Chemical, Nalco Chemical, American Cyanamid	42–48 lb/CF	Eye and skin irritant

Chemical	Materials of construction	Purity/Form	Solubility	Suppliers	Density	Remarks
Chlorine gas	Steel, copper for gas, Hast.C, PVC, Vitor for solution	99.8% Purity	0.98 @ 10°C	Dow Chemical, Jones Chemical, McKeeson Chemical	$d = 3.214$ (air = 1)	Poisonous and very corrosive when wet
Sodium hypochlorite	Rubber, PVC, Vinyl, Hast.C	Liquid with 15% NaOCl	Infinite	Delta Chemical, Farm & Industrial Chemical, Jones Chemical	10.2 lb/gal (15%)	Strongly alkaline, short storage life
Calcium hypochlorite	Rubber, PVC, Hast.C	Dry form with 70% Cl	22% at 10°C	Delta Chemical, Jones Chemical, Pennwalt Corporation	70 lb/ft³ (granules)	Hygroscopic
Fluosilicic acid	Rubber, PVC, Hypalon, Carp. 20	20–30% (16–24% F)	Infinite	Essex Chemical, Thompson Chemical	10.5 lb/gal (30% solution)	Corrosive, toxic, attacks glass
Sodium fluoride	Rubber, 316 SS, PVC, Hypalon	95–98% (43–44% F)	4% @ 10°C 4.1% @ 25°C	Allied Chemical, Ashland Chemical, J. T. Baker Chemical	90–105 lb/CF (crystal)	Poisonous
Sodium silicofluoride	Rubber, 316 SS, PVC, Hypalon	98.5% (60% F)	0.46 @ 10°C 0.76 @ 25°C	Ashland Chemical, Delta Chemical, Olin Corporation	85–105 lb/CF	Poisonous
Potassium permanganate	Steel, 316 SS, FRP, PVC	97–99%	3.3 @ 10°C 5.0 @ 20°C	Ashland Chemical, Carus Chemical	86–100 lb/CF	Hygroscopic, oxidant, toxic
Sodium hexametaphosphate	316 SS, Rubber, PVC, RFP	67% P_2O_5 (minimum)	Infinite	Ashland Chemical, FMC Corporation, Monsanto Company	64–100 lb/CF, glass or flake	Hygroscopic, sequestering agent

TABLE 4.1-1 (Continued)

Name	Suitable Materials	Available Strengths	Solubility (g/100 mL)	Major Suppliers	Bulk Weight or Dry Chemical in Solution (lb/gal)	Remarks
Zinc ortho-phosphate	RFP, PVC, 316 SS	8% Zinc in sulfate type, 25% zinc in chloride type	Infinite	Technical Products Company, Monsanto Industrial Chemicals, Conray Chemical	10.5 gal/gal, S.G. = 1.27	Acidic solution, avoid skin/eye contact
Copper sulfate	Rubber, PVC, RFP, 316 SS	99%	25.2 @ 15°C	Ashland Chemical, Jones Chemical, Westco Chemical	75–90 lb/CF	Poisonous, algaecide
Carbon dioxide	Steel, 316 SS, PVC	99.5% Liquid CO_2, 10–14% in combustion	0.35 @ 10°C, 0.15 @ 25°C	Union Carbide, Arco Chemical	$d = 1.53$ (air = 1)	Liquified CO_2 in cylinder, recarbonation use
Sulfur dioxide	PVC, Ceramic, Hypalon, Carp. 20	100%	16.2 @ 10°C, 11.3 @ 20°C	Delta Chemical, Jones Chemical, Stauffer Chemical	$d = 2.26$ (air = 1)	Liquified SO_2 in cylinder, dechlorination use
Air (reference)	—	—	14 mg/L @ 10°C, 10 mg/L @ 25°C		$d = 1.29$ g/L (dry)	
Ozone gas	Ceramic, Teflon, 316 SS, glass	2–8% Depending on generator	6.5 mg/L @ 10°C, 3.5 mg/L @ 25°C	On-site generation only	2.14 g/L @ 0°C, $d = 1.66$ (air = 1)	Toxic, fire hazard, strongest oxidant

Chemical	Form	Materials of construction	Solubility	Suppliers	Density	Hazards
Hydrogen peroxide	35, 50, and 70% Solutions	Aluminum—type 5254, polyethylene, porcelain, 316 SS (fair)	Infinite	FMC, DuPont, Interox	S.G. = 1.29 (70% solution)	Potentially a fire/explosion hazard, eye and skin irritant
Ammonia gas	99% NH_3	Steel, 316 SS, neoprene	40 @ 10°C	Delta Chemical, Monsanto Industrial Chemicals, Olin Corporation	0.77 g/L, $d = 0.56$ (air = 1)	Pungent, toxic, explosive
Aqua ammonia	30% NH_3 (26 Be)	Iron, 316 SS, PVC	Infinite	Allied Chemical, Hills Brothers Company, Jones Chemical	S.G. = 0.897	Unstable, strongly alkaline
Sodium chlorite	80%	Glass, PVC, FRP, Hast. C	35 @ 10°C	ICD Industry, Olin Corporation, Westco Chemical	70 lb/ft^3	Hygroscopic, combustible with organic matter
Quick lime—CaO	80–95%	Iron, concrete, PVC	22 @ 10°C	AC Industry, J. T. Baker, Kraft Chemical	60 lb/ft^3	Emits heat upon contact with water
Hydrated lime—Ca(OH)	82–95%	Iron, concrete, PVC	22 @ 10°C	AC Industry, Asher-Moore Company, Rugger Chemical	30 lb/ft^3	Dusty, irritant
Sodium hydroxide (caustic soda)	Dry or 50% liquid	Steel, iron	51 @ 10°C	Stauffer Chemical, Allied Chemical, FMC Corporation	60 lb/ft^3 or 2.7 lb NaOH in 25% solution	Toxic, dangerous to handle

TABLE 4.1-1 (Continued)

Name	Suitable Materials	Available Strengths	Solubility (g/100 mL)	Major Suppliers	Bulk Weight or Dry Chemical in Solution (lb/gal)	Remarks
Sodium carbonate (soda ash)	Steel, rubber, PVC	99% pure	12.5 @ 10°C	Hacros Chemical, Ashland Chemical, Stauffer Chemical	40–60 lb/CF	Hygroscopic
Sulfuric acid	Steel, PVC for 92% acid, Carp. C, Viton for <92%	66 Be:93% H_2SO_4 60 Be:78% H_2SO_4 50 Be:62% H_2SO_4	Infinite	Allied Chemical, Ashland Chemical, Monsanto Chemical	15.2 lb/gal (66 Be), 14.2 lb/gal (60 Be)	Very corrosive, hygroscopic
Activated carbon	316 SS FRP Rubber Hastelloy C Epoxy paint Saran	Power (PAC) 200 mesh size or Granules (GAL) E.S.: 0.6 to 0.9 mm UC: 1.6 to 2.4	— —	Calgon Corp Amer. Novlt. Co. Trans Pacific Carbon Corp, Calgon	13–46 lb/cf (20 avg) 25–26 lb/cf	Dusty, can be explosive

Disinfectants

Chlorine	Liquid form	$250/ton
Chloramines (Cl_2 + NH_3)		$300/ton for NH_3 alone
Chlorine dioxide (Cl_2 + $NaClO_2$)		$30,000/ton for dry Na-$ClO_2$ alone
Ozone (on-site gas generation)		$2,000/ton for cost of generation

pH Control

Lime—$Ca(OH)_2$	Dry form	$70/ton
Caustic soda	50% Liquid	$600/ton as NaOH
Soda ash	Dry form	$150/ton
Sulfuric acid	98% liquid	$200/ton

Taste and Odor Control

Powdered activated carbon (PAC)	Dry form	$1,100/ton
Granular activated carbon (GAC)	Dry form	$1,700/ton
$kMnO_4$	Dry form	$1,800/ton
Ozone (on-site gas generation)		$2,000/ton

Corrosion Control

Sodium metaphosphate	Dry form	$1,800/ton
Zinc orthophosphate	Liquid form	$1,500/ton
Caustic soda	50% Liquid	$600/ton as NaOH

Fluoridation

Hydrofluosilicic acid	23% Liquid	$850/ton as F
Sodium fluoride	Dry	$400/ton
Sodium silicofluoride	Dry	$1300/ton as F

Softening

Lime—CaO	Dry	$60/ton
Caustic soda	50% Liquid	$600/ton
Carbon dioxide	Liquid	$70/ton

Algae Control

Copper sulfate	Dry	$1,200/ton

Considerations Prior to designing the chemical feed system, the engineer should be familiar with the raw water quality, the desired finished water quality, and the treatability of the water. These factors determine the types of chemical that can be employed and their dosages. The design engineer may then address the following issues: the characteristics of each chemical, local conditions, mode of delivery, availability of the chemicals, cost, local and federal safety codes, and procedures for handling the storage.

The chemical characteristics that should be analyzed are the stability (storage life), corrosivity, crystallization temperature (freezing), safety, potential hazard to health, hygroscopicity, and flammable or explosive nature. Similar to the chemical characteristics, the local conditions affect the cost and the form in which the chemicals are transported and stored. The local conditions may also affect the reliability of the chemical supply. Additional design considerations are the ambient temperature, humidity, amount of snowfall, potential for flooding, distance from major freeways and rail lines, and the proximity of the plant site to residential and commercial areas.

The method of chemical transportation is an important issue because it determines the design of the receiving facility. The treatment chemicals may be delivered in the form of bags via trucks or in bulk form by semitrailer, rail, or barge. If the chemicals are delivered by trucks, the access road must be wide, with an adequate turning radius. The engineer must also establish the proper grade and the type of pavement (access road) in addition to researching the weight limits of the public roads and bridges.

The availability of the water treatment chemicals is a very important consideration for plants located in remote areas and developing countries. For example, polymers and liquid chemicals, such as liquid alum and liquid caustic soda, are seldom available in these regions. Engineers are therefore advised to research the local availability of particular chemicals prior to designing the system.

An issue related to chemical availability is cost. Depending on the availability, the difference in cost between two chemicals may be quite significant. For example, caustic soda is generally 10 times more expensive than lime. Chemical cost is also affected by the manner in which it is purchased; bulk versus small amounts. Engineers are encouraged to study market trends in order to predict possible shortages in chemical supply, which will substantially increase their cost.

With regard to the handling and storage of water treatment chemicals, the design engineer must be thoroughly familiar with the local, state, and federal safety regulations. In compliance with the OSHA regulations, plants should have emergency showers, eyewashes, exhaust fans, and special clothing and gear for the safety of the workers. Additionally, chemical storage tanks must be located within a bermed area—to contain chemical spills—and acid holding tanks must be isolated from alkali chemical storage tanks by a separation berm. OSHA also demands that oxidants, such as chlorine and potassium permanganate, be stored in a separate room from organic chemicals such as polymers and activated carbon.

The handling and storage of treatment chemicals must be based on the nature of the chemicals. Hygroscopic chemicals such as ferric chloride, ferric sulfate, potassium permanganate, soda ash, calcium hypochlorite, and alum are generally shipped and stored in their dry form and therefore require special packaging and storage to protect them from moisture. The powdered form of all substances is potentially hazardous. If the dust concentration is allowed to reach a certain level, one spark will cause an explosion. Ozone, oxygen, calcium oxide (quick lime), sodium chlorite, and activated carbon are all considered to be fire hazards; some are also explosive and toxic. They must therefore be handled and stored in strict accordance with the safety codes and rules provided by the manufacturer.

There are two other miscellaneous but important design concerns: clogging of the feed line and the mode of chemical feeding. Clogging of the chemical feed pipe is a common, yet serious, problem.

Alkali chemicals that have hydroxide in their chemical formula and metal coagulants quite often produce heavy scale in the feed lines and diffusers. Many treatment plants dilute the chemicals downstream of the feeder in order to obtain better diffusion of the chemical within the process water, to minimize the feed lag time, and to avoid having the slurry settle in the feed line. However, extreme caution must be taken when diluting metal coagulants such as alum and ferric chloride due to their pH and solubility relationship (see Figure 3.2.3-1). In the case of alum and ferric chloride, solutions up to 1% and 5% (using ordinary filtered water) may be the maximum safe dilution, respectively. Alkaline chemicals such as lime, caustic soda, and aqua ammonia must not be diluted by carrying water unless the total hardness of the carrying water is less than 20–30 mg/L; otherwise, calcium carbonate scales will rapidly form due to the softening reaction.

Regardless of the metered solution concentration of the chemicals, the diffuser pipes can still potentially become clogged as the result of the instant metal hydroxide formation or softening reaction that occurs when these chemicals hit the processing water, provided that the hardness is moderate or high. For this reason, they should be designed for easy cleaning and if the diffusers are not easily accessible, an alternative diffuser should be provided as a backup. The ideal design is to locate the feeders as close as possible to the application points.

It is worth restating the importance of the continuous chemical feed versus the on/off mode. The pulsation pump is not the primary choice as a feeder unless a pulsation damper can be installed to smooth out the discharge flow: noncontinuous chemical feeding will overdose one particular part of the process water while leaving the rest devoid of chemicals. Certain chemicals such as alum undergo instantaneous and irreversible reactions.

From the view of the plant operators, chemical feed pipes should be installed at every possible location. The convenience of changing chemical feed locations will offset the initial cost of installation. If there is a possibility for future plant expansion, the design should also include the space and capacity required by the additional feeders, feed lines, and appurtenances.

Overview of the Chemical Feed Systems

Three types of chemical feed system are used in water treatment: liquid chemical, dry chemical, and gas feed. Of the three types, the liquid chemical feed system is preferred because it is clean and compact, is not labor intensive to operate, has easy automatic controls, and does not have problems with chemical dust or fumes.

Liquid Chemical Feed System   The simplest, proven, and most cost effective liquid chemical feed system has a pump capacity that is at least 10% larger than the maximum feed rate (Alternative 1, Figure 4.1-1). This system contains a rotameter, flow control valves, shut-off valve(s), and a recycling pipe line; the recycling line can return the excess flow either to the tank or to the pump suction line. This type of system is not suggested for use with chemicals that coat the interior of the rotameter (glass tube). Chemicals classified under this category are caustic soda, ferric chloride, potassium permanganate, and certain types of polymer. It is recommended that the rotameters not be installed in locations exposed to sunlight, since algae will grow within the glass tube. A quick dismounting connection such as a union coupling should be provided at one side of the rotameter connection line to facilitate simple cleaning of the glass tube. One major drawback of this system is that there is no simple way to pace the chemical feed rate to changes in the plant flow rate. However, this problem can be avoided by using a variation of Alternative 1—installing a rotodip type of feeder (Alternative 2 of Figure 4.1-1).

The rotodip feeder receives a 4–20 mA signal from the plant flow meter and is therefore capable of pacing itself to the plant flow rate. The feeder has a set of revolving cups and contains a unique feature, a wide turn-down ratio. A radio of 1500:1 may be achieved by changing the revolving speed, the submergence of the cups, and the speed reduction gears. It requires minimal maintenance and can easily last over 20 years under normal operating conditions.

The majority of modern water treatment plants use metering pumps. Most of these pumps have a 150:1 feed range, including adjustment of the pump stroke length (15:1) and motor speed variation (10:1). The advantages of the metering pump are its simple pacing of the chemical feed rate to the plant flow rate, a very high feeding accuracy (1%), a discharge pressure up to 150 psi (1035 kPa) or more, and a compact size. Refer to Figure 4.1-2 for an illustration of this second type of feed system.

A third type of chemical feed system combines a magnetic flow meter with a needle valve (for flow control) and a pressurized storage tank (Figure 4.1-3). The pressure level within the air padded tank depends on the back pressure from the processing water line. For example, if a chemical is fed into an open channel, an air padding pressure of 10 psi (70 kPa) will be sufficient for the storage tank. This system is simple and has proved to be effective, but it has two major disadvantages: (1) a limited turn-down ratio

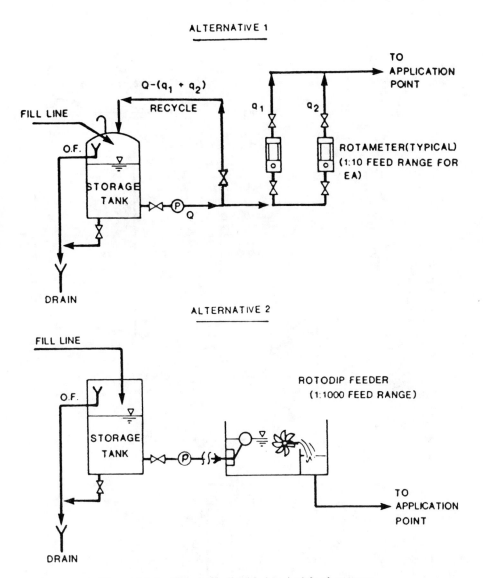

Figure 4.1-1 Two basic liquid chemical feed systems.

and an overall practical limit of 15:1, and (2) a potential for clogging of the flow control valve and meters. A variation of this system uses a recycling pump, rather than the pressurized storage tank, in much the same manner as the first feed system containing the rotameter.

Since exposure to some of the water treatment chemicals is hazardous to human health, the plant design must conform to the local safety codes and OSHA regulations. All storage and handling facilities should therefore have

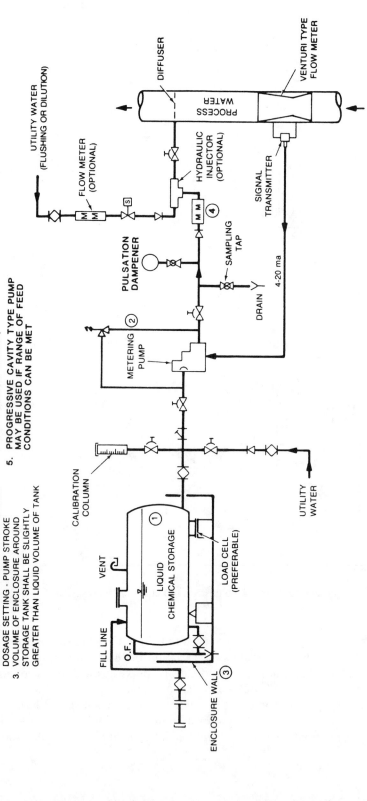

Figure 4.1-2 Liquid chemical feed system (metering pump system).

326

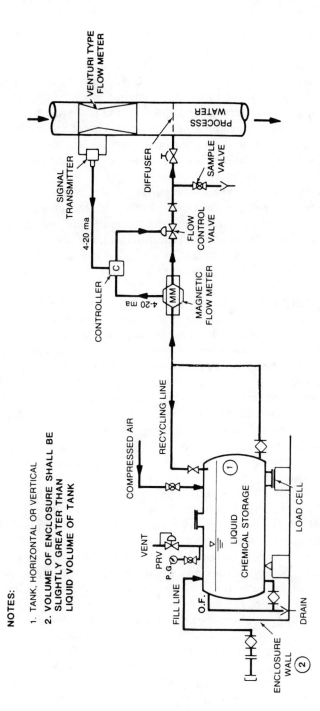

Figure 4.1-3 Flow pacing solution feed system (closed-loop control with magnetic flow meter).

eyewashes, showers, clear warning signs, berms for containment, and adequate lighting and ventilation. All storage tanks must have tags labeling their contents, an access hole, fill line, drain, overflow, discharge valve, vent, and content indicators, such as a load cell or a differential pressure cell, or sonic level indicators. Optionally, the tanks may contain a mixing device. The use of glass-sided gauges is not recommended due to the potential for breakage and because they become fogged by the chemicals. Use of the bubbler tube system should be avoided as it may introduce oxygen, carbon dioxide, or bacteria into the stored solutions and therefore degrade certain chemicals such as caustic soda and polymers.

For any of these systems, the most important design consideration is the selection of materials that are most compatible with the chemicals to be handled—materials used in constructing the metering unit, valves, piping, and storage tank. Other design considerations include the ease of operation, reliability, safety, cost, and the method by which the chemicals are fed: it is essential to use continuous feeding and not intermittent feeding of chemicals.

Dry Chemical Feed System   The dry chemical feed system is a well established and reliable method of adding dry or solid chemicals. It has been successfully used in industrial and water treatment applications for many years. This type of system may easily be automated and problems associated with dust control can be minimized with the installation of an appropriate type of control system. Yet, the dry chemical feed system is more labor intensive (operation and maintenance) than the liquid chemical feed system.

The dry feed system is usually not preferred for most plant operations. However, certain chemicals such as lime, potassium permanganate, PAC, and most anionic and nonionic polymers are only available in their dry form. Moreover, a few plants can only obtain chemicals in their dry or solid form due to problems associated with the liquid form: freezing, high delivery cost, degradation of the liquid form of the chemicals when stored for long periods of time, or a limited selection (polymers) when compared to the dry form.

The design considerations for the dry chemical feed system primarily focus on the storage silo and dissolving tank. The major concerns of the engineer are explosion-proof handling of dusty material and the health and safety of the plant operators. An appropriate type of dust collector unit and an exhaust fan must be installed in the feed rooms. Engineers should realize that all bag type dust collectors require the dust to be completely dry. If this cannot be accomplished, a water scrubber type of collector should be specified.

For any type of solid material, especially the powdered form, the angle of the hopper bottom is an important issue. The bottom of the hopper should have an angle of no greater than 30° from the vertical and should be fitted with ordinary vibrators. However, if a "bin activator," manufactured by Vibra Screw, Inc., is installed in the bottom, the hopper angle may be more shallow due to the effective control in bridging and flashing the dry material.

Since the surface of the material stored in the silo is generally not level, it is very difficult to select a reliable inventory indicator for dry materials. Currently, the best and most reliable method of keeping inventory is the installation of load cells to weigh both the silo and its contents.

In the case of hygroscopic chemicals, the storage silo should be made as hermetic as possible; this will minimize caking caused by atmospheric moisture. The silo should also have a discharge bin gate to isolate the feeder for service. This gate should be followed by a flexible connection, which isolates the vibrations of the feeder and facilitates the simple connection of the feeder to the bin outlet.

If a pneumatic system is used to load the dry chemicals into a bulk storage silo, the fill connection must be of a suitable size and type so that it is compatible with the delivery trucks of the anticipated chemical suppliers. Other essential items that should be incorporated into the design are the use of elbows with a long radius in the fill line and a dust collector, which should be installed in an easily accessible location.

Engineers are also advised to investigate commercially available types of packaged bulk storage which are combined with feed and mixing systems. One example is the "Chem-Tower" by Smith and Loveless, BIF, and Advance Industries, Inc.

A dissolver or slurry tank is generally required to create a chemical solution or slurry prior to application to the processing flow—for better chemical dispersion. Thus, the dry chemical feed system—storage bins or bag storage area, dust collectors, feeders, dissolvers, duct or pipe works, and grit or waste bins—occupies a considerable amount of floor space and height.

Selection of the dissolving or slurry tank should be based on the characteristics of the chosen chemical. The necessary design considerations include the solubility of the chemical at the lowest water temperature, the time required to dissolve the chemical at the maximum feed rate, the appropriate ratio of water to chemical which will create a solution or slurry with the proper concentration, the amount of heat generated during hydration, the possibility of flow short-circuiting within the tank, and installation of overflow and drainage pipe connections that are adequate in size and in the proper location.

The feed line should provide clean-out fittings, in addition to a connection for flushing water, so that any solids or grit will be flushed out before they clog the feed pipe. The flow velocity in the feed line should be approximately 5 ft/s (1.5 m/s). If the slurry form of the chemical is to be fed to the processing water, the pipeline should avoid sharp turns and steep rises and falls because these features promote caking and increase the headloss. In other words, the design should minimize erosion and the sedimentation of slurry within the pipeline.

Figure 4.1-4 is a schematic illustration of the general dry chemical feed system. Figure 4.1-5 is a lime feed system that is installed directly above the application point. Figure 4.1-6 is a dry polymer feed system, which may also be used as a liquid polymer feed system. This last system is proven and performs well with minimal operation and maintenance effort. However, the

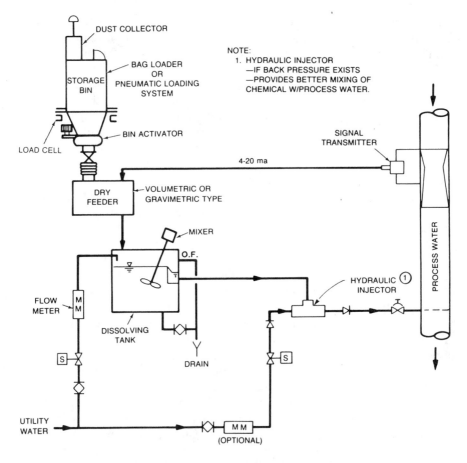

DRY CHEMICAL FEEDING SYSTEM WITH DISSOLVING TANK

Figure 4.1-4 Dry chemical feed system with dissolving tank.

skid-mounted, self-contained system generally costs about $25,000. There are a few types of inexpensive dry polymer feed system that utilize high-speed blenders without an aging tank. However, these systems are not effective due to the high shear forces generated by the blender.

The basic types of dry chemical feeder are the gravimetric and volumetric feeders. The gravimetric feeder is preferred because of its wide turn-down ratio (100:1), high accuracy (1%), easy automation, and ability to feed information directly to a computer. Moreover, plant operators do not have to refer to a calibration chart to set the chemical dosage. However, these units are significantly more expensive than the volumetric dry feeders. Typical gravimetric dry feeders are the belt type and loss in weight type.

Volumetric feeders include the oscillating hopper, vibrating feeders, grooved disk feeders, screw (helix) feeders, and belt feeders. The first two types should

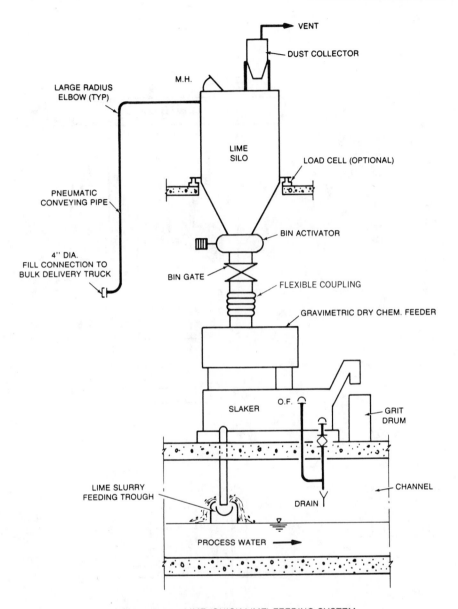

LIME (QUICK LIME) FEEDING SYSTEM

Figure 4.1-5 Lime (quick lime) feed system.

not be considered for use in the water treatment field because of their low feeding accuracy (10%). The grooved disk feeders are highly accurate but have a limited capacity. Their application is therefore limited to small sized plants. The screw type and belt feeders are reasonably accurate (± 1–3%). The feeding range of the screw feeder is 20:1 and 10:1 for the belt feeder.

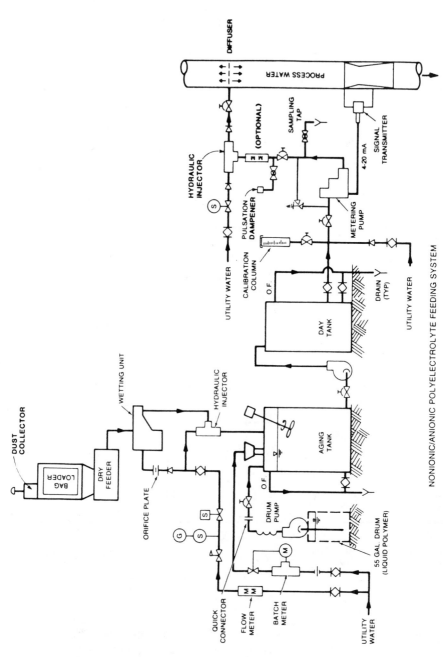

Figure 4.1-6 Nonionic/anionic polyelectrolyte feed system.

The last two types are most commonly used in water treatment because of their lower cost compared to the gravimetric feeders. Yet, the volumetric feeder also has a disadvantage: the feed rate setting must be calibrated frequently due to variations in bulk density from batch to batch. This difference is the result of purchasing the same chemical from different suppliers, since the purchase of the chemicals is based on bids. The variations may also be attributed to the particle size, density, shape, and moisture content of each batch. If different types of chemical are to be fed by means of a volumetric feeder, each chemical must have its own calibration chart so that its dosage can be set for the feeder.

Gas Feed System With the exception of ozone and carbon dioxide, the gaseous form of a chemical is usually fed to the process water as a solution, provided that the gas is reasonably soluble in water. Although chlorine, anhydrous ammonia, and sulfur dioxide are shipped and stored in steel cylinders as liquified gas, they are commonly made into a solution by means of a hydraulic eductor and fed to the process water as a solution. However, the liquified gas is generally evaporated to form a gas, then injected into the process water by means of a vacuum, pressure, or direct gas feed system.

The gas feed systems are generally simple in mechanism and easy to operate. The solution feed vacuum type of feeder is manufactured as a proprietary type of feeder and is commonly used in chlorination, ammoniation, dechlorination, and in feeding chlorine dioxide. This type of gas feeder automatically shuts off the gas with a loss in vacuum, loss in operating water pressure, or when the solution stops flowing from the discharge line.

Direct gas feeders are seldom used but may be utilized in situations where pressure water and/or electricity is not available. It is therefore used primarily in very small facilities such as campsites or small plants located in remote areas or in developing countries.

Both carbon dioxide and ozone, particularly ozone, have extremely low solubility in water (refer to Table 4.1-1). These gases must be fed to the process water via a rotameter, where they are released as fine gas bubbles from the bottom of the contact channels. The contact channels have a water depth ranging from 12 to 18 ft (3.5–5.5 m). The ozone contact channel must have a water depth of over 18 ft (5.5 m) due to the very poor solubility of ozone in water. The chlorine gas feed system is a typical gas feed system and is illustrated in Figure 3.2.8-2. The ozone feed system is discussed in detail in Section 4.8.

Required Capacity for the Chemical Feeder and Storage The capacity of the feeder should be able to cover both the maximum and minimum chemical feed requirements. If the turn-down is not large enough to meet the maximum and minimum requirements, two sets of feeders may be needed to adequately cover the entire chemical feed range. The feed range may be computed by the following method (lb/day):

$$\text{Maximum feed rate} = [(\text{maximum dosage} \times 8.34) \times Q_{\text{max-day}}] \div 24$$

$$\text{Average feed rate} = [(\text{average dosage} \times 8.34) \times Q_{\text{ave-day}}] \div 24$$

$$\text{Minimum feed rate} = [(\text{minimum dosage} \times 8.34) \times Q_{\text{min-day}}] \div 24$$

$$\text{lb/hr} = [(\text{mg/L} \times 8.34) \times \text{mgd}] \div 24$$

Notes

1. Minimum plant flow occurs during winter nights or during the period of plant start-up.
2. Many chemical feeders have a limited feed range of 10:1 to 15:1.
3. In the United States, the capacity of the chemical feeders is commonly expressed as pounds per hour or gallons per hour.

The quantity of water treatment chemicals which should be stored at the plant site is influenced by the local weather conditions, the raw water quality, and the local supply situation. Treatment plants are generally considered to be adequately stocked if they have a 15 day supply; this figure is based on the maximum dosage and the average daily flow rate. Remote areas or regions with frequent flooding or heavy snowfall should have a much larger supply.

$$\text{Storage capacity} = [(\text{maximum dosage} \times 8.34) \times Q_{\text{ave-day}}] \times 15 \text{ days}$$

$$\text{lb} = [(\text{mg/L} \times 8.34) \times \text{mgd}] \times 15 \text{ days}$$

If the maximum dosage and the maximum daily flow rate are expected to coincide for more than 1 week, the following equation may be used to calculate the storage requirement:

$$\text{Storage capacity} = [(\text{maximum dosage} \times 8.34) \times Q_{\text{max-day}}] \times 15 \text{ days}$$

Specific Design Issues   Due to the different natures of the water treatment chemicals, each type of feed system has a few specific design issues that must be addressed by the engineer. This section discusses the special considerations associated with the polymer, PAC, potassium permanganate, and lime feed systems.

Polymer Feed System   Cationic polymers are generally shipped as a solution in drums or by bulk delivery; bulk deliveries are approximately 3000 gal, but trucks with 1000 gal tanks are also available in some areas. Nonionic and anionic polymers may be shipped in either liquid or dry form. The dry form is preferred by some plants because it occupies less storage area and can be stored for a longer period of time, compared to a maximum effective storage of 1 year for the liquid form.

The system should be designed to feed both liquid and dry polymers as shown in Figure 4.1-6, especially in cold weather regions and remote areas. This design gives the plant operators maximum flexibility in feeding the polymer.

All polymers are very viscous and slippery if spilled on wet concrete floors. Therefore, a concrete curb should be provided around the polymer feed system, in addition to hose bibs and floor drains to contain and wash away the spilled material.

Mechanical mixers with relatively low speeds should be used in preparing the polymer solutions. If the mixers are operated at 400 rpm or less, the possibility of breaking the polymer chains is decreased.

Dilute polymer solutions (1% or less) are subject to rapid degradation. Thus, the capacity of the working solution (day tank) should be sized to require a new batch every 1–2 days.

Powdered Activated Carbon (PAC) Feed System   There are two basic PAC feed systems.

Alternative 1	Dry feeder with bag-loading hopper, extension hopper, dust collector, and either a dissolving tank or a vortex mixer tank.
Alternative 2	PAC slurry storage tank with a metering pump.

If PAC feeding is infrequent and the feed rate is less than 150 pounds per hour (pph), Alternative 1 should be considered because the capital investment is much less. However, if the system is used frequently and the feed rate exceeds 150 pph, Alternative 2 is the proper choice.

Two major problems have been associated with Alternative 2: solidification of the PAC slurry when it is not mixed frequently and the inability of hydraulic eductors to convey the PAC slurry. The mixer for the PAC slurry tanks should be designed to prepare 1.5 lb/gal of slurry (1 lb/gal is standard).

PAC adsorbs organic compound, including airborne gases. This characteristic may reduce its effectiveness and engineers must incorporate special storage considerations into the plant design. A water scrubber type of dust collection system should be installed for moist PAC and a bag type dust collection system should be used exclusively for dry PAC.

Carrying water is added to the PAC feed line so that a flow velocity of 5 fps (1.5 m/s) is maintained. The feed system should also have clean-outs and large radius elbows.

PAC particles are very abrasive and wet PAC is extremely corrosive. Thus, the design of the feed system should take these characteristics into account. The rotodip feeder with a hydraulic eductor system has a very successful operational history and this system should be considered whenever applicable.

50% moistured PAC is exclusively used all major water treatment plants in Japan and practically no dust problems. The product has a consistency of natural table salt and easy to handle.

Potassium Permanganate Feed System   There are two basic potassium permanganate feed systems.

Alternative 1	Volumetric dry feed with a bag-loading hopper, extension hopper, dust collector, and dissolving tank (no metering pump).
Alternative 2	A batch system consisting of two large dissolving tanks and a metering pump.

Like other chemicals, the optimum dosage of potassium permanganate usually covers a wide band and a deviation of $\pm 5\%$ from the desired dose does not generally affect the result. For this reason, Alternative 1 is the preferred system. Also, the floor area required by Alternative 1 is usually half that required by Alternative 2 because potassium permanganate has a very low solubility, 3 g/100 mL at 10°C and 2.8 g/100 mL at 5°C. Alternative 2 usually requires large dissolving tanks.

Lime Feed System   The most effective form of adding calcium hydroxide to the process water is quick lime (CaO). Quick lime with a purity of 85% and hydrated lime ($Ca(OH)_2$) with a purity of 90% are easily obtainable. When compared to hydrated lime, both the storage and dosage requirements of quick lime are lower because 1 mole of CaO (56 g) produces 1 mole (74 g) of $Ca(OH)_2$ after hydration with water. Thus, quick lime should be considered if lime is to be used frequently at a minimum dosage of 1 ton/day.

Quick lime may be shipped as pebbles or in granular form; the approximate bulk density is 60 lb/ft^3. In contrast, slaked lime has a bulk density of only 30 lb/ft^3 but is only available as a light powder. The quick lime storage silo may therefore be much smaller and "flushing" and "bridging" problems associated with silo-stored slaked lime may be avoided. Yet, quick lime must be slaked prior to use and grit must be removed from the slaker on a routine basis.

The lime feed system should consist of a storage silo with mechanical pneumatic loading, dust collector, bin activator such as those produced by Vibra Screw, bin gate, content indicator (load cells), volumetric or gravimetric dry feeder, slaker (for quick lime), or dissolving tank with mechanical mixer (for slaked lime). For optimal results, the feeder should be located directly above the application point. This arrangement will prevent "caking" within the feed pipes.

Design Criteria   Depending on the form of the chemical, the design parameters for the feeder, storage, feed line, and diffuser may be different. This section discusses these issues for the liquid chemical, dry chemical, and gas feed systems. The design criteria for the treatment plant access road and loading dock are also presented.

Liquid Chemical Feed System

Feeder
Type	Metering pumps, rotodip, magnetic flow meter or rotameter.
Feed range	100:1—this figure includes a range of 10:1 for the dosage setting and a range of 10:1 for plant flow pacing. If the feeder does not have a 100:1 feed range, in most cases, more than one feeder must be provided to cover the range.
Control	Ideally, the plant will have automatic flow pacing and remote dosage setting control, but this is not absolutely necessary. However, local manual control must be provided.
Accuracy	±1.0% is preferred, based on the maximum feed rate.

Storage
Type	Noncorrosive material. A steel tank may alternatively be used if it has a protective lining. The engineer must also incorporate earthquake, freezing, safety,and aesthetic considerations into the design.
Capacity	Normally 15 days of storage, based on the maximum dosage and average daily flow rate.
Control	A content indicator should be provided for each tank and possibly another indicator for the main control board.
Feed line	Noncorrosive materials such as PVC or 316 SS are commonly used. If pulsation pumps are used, the feed line should be sized to accommodate the peak pulsation flow rate—this rate is approximately three times higher than the rated pump capacity—regardless of whether a pulsation damper is used or not. Flushing water connections and clean-outs are also necessary items.
Diffuser	The type of diffuser depends on the type of chemical used. The perforated pipe diffuser should be avoided if the chemical has scale forming characteristics and Types C, D, and E, of Figure 3.2.3-3 in Chapter 3, should be considered. Non-scale-forming chemicals will be well dispersed by the perforated pipe diffuser; refer to Types A and B of Figure 3.2.3-3. The diffuser design is based on Figure 3.2.3-4.

Dry Chemical Feed System

Feeder
Type	Belt feeders for large feed capacities; screw or helix feeders for small capacities; grooved disk feeders for small capacities; loss in weight type or vibration feeders in special cases.
Feed range	Each type of feeder has an inherent feed range and accuracy. The following table summarizes the data.

	Capacity	Feed Range	Accuracy
Belt Feeder			
Volumetric	Minimum, 0.05 cfm	10:1 Normal	±1% of set rate
	Maximum, 60 cfm	10:1 Optional	
Gravimetric	Minimum, 0.05 lb/min	100:1 Normal	±1% of set rate
	Maximum, 2000 lb/min		
Screw (helix type)	Minimum, 0.02 cfh	20:1 Normal	±1% by volume
volumetric feeder	Maximum, 40 cfh		±3% by weight
Grooved disk type	Minimum, 0.17 cfh	100:1 Normal	±1% by volume
volumetric feeder	Maximum, 17 cfh		±3% by weight
Miscellaneous			
Loss in weight type	Minimum, 20 lb/h	10:1 Normal	±1% of set rate
(gravimetric)	Maximum, 500 lb/h		
Vibration feeder	Minimum, 0.7 cfh	40:1 Normal	±10% by weight
(volumetric)	Maximum, 40 cfh		

Control	The same as liquid feeders.
Accuracy	The same as liquid feeders.
Storage	This is basically the same as the liquid feed system with respect to type, capacity, and control. The design considerations are covered in the previous section.
Feed line	This is fundamentally the same as the liquid feed system, except that a removable top channel with gravity flow is often used for lime slurry.
Diffuser	The design of the diffuser is the same as the liquid feed system. Type E of Figure 3.2.3-3 illustrates a trough type diffuser.

Gas Feed System

Feeder
Type	Rotameter system, Venturi flow meter system.
Feed range	20:1 is preferred, with a minimum of 10:1.

Control	Automatic flow pacing is required. A closed-loop control with residual analyzer and plant flow meter is preferred.
Accuracy	$\pm 2\%$ based on the maximum feed rate.
Storage	
Type	A high-pressure steel cylinder is standardly used in storing most liquified gas. For large installations, a larger capacity tank mounted on a trailer or a tank rail car (stationary storage tanks) may be used. Ozone must be generated on site.
Capacity	This is basically the same as the liquid feed system.
Control	The same as the liquid feed system.
Feed line	Fundamentally, the same as the liquid feed system. However, in the case of ozone, 316 SS is the only material that may be used.
Diffuser	Similar to the liquid feed system, but Type E of Figure 3.2.3-3 must be used for carbon dioxide and ozone.

Design Criteria for the Access Road and Loading Deck   The dimensions of typical delivery trucks are presented below:

	Van	Straight Body	Conventional Semitrailer
Length (L)	15' to 20'	17' to 35'	55'
Width (W)	7'	8'	8'
Height (H)	7'	13'-6"	13'-6"
Floor height (FH)	2' to 2'-8"	3' to 4'	4' to 4'-4"
Track (T)	5' to 5'-4"	5'-10"	6'-6"
Rear axle (RA)	—	1'-3" to 12'	4' to 12'

The allowable maximum gross weight (lb) may be calculated from $1025(D + 24) - 30D^2$, where D is the distance (in feet) between the first and last axle of the vehicle or combination of vehicles (AASHO). Generally, the maximum cargo weight is 18 tons.

The required turning radius for a 55 ft semitrailer and truck combination is illustrated below.

Example Design Calculations

Example 1

Given   A treatment plant with a daily average flow rate of 20 mgd is located in North Carolina and requires an alum feed and storage system. The maximum, average, and minimum alum dosages are 50, 15, and 8 mg/L, respectively.

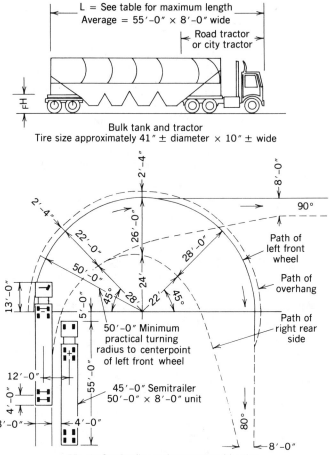

Bulk tank and tractor
Tire size approximately 41″ ± diameter × 10″ ± wide

55′-0″ Semitrailer and tractor combination
Minimum practical turning radius of 50′-0″

Determine

(i) The number and capacity of each feeder.
(ii) The number and capacity of the storage tank(s).
(iii) The major design considerations.

Solution   The maximum, average, and minimum plant flow rates should be determined first.

$$Q_{max} = 1.5 \times 10 = 30 \text{ mgd}$$

$$Q_{ave} = 20 \text{ mgd}$$

$$Q_{min} = 0.25 \times 20 = 5 \text{ mgd}$$

Commercially available liquid alum is used for operator convenience. Since liquid alum contains 5.6 lb of dry alum per gallon,

$$\text{Maximum dosage rate} = 50 \times 8.34 \times 30 = 12{,}510 \text{ ppd} \quad \text{or} \quad 93 \text{ gph}$$

$$\text{Average dosage rate} = 15 \times 8.34 \times 20 = 2502 \text{ ppd} \quad \text{or} \quad 18.6 \text{ gph}$$

$$\text{Minimum dosage rate} = 8 \times 8.34 \times 5 = 334 \text{ ppd} \quad \text{or} \quad 2.5 \text{ gph}$$

$$\text{Turn-down ratio} = 93{:}2.5 \text{ gph} \quad \text{or} \quad 38{:}1$$

(i) The system should have a hydraulic diaphragm type of metering pump with a 100 gph capacity, a 100:1 turn-down unit, and a second metering pump with a capacity of 50 gph. A third metering pump (50 gph) is specified as a spare since alum is one of the most important chemicals in the treatment process. Thus, one 100 gph and two 50 gph metering pumps are specified.

(ii) The storage capacity is calculated to be $[(50 \times 8.34 \times 20) \times 15] \div 5.6 = 22{,}340$ gal. Thus, one 23,000 gal fiberglass storage tank should be provided. Freezing is not a likely problem in this location.

(iii) The major design considerations are listed below.
  - The alum dosage should always be based on dry alum even when liquid alum is used.
  - Alum is a very corrosive chemical. PVC, FRP, 316 SS, or Hasteloy C SS are therefore the most appropriate types of material.
  - Alum should be fed as a continuous flow.
  - An effective flash mixing system is essential at the application point.
  - In order to avoid clogging both the feed lines and the diffuser, it is recommended that neat alum be fed to the process water.
  - The alum fill line should be a 316 SS pipe instead of PVC because liquid alum may be delivered at temperatures over 130°F. The exits of the fill line should be above the highest liquid level in the tank because there is a tendency for the liquid alum to shoot out from the air vent at the last moment during the unloading operation; this is a function of the massive air supply from the delivery truck blower.
  - The feed system should be designed in accordance with the safety codes and rules, including OSHA.
  - The design should allow for ferric chloride or ferric sulfate (alternative coagulants) to be used in the alum feed system. Since the ferric salts are more corrosive than alum and may contain some grit, engineers must take these issues into account when designing the feed system. Examples are the use of a hopper type tank bottom to remove grit, Hasteloy C stainless steel as the metal, and Kynar for the plastic parts. Ferric chloride solution freezes at 0°C if the solution strength is over 45%, but a 40% solution will not freeze

until the temperature reaches $-12°C$—approximately the same freezing point of commercial liquid alum (49% solution).

The concept of using an alternative coagulant is important because any treatment plant that achieves alum flocculation at a pH of 7.5 will often have a high concentration of residual aluminum, equal to or above 0.2 mg/L, in the process water. The general public and the EPA are concerned about high levels of aluminum in potable water due to its possible connection to Alzheimer's disease. It should also be noted that ferric coagulant is much more effective for flocculation and clarification in cold months than alum.

Example 2

Given A treatment plant with a maximum daily flow rate of 50 mgd is located in California and requires a caustic soda feed system. The maximum, average, and minimum dosages are 10, 5, and 2 mg/L, respectively.

Determine

(i) The number and capacity of each feeder.
(ii) The number and capacity of the storage tanks.
(iii) The major design considerations.

Solution The maximum, average, and minimum flow rates are as follows:

$$Q_{max} = 50 \text{ mgd}$$

$$Q_{ave} = 50 \div 1.5 = 34 \text{ mgd}$$

$$Q_{min} = 0.25 \times 34 = 8.5 \text{ mgd}, \quad \text{say 8 mgd}$$

Commercially available 50% liquid caustic soda is purchased and stored as a 25% solution in order to avoid crystallization during the winter. The 25% solution contains 2.7 lb of dry caustic soda per gallon.

Maximum dosage rate = $10 \times 8.34 \times 50 = 4,170$ ppd or 65 gph

Average dosage rate = $5 \times 8.34 \times 34 = 1428$ ppd or 22 gph

Minimum dosage rate = $2 \times 8.34 \times 8 = 133$ ppd or 2 gph

Turn-down ratio = 65:2 or 22:1

(i) The capacity of the pump selection is the same as the previous example, except that a spare pump is not specified because this chemical is not absolutely essential in water treatment.
(ii) The storage capacity is calculated to be $[(10 \times 8.34 \times 34) \times 15] \div 2.7 = 15,753$ gal of 25% solution. Provide two identical 10,000 gal

tanks because a dilution step is required even during the mild California winters.

(iii) The major considerations for the caustic soda feed system are as follows.

- The dosage is based on the dry weight and expressed in mg/L.
- A 50% solution of caustic soda will freeze at 13°C (55°F); however, a 25% solution will freeze at −18°C (0°F); refer to Figure 4.1-7.

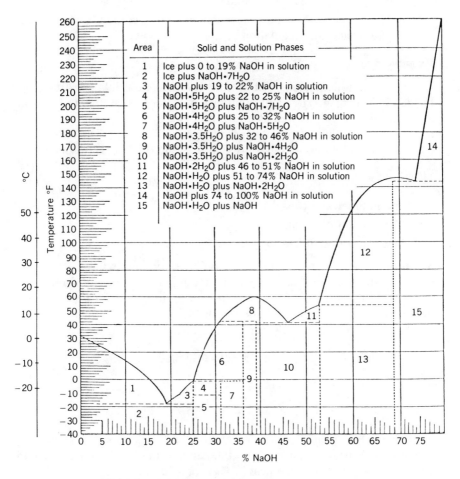

Area	Solid and Solution Phases
1	Ice plus 0 to 19% NaOH in solution
2	Ice plus NaOH·7H₂O
3	NaOH plus 19 to 22% NaOH in solution
4	NaOH·5H₂O plus 22 to 25% NaOH in solution
5	NaOH·5H₂O plus NaOH·7H₂O
6	NaOH·4H₂O plus 25 to 32% NaOH in solution
7	NaOH·4H₂O plus NaOH·5H₂O
8	NaOH·3.5H₂O plus 32 to 46% NaOH in solution
9	NaOH·3.5H₂O plus NaOH·4H₂O
10	NaOH·3.5H₂O plus NaOH·2H₂O
11	NaOH·2H₂O plus 46 to 51% NaOH in solution
12	NaOH·H₂O plus 51 to 74% NaOH in solution
13	NaOH·H₂O plus NaOH·2H₂O
14	NaOH plus 74 to 100% NaOH in solution
15	NaOH·H₂O plus NaOH

GRAPH OF FREEZING POINTS AND SOLUBILITY

Figure 4.1-7 Graph of freezing points and solubility for a caustic soda solution. Complete solution prevails at points on or above the freezing point line. Below this line, some of the caustic soda exists as a solid, anhydrous or hydrated, depending on the temperature and composition. Numbered areas and table inset indicate the nature of the mixtures. (Courtesy of Stauffer Chemicals, New York, *Caustic Soda Bulletin*, 1965.)

- Liquid caustic soda is not corrosive to steel.
- Caustic soda is toxic and dangerous to handle.
- Scale formation will occur when caustic soda is diluted with water unless the water is very soft. Also a heat generation by dillution.
- Caustic soda generally costs eight times more than lime, but the required dosage for pH control costs approximately the same.
- Plant operators usually prefer to use liquid caustic soda rather than lime since it is easier to handle.
- The design of the feed system should be in accordance with current safety rules and regulations.

Example 3

Given A treatment plant with a daily average flow rate of 20 mgd is located in Michigan. The plant requires a polymer feed system to function as a flocculation aid, as well as a filtration aid. Pilot tests confirm that an anionic or nonionic type of polymer feed system is needed. The maximum, average, and minimum dosages are 0.5, 0.2, and 0.05 mg/L, respectively.

Determine

(i) The number and capacity of each feeder.
(ii) The amount of required storage.
(iii) The major design considerations.

Solution Commercially available anionic or nonionic polymers are primarily of the dry form; however, a limited number of these polymers are available in liquid form. The system design should allow for both dry and liquid forms to be used, even though the severe winter temperatures are not favorable for liquid polymer because of the potential for freezing.

$$Q_{max} = 1.5 \times 20 = 30 \text{ mgd}$$

$$Q_{ave} = 20 \text{ mgd}$$

$$Q_{min} = 0.25 \times 20 \times 5 = 5 \text{ mgd}$$

Most dry polymers are very viscous at concentrations over 1%. Therefore, a 0.5% concentration is usually recommended; the 0.5% solution should contain 0.042 lb of dry chemical in 1 gal of solution.

$$\text{Maximum feed rate} = 0.5 \times 8.34 \times 30 = 125 \text{ ppd}$$
$$= 1.25 \div 0.042 = 2976 \text{ gpd} \quad \text{or} \quad 124 \text{ gph}$$
$$\text{Average feed rate} = 0.2 \times 8.34 \times 20 = 33 \text{ ppd} \quad \text{or} \quad 33 \text{ gph}$$
$$\text{Minimum feed rate} = 0.05 \times 8.34 \times 5 = 2 \text{ ppd} \quad \text{or} \quad 2 \text{ gph}$$
$$\text{Turn-down ratio} = 124{:}2 \quad \text{or} \quad 62{:}1$$

(i) A hydraulic diaphragm type of metering pump with a 150 gph capacity with a 100:1 turn-down and another unit with a 50 gph capacity should be installed.

(ii) The amount of storage is calculated to be $[(0.5 \times 8.34) \times 20] \times 15 = 1251$ lb of dry polymer. An ordinary bag of polymer weighs 30 lb. Therefore, $1251 \div 30 = 42$ bags. Let us say 40 bags. Twenty bags should be stacked on a pallet for easy hauling and to protect the bags from the wet floor. Additionally, floor space for four 55 gal drums should be provided for the liquid form of the polymer.

(iii) The design considerations for the polymer are as follows.

- The dosage is based on the dry weight of the chemical. In the case of liquid polymer, the dosage should be determined by running jar tests, assuming that the stock solution is a 100% solution, for operator convenience.

- It is difficult to produce a polymer solution that forms "fish eyes" when making solution from dry polymers using ordinary dissolving tanks.

- Polymers that are EPA approved for the treatment of drinking water should be selected and the maximum dosages should abide by the criteria set by the EPA.

- Polymers act selectively; thus, care must be taken to choose a polymer that performs best under the given conditions.

- The application sequence plays an extremely important role in the effectiveness of the polymer.

- Never overfeed polymers; their optimum dosages are located in a very narrow range. A minute amount of overfeeding will cause the turbidity colloids to restabilize, thus resulting in a sharp increase in turbidity.

- Since diluted polymer solutions deteriorate quickly, every batch should be used within 2 days (during hot weather).

- Some polymers are corrosive in nature.

Operation and Maintenance The major issues in operation and maintenance are safety programs and training, good housekeeping, and good recordkeeping. Each of these issues is discussed in detail.

Safety Programs and Training The chemicals used in the water treatment processes are, for the most part, harmful to human health. Thus, the water production manager and water supply department must establish a formal safety program and conduct periodic hands-on training. It is essential that every treatment plant have a custom made operation and maintenance manual for each type of chemical feed system. Furthermore, the necessary type and number of special clothing, emergency showers, and eyewashes should routinely be checked.

Good Housekeeping   Good housekeeping methods are an important issue since the chemicals are essentially hazardous in addition to being toxic. Any chemical spill should be cleaned immediately and the accumulation of gas and dust must be avoided through the frequent use of exhaust fans and a vacuum cleaner. Polymer solution spilled on the floor will result in extremely slippery conditions; all spills should therefore be immediately washed out by means of the floor drains. Operators should note that wet concrete floors are also very slippery. The appropriate hazard signs should be posted in highly visible locations at all times. Lastly, regular maintenance for all mechanical equipment must be performed on time and the chemical feed lines and diffusers should be checked periodically for clogging.

Recordkeeping   Recordkeeping is a vital task because it is required by both local and federal regulatory agencies. Hourly, daily, and monthly records must be kept with respect to the type and dosage of each chemical used. The chemical inventory must also be known at all times in order to maintain an adequate supply of treatment chemicals on site. Moreover, the operating performance of each piece of equipment must be logged on a daily basis and all pertinent information should be conveyed to the next shift operator through verbal communication and in writing (on the daily operational data sheet). Computer-assisted instrumentation and control can greatly assist in recordkeeping.

BIBLIOGRAPHY

AWWA, *Water Fluoridation—Principles and Practices*, AWWA Manual M-14, 1977.

AWWA, *Water Quality and Treatment—A Handbook of Public Water Supplies*, 3rd ed., McGraw-Hill, New York, 1971.

BIF, *Chemicals Used in Treatment of Water and Wastewater*, Engineering Data, August 1982.

Williams, R. B., and Culp, G. L., eds., *Handbook of Public Water Systems*, Van Nostrand Reinhold, New York, 1988.

4.2   HANDLING FACILITY FOR WASTE WASHWATER

Purpose   Conventional water treatment processes ordinarily produce 2–3% of the treated water as filter waste washwater. Prior to 1965, it was common practice to discharge this waste back to the water source or other public water course. However, due to the passage of Public Law 92-500 and the Water Pollution Control Act Amendments of 1972, this practice is now prohibited. Thus, both the waste washwater and sludge from the clarifiers are categorized as industrial wastes and a permit must be obtained from the local regulatory agency to discharge them; the agency is governed by the National Pollutant Discharge Elimination System (NPDES).

Since the passage of Public Law 92-500, many treatment plants have been recycling the wash waste with or without clarification, and this practice has not had adverse effects on the efficiency of the process or the quality of the finished water. However, a recent EPA proposal seeks to restrict the rules governing surface water treatment by requiring a high degree of treatment— including flocculation, sedimentation, and disinfection—before recycling. Furthermore, the recycling must be directed to the head of the plant and not the filter influent.

This section describes how to design a waste washwater handling facility so that it meets the recent requirements set by the regulatory agencies. The design of a waste recycling system, where the waste is routed to the head of the plant without a purification process, will also be discussed because this practice is still applicable to some plants in the United States and to most water treatment plants outside the United States.

Considerations The quality and quantity of the wash waste, the desired quality of the recycled or discharged (treated) waste, and the type and capacity of the treatment process are the major design considerations for the waste washwater handling facility. Alternatives such as lagooning, flocculation and sedimentation, and recycling without a purification process are also important issues and are discussed as well.

Quality and Quantity Many design engineers have chosen to mix together the waste washwater and the sludge from the basins. Yet, this practice is undesirable because the solids concentrated in the sludge are allowed to disperse in the waste water. One-half to 1.5% of the sludge is composed of concentrated solids: microorganisms, organic compounds, and heavy metals that have been extracted and settled from the raw water. In contrast, waste washwater has a solid concentration of 0.01–0.04%. If the two are mixed, the turbidity of the water increases and more chemicals are required to treat the sludge and waste mixture. This can potentially lead to the recycling of objectionable compounds and the impartation of undesirable taste and odor (of the sludge) to the water. Thus, the sludge from the basins should not be mixed with the filter wash waste (which is recycled), but should be discharged to the sludge handling system.

The quality of the wash waste depends on the raw water quality and the pretreatment unit processes. The typical properties of the waste, for a conventional water purification plant using alum as the coagulant, is shown below.

Turbidity (NTU)	BOD (mg/L)	COD (mg/L)	pH	TDS (%)
150–250	2–10	30–150	6.7–7.5	0.01–0.05

The quantity or production rate of the wash waste is also a function of the raw water quality and the type of pretreatment processes, in addition to the efficiency of the processes. The percentage of waste to the amount of water treated is usually about 2% during the warm months and 3% during the cold. Direct filtration and in-line filtration plants generally produce higher percentages of waste, and figures greater than 10% have been reported during episodes of algal bloom and turbidity spikes in the water source. Engineers are strongly advised to recycle the waste washwater rather than discharging it after the clarification process because the cost of water is often rather significant.

Quality Goal of the Recycled or Discharged Treated Waste   The quality goal of the recycled treated waste should be equal or better than the average raw water quality. However, if the waste washwater is to be discharged to a public water course, the quality of the treated waste should basically be equal to or better than the quality of the receiving body. Nonetheless, the NPDES permit may include other special requirements.

Type and Capacity of the Treatment Process   The type of the waste treatment process is fundamentally the same as a conventional flocculation and sedimentation process with disinfection. Proprietary reactor-clarifiers, which have a large internal flow recirculation, have good process efficiency and are tolerant of shock loadings. Other advantages include the ability to thicken the sludge, easy sludge removal, and a small site requirement. Regular horizontal flow, rectangular tanks are also effective waste handling facilities, although they may require a larger site area than the proprietary types due to their lower allowable hydraulic loading rates.

No matter the type of unit, the best process performance is always achieved if the units are operated on a continuous basis and not in an on-and-off fashion. This ideal condition may be attained by providing a large waste holding tank and by pumping the waste into the treatment unit at a constant flow rate.

The capacity of the waste treatment facility should be based on both the maximum waste production rate and the tolerable recycling rate of the treatment plant. The practical maximum recycling rate is 20% of the plant flow rate, yet the design should be based on a recycling rate of 10% or less so that the hydraulic loading of each process unit is not significantly increased. The sludge produced by the waste treatment facility, from the sedimentation basins, should be discharged to a sludge handling facility for further treatment.

Alternatives   Under normal conditions the suspended matter in the raw water is already well coagulated and flocculated prior to filtration. Furthermore, additional effective flocculation usually occurs in the filter bed during filtration. Consequently, the backwashing waste water contains well flocculated particles that settle very well. Figure 4.2-1 presents the batch settling characteristics of wastewater without additional flocculation with coagulant. The figure shows that the supernatant reaches nearly 10 NTU, from an original turbidity of 210 NTU, after 30 min of settling. A settling time of 1–2 h yields

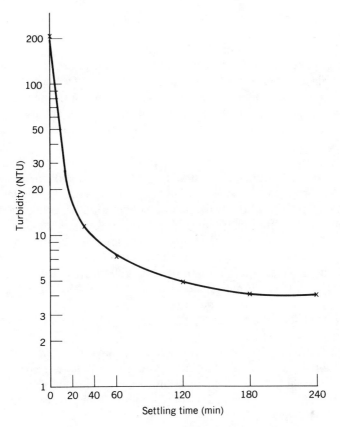

Figure 4.2-1   Settling characteristic of waste washwater.

a supernatant of nearly 5 NTU. Knowing the quick settling characteristics of the waste washwater, two basic treatment alternatives may be used: (1) lagooning with disinfection and (2) ordinary flocculation, sedimentation, and disinfection. Moreover, the wash waste may be recycled to the head of a conventional complete process without settling, as long as the waste is fresh; this practice is still a viable alternative under certain conditions.

LAGOONING   Lagooning is an effective method of clarification if the plant site has adequate space and if the filter backwash intervals are reasonably spaced. The lagoon should be large enough to hold the wash waste of approximately 10 backwashes, or a series of three or more smaller lagoons, each holding three or four filter wash wastes, may be used.

All lagoons should be elongated in shape in order to provide the maximum distance between the inlet and outlet. This configuration minimizes the effect of the suspended sludge, which is caused by the stirring action of the incoming backwash waste. The inlet of each lagoon should have an energy dissipator.

The outlet should be designed to decant, as well as drain the lagoon, and should also act as an overflow facility. The side walls of the lagoon must be protected from erosion, rainfall, and wave action. Each lagoon should also have an adequate access road, which allows vehicles to remove accumulated sludge. Lastly, a disinfection facility, together with a recycling pump, should be provided at the effluent of the lagoon. Furthermore, engineers should realize that the lagoons tend to breed mosquitos and other aquatic organisms, in addition to taste and odor problems characteristic of certain regions.

FLOCCULATION AND SEDIMENTATION PROCESSES   A conventional flocculation and sedimentation pretreatment unit, with a disinfection facility, is required whenever the site does not have adequate space for the lagoons or when the settleability of the waste is unacceptable unless conventional flocculation and sedimentation are implemented. As previously discussed, either the upflow type of reactor-clarifier or a conventional horizontal flow, rectangular flocculation and sedimentation basin may be used. In most cases, the preferred chemical is either 30 mg/L of alum or 2 mg/L of cationic polymer. The process will have 20 min of flocculation, followed by a surface loading of 0.8–2.5 gpm/ft^2 (2–6 m/h) and a settling time of 0.5–2 h, depending on the type of unit provided. A disinfection facility should also be included in the design.

The conventional pretreatment process should have a backwash waste holding tank. This tank must be large enough to accommodate the waste of two to three backwashes, depending on the anticipated frequency of filter washing. Floc removal should be achieved during the subsequent process (clarification) not in the holding tank.

The holding tank may be either square or circular rather than elongated. These shapes will prevent the floc from settling since the inflow creates a stirring effect. Experience has shown that the ideal holding tank is cylindrical in shape and has an inlet pipe that produces a tangential flow direction, thus creating a swirling or vortex flow pattern within the tank. In order to be able to completely drain the tank and collect filter medium, which is very abrasive to the pump impeller, a sump must be provided at the center of the floor of the holding tank. Furthermore, an overflow facility should be included in the design.

Tanks that have the transfer pumps located in the center of the circular tank have been observed to be virtually free of sludge accumulation. The capacity of the transfer pump used in the flocculation and sedimentation units is dictated by the anticipated maximum filter washing frequency; the waste flowing into the holding tank must be pumped out to accommodate the next filter washing. Likewise, the rated capacity of the treatment process is determined by the maximum pumping rate of the transfer pumps.

Ideally, the waste handling facility is located at a higher elevation than the main treatment plant and the treated water is recycled back to the head of the main plant by gravity. If this scheme cannot be adopted, then recycling pumps must be installed. Yet, the recycling pumps should only be used as a last resort because the double pumping scheme tends to have "jogging" problems from the transfer pump to the recycling pump—repeated starting and stopping of the pump at frequent intervals.

RECYCLING OF THE WASTE WITHOUT A PURIFICATION PROCESS   This alternative has been applied to many existing plants and has proved to be effective provided that the recycling rate and other factors are correctly chosen. This is especially true for conventional treatment processes that must produce large and easily settleable floc.

There is an abundance of clear evidence which confirms that recirculation of preformed floc remarkably improves both flocculation and sedimentation; the majority of companies that design and manufacture reactor-clarifier units are exploiting floc recirculation.

The key design guides are provided by bench scale tests. The test results obtained from a study conducted by the author are presented in Figure 4.2-2. The waste water used in the experiment was not the supernatant but the well stirred turbid washwater with a turbidity of 210 NTU (see Figure 4.2-1). The test results demonstrate that both flocculation and sedimentation were dramatically improved as the amount of wash waste added approached 30% of the total water volume. However, in actual treatment plants the practical blending ratio of unsettled waste washwater is up to 10% due to the effect of hydraulic shock loading and the limited detention time of the units. Fresh settled floc (sludge) also improves the settling efficiency if the blend volume ratio is no more than 2%. However, it should be remembered that the danger associated with sludge recycling lies in the potential for recycling objectionable taste, odor, and other undesirable substances. Moreover, sludge that is over 2–3 days old often loses its effectiveness completely (see Figure 4.2-3).

In summary, the waste washwater should be stored in a holding tank in the manner previously described and should be recycled, without clarification, at a blending rate that is 5–10% of the plant inflow rate. It is always a good practice to disinfect the recycling waste.

Design Criteria   The design criteria for proprietary reactor-clarifiers and regular horizontal flow rectangular tanks will not be fully discussed in this section because they were covered in Sections 3.2.3–3.2.5. However, the criteria for the alternative system of lagooning are presented, and the flocculation and sedimentation processes and the disinfection facility are briefly described.

Lagooning

Number of lagoons	A minimum of two to satisfy the cleaning requirement.
Size of each lagoon	Large enough to hold the wastewater of a minimum of 10 filter backwashes.

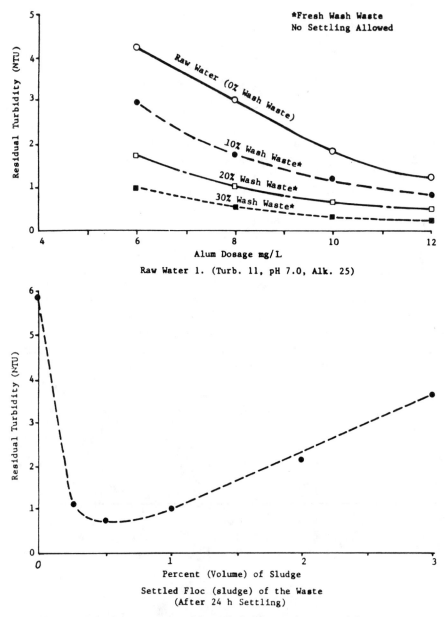

Figure 4.2-2 Improvement of flocculation by wash-waste mixing.

	Condition	Turbidity	Color	pH	Alkalinity
Raw water (reservoir)	Ⓐ 100% raw water	6	7.5	6.8	16.5
	Ⓑ Fresh wash waste	230	220	6.65	27
	Ⓒ 12 h old waste				
Filter wash waste	Ⓓ 1 day old waste				
	Ⓔ 2 day old waste				
	Ⓕ 4 day old waste				

Note: Experimental conditions Ⓑ through Ⓕ were evaluated using 10% addition of "aged" wash waste to raw water.

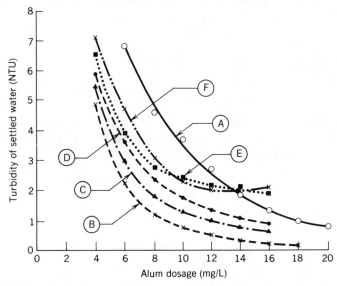

Figure 4.2-3 Settled water turbidity as a function of wash-waste admixture age.

Shape of the lagoon	Free form or rectangular. The length to width ratio should be 4:1 or greater.
Depth of the Lagoon	A minimum of 10 ft (3 m.
Inlet	Isolation valve and an energy dissipator.
Outlet	Should function in decanting, draining, and overflow protection.
Miscellaneous	The side wall should be protected from erosion and periodically checked for seepage. If the area is open to the public, provide a safety fencing around the lagoons.

Flocculation and Sedimentation Processes Despite the high turbidity of the waste washwater, the waste is easily flocculated and clarified. The design criteria of each unit process are fundamentally the same as presented in Sections 3.2.3, 3.2.4, and 3.2.5.

If the site area for the waste reclamation facility is very limited, the Densadeg reactor-clarifier should seriously be evaluated. This unit is manufactured by Infilco Degremont, Inc., and provides a surface loading rate of as high as 5–6 gpm/ft^2 (12.5–15 m/h) for the clarifier and is capable of producing an alum sludge composed of 2–3% solids by means of a rather high polymer dosage.

Disinfection Facility Chlorination is the prime candidate for the disinfection of recyclable water. Since the recycling rate is generally only 2–3% of the plant flow rate, the THMs produced by chlorination should not be a problem, unless the raw water characteristics pose a threat. Refer to Section 3.2.8 for the design criteria for the disinfection facility.

Example Design Calculation

Example 1

Given The filters in Example 2 of Section 3.2.6, Filtration, are part of a 50 mgd filtration plant. The treatment plant requires a waste washwater treatment facility and a recycling facility.

Determine

(i) The type of waste reclamation facility.
(ii) The treatment process.
(iii) The basic design criteria for each unit process.

Solution

(i) A conventional pretreatment process should be selected since the treatment plant is located in a good residential area and the cost of the land does not permit the plant to purchase large, extra parcels of property. Furthermore, the residents will probably not tolerate having a sludge lagoon in their neighborhood due to aesthetic and health considerations and the potential mosquito problem.

(ii) The treatment process should have a waste holding tank located near the filter bank, waste transfer pumps on the holding tank, and conventional pretreatment units at an elevation that is several feet higher than the water level at the head of the main plant. The clarified water should be chlorinated and transported to the head of the plant via gravity flow. The head of the plant should have a flow meter in the discharge pipe line.

(iii) The six basic design criteria for the treatment process units are presented below.

DATA ON THE FILTERS

Total number of filters	Six filters, each with a filter bed area of 960 ft^2
Backwash condition	22 gpm/ft^2 for 7 min (conservative)
Surface wash condition	3 gpm/ft^2 for 3 min

WASTE WASHWATER HOLDING TANK

Number	One tank with a 62,800 ft^3 or 470,000 gal capacity
Size	80 ft in diameter with a water depth of 12.5 ft, to hold three filter washings

TRANSFER PUMPS

Type	Vertical turbine pump
Number	Three identical pumps (one pump as a standby)
Capacity	1000 gpm each

TREATMENT PROCESSES

Total capacity	2000 gpm
Flash mixing	10 in. in-line static mixer
Flocculation tanks	Two identical baffled channel types— round the end type
Size of each	20 ft (W) × 20 ft (L) × 10 ft (swd)
Number of channels	Four channels with a total of 16 baffles for each tank
Volume of each	4000 ft^3 or 30,000 gal
Mixing time	30 min at 1000 gpm
G	An average of 30 s^{-1}—tapered mixing

SEDIMENTATION TANKS

Number	Two identical tanks with tube settler modules
Size of each	20 ft (W) × 35 ft (L) × 10 ft (swd)
Tube settler	500 ft^2 of each tank is covered by the settlers

Surface load-ing	2 gpm/ft² for the area covered by the settlers
Launders	Two launders per tank, each 25 ft long and a loading rate of 10 gpm/lf
Sludge collec-tor	One chain-and-flight type for each tank

TOTAL VOLUME OF THE WASTE WASHWATER

$$(22 \times 960 \times 7) + (3 \times 960 \times 3) = 156{,}480 \text{ gal/wash}$$

Since the water source is a large lake, the potential for algal bloom exists. Therefore, the worst scenario would be that each filter is washed three times a day. The average wash cycle is assumed to be 24 h for the filters. The waste holding tank should be sized to hold three backwashes because two to three filters may become clogged at the same time, thereby requiring the filters to be washed in a consecutive manner.

Total required volume: $156{,}480 \times 3$

$$= 46{,}9440 \text{ gal}\quad \text{or}\quad 62{,}800 \text{ ft}^3$$

The tank should be cylindrical in shape with a maximum water depth of 12.5 ft.

Diameter of the tank: $D = \left(\dfrac{62{,}800 \div 12.5}{0.785}\right)^{0.5} = 80 \text{ ft}$

Under the worst conditions, a total of 18 filter washes will be performed per day; thus, 24 h $\div$ 18 = 1.33 h or an average of 1 h and 20 min between filter washings. In other words, the total volume of wash waste produced by one filter washing must be pumped out within 1.33 h. The required capacity of the transfer pump is

$$156{,}480 \div 80 = 1956 \text{ gpm}$$

Make this 2000 gpm or 2.88 mgd.

Under normal conditions, each filter is washed once a day. Since the raw water is obtained from a large lake, it is assumed that the treatment mode of the plant, for the majority of the year, is a direct filtration simulation process. In this case, the average time between the filter washing is 24 $\div$ 6 = 4 h. The required recycling pump capacity is therefore

$$156{,}480 \div 240 = 652 \text{ gpm}$$

However, the pumping capacity should be roughly 50% larger (1000 gpm) since plant operators generally prefer to wash as many filters as possible during the day shift; the graveyard shift has only the absolute minimum number of operators and in many cases it is only one or two persons.

Provide three identical transfer pumps, each with a 1000 gpm (1.44 mgd) capacity, at the center of the circular tank. One of these pumps should act as a standby.

The total design capacity of the treatment facility should be 2000 gpm. Therefore, provide two parallel units, each consisting of a 1000 gpm process train.

Since the recycling rate is rather constant, that is, only one pump is operated regardless of the water level in the holding tank, a baffled channel flocculation process is the most cost effective method for producing good floc; it is also cost effective to build and operate. An alternative to the baffled channels is the vertical shaft flocculator; refer to Section 3.2.4 for further information.

Each sedimentation tank should be designed with both a tube settler module and chain-and-flight collector units. The tube settler modules are used for their space-saving feature and their effectiveness.

The blending ratio of the recycled wash waste should always remain below 10% of the anticipated normal operational conditions. The blending ratio is generally 5–8%; thus, there is little concern for hydraulic shock loading to the main process units. Figure 4.2-4 illustrates the general features of the waste holding tank and the waste treatment processes.

Operational Issues   An issue that is important but often neglected is the monitoring of filter medium accumulation in the wastewater holding tank. Plant operators should drain the tank every 4 months and estimate the volume of accumulated sand and/or anthracite coal at the bottom of the tank. If the operator finds an excessive amount of medium, it is clearly evident that the filters are losing a significant amount of medium during the washing process. The operator should then modify the filter washing conditions or adjust the height of the wash troughs. Accumulation of filter medium should be detected at an early stage because the medium is very abrasive and will erode the impeller of the transfer pumps.

With regard to the treatment of recycled water, the coagulant dosage of the main process would need not be adjusted if the blending ratio is less than 10%, regardless of the turbidity of the recycled water. The turbid form of wash waste contains a great deal of floc and this generally improves the efficiency of the flocculation and sedimentation processes due to the seeding effect (Figure 4.2-2). Moreover, the dosages of other water treatment chemicals usually do not have to be readjusted as long as their blending ratios remain under 10%.

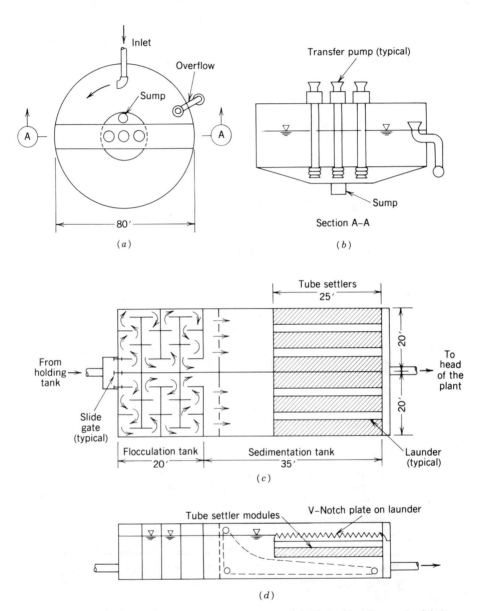

Figure 4.2-4 Waste washwater treatment process. (a) Plan of holding tank. (b) Section A–A. (c) Plan. (d) Longitudinal section.

The operation and maintenance of the pumps and the flocculation and sedimentation units do not require special operator considerations. Refer to Sections 3.2.4, 3.2.5, and 4.7 for further information on the operation and maintenance aspects.

BIBLIOGRAPHY

Kawamura, S., "Effect of Filter Wash Waste Recycling on Both Alum Flocculation and Cost of Plant Operation," *J. Jpn. AWWA*, 288:28(September 1958).

Kawamura, S., "Considerations on Improving Flocculation," *J. AWWA*, 68:328(June 1976).

4.3 SLUDGE HANDLING AND DISPOSAL

Purpose   The sludge produced by water treatment plants is categorized as chemical sludge and primarily consists of inert materials. It is a by-product of the water treatment process and, until recently, treatment plants would return it to the water source downstream of the intake. Currently, federal and state legislations classify most types of sludge as pollutants and prohibit their disposal into the public water supply.

The majority of water treatment processes employ aluminum sulfate as a coagulant and sludge is therefore commonly referred to as alum sludge. Alum sludge is difficult to dewater and dried sludge can revert to slurry when wet and disturbed, due to its thixotropic nature. This section presents some design guides for creating the most cost effective and feasible sludge treatment (conditioning, thickening, and dewatering) and disposal methods.

Considerations   The major considerations facing every design engineer are the Water Pollution Control Act, estimation of the sludge production rate, characteristics of the sludge, methods of minimizing the sludge production rate, dewatering methods, and the ultimate disposal of the sludge. This section addresses the first five issues and presents several sludge handling methods. The ultimate disposal of the sludge is discussed in a separate section.

Water Pollution Control Act   Public Law 92-500, the Water Pollution Control Act Amendment of 1972, categorizes sludge from water treatment plants as an industrial waste requiring compliance with the provisions stated in this act. Under the NPDES (National Pollutant Discharge Elimination System) provisions, a permit must be obtained to discharge materials such as sludge and waste washwater. Wide variations are found in the limitations of the permit among the various states. The parameters include the pH, BOD, turbidity, temperature, flow of discharge, and the total concentration of the following substances: suspended solids, settleable solids, iron, manganese, aluminum, chlorides, sulfate, residual chlorine, floating solids, and foam. According to this amendment, no utility is allowed to discharge waste after 1985, unless it can show good cause for not recycling the treated waste water.

Thus, engineers must grapple with the problem of the ultimate disposal of sludge.

Estimation of the Sludge Production Rate   It is rather difficult to estimate the sludge production rate since there are no concrete methods for predicting turbidity spikes and the chemical dosages required to treat them. However, there are several empirical formulas that may be employed. It is recommended that engineers select parameters that are similar to the operational conditions and use the basic formulas, such as the ones shown below.

Dry alum sludge production rate (pounds per million gallons of water treated):

[Alum dosage (mg/L) $\times$ 2.2]
$$+ \text{[raw water turbidity (NTU)} \times 1.3^* \times 8.34]$$

Dry ferric hydroxide sludge production rate

(pounds per million gallons of water treated):

[Ferric salt dosage (mg/L) $\times$ 3.2]
$$+ \text{[raw water turbidity (NTU)} \times 1.3^* \times 8.34]$$

* 1.3 is the ratio between total suspended solids (mg/L) and turbidity (NTU). This ratio ranges from 1.0 to 2.0.

In the case of sludge produced by lime softening, the rate of sludge production is largely a function of the hardness of the raw water and the hardness level of the softened water. In general, 2–3 mg/L of solids will be produced for each mg/L of hardness removed. Thus, average lime sludge production rate may be estimated as follows:

Dry lime sludge production rate (pounds per million gallons of water treated):

$$2.5 \times 8.34 \times \text{total hardness removed (mg/L as } CaCO_3)$$

Sludge Characteristics   The characteristics of water treatment sludge differ from location to location because they are largely dependent on the raw water characteristics, the type and amount of coagulant, and other types of chemical aids used in treating the raw water. In general, dried alum is composed of 30% aluminum and 50% inert solids, such as silica and calcium. A typical sample of alum sludge contains the components listed in Table 4.3-1. The typical components of lime softened sludge are presented in Table 4.3-2.

The characteristics of lime sludge vary as a function of magnesium hydroxide content. The drying characteristics of lime sludge are similar in nature

TABLE 4.3-1   Alum Sludge Characteristics

BOD (mg/L)	COD (mg/L)	pH	Total Solids (%)	Al_2O_3 (%)	SiO_2 and Inert (%)	Organics (%)
30–300	30–5000	6–8	0.1–4	15–40	35–70	15–25

TABLE 4.3-2   Lime Sludge Characteristics

$CaCO_3$ (%)	$Ca(OH)_2$ (%)	$Mg(OH)_2$ (%)	SiO_2 and Inert (%)
85–93	0–1	0.5–8	2–5

to alum sludge. The concentration of magnesium hydroxide may range from insignificant to 30% by weight. Lime sludge containing low levels of magnesium hydroxide may be dewatered to 50–60% solids through the use of drying beds. Yet, these figures will be reduced to 20 and 25% solids if the sludge has higher concentrations of magnesium hydroxide.

The characteristics of sludge are important to know because they greatly influence the methods of handling and disposal. For example, semidried alum and ferric sludges are thixotropic; the consistency of the sludge changes with physical force. This characteristic makes alum and iron sludge very difficult to handle.

The consistency of drying sludge differs primarily due to the amount of metal hydroxides and the applied polymer (coagulant) dosage. The general expression of sludge consistency is as follows:

Solids Content (% of solids)	Sludge Consistency	
	Alum Sludge	Lime Sludge
0–10	Liquid	Liquid
10–15	Viscous liquid	Viscous liquid
15–20	Pasty	Pasty
20–25	Semisolid	Pasty to semisolid
25–30	Soft solid	Soft solid
30–50	Crumbly cake	Crumbly cake

Reduction of Sludge Production Rate and Volume   In order to minimize capital expenditure and the operational costs of sludge handling and disposal, the first step to be considered is the reduction of the sludge production rate. The sludge production rate may be cut by 30–80% from the amount produced by a conventional complete treatment process using alum (coagulant) if either of these two methods are implemented: the simultaneous use of a polymer and alum (as coagulant) or adopting a direct filtration type of process to replace the conventional complete treatment process. These methods should be investigated prior to design work.

The actual data from many operational water treatment plants demonstrate that if 0.1–1.5 mg/L of cationic polymer are applied in conjunction with alum, the alum dosage can be reduced by 30–50%. The combined application of 0.2–0.3 mg/L of anionic or nonionic polymer and alum often improves flocculation dramatically, to the extent that alum dosage may be reduced by 15–30%. In the case of lime softening, the use of caustic soda may decrease the production of calcium carbonate sludge by 35–50%. However, the cost of caustic soda often outweighs the benefits associated with a decrease in sludge production.

The generation of sludge is most dramatically reduced when the conventional complete treatment process is replaced by two-stage, direct, or in-line filtration. These processes may be used to treat raw water obtained from large lakes or reservoirs and require only 4–6 mg/L of alum or the combined application of 2–4 mg/L of alum and 0.5–1 mg/L of cationic polymer. In contrast, 12–16 mg/L of alum are required by the conventional complete process for the same raw water. Thus, it is apparent that the sludge production rate may be reduced to levels that are only 10–20% of the conventional process. If the treatment plant is forced to adopt a conventional process due to occasional turbidity spikes or algal bloom, the plant design could possibly have a bypass channel across the basins so that plant operators have the ability to treat the water in a direct filtration or in-line filtration mode for most of the year.

The second step in minimizing the costs of the sludge handling and disposal processes is to decrease the volume of the sludge. The volume of the sludge may be minimized through effective dewatering techniques and by-product recovery.

It is important for engineers to select an appropriate type of clarifier and sludge handling facility. For example, the composition of alum sludge in rectangular sedimentation tanks is generally 0.3–1.5% solids. However, the composition is only 0.2–0.5% if a sludge blanket type of clarifier is used. Furthermore, basins lacking a continuous mechanical sludge removal unit usually have the sludge accumulating to heights of 10–12 ft (3–3.5 m) before the basins are cleaned. Thus, the sludge is compressed to the extent that it is composed of 4–6% solids, whereas sludge from sedimentation tanks containing mechanical sludge removal units has a composition ranging from 0.5 to 1.5% solids, depending on the type of sludge removal unit and the scraping cycle.

Under normal conditions, approximately 0.1–0.5% of the water treated at a water purification plant will be wasted by the sludge withdrawal operation unless recycled. The relationship between the suspended solid content (mg/L) and the measured turbidity (NTU) is a function of the type, size, shape, and color of the suspended solids. Thus, a definite relationship cannot be derived. The ratio of the total suspended solids to the turbidity usually ranges from 1 to 2, with 1.3 being an average figure.

The solids content of lime softened sludge is much greater than alum sludge; it ranges from 1 to 10%, depending on the type of clarifier unit. Sludge removed from clarifiers may be 3–4% solids, and 0.3–1.5% of the lime softened water is used to the sludge withdrawal cycles.

Sludge Handling   A certain level of handling and treatment is required prior to the ultimate disposal of the sludge. The degree of handling depends on the requirements governing the ultimate disposal.

Mechanical dewatering systems may be used in situations where drying beds cannot be constructed, due to the lack of land or lack of capital, and/or a high solids content is required. These systems include gravity thickening, vacuum filters, filter presses, centrifuges, and freeze-and-thaw techniques. Yet, in regions where the weather is mild to hot and evaporation exceeds rainfall, the most simple and cost effective methods are the sludge lagoon drying beds and the shallow depth sludge drying beds. If properly designed and operated, these beds are capable of drying both alum and lime sludge containing 45–50% solids.

The drying time for sludge drying beds may be drastically reduced if the beds are continuously aerated at a rate of 1–2 cfm/ft^2 for the first 3 weeks of the drying cycle, until cracks appear on the surface of the sludge. One Japanese equipment manufacturer has demonstrated that the drying time, for alum sludge, may be shortened from 4 months (using ordinary weather drying) to 6 weeks through the implementation of sludge aeration. Rather than using compressed air, a vacuum may be applied to the sludge beds. However, this practice causes the underdrainage system to become frequently clogged. Similarly, sand drying beds may significantly cut the drying time of both alum and lime sludge; however, in many cases the beds must be reconstructed every 1–2 years due to clogging.

In locations where severely cold temperatures are predominant for one-third of the year, freezing is a very effective method of sludge dewatering. Freezing can produce alum sludge that is 25% solids, and the sludge is no longer thixotropic in nature. The application rate of the sludge is approximately 2.5 lb/ft^2 and the wet sludge is usually kept below a depth of 12 in. A large-scale mechanical freeze-and-thaw facility was built in Chiba and Sendai, Japan, to treat alum sludge. However, the widespread use of this facility has been limited by the high initial cost and high operational cost ($100/ton).

Gravity Thickening   Gravity thickening may be attained through the use of lagoons or mechanical thickening tanks, provided that a polymer is added to the liquid sludge. Alum sludge treated with 2–3 mg/L of cationic polymer can produce a sludge containing 3–5% solids through lagooning. Lime conditioning of alum sludge yields a sludge composition of 6–9% solids when processed by a thickener tank at a feed rate of 0.07 to 0.14 gpm/ft^2 and approximately 5 to 10 lb of solids per square foot of thickener surface per day. Lime sludge may be thickened to a greater degree than alum through the use of thickeners; a loading rate of 30 to 60 lb of solids per square foot of thickener surface area per day may yield lime sludge that is 15–35% solids. In case of alum or ferric hydroxide sludge, the standard solid loading rate for sludge thickener is 10 lb of solids per square foot per day or 0.12 to 0.15 gpm/ft^2 hydraulic loading.

Centrifugation   Among the several types of commercially available unit, the most applicable units are the scroll-discharge, solid-bowl, and basket-bowl. Basket centrifuges produce alum sludge that is 10–11% solids without the use of polymers. However, the sludge composition can be improved to 15% solids if the sludge is preconditioned with polymer (4–5 lb per 1 ton of dry solids). Optimum preconditioning of alum sludge calls for the application of 1–2 lb of polymer per ton of solids and results in a 25–30% cake.

Bowl centrifugation allows lime softened sludge to be more easily dewatered than alum sludge. This technique produces a cake that is 30–70% solids with or without polymer preconditioning. Yet, the disadvantage is that it may require a high level of maintenance and has a high operating cost.

Drum Type of Dehydrator (Dehydrum)   The drum type dehydrator is a proprietary unit that has been used successfully in Japan. It transforms the sludge into pellets that are 20–35% solids, provided that polymer is the sludge conditioner. The shape and motion of this unit resemble a concrete mixer truck. Mechanically, the system is simple and therefore has the lowest operation and maintenance costs of all the mechanical dewatering systems.

Filter Press (Pressure Filtration)   The filter press may be used to dewater both alum and lime sludge. However, it is used primarily to dewater alum sludge that has been preconditioned with polymer or lime or precoated with diatomacious earth. The solid content of the cake is generally 25–50%, with a pressing cycle of 8 h. The life cycle of the filter cloth is normally 1.5 years. The filter press is a batch process that requires a considerable outlay of capital and has a high operational cost. Yet, if the sludge is required to be over 40% dry alum solids, this process is the most reliable and quickest available method.

Belt Press   The belt press is a relatively small unit requiring minimal attention and maintenance once the system is optimized. It has been used to process coagulant sludges and has yielded alum cakes that are 20–25% solids as long as polymer is added to the feed sludge.

The Skinner Filtration Plant, operated by the MWD of Southern California, has been using two sets of belt presses with a processing capacity of approximately 250 tons/h for the past several years. This plant preconditions the alum sludge, which is approximately 3% solids, with 2.5–3.2 lb of cationic polymer (Telacol) for every ton of dry solids. The product (cake) is 22–25% solids. The system requires only one operator for a period of about 2–3 h/day.

TABLE 4.3-3   Solids Content (%) of Sludge by the Dewatering Processes

	Gravity Thickening	Drying Beds	Centrifugation	Filter Press	Belt Press	Freezing
Alum sludge	2–5	30–50	10–25	25–50	20–25	25–30
Lime sludge	8–12	40–60	30–70	40–50	—	—

The belt cloth costs $2000 per unit and lasts approximately 2000 operating hours before any major repair work is required. Other treatment plants on the East Coast of the United States report the belt press as producing cakes that are 15–20% solids, requiring 10–15 lb of polymer per ton of solids.

Vacuum Filters   Vacuum filters are primarily used in processing lime sludge. The typical yield is a cake that is 40–50% solids. For low magnesium lime sludge, the loading rate is 40–90 lb/ft² per hour, but only 10–20 lb/ft² per hour if the magnesium level is high. Vacuum filters are generally ineffective in processing alum sludge.

Table 4.3-3 summarizes the typical performance of several dewatering processes.

Filtrate Disposal   The liquid waste from all the mechanical dewatering processes pose a problem because of their chemical characteristics: they contain a high concentration of polymer, alkali chemicals, aluminum, and organic compounds and a significant level of heavy metals. Recycling the liquid waste back to the head of the plant, alone or combined with the filter wash waste, is not recommended because it will adversely affect the flocculation process and will significantly increase the level of unfavorable compounds in the processing water. Moreover, the Department of Health Service (for certain states) discourages this practice because of the uncertainty about the effect of trace metal and other unfavorable compounds on human health. Possible methods of disposal include discharge into the sanitary sewer, purification treatment, and discharge to sludge drying beds if they are on site.

Ultimate Disposal   Four practical options are available for the ultimate disposal of sludge: direct discharge to the sewer system, hauling to a landfill, on-site disposal, and recovery of by-products.

Discharge to the Sewer System   The easiest solution for the treatment plant is direct discharge of the sludge to a sanitary sewer. However, the waste sludge flow must be equalized prior to discharge because the ordinary sewer lines are incapable of handling the flow rate (directly from the basins) and the solid content is often too thick. The equalized sludge may be pumped from the holding tank to the sewer line at a constant rate that does not flood the sewer system. In several cases, the discharge of waste into the sewer lines was restricted to night, due to the limited capacity of the existing sewer system. The actual effect of this practice is usually not noticeable, unless the sludge from the water treatment plant exceeds the total suspended solids received by the wastewater treatment plant (in the form of sewerage) by more than a few percent. On certain occasions, the alum sludge may even be beneficial to the wastewater treatment processes; it may increase the removal of phosphate or may aid in the removal of suspended solids due to the aluminum hydroxide. Water treatment plants using a high proportion of organic polymer as coagulant will yield sludge that tends to be more biodegradable. The negative effect of alum sludge on wastewater treatment plants is the reduction in methane gas, caused by excessive amounts of inert materials, such as aluminum hydroxide and clay materials, being brought to the sludge digesters. Unless both the water and wastewater treatment plants are under the same administrative jurisdiction, a connection fee will be imposed for the direct discharge of sludge to the sewer system. Following is an example of the fee structure used in southern California.

$$\text{Fee} = \$375 \times [0.0034 \ (gpd) + 0.115 \ (ppd \ COD) + 0.3595 \ (ppd \ SS)]$$

where gpd = sludge flow in gallons per day,
ppd COD = pound per day of COD,
ppd SS = pound per day of suspended solids.

In addition to the initial connection fee, the sewer authority will assess an annual surcharge based on the established unit costs. An example is presented below:

Cost per year per million gallons discharge	$200
Cost per year per 1000 lb of COD	$ 12
Cost per year per 1000 lb of SS	$ 33

In contrast to the difficulties encountered in the United States, the discharge of sludge to sanitary sewers is a popular method of disposal in Great Britain and the Greater Paris area in France. It has been claimed that this practice has very little effect on the wastewater treatment processes or on subsequent sludge digestion.

Hauling Sludge to Landfills Selection of a sanitary landfill that is near a city is becoming increasingly difficult. In recent years, residents have become more concerned about environmental issues and issues that may affect land value. To further complicate matters, many existing sanitary landfills are reluctant to accept liquid and semiliquid sludge in order to comply with the federal landfill guideline PL 91-512 (Solid Waste Disposal Act) and because of their operational inconvenience. The Solid Waste Disposal Act recommends that landfill sites accept water treatment sludge only if it does not contain free moisture. Free moisture is defined as liquid that will drain freely from solid material by gravity. Currently, most landfill sites only accept sludge with a minimum solids content of 25%; some sites demand a solids content of over 40%. It is important to note that alum sludge containing less than 20% solids generally cannot be handled by sludge handling equipment nor hauled from the plant in trucks.

Landfills generally charge a disposal charge of $5–6 for each ton of dried sludge. If the sludge is located in a remote area, the hauling charge becomes

significantly higher. Design engineers should also investigate alternative landfill sites in case the nearest site becomes unavailable.

If the sludge hauling trucks must pass through respectable residential areas
or busy commercial zones, the option of using landfills may possibly be eliminated by public hearings. Environmental impact studies may also prevent
the implementation of this alternative.

On-Site Disposal   On-site disposal may be considered as an alternative method
of sludge disposal only if the treatment plant has an adequate site area, has
an old quarry on the premises, produces very small amounts of sludge, or
has a combination of these conditions. If a plant is a candidate for on-site
disposal, the following issues must carefully be evaluated prior to design work:
the potential for contaminating the finished water and/or groundwater and
the aesthetics and safety of the facility.

The disadvantages of on-site disposal may be illustrated by the following
example: application of alum sludge to the ground surface tends to clog the
soil, thus preventing the passage of air to the roots of vegetation. However,
lime sludge offers some advantages: lime sludge has been shown to improve
the condition of clay soil and neutralize acidic soil.

Recovery of By-products   Acidic alum recovery was first tried on a large-
scale basis in Japan: 15 plants were built between 1965 and 1972. Alum
recovery can generally be attained at a rate of 80% or more, through sulfuric
acid extraction at a pH of 2.5. The recovered alum is nearly as effective as
commercially available alum. However, the quality of the alum often does
not meet water treatment chemical standards because other substances are
also extracted during the acidification process: iron, manganese, heavy metals,
and a wide variety of organic compounds. Consequently, the recovered alum
is used in conjunction with commercially available alum. Due to the possible
harmful effects of recycling and the accumulation of heavy metals, this practice
is now discouraged. The recovered by-products of alum sludge have also been
used in the production of low-grade bricks (in Brazil), as road stabilizer, and
as aggregates for low-grade road material. In Japan, alum sludge is used a
raw material for cement manufacturing industry, as soil conditioner for certain
type of vegetation and high grade bricks.

Many locations have practiced lime recovery by burning the lime sludge
in kilns. However, very few plants now practice this alternative because of
recent concerns associated with air pollution and because of the economic
feasibility.

The Midwest (United States) has successfully employed lime sludge to
neutralize acidic soil. Lime sludge may also be used to increase the porosity
of tight soils, thus allowing them to be more workable and productive for
agricultural purposes.

Design Criteria   The design criteria for the properties of alum sludge, lagoon drying beds, sand drying beds, and mechanical dewatering systems are now presented.

Properties of Alum Sludge

1. The solids content of wet alum sludge obtained from ordinary rectangular sedimentation tanks containing mechanical sludge collection systems is 0.5–1.5%.
2. Alum sludge from sludge blanket type clarifiers has a solids content of 0.1–0.5%.
3. Sludge lagoons are primarily used in gravity thickening. The solids content of sludge processed in this manner, with 3–4 months of continuous sludge filling and decantation operation, may be 3–4%.
4. A regular sludge thickener with stirring rakes can yield alum sludge that is 4–6% solids with the addition of very little polymer.
5. Sludge containing 4–5% solids at the beginning of a drying cycle will shrink to approximately 15% of the original volume after drying: the dried sludge will thus be 45% solids.
6. The initial depth of the sludge, prior to drying, has a drastic effect on the length of the drying period. Thus, the initial depth should not exceed 2–3 ft if the drying cycle is established to be 3–4 months.
7. The supernatant must be decanted continuously during the drying cycle in order to facilitate the formation of cracks on the surface of the drying sludge; these cracks will speed up the drying time by exposing more surface area for evaporation.
8. Figure 4.3-1 plots the unit volume weight (bulk density) of the alum sludge at various stages of the drying cycle. This chart should be used as a reference when designing the drying bed.

Lagoon Sludge Drying Bed

Sizing guide	The required net bed area for alum sludge is estimated to be 8 lb/ft² (40 kg/m²) for wet regions and 16 lb/ft² (80 kg/m²) for dry regions; these figures are based on dry sludge weight.
Minimum number of lagoons	Three, but four are preferred.
Size of each lagoon	Each lagoon should be capable of holding the amount of sludge produced (average) by the treatment plant over a period of 3–4 months.
Depth of the liquid sludge	A maximum depth of 6 ft during the filling stage; however, a depth of 4 ft is preferred.
Bottom slope	A downward slope of 0.5–1% toward the outlet.

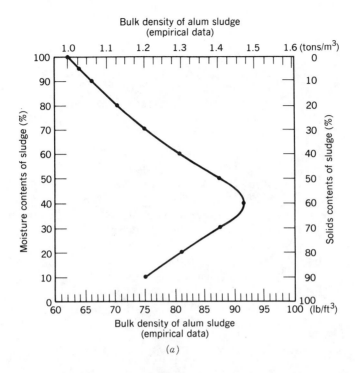

Figure 4.3-1 Unit volume weight (bulk density). (a) alum sludge. (b) Lagoon type sludge. (c) Drying alum sludge.

Shape of each lagoon	Rectangular or free form with a minimum length to width ratio of 4:1.
Underdrainage system	This is an optional item. The system is composed of perforated pipe laterals with gravel packing and 1 ft of sand above the gravel bed. Note that the underdrainage system tends to become clogged and is often ripped away by the sludge removal equipment.
Lining	The entire surface, including the bottom of the drying bed, should be covered by 2–3 in. of Gunite, asphalt cement, or soil cement to facilitate easy sludge removal (by means of mechanical equipment), to prevent contamination of the groundwater, to provide protection from erosion, and for weed control.
Distance between inlet and outlet	A minimum of 100 ft (30 m) should be provided to prevent flow short-circuiting.
Inlet pipe	Isolation valve and a simple energy dissipator.
Outlet	Multilevel (9–12 interval) decantation facility, overflow protection, and complete drainage of the lagoon.
Vehicle access	A ramp should be provided for each bed to facilitate the removal of dried sludge.
Access road	A good access road, both to and around each lagoon, should be included in the design.
Flow velocity in the pipe	The flow velocity of liquid sludge should be higher than 2.5 fps (0.75 m/s).

Sand Bed Type of Drying Bed The design criteria for this type of drying bed are fundamentally the same as those listed for the lagoon drying beds, with the exception of the following items:

Depth of the liquid sludge	A maximum of 18 in. (12 in. preferred).
Underdrain laterals	Perforated laterals with 6 in. diameters and 10 ft on center.
Gravel bed	Three layers with a total depth of 14 in: 3 in. layer of $\frac{1}{8}$–$\frac{3}{8}$ in. gravel; 3 in. layer of $\frac{3}{8}$–$\frac{1}{2}$ in. gravel; and 8 in. layer of $\frac{3}{4}$–1.5 in. gravel.
Sand bed	Sand ranging from #10 to #30 mesh at a depth of 12 in.

Note: Lime sludge is generally easier to dewater by the sand drying bed than alum sludge; this method yields a sludge that is 25% solids and is therefore spadeable (weighs 120 lb/ft^3).

Mechanical Dewatering Systems   The belt press, filter press, vacuum filter, centrifugation, and freeze-and-thaw processes are the mechanical dewatering systems that are applicable to water treatment. The first two units are most frequently employed. Since these systems are proprietary items, the project engineer only needs to determine the functional specifications: the solids content of the dewatered sludge, floor space requirements, and limitations on power and chemical usage.

A pilot study must be conducted prior to writing the specifications, to confirm the effectiveness of the system and to evaluate the type and required chemical dosage for preconditioning the sludge.

Example Design Calculations

Example 1

Given   A water treatment plant with a daily average flow of 20 mgd is located in southern California and requires a lagoon type of sludge drying bed. The average turbidity of the raw water is 12 NTU and the average alum dosage is 10 mg/L.

Determine

(i) The required number of lagoons.

(ii) The depth of each lagoon.

(iii) The size of each lagoon.

Solution   By using the formula presented in the "Considerations" section, the sludge production rate may be calculated:

$$[(10 \times 0.26 \times 8.34) + (12 \times 1.3 \times 8.34)] \times 20$$

$$= 3035 \text{ lb of dry sludge per day}$$

(i) Provide four lagoons with identical capacities. Each lagoon should be sized to hold sludge with an average of 4% solids for a duration of 3 months.

(ii) The maximum depth of the liquid sludge, during the filling stage, should be set at an average of 6 ft. When the drying cycle commences, continuous decantation should be initiated so that the depth of the sludge is reduced to 2–3 ft prior to the actual drying process.

(iii) The required net area of one lagoon may be estimated from the sizing guide listed in this section, under design criteria. Since southern California is sunny for most of the year, with a warm to hot climate, use a dry solid weight of 15 lb/ft^2 of required area to determine the required lagoon area.

Based on a 3 month supply of sludge, the average amount of dry solids is

$$(3 \times 30) \times 3,035 = 273,150 \text{ lb}$$

The required net area for the lagoon is therefore

$$273,150 \text{ lb} \div 15 \text{ lb/ft}^2 = 18,210 \text{ ft}^2$$

Thus, make the lagoon rectangular in shape, 60 ft wide and 300 ft long, although a free shape, such as a kidney bean shape, is aesthetically better.

Check the calculations. Based on Figure 4.3-1, the volume of the 4% solids sludge is approximately 63.8 lb/ft³. Therefore, the total volume of the sludge may be calculated to be

$$V = (273,150 \div 0.04) \div 63.8 = 107,034 \text{ ft}^3$$

Since the depth of the liquid sludge at the end of the filling cycle is 6 ft, the area required for 136,480 ft³ of 4% solids sludge is

$$107,034 \div 6 = 17,840 \text{ ft}^2$$

which is essentially the same as 18,000 ft².

In summary, provide four lagoons that have a net area of approximately 72,000 ft² (1.7 acres) and a depth of 6 ft. Other considerations such as vehicle access, the building of dikes around the lagoons (side wall slope of 2.5:1), and a vehicle access road that is 10 ft wide at the top of the dikes, require the land to be almost twice the net area required for the lagoons (3.5 acres).

Example 2

Given A 10 mgd lime softening plant is located in the Midwest. The plant requires lime sludge drying beds to be constructed. The average total hardness of the river water is 360 mg/L, with insignificant amounts of magnesium. The finished water hardness is, on the average, 80 mg/L.

Determine

(i) The number of sludge drying beds.
(ii) The filling depth of the sludge.
(iii) The size of each bed.
(iv) Provide a sketch of the sludge bed section.

Solution There are several general considerations that must be determined prior to making the four determinations. The following assumptions are made for this example.

1. For every 1 mg/L of hardness removed, 2.5 mg/L of dry solids (sludge) is produced.
2. Lime sludge containing insignificant amounts of magnesium hydroxide is easily filtered through the sand bed and dries faster than alum sludge.
3. The drying beds will yield sludge that is 50% dry solids (80 lb/ft^3). Sludge that is 25% solids contains 72 lb of dry solid per cubic foot of sludge; this sludge is spadeable and is also easily handled by the mechanical sludge removal equipment.
4. Assume that the average daily flow is 67% of the plant capacity of 10 mgd.
5. The hardness removed is averaged to be $360 - 80 = 280$ mg/L.

The average sludge production rate (dry sludge) is 39,115 lb/day:

$$[(2.5 \times 280) \times 8.34] \times 6.7 = 39,115 \text{ lb/day}$$

Now assume that the process unit yields a liquid lime sludge that is 5% solids with a unit weight of 65 lb/ft^3. The volume of the 5% solids sludge produced each day may be calculated as

$$V = (39,115 \div 0.05) \div 65 = 12,035 \text{ ft}^3/\text{day}$$

The volume of the 25% solids sludge is $39,115 \div 72 = 543$ ft^3/day.

(i) A total of eight sludge beds are needed. Each bed should be able to hold the amount of 25% solids sludge produced over an average of 15 days so that a total of 4 months storage is provided.
(ii) The maximum depth of accumulated sludge during the filling cycle is 12 in.
(iii) The size of each bed must be calculated as follows. The estimated volume of the semidried sludge (25% solids) for 15 days of storage is

$$V = 543 \times 15 = 8145 \text{ ft}^3$$

The depth of the bed is limited to 1 ft. If the width of each bed is designated to be 40 ft, then the required length is

$$(8145 \div 1) \div 40 = 203 \text{ ft}, \quad \text{say 200 ft}$$

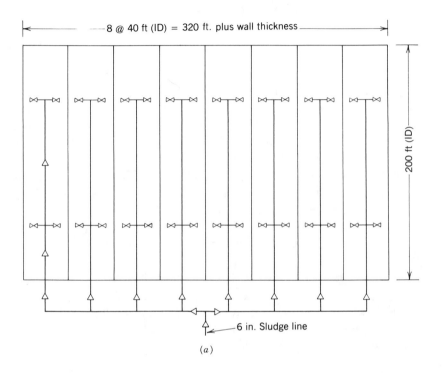

(a)

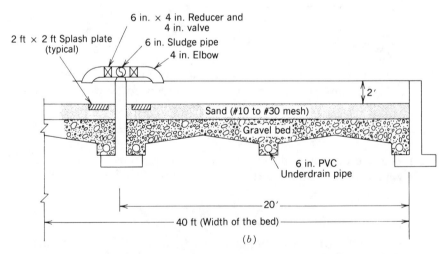

(b)

Figure 4.3-2   Lime sludge drying bed. (a) Plan. (b) Partial cross section.

Therefore, the drying bed is 40 ft (W) × 200 ft (L) per sand bed. Figure 4.3-2 is a sketch of the designed sludge bed.

Notes

1. The sludge beds tend to become clogged and the beds are easily disturbed by the heavy equipment used during sludge removal.
2. The sludge lagoon drying bed is the most popular and practical method of sludge handling.
3. The design criteria for the sludge lagoons are basically the same for lime sludge and alum sludge.
4. A conventional lime sludge lagoon system has a minimum of three lagoons, each having a depth of 10 ft and the capacity to hold nearly a 1 year production of wet sludge. The land requirement is approximately 1 acre per 100 mg/L hardness removed for each million gallons treated. A decantation facility is essential for proper sludge drying.

Operation and Maintenance   Plant operators must pursue three major issues: reduction of the sludge production rate, withdrawal of highly concentrated sludge from the process, and optimization of the sludge dewatering operation.

Reduction of the Sludge Production Rate   In order to decrease the sludge production rate, operators must select the appropriate coagulant, optimize the chemical dosage, and possibly optimize the flocculation process. Since 26% of the added alum produces aluminum hydroxide (i.e., becomes part of the sludge), the primary goal of the operator should be to reduce the alum dosage without degrading the finished water quality. The operator should first check to see if the polymers are being effectively employed. Since the cost of most cationic polymers is approximately six times higher than alum, there will be no difference in the total chemical cost if 1 part of polymer is used to replace 6 parts of alum. There are many case histories which prove that a cationic polymer dosage of 0.5 mg/L, together with alum, reduces the alum dosage by 3–5 mg/L when compared to flocculation by alum alone. The use of anionic and nonionic polymers as flocculation aids can also reduce the required alum dosage, as illustrated in Table A7-1 in Appendix. In the case of direct or two-stage filtration, the use of cationic polymers virtually replaces alum as the coagulant. Since there is a minimal amount of aluminum hydroxide in the sludge, there is a reduction in sludge production and the sludge is more easily handled.

The adoption of a direct filtration process during periods of good raw water quality is an alternative method of drastically reducing sludge production. Many water treatment plants that draw their water supply from large reservoirs or lakes should be able to implement this mode of treatment, but only at certain times of the year. Rather than feeding 15 mg/L of alum to raw water containing 3 NTU of turbidity to form settleable floc, the direct filtration process requires an alum dosage of only 6 mg/L or a combination of alum (3 mg/L) and cationic polymer (0.5 mg/L) may be used to coagulate the turbidity. The flocculation and filtration processes subsequently occur in the filter bed. To reiterate, there are many actual case histories of successful treatment of water by this mode; technical articles on this subject may be found elsewhere.

The third item that the operator must be aware of is the effect of feeding lime with alum at the head of the plant. Many old textbooks often illustrate this practice without regard to the alkalinity and pH of the raw water, and a number of treatment plants consider this practice as a given. Unless the alkalinity of the raw water is below 15 mg/L and the required alum dosage exceeds 15 mg/L, there is no need to add extra alkalinity. Flocculation in a pH range of 5–7 is actually ideal, except for the potential for corrosion; however, the pH may be adjusted in the later stages. If lime is not added with alum to the head of the plant, both residual alum in the treated water and sludge production will be reduced.

All cases described in this section require bench scale tests to be performed to verify and obtain the optimum chemical dosage.

Withdrawal of thick sludge   If the sludge delivered to the dewatering system is thick, then the process is more cost effective and more effective in concentrating the solids, thus allowing the ultimate disposal operation to be more efficient. For example, the operator may adjust the sludge collection cycle of the clarifier by observing the consistency of the liquid sludge, provided that the facility allows the operator to do so. Excessive sludge withdrawal will result in an extremely low solids content.

Thus, operators should adjust plant operations so that the sludge will have a solids content of at least 0.5%; alum sludge obtained from the clarifier preferably has a solids content of 1%. A word of caution: do not allow the sludge to remain in the clarifier for a long period of time in an attempt to increase sludge concentration because the sludge may become septic and may impart an objectionable taste and odor to the treated water.

Although it is common practice to discharge the sludge from the clarifiers to the filter waste-wash holding tank, it is highly discouraged for two reasons: (1) the sludge becomes diluted and (2) unfavorable substances such as organic compounds, microorganisms, heavy metals, and taste- and odor-causing compounds may potentially be recycled back to the treated water. Sludge from the clarifiers should be directed to either the sludge beds or the sewer system.

Optimization of Sludge Thickening and Disposal Operations   There are many areas in which the plant operator can implement change; however, these depend on the type of thickening system. Plant operators have the ability to optimize the sludge loading rate to the thickening process, the thickness of the feed sludge, the belt speed, and the pressure level to the filter press. However, the efficiency of the sludge thickening process may greatly be improved by adding an appropriate type and proper dosage of polymer to the liquid sludge, regardless of the type of sludge being processed.

In regions where lime is readily available (at a reasonable cost) and where the cost of personnel is also reasonable, the application of an additional 15% (or more) of lime to alum sludge raises the pH of the sludge to 12. Experience has shown that this practice dramatically improves the dewatering characteristics of the alum sludge.

BIBLIOGRAPHY

AWWA, *Water Quality and Treatment*, 3rd ed., McGraw-Hill, New York, 1971.

Bishop, M. M., et al., "Testing of Alum Recovery for Solids Reduction and Reuse," *J. AWWA*, 79(6):76(June 1987).

Cornwell, D. A., et al., "An Overview of Liquid Ion Exchange with Emphasis on Alum Recovery," *J. AWWA*, 71(12):741(December 1979).

Culp, R. L., et al., "Is Alum Sludge Advantageous in Wastewater Treatment?," *Water Waste Eng.*, p.16(July 1979).

Garbarek, R. J., et al., "Silvicultural Application of Alum Sludge," *J. AWWA*, 79(6):84(June 1987).

Japan WWA, "Design Criteria on Handling Facilities for Water Treatment Wastes," *J. Jpn. WWA*, 489:103(June 1975).

Knocke, W. R., "Effects of Coagulation on Sludge Thickening and Dewatering," *J. AWWA*, 79(6):89(June 1987).

Novak, J. T., et al., "Use of Polymers for Chemical Sludge Dewatering on Sand Beds," *J. AWWA*, 69(2):106(February 1977).

"Research Needs for Alum Sludge Discharge," Committee Report, *J. AWWA*, 79(6):99(June 1987).

"Water Treatment Plant Sludges—An Update on the State of the Art," Parts 1 and 2, *J. AWWA*, 70(9):498 and 70(10):548(September and October 1978).

Wilhelm, J. T., et al., "Freeze Treatment of Alum Sludge," *J. AWWA*, 68(6):312(June 1976).

4.4 INTAKE SYSTEM

Purpose A water treatment plant ceases to function when the intake system fails to supply water. If the plant is relying on a single intake system, malfunction of this unit results in the interruption of the water supply to the community. Therefore, the intake must be located in an easily accessible location and designed and built to supply a specified quantity of the best available quality of water; the source should be reliable, that is, no interruption under any condition. The intake system is often combined with the grit chamber and raw water pumping station. However, this section only discusses the intake system.

The intake structure is constructed at the water source for the sole purpose of extracting water for water treatment and water supply. The capacity of the intake should be the maximum daily demand anticipated for the next 50 years; the permits required to build the intake and the cost of construction are both difficult and costly tasks. If an intake is twice the size required to satisfy present needs, it would cost an additional 20–30%. However, a well designed intake system, for middle to large sized plants, may cost 15–20% of the total cost of constructing the entire water treatment plant.

Considerations   The essential factors for the intake system are reliability, safety, and minimal operation and maintenance costs. Thus, the design engineer must conduct extensive studies and use ingenuity in selecting the site and in designing the intake works.

Numerous studies must be performed prior to the selection of the intake site: water rights, the quality of the water source, climatic conditions, fluctuations in the flow rate and water level, regulations set by the Department of Fish and Game, water navigation, geographical and geological information, and economic issues.

The intake may be located on a river, lake, or reservoir or may be designed to extract groundwater. The basis for site selection varies according to the water source.

River Intakes   Site selection for the intake system should be based on (1) obtaining the best quality water through the implementation of procedures that avoid polluting the water source; (2) forecasting possible changes in the flow and course of the river; (3) minimizing the effects of icing, flooding, floating debris, river navigation, and surges in flow; (4) providing easy access to the intake for maintenance and repair; (5) providing adequate maneuvering space for vehicles; (6) allowing for future additions to the facility; (7) maintaining a safe quantity of water flow during periods of drought; (8) minimizing the effects of the facility on aquatic life; and (9) obtaining the best or at least good, geological conditions.

If the intake site is located at a bend in the river, the outside bank of the curve should be selected since it offers better conditions. The inside bank will most likely be troublesome because of the shallow water depth, and the presence of sand bars and accumulated debris.

Most river intake structures are constructed as shore intakes: an intake system and raw water pumping station combined. Yet, several alternatives are available: submerged intake, intake tower, suspended intake, siphon intake, floating intake, and movable intake. The last two types are worth evaluating if the water level of the river bed tends to fluctuate widely, if there is a frequent flooding problem and if the intake capacity is rather small—less than 5 mgd or 0.2 m³/s.

The infiltration galleries or pipes are a unique type of intake system. This system is a gravel-packed gallery containing many openings (perforated pipes). It may be constructed under the river bed parallel to the shore or in a gravel-packed well located on the river bed. The advantages are that it avoids problems associated with ice formation and the water quality is substantially better than the river water itself. The disadvantages are the potential reduction in yield due to clogging by the sand and gravel layer and its limited capacity.

The design of the intake system should include multilevel intake ports and bar screens or racks. Additionally, engineers must provide protection against ice, floating debris, boats, and barges, and the design must meet the requirements established by the Department of Fish and Game.

Lake and Reservoir Intakes   Tower intakes and submerged intakes are two typical types of lake and reservoir intake. Submerged intakes are generally less expensive to build than the intake towers. These units also have no aesthetic impact and do not interfere with river navigation. Submerged inlets that are properly designed and located will have minimal problems with ice as long as the water depth exceeds 20–30 ft. The characteristics of tower intakes may be found in the next section, "Design Criteria."

An alternative to the tower and submerged intakes is the shore shaft. The shore shaft is the part of the intake system that connects to the submerged inlet structure by means of a pipe or conduit. It functions as a screen chamber, as well as a pump suction well. Thus, the shore shaft should be located at a depth that allows for the safe withdrawal of the required maximum flow rate when the lake or reservoir is at its lowest level.

Site selection for the lake and reservoir intake system should be based on the considerations listed for the river intakes and the following items: (1) the stratification characteristics of the water due to seasonal changes in water temperature; (2) the water quality of minor streams, algal growth characteristics, and the growth cycle of other microorganisms; (3) shore conditions, wind direction, and velocity; (4) the conditions of the catchment area, including potential sources of pollution; (5) conditions conducive to siltation in the reservoir; (6) the ultimate purpose of the reservoir—multipurpose, such as hydroelectric power, irrigation, water supply, recreation—and the impact of joint users on the inlet location; (7) the level of sanctioned recreational activities and the degree to which body contact sports will be allowed; and (8) the flood level. In addition to these considerations, the design engineer must also contend with issues such as a cooperative design arrangement with the joint users (if the intake is for multipurpose use), the possible incorporation of an artificial water mixing device to destratify the water in the intake design, and special ice-breaking measures (air agitation) at the intake location for cold weather regions.

Refer to Figures 4.4-1a and 4.4-1b for the various types of intake structure.

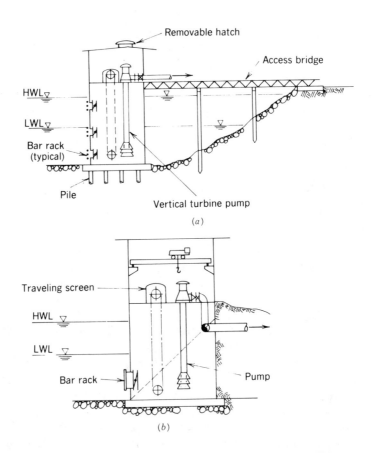

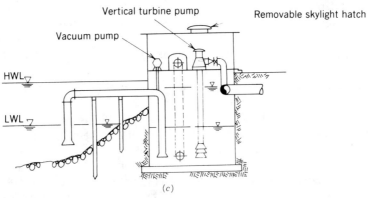

Figure 4.4-1a Various types of intake structures. (a) Intake tower. (b) Shore intake. (c) Siphon well intake.

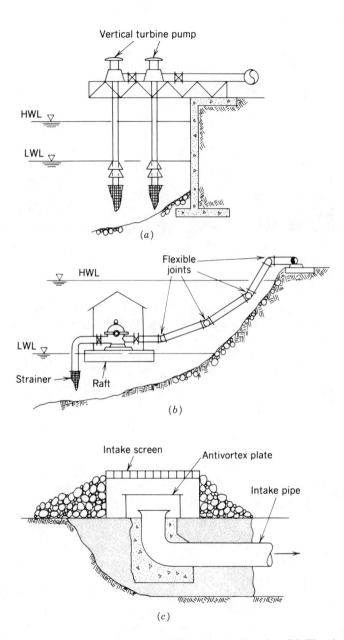

Figure 4.4-1b Various types of intakes. (a) Suspended intake. (b) Floating intake. (c) Intake crib.

Groundwater Intakes The term groundwater includes artesian wells, deep infiltration galleries or pipes, shallow wells, and deep wells. In all cases, the quality and quantity of the water should be investigated by a test well or other methods.

The first step of the design project is to find a reliable and suitable aquifer to function as the source of groundwater. The initial study must also investigate (1) the relationship between the level of the nearby river and the groundwater level of the site during the dry season; (2) the quality of the groundwater during the dry season, providing that the wells are close to the seashore; (3) the quality, yield, and geological condition of wells and infiltration galleries in the immediate vicinity; (4) the temperature and quality of the water and, if artesian wells and springs are targeted as the potential sources of water, the yield; and (5) the effect of industrial wastes and the possible seepage of wastewater (nitrates) into the aquifer. Site selection for groundwater intakes must also consider the potential intrusion of seawater, the effects of having the groundwater level being drawn down by a cluster of wells, and the potential migration of pollutants from remote dump sites harboring industrial toxic wastes.

After a well site has been selected, a test well should be drilled and pumping tests must be performed to obtain a safe yield from the test well and to check the quality of the well water. A safe yield is assumed to be approximately 60–70% of the maximum yield obtained during the pumping test.

Since the quality of the groundwater is generally very good, all conventional treatment processes, except chlorination, may be eliminated. Thus, the design of the water supply well must be in strict accordance with all requirements dictated by the regulatory agencies.

Design Criteria This section presents the design criteria for the intake tower, shore intake, intake crib, intake pipe, and infiltration gallery.

Intake Tower

Location	As close to the shore as possible but at a minimum water depth of 10 ft (3 m), with the exception of small intakes.
Shape and size	The standard cross section of the tower should be circular or oval. In the case of a river intake, the major axis of the oval must be parallel to the flow direction. The top of the tower should be at least 5 ft (1.5 m) above the highest water level and the access bridge to the tower must also have the same clearance. The interior diameter of the tower must be large enough to be able to install and service the intake gates and pumps (if pumps are installed).

Structure	The materials used to construct the tower should be strong and durable, such as reinforced concrete, and should be built on a solid foundation so that it may resist the force of flooding.
Intake ports	Gated intake ports must be provided at various depths; the lowest of the ports should be approximately 2 ft above the bottom. The vertical interval of the ports should be 10–15 ft (3–4.5 m). The flow velocity through the gross area of the ports, at the same elevation, should generally not exceed 1 fps (0.3 m/s). In regions where the hazards of icing are to be expected, a velocity of 0.5 fps (0.15 m/s) or less is desirable. Sluice gates or butterfly valves must be specified for the intake ports on the interior or exterior of the tower. The gates are used for isolation but butterfly valves are preferred since the sluice gates have a tendency to freeze up if they are not used on a regular basis.
Bar screen	Bar screens (racks) should be provided at each port exterior to the valve. The bars should be composed of $\frac{1}{2}$–$\frac{3}{4}$ in. steel and spaced 2–3 in. on center. Under normal conditions, the flow velocity through the net opening area of the bar screens must not exceed 2 fps (0.6 m/s). In certain cases, the velocity is limited to under 0.35 fps (0.1 m/s) by the Department of Fish and Game to protect small fish from being sucked in.
Fine screen	If screens are not installed downstream of the tower, it will be necessary to install fine screens to remove small floating objects and to protect fish. In most cases, the clear openings of these screens should be approximately $\frac{3}{16}$–$\frac{3}{8}$ in. (5–9.5 mm), and the maximum flow velocity through the net openings in the screens should be no more than 2 fps at the maximum designed flow rate. The use of an automatic hydraulic cleaning device is highly recommended. Moreover, in regions with very cold climate, the intake tower and screens should be protected from frazil and anchor ice formation. The two methods that are often used are compressed air and steam injection.

Shore Intake

Location	Minimum water depth of 6 ft (1.8 m).
Type	Typical shore intake—siphon well, suspended, or floating type—depending on the situation.

Structure	This is determined by the type of intake but is basically the same as the intake tower.
Intake bay	The cross section of the intake opening should be capable of accommodating a maximum flow velocity of 1.5 fps (0.45 m/s). The flow velocity should be reduced to below 1 fps (0.3 m/s) if an unusual amount of debris or icing is to be expected.
Bar screen	These should be provided at the intake bay and set approximately 60° from the horizontal. The bar should be composed of $\frac{1}{2}$–$\frac{3}{4}$ in. steel and spaced 2–3 in. on center. Under normal conditions, the flow velocity through the screens should not exceed 2 fps (0.6 m/s). The bar screen has a trapezoidal cross section, which prevents the accumulation of solids, with a reciprocating action, that is, a climbing screen inclined 80° to the horizontal.
Fine screen	Fine screens are often installed just downstream of the bar screen or may be installed in the grit chamber. The clear openings in the screen must be $\frac{3}{16}$–$\frac{3}{8}$ in. (5–9.5 mm) and, under normal conditions, the maximum flow velocity through the net opening should occur at minimum submergency. As with the tower intake, regions with cold weather must protect the fine screens from icing hazards through the use of compressed air or steam injection.

Intake Crib

Location	Deeper than 10 ft (3 m) from the surface and in a location where it will not be buried by sediment, washed away, or hampered by problems associated with ice formation.
Structure	In areas where the water depth exceeds 10 ft, the top of the intake should be 3 ft (1 m) above the river bottom. However, if the water depth is less than 10 ft, the crib must be buried anywhere from 1 to 3 ft (0.3 to 1 m) below the river bed. Cribs are usually polygonal and built with a firm frame of lumber or reinforced concrete.
	All sides of the crib should be protected by riprap or a concrete slab. Crib ports must be sized to provide a maximum velocity of 0.25–0.5 fps (0.08–0.15 m/s). The crib should surround a bell-mouth pipe that is connected to the intake conduit.

Intake Pipe or Conduit

Size	In order to minimize the accumulation of sediments, the pipe or conduit must be capable of providing a maximum flow velocity of 3–4 fps (0.9–1.2 m/s).
Protection	If the pipe or conduit crosses a river or lake to connect with the shore shaft, it should have 3–4 ft of cover over the top of the crown. Additionally, crushed rocks should be placed above the trench; a rule of thumb is to use 1 yd^3 of crushed rock per linear foot of pipe or conduit.
Slope	In order to avoid creating an air blockage in the line, the pipe or conduit must be laid on a continuously rising or falling grade.
Permit	A permit must be obtained from the appropriate authorities prior to the design phase if the intake pipe or conduit is designed to cross the bottom and bank of a river, lake, or reservoir.

Infiltration Gallery

Direction	At right angle to the river or parallel with the river flow, depending on the underflow pattern, the anticipated difficulties in construction, and the cost of constructing the gallery.
Depth	The common depth is generally 15 ft (4.5 m) below the bottom of the river or lake. Yet, the actual depth should be determined through hydrological studies.
Orifice size	The diameter of each perforation is generally $\frac{3}{8}$–$\frac{3}{4}$ in. (10–20 mm). The number of orifices is 2–3 per square foot (10–30 per m^2) of the collector surface.
Length	The length of the gallery is based on the assumption that the flow velocity through each orifice, at maximum intake flow rate, must be 0.1 fps (0.03 m/s).
Structure	The gallery must be constructed with perforated reinforced concrete pipe or conduit. A stainless steel well strainer may be used for small-scale installations.
Slope	The gallery can be laid on level ground but a slight slope of 500:1 will minimize the problem of air blockage.
Velocity	The flow velocity must not exceed 3 fps (0.9 m/s) at the outlet of the gallery.

Backfilling	The collector is packed with cobblestone, gravel, and sand similar to that used in filter bed gradation. The thickness of each layer must be a minimum of 20 in. (0.5 m). The trench of the gallery collector is backfilled with the specified sand or native sand may be laid on top of the gravel and sand packing.
Junction well	This feature must be provided at the junction points of the gallery and at the end of the gallery line. The diameter of the junction well must be a minimum of 3 ft (1 m) to facilitate the inspection of the gallery.

Shallow Wells and Deep Wells   The design and construction of the wells are addressed in many books. Therefore, no attempt is made to discuss this subject (see the Bibliography). The regulatory agencies and the Ten States Standards provide detailed instructions on the design of the wells.

Operation and Maintenance   Under normal conditions, a properly designed and constructed intake system requires very little in the way of operation and maintenance efforts. The usual operational tasks include regular maintenance of mechanical equipment—traveling screen, intake pumps, and appurtenances—and the removal of large floatables collected by the bar racks.

If the intake is provided with multilevel ports, the valves of each port must be operated on a regular basis. There have been too many cases where the sluice gates attached to various levels of intake ports became inoperable because they were not run for the sake of maintaining performance and severely corroded valve operating stems as well as the gates.

Floods are a potential problem for the intake system because logs and other large objects may jam the intake bar racks. In regions of cold weather, ice floes driven by the wind may potentially block the ports of the shore intake. Operators must keep careful watch for these problems and appropriate actions must be taken before any damage occurs to the intake or before there is a drastic reduction in the capacity.

After each flood or earthquake, it is vital to send a diver down to examine the condition of the underwater intake structures, to determine if the ports have become blocked by sand migration or if the foundation of the intake has been eroded or damaged.

When the intake is located in a large reservoir, the plant operator and laboratory personnel should periodically check the quality of water obtained from various depths of the lake. If the quality is better at a particular level, the intake level should be adjusted accordingly. This practice reduces the chemical cost of water treatment and also saves a considerable amount of extra labor by avoiding conditions associated with algal bloom and taste and odor problems.

In some cases, the raw water must be conveyed to the treatment plant over a long distance by means of a pipeline or conduit. One of the problems associated with this practice is the growth of slimy organisms or freshwater mussels in the interior of the pipeline. The growth of these organisms may dramatically reduce the carrying capacity of the raw water aqueduct. A common solution is to chlorinate the water at the intake site, leaving approximately 0.5 mg/L of free residual for 15–30 min. Chlorination may be continuous or a shock treatment, depending on the situation; operators must give special consideration to the potential for THM formation. Regardless of the potential for THM formation, it is a good practice for engineers to incorporate a chlorine feed system, including a chlorine diffuser, during the design phase. If no such facility is provided, plant operators should consider adding a chlorination system before the situation becomes uncontrollable.

BIBLIOGRAPHY

Anderson, K. E., *Water Well Handbook*, 4th ed., Missouri Water Well & Pump Contractors Association, Belle, MO, 1971.

AWWA, *Groundwater*, AWWA Manual M21, AWWA, New York, 1973.

Davis, C. V., ed., *Handbook of Applied Hydraulics*, McGraw-Hill, New York, 1969.

DHV Consulting Engineers, *Shallow Wells*, DHV Consulting Engineers, Anersfoort, The Netherlands, 1979.

Fair, G. M., Greyer, J. G., and Morris, J. C. *Water and Wastewater Engineering*, Wiley, New York, 1966.

Gibson, U. P., *Water Well Manual*, Premier Press, Berkeley, 1971.

Japan WWA, *Design Criteria for Waterwork Facilities*, Japan WWA, Tokyo, 1969.

National Water Well Association, *Design and Construction of Water Wells*, Van Nostrand Reinhold, New York, 1988.

Richardson, W. H., "Intake Construction for Large Lakes and Rivers," *J. AWWA*, 61(8):365(August 1969).

Urquhart, L. C., ed., *Civil Engineering Handbook*, McGraw-Hill, New York, 1959.

Water Well Specifications, Committee on Water Well Standards, National Water Well Association, Premier Press, Berkeley, 1981.

4.5 GRIT CHAMBER

Purpose Grit is defined as a combination of silt, sand, gravel, shells, and other abrasive materials. The grit chamber is a plain sedimentation tank that removes grit by simple gravity sedimentation. The purpose of this chamber is to protect moving mechanical equipment (such as pumps and mixers) from the abrasive substances and to prevent the accumulation of grit in the raw water line and the pretreatment processes, including the ozone contact tanks and basins.

The grit chamber should be situated upstream of the raw water pumps and as close to the intake structure as possible. This arrangement prevents problems associated with siltation and simplifies grit handling. As described in Section 4.4, the grit chamber and inlet structure are often constructed at the intake site and screens are often installed at the inlet of the grit chamber. Hydraulic scale model conducted by the author clearly demonstrated if the screen was properly installed: it functioned as an inlet diffuser wall.

Engineers should note that the grit chambers for water treatment processes and sewage treatment processes are distinctly different from those used in sewage treatment in regard to one function—the settling of undesirable compounds. Sewage plants are primarily interested in removing the large inert compounds of suspended matter. Thus, most of the organic decomposable solids are permitted to flow through the tank. The tank design therefore does not need to provide good flow characteristics for settling and the tank is often aerated to produce a spiral flow pattern. For this reason, it is obvious that engineers should not use the blueprints of a sewage treatment grit chamber when designing a water treatment grit chamber.

Considerations The four basic considerations in designing a grit chamber are the location of the tank, the number of required tanks, the shape of each tank, and the size of the grit to be removed. The first two issues are only briefly discussed.

The most desirable location of the grit chamber is at the intake site since this will maximize the purpose of the tank. A single divided tank or two separate tanks are generally used to meet minimum operational requirements. This arrangement allows one tank or side to be drained for cleaning or repair while still allowing the plant to remain operational. In situations where sand carry-over is not a major problem, a single tank with a by-pass channel or pipe is considered to be satisfactory.

Grit chambers are available in two basic shapes: rectangular and square. The rectangular tank is identical to the rectangular horizontal flow sedimentation tank. The square tank, also known as a detritus tank, is commonly used in the treatment of sewage (see Figure 3.2.5-9). The square tank is satisfactory for the purpose of wastewater treatment but does not perform well as a water treatment grit chamber.

The main problem with the detritus tank is a function of its physical characteristics. The short tank length and the uneven inflow to the tank cause severe flow short-circuiting. Figure 4.5-1 illustrates the result of a tracer test using the square tank. The graph clearly demonstrates the poor flow characteristics in the tank. In order to provide good flow characteristics for grit settling, the chamber should be rectangular with a contracted inlet.

If there is a significant amount of grit carry-over, the grit may be removed by chain-and-flight or traveling bridge scrapers since these will continuously remove the grit. If these conditions do not exist or are not expected to occur, a gutter should be provided at the center of the tank bottom (longitudinally)

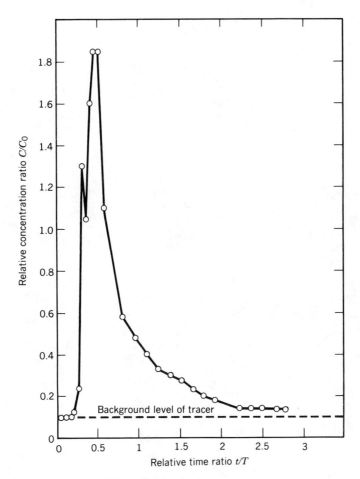

Figure 4.5-1   Grit basin tracer curves.

to facilitate the manual cleaning of grit (by hosing) and removal by a grit pump or a hydraulic eductor system. The lower portion of the bank may be longitudinally sloped 100:1 toward the sump and transversely sloped 1:50, or steeper, toward the center gutter.

A screen system may be constructed at the downstream portion of the inlet diffuser wall at an angle of 70° from the horizontal. Since the screen acts as a diffuser wall and the direction of flow (just past the screen) is perpendicular to the screen angle, an angle less than 70° would direct the flow to the tank bottom and scour the sediment (Figure 4.5-2). It is also important to provide isolation valves at both ends of the tank and to install a weir type effluent so that a minimum water depth can be maintained in the tank. The diffuser wall should have a total orifice area of approximately 15% of the cross-sectional area of the tank. These criteria are based on the hydraulic scale model test

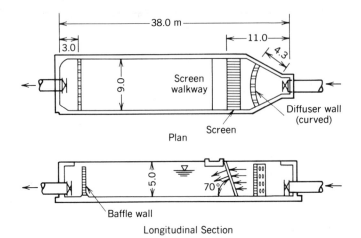

Figure 4.5-2   Grit chamber.

of a rectangular grit chamber conducted by the author to optimize the diffuser wall design and screen angles (see Bibliography).

Generally, the grit to be removed is sand or silt with diameters exceeding 0.1 mm or larger than #100 mesh (0.15 mm). Grit any smaller than this does not pose a major threat to the pumps and pipelines. Figure 4.5-2 is a grit chamber designed to treat 35 mgd (1.5 m³/s) of flow; the design is based on the hydraulic scale model study.

Design Criteria

Number of tanks	Two tanks; one with a by-pass line may be used as an alternative.
Water depth	10–13 ft (3—4 m) with a grit remover; 11.5–16 ft (3.5–5 m) without a grit remover.
Length to width ratio	Minimum of 4:1
Length to water depth ratio	Minimum of 6:1
Mean flow velocity	10–15 fps (3–4.5 m/s)
Detention time	6–15 min
Surface loading	4–10 gpm/ft² (10–25 m/h)

Note: If a gravity flow system is planned, the highest water level in the tank must be lower than the lowest water level at the intake.

Operation and Maintenance   A properly designed grit chamber does not require much operator attention. The inlet diffuser vanes of square tanks, such as the detritus tank, require periodic adjustment in order to maintain uniform inlet flow conditions. Yet, it is virtually impossible to create a good inflow condition based solely on the adjustment of the diffuser vane.

Plant operators must observe the conditions of the grit chamber on a daily basis. However, more attention should be given to the intake and the grit chamber after heavy rainfall, and tasks such as trash removal, adjustments in the grit removing cycle, and monitoring of the quality of the incoming raw water must be performed before the water is permitted to flow to the treatment processes. If the plant design does not include an automatic screen cleaning system, the plant operator may be quite busy removing debris from the screen. Plant operators should also consider installing a turbidimeter in the grit tank for early detection of turbidity spikes if the design has failed to provide one.

Regardless of the grit chamber conditions, the interior of the tanks should be inspected on an annual basis, during periods of low water demand. All required maintenance should also be performed at this time. It is always a good practice to collect grit samples from the tank because sieve analysis of the samples provides operators with valuable information on tank performance. An example design calculation of a grit chamber design is shown in chapter 3.2.5 as the Example 2 of the example calculations.

BIBLIOGRAPHY

Degremont, *Water Treatment Handbook*, 5th ed., Halstead Press, New York, 1979.

Kawamura, S., "Hydraulic Scale Model Simulation of the Sedimentation Process," *J. AWWA*, 73(7):372(July 1981).

Metcalf & Eddy, Inc., *Wastewater Engineering: Treatment, Disposal, Reuse*, 2nd ed., McGraw-Hill, New York, 1979.

Montgomery, J. M., Consulting Engineers, *Water Treatment—Principles and Design*, Wiley, New York, 1985.

4.6 OPERATIONS BUILDING

Purpose The operations building of many of the older, small to mid-sized plants were designed with little attention to architectural and human consideration. More emphasis should be placed on designing a functional and pleasing environment since this building is considered to be the human-oriented structure of the entire water treatment plant and often dictates the theme of all surrounding structures.

Considerations The three basic issues that must be addressed are the functional, architectural, and structural considerations. The issues range from building configuration and exterior finish to design loads.

Functional Issues The operations building should have an administrative zone, operational management zone, product quality control zone (laboratory), and a mechanical/workshop zone. Additionally, the chemical feed system is often included in the operations building.

The administrative zone is composed of a lobby with an area for the receptionist or secretary, a storage room for records and office supplies, toilet facilities for visitors, a conference room, and offices for the water production manager, chief operator, and supervisor.

The zone of operational management minimally contains the control room with its monitors and data loggers, a process control laboratory, and a lunch room. The laboratory should at least have a jar tester, turbidimeter, pH meter, and a water sampling sink. A large meeting room suitable for training operators and accommodating a large group (30 people) of visitors is preferably included in the design. By providing an accordion type door, part of this meeting room may be sectioned off for use by shift operators and maintenance personnel who are stationed at the facility.

A lunch room with vending area is an essential part of modern treatment plants, for the morale and well-being of the plant personnel. Locker room facilities, with showers, must also be provided for both men and women operators. Modern supervisory and control systems produce a large quantity of records; thus, an adequate room must be provided for the storage of operational records and computer supplies.

The Safe Drinking Water Act was recently amended to increase the number of water sampling and analysis items from the present 49 to 183 items by the year 2000. Based on this knowledge, all major treatment plants should have a rather large laboratory zone, unless the water samples are sent to an outside laboratory.

The laboratory should have four distinct areas based on specialty: general chemistry lab, instrumentation lab, bacteriological lab, and an office for the water quality control superintendent. One of the laboratories must have a sampling sink with sampling taps from the head of the plant, settled water, filtered water, finished water, and other necessary phases of the processing water. A minimum total lab space (floor area) of 1500–2000 ft^2 would be required to process the drastic increase in future water analysis.

The fourth zone in the operations building is the mechanical equipment room and the machine (maintenance) shop. The function of this zone is to house the mechanical equipment—the heating, air conditioning, and ventilation—necessary for the entire operations building and the workshop for the process equipment. Engineers must design one room in this zone with a separate air conditioning system. This room does not need to be large, generally 200 ft^2, but it must have a dust-free environment so that it may be used as an electronics and computer workshop.

The chemical storage and feeding system may be designed as part of the operations building. It is usually located at the rear of the administrative building, behind the administrative and operational management zones for reasons of safety and traffic control. The advantage of having a separate building for the chemical feed systems are improved and convenient access for delivery trucks, shorter chemical feedlines and suction pipelines for the liquid chemical metering pumps, better housekeeping, better climate control, and safety—in case of a severe chemical leakage.

Architectural Issues   There are eight major architectural issues in the design of the operations facility. Some of these are common sense; however, many are a function of owner preference and state and federal laws.

ENVIRONMENTAL ISSUES   This category includes the topography of the site, weather conditions, the type and the quality of the surrounding neighborhood, and the view of the process units from the control room. Most of these data will already have been collected during the plant site evaluation phase.

SPACE REQUIREMENTS   The size, type, and number of rooms are strongly influenced by owner preference and the real needs. The limits on the occupancy rate affects the layout of the office, process, and maintenance zones. Two areas within the administrative zone—the toilet/locker room and the laboratory—require larger than average space. Handicap toilet facilities and separate locker rooms for both men and women are mandated by law and the need for a larger laboratory space is a result of the Safe Drinking Water Amendment.

Based on past experience, mid-sized plants (30–60 mgd or 1.3–2.6 m^3/s capacity) should have the following major rooms and functional sizes:

Lobby/reception area	18 ft × 20 ft
Control room	15 ft × 20 ft
Conference room	15 ft × 20 ft
Offices	
Superintendent	15 ft × 15 ft
Plant supervisor	15 ft × 15 ft
Supervisor	12 ft × 15 ft
Chemist	12 ft × 15 ft
Maintenance	12 ft × 15 ft
Lunch room	12 ft × 12 ft
Laboratory	25 ft × 50 ft
	(minimum)
Toilet/locker room	16 ft × 32 ft
HVAC equipment room	15 ft × 16 ft
Electrical equipment room	12 ft × 16 ft
Machine shop	25 ft × 30 ft
Warehouse	24 ft × 30 ft
Chemical feed room	40 ft × 62 ft
Chlorine feed room	15 ft × 20 ft
Chlorine storage room	15 ft × 44 ft

The other rooms should be sized according to the type of equipment, usage, and the requirements set by the manufacturer.

CONFIGURATION OF THE ROOMS   Factors such as the location of the visitor parking lot, location of plant facilities, and the view of the treatment plant often influence the configurations of the rooms in the operations building. In many cases, the plant superintendent and control room operators demand that their rooms have a view of the main treatment process units or the plant entrance. Thus, the following is a list of several design criteria which may be used to arrange the space within the operations building.

- The control room should be adjacent to the lobby. This will allow operators to monitor and control visitor entry.
- Ideally, the control room should be located in the interior of the building in order to avoid exposing computer equipment to the effects of direct sunlight and fluctuations in temperature. Moreover, the computer terminals are easier to view if they are shielded from window glare.
- Access to the conference room, offices, and so on should be convenient for the employees.
- Visitor access to the toilet facilities must be simple. A separate toilet facility should be provided for all visitors and this room should be kept clean at all times.
- Separate rooms should be provided for meetings and lunch breaks.
- The lunch room should have the same appliances and storage space as those found in a small residential kitchen.
- If the treatment plant is located in a residential area, the operations building should be a one-story structure to minimize the visual impact to the neighborhood. However, the height of the building is a function of the situation and owner preference.

EXTERIOR FINISH   The exterior of the building should be attractive, durable, and require minimal or no maintenance. The walls should be constructed of a material that may be used for both the interior and exterior of the building; concrete or masonry meets these requirements. Masonry is relatively inexpensive and is more attractive than concrete. The roof should be pitched to minimize the potential for leaking and also to create an architectural feature. Many options exist for roofing material: clay tile, shingle, or metal. Lastly, the windows and window wall units should be specified to be insulated glass. Engineers may even wish to stipulate that all rooms requiring climate control shall have tinted windows.

INTERIOR FINISH   The materials selected for the interior of the building should be both durable and attractive. However, owner preference often strongly influences the selection. The following is a typical room finish schedule.

Lobby and corridors	Quarry tile	Wallpaper	Acoustic tile
Control room	Sheet vinyl or tile	Wallpaper	Acoustic tile
Offices	Carpeting or sheet vinyl	Wallpaper	Acoustic tile
Toilet/locker rooms	Mosaic tile	Ceramic tile	Painted gypsum board
Laboratory	Sheet vinyl	Painted gypsum board	Acoustic tile
Electric room	Concrete	Painted gypsum board	Painted gypsum board
Process room	Concrete	Sealed masonry	Acoustic tile or painted gypsum board

BUILDING CODE   The building code divides the operations building into an occupancy zone and a factory type zone. The codes also dictate certain design features based on the occupancy rate and safety considerations; several of these considerations are listed below.

Number of exits
Fire-rated corridor walls
Fire-rated doors and windows
Area separations
Fire-rated ceilings
Flame spread classifications
Smoke detectors

The factory zone must be separated from the occupancy zone by a fire-rated wall as required by the Uniform Building Code (UBC). The heating, ventilation, and air conditioning room (HVAC) requires a fire-rated wall containing sound-insulating material. Additional issues are the amount of insulation in the walls and ceiling. Refer to the UBC for the specific requirements related to each of these considerations.

BUILDING ARRANGEMENT   It is in the best interest of all parties if the building arrangement issue is developed by a qualified architect who is familiar with the functions of a water treatment plant or an architect who is guided by a project engineer. Figure 4.6-1 is a sample building arrangement.

Structural Issues   The structural system of the operations building is directly affected by the occupancy groups in the building. Each occupancy group has a maximum allowable floor area; refer to the UBC for the maximum allowable

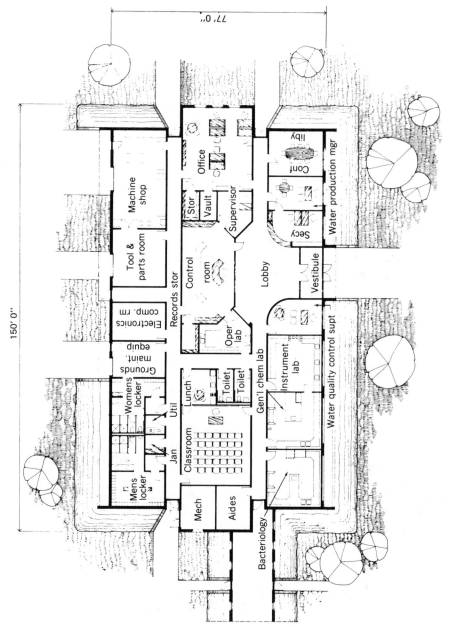

Figure 4.6-1 Example of a building floor plan.

floor area for the five different types of construction. Note that the washdown and durable materials used in the construction of water treatment plants often give the operations facility Type II or III building characteristics.

The goal of the structural design, for hydraulic structures, is to produce strong units that will remain serviceable and leak-free for the life of the plant. This design should follow the recommendations of the UBC and ACI-350, "Concrete Sanitary Engineering Structures." Regardless of the specific application, the design and details of the structure must minimize cracking due to shrinkage and limit direct cracking at construction joints and expansion–contraction joints.

Since geotechnical conditions of the plant site also affect the structural design of the plant facilities, recommendations should be obtained from a geotechnical consultant in regard to the subgrade and groundwater level of the property. Advice should also be obtained on shoring, earth pressures, sheet piles, and so on.

Design Criteria   This section presents very general design criteria for the materials and the design loads used in constructing an operations building. Specific recommendations cannot be presented because adjustments must be made according to the local and specific conditions of each plant site.

Construction Materials

Concrete	4000 psi compressive strength
Reinforcing steel	ASTM A615, grade 60
Structural steel	ASTM A36
Masonry	ASTM C90, grade N-1, lightweight block, 1500 psi combined compressive strength
Aluminum	Alloy 6061-T6
Stainless steel	Type 304

Design Loads

Office, laboratory, and control room	100 psf
Areas with process units and mechanical equipment	150 psf
Areas with heavy equipment	Equipment weight plus 200 psf
Electrical control room	250 psf
Traffic loading	HS-20 (AASHTO)
Roof loads (UBC requirement)	20 psf (basic)
Wind loads (UBC requirement)	100 mph
Seismic loads	Recommendation of the geotechnical expert

4.7 PUMPING SYSTEM

Purpose Although pumping is not a water treatment process per se, it is a major ancillary item. The water treatment design engineer is responsible for selecting and specifying the appropriate type of pump, for setting the system requirements, writing the pump specifications, and developing all other necessary information. Since the pumping system is usually not the design engineer's area of expertise, he/she is strongly advised to refer to the literature published by pump manufacturers and to consult with mechanical engineers in order to obtain more detailed information and data. This section only presents basic guidelines for pump selection, designing a pumping facility, preparing pump specifications, and other special considerations.

Considerations Several items must be addressed when designing a pump system. These items, in addition to those used in selecting pump drive units and determining the number of pumps and standby generators, are briefly covered in this section.

Type of Fluid The primary concerns that must be addressed during pump selection are both the chemical and physical characteristics of the fluid to be pumped. The main characteristics that strongly influence pump selection and system design are viscosity, temperature, density, corrosivity, chemical stability, lubricating property, and the type and amount of suspended material (in water).

System–Head Curve The design engineer should establish the system–head curve by analyzing the pumping system; the preliminary design of the system should include a pump layout, piping scheme, and instrumentation diagram. A clear and accurate curve of the pumping equipment characteristics, in the expected range of operation, may be obtained from the manufacturer. Engineers should remember that hydraulic headloss is a function of the flow rate and the physical characteristics of the piping system. Therefore, a clear piping scheme will reduce the chance of additional headloss and ultimately minimize the operational cost of pumping. The details for establishing the system–head curve may be found in most engineering and hydraulic textbooks.

Potential System Modifications In the majority of cases, the pumping system designed during the initial design phase is later expanded or modified. If future modification of the pumping station is required by the master plan, the pumping system should be designed to accommodate them. This will allow the original design to still be effective and will not hinder existing plant operations during construction.

Operational Modes Important considerations for designing a pumping system are the degree of flow or head fluctuation and the mode of operation, such as continuous or intermittent. The operational modes dictate the number of pumps and their capacities and determine if booster pumps are required in a flow path.

Required Margins   Engineers must specify a pumping unit that has adequate margins so that the equipment will be able to withstand infrequent, short-lived, abnormal conditions: fluctuations in pressure, dips in electric voltage and frequency, hydraulic surges, or the loss of cooling water. With respect to small electrical disturbances, the margins may be added to the pump head and capacity. Yet, the pumps should only be specified to be 15–20% over the design points because oversized pumps may develop mechanical and hydraulic problems.

Pump Selection   Different types of application require specific types of pump. Pump selection is based on factors such as the fluid characteristics, turn-down ratio, discharge pressure and system requirements, availability of space, layout, life, energy and pump costs, code requirements, and the materials used in construction.

Reciprocating pumps, such as the plunger or diaphragm type, should be selected for liquid chemical metering and injection applications. These types of pump deliver a constant capacity of liquid over a wide range of system–head variations. However, the capacity of these pumps is rather small and the pumps are characterized by a pulsating flow.

Positive displacement pumps, also known as rotary pumps, may be used for both liquid and gas pumping. Although this type of pump does not exhibit a pulsating type of flow and may also be used as a slurry or metering pump, its application is limited to a low to medium pressure range.

The centrifugal pumps are used for a wide variety of hydraulic head and over a wide range of capacity requirements. However, in water treatment applications they are generally used in situations requiring low to medium capacity with medium to high pressure. These pumps require priming unless installed below the water surface in a dry well.

Axial flow pumps are best suited for applications requiring low hydraulic head and high flow conditions. The mixed flow type is used in situations with intermediate conditions, between the axial flow and centrifugal pump. Refer to Figure 4.7-1 for the pump characteristics.

Vertical turbine pumps require much less space than the centrifugal type of pump and are also self-priming. Yet, this type of pump generally requires more head room. Thus, if the pumps are installed in a housing, a removable skylight or hatch must be provided directly above each pump to facilitate repairs. Nonetheless, vertical pumps such as turbine pumps, axial flow pumps, and mixed flow pumps are often installed without a housing. Rather than using a pump well construction, vertical turbine pumps may use a "canned" suction type of design in the suction pipeline to help minimize the construction costs of a pumping station.

For further information of the various aspects of pumps, refer to the *Hydraulic Institute Standards*, 14th edition.

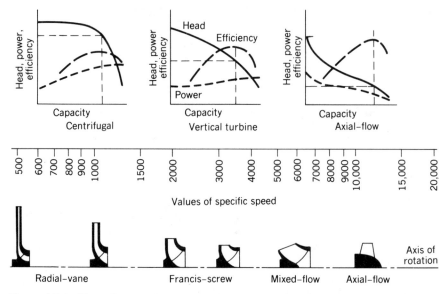

Figure 4.7-1 Type characteristics and specific speed scale for various impeller designs. (Adapted from R. P. Walker, *Pump Selection*, Ann Arbor Science Publishers, Ann Arbor, MI, 1973.)

Drive Selection The available energy source, the cost of operation, and various local conditions all affect the selection of the pump drive unit. Pumps may be driven by a variety of means:
electric motor, internal combustion engine, or steam and may either have constant or variable speed drives.

The internal combustion engine can be the drive unit due to local conditions and to avoid increasing the energy demand during peak hours of electrical consumption. An electric motor may be selected as part of the dual drive to act as a redundancy drive. Yet, the majority of pumps are driven by squirrel-cage induction motors due to their versatility and availability.

Other miscellaneous types include synchronous motors, used in applications requiring large horsepower and where a power factor correction is required; wound rotor motors, which are occasionally used in applications requiring variable speed drives; and reduced voltage starting and low in-rush current motors. The last two types may be used as high power motors in situations where an initial transient reduction in-line voltage would adversely affect other equipment.

Number of Pumps and Standby Generators Reliability considerations often tend to favor the use of multiple-pump schemes. A three-pump system employing identical electric motor driven pumps, each having a capacity that is 50% of the maximum demand, is commonly selected. This type of system only requires the power supply to run two of the pumps at any one time and to maintain the third pump as a standby.

If frequent power outages are anticipated, an engine driven pump or standby diesel generator should be incorporated into the pumping system design. If a standby generator is selected, the system will require reduced voltage starting and the engine generator should be capable of supplying a starting current that is at least three times the normal full load current. However, a smaller emergency generator will be sufficient if sequence starting is used, since each pump may be started in a stepwise fashion.

The size of the pumping system is also an important factor in determining the number of pumps. Smaller systems may opt to utilize two to three pumps of the same capacity in order to make the pumping system more flexible. This scheme will enable the system to meet the water demands, will facilitate easier maintenance, and will allow interchangeability of parts. However, larger installations experiencing high peak demands would benefit from a two-bank system. In this scheme, one bank of pumps (constant speed drives) supplies the base flow to satisfy the normal demand. The second bank supplies the extra flow necessary to satisfy the peak demands. The pumps of the second bank may be driven by internal combustion engines, such as diesel engines, so that the expensive peak power costs (including the potential penalty charge) may be avoided.

Pump Specifications   The two basic types of pump specification are those concerned with pump construction and those related to performance. The first specifies details such as the type of design, the design method, and pump construction. The performance specifications dictate the performance standards for the pump and contain a few details related to pump construction, such as the pump construction material.

Since the owner is primarily responsible for the performance of the pump, it is more than likely that the plant owner, not the manufacturer, will be held legally responsible if pump malfunctions develop, provided that the specifications describe the pump design and construction. Thus, most water treatment plant projects use the pump performance specifications when constructing the pump system. The project engineer can always seek help from pump manufacturers when developing the clarifying specifications. However, the engineer should remember that each supplier is partial to his own product. It is always a good practice to include a series of alternative, in addition to the basic, specifications because it allows the pump suppliers to present their best offers and provides the owner with the advantage of obtaining bids from several manufacturers.

The pump specifications should give reference to the quality codes and standards of the materials to be used. Industrial standards such as ANSI and ASTM are good sources for basic requirements relating to metallurgy, dimensions, tolerance, and flange details.

It is important to note that a number of design engineers have problems with selecting the appropriate type of metal for pump construction. In the majority of cases, stainless steel is used with excellent results. However, if the plant is pumping anaerobic sludge or deep well water or pumping out the bottom portion of deep lakes, type 316 stainless steel will become severely pitted under these anaerobic conditions. Therefore, other types of stainless steel, such as type 301 or 304, should be specified having manufacturer's assistances. Another area of concern is the problem of abrasion—when stainless steels of the same type rub together. As a rule of thumb, the hardness of the two adjacent metals should have a difference of 50 Bhn on the Brinell scale of hardness.

The bidding documents are usually divided into two major parts, the Technical Specifications and General Information (commercial term). The Technical Specifications should include the following items:

1. Scope of the work.
2. Work not included in the document: for example, foundations, anchor bolts, labor.
3. Ratings and service conditions such as the characteristics of the liquid, temperature of the liquid, flow rate, total dynamic head, preferred speed range, efficiency at the design point, load conditions, NPSH, overpressure, runout, and operating conditions.
4. Design and construction brief detailing the codes, standards, material used, type of casing, stage arrangement, balancing, nozzle orientation, type of support, vents and drains, type of bearings, shaft seals, baseplates, connecting pipes, and instruments.
5. The lubricating oil system.
6. Details of the drive unit, such as motor voltage, power supply and regulation, local panel requirements, wiring standards, terminal boxes, and electrical devices; and the type of fuel, number of cylinders, cooling system, speed governing, starting requirements, couplings, and exhaust muffler for the internal combustion drives.
7. Cleaning and painting.
8. Performance testing.
9. Drawings and supporting data.
10. Special tools.
11. Criteria for evaluation.

The General Information should include the name and address of the owner, the applicable codes and standards, information on the location and condition of the plant site, and a schedule that explains the drawing requirements, bid dates, manufacturing time, and equipment delivery. The clause concerning the plant site should specify, among other things, the elevation of the plant site, the range of ambient temperatures, and humidity. It is recommended that the General Information also clarify the key people associated with the project: for example, owner, engineer, and seller.

Performance testing is generally not an important consideration for small pumps: those with a discharge size that is less than 6 in. Yet, it is vital for larger pumps because a loss of efficiency of 1–2% from the specified value results in a significant increase in the annual power cost. It is very important that the owner and supplier agree on the method of testing and whether the testing of the pump should be witnessed.

Special Considerations In addition to the previously mentioned considerations, the design engineer should address five other issues: the pump starting conditions, the use of high-speed versus low-speed pumps, the method used in controlling hydraulic surcharge, the design of the pump suction well, and selection of the sludge pump.

Pump Starting Conditions The discharge valve of centrifugal pumps must be closed at a particular time during start-up in order to prevent hydraulic surges. Under these circumstances, the pumping unit will not require additional horsepower. As long as an adequate net positive suction head is provided, this practice also removes the concern over cavitation. This is also true for mixed flow pumps, up to an approximate specific speed of 5000.

Propeller type pumps, including most types of mixed flow pump, require very high horsepower at start-up. Thus, the pump discharge valve should be sufficiently opened after the drive unit has operated at full speed for a few seconds. If the discharge valve is allowed to remain in a closed position, the water in the bowls will become heated, thereby allowing steam to displace the water in the column. For this reason, valve control during pump start-up must be evaluated carefully.

High-Speed Versus Low-Speed Pumps The pumping requirement for a particular flow rate at a specific total dynamic head may be met by two different methods. The first is to use the combination of a small sized impeller, capable of high speeds, such as 3600 rpm, with a high-speed motor. The second alternative is to use a slower speed drive (800 to 1200 rpm) with a larger impeller.

Although pump selection is influenced by owner preference and plant conditions, the high-speed pumps are cheaper to manufacture and the owner therefore has the option to select the lowest bidder. Despite their lower cost, the high-speed pumps are generally maintenance intensive due to problems associated with wear, erosion, misalignment, and vibration. The high-speed units also tend to produce a high-pitched noise. It is the responsibility of the engineer to carefully evaluate the various offers, including the economic life span of the pump and the associated maintenance costs.

Hydraulic Surge Control Two types of hydraulic surge have been associated with the pumping system. The first occurs during pump start-up and in small plants; this type of surge may be eliminated by installing a surge control valve between the pump and the check valve (Figure 4.7-2). The second type, the water hammer, is produced if the pump is suddenly stopped by the operator or if all the pumps suddenly fail due to a power outage. If the facility is expected to have a significant problem with water hammer due to power outages, a surge tank should be installed to alleviate this problem.

A special technique in controlling hydraulic surges is surge analysis. Although the technique requires both knowledge and experience to be used effectively, a rule of thumb is to close the surge control valve (during pump start-up) and to open the valve (during pump topping) after the shock wave has reflected. The required time may be calculated from the formula $2L/a$, where a is the velocity of the shock wave (4000–5000 ft/s) and L is the length of the pipeline in feet.

Design of the Pump Suction Well A critical part of the pumping system design, for the engineer, is the pump well, because the pump impeller requires a particular type of flow condition to exist. The engineer must verify that the flow conditions are uniform and ascertain that the design of the pump suction well will provide the required type of flow.

The pump well may be an open pit or a large sized pipe. Detailed design criteria for optimum pump suction configurations may be found in the *Hydraulic Institute Standards* (14th Edition, 1983). Figures 4.7-3 and 4.7-4 illustrate some of these design guides.

Sludge Pumps Three types of pump are best suited for sludge pumping: the positive displacement type, such as the plunger and progressive cavity pumps, the centrifugal type, and the torque flow type. Regardless of the type of pump chosen, the sludge piping should be no smaller than 6 in. in diameter and the velocity should be 5–6 ft/s to prevent clogging problems.

The plunger type of pump has a satisfactory history of operation. These pumps are self-priming and the pulsating action tends to concentrate the sludge in the hoppers (located prior to the pumps) and resuspends solids in the pipelines during flows of low velocity. The main disadvantages of the plunger pumps are their bulky size and cost. For these reasons, this type of pump is rarely used today.

The use of progressive cavity pumps, such as the Moyno pump, has been particularly successful in pumping concentrated sludge. The Moyno pumps are progressive cavity pumps and are self-priming at suction lifts up to approximately 20 ft. However, the pump is usually installed under flooded suction conditions to prevent the rubber stator from burning out.

Specially designed centrifugal pumps, such as screw feed, bladeless, and torque flow, have also been used successfully. The torque flow type of pump, such as the Wemco pump, has fully recessed impellers and is very effective

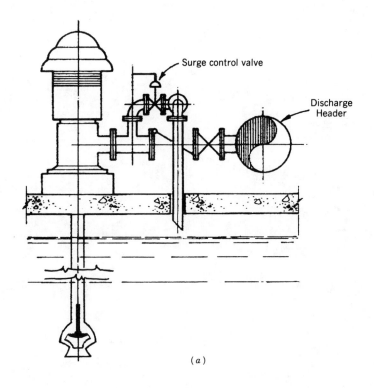

(a)

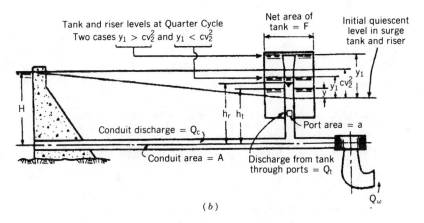

(b)

Figure 4.7-2 Surge control valve and surge tank. (a) Surge control valve. (Adapted from R. P. Walker, *Pump Selection*, Ann Arbor Science Publishers, Ann Arbor, MI, 1973.) (b) Surge tank installation. (Adapted from *Handbook of Applied Hydraulics*, C. V. Devis, ed., McGraw-Hill, New York, 1952.)

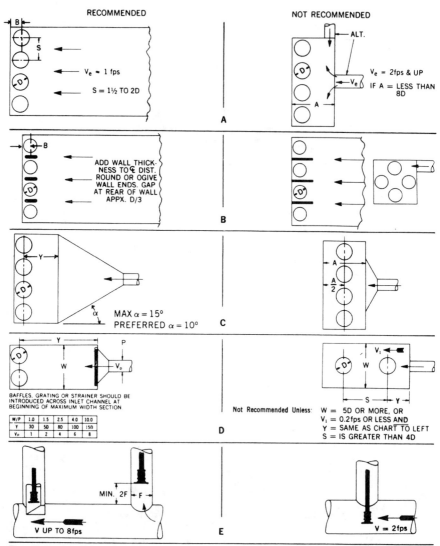

The Dimension D is generally the diameter of the suction bell measured at the inlet. This dimension may vary depending upon pump design. Refer to the pump manufacturer for specific dimensions.

Figure 4.7-3 Multiple pump pits. (Adapted from *Hydraulic Institute Standards*, 14th ed., 1983.)

in pumping sludge but has very low pump efficiency. Engineers should refer to *Wastewater Engineering* (2nd edition) by Metcalf and Eddy, Inc., for more detailed information on sludge pumps.

Operation and Maintenance As described earlier, the starting and stopping of the pump requires special consideration. These procedures are dependent on the pump type and the associated accessories.

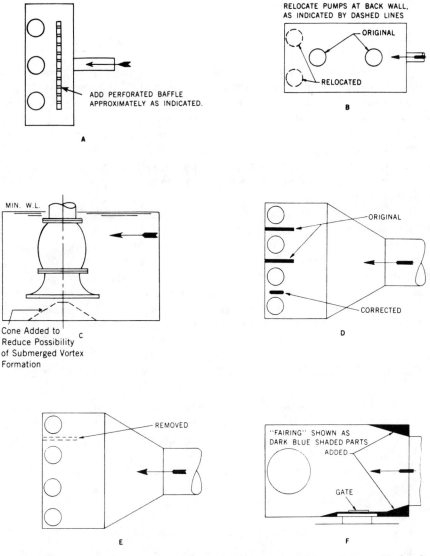

Figure 4.7-4 Correction of existing sumps (part one). (Adapted from *Hydraulic Institute Standards*, 14th ed., 1983.)

Pumps with low hydraulic head, mixed flow, and of the propeller type should be started with the discharge valve wide open against a check valve to prevent backflow. However, pumps with medium and high hydraulic head should be started against a closed discharge valve to reduce the starting load on the motor. If the discharge line is already pressurized by another pump, a check valve will be equivalent to a closed valve. It should be remembered that whenever a pump is started against a closed discharge valve the recirculation bypass line must be opened to prevent overheating.

The pump stopping procedure is also dependent on the pump type and the accessories. The plant operator is therefore advised to strictly follow both the operational manual provided by the manufacturer and the plant operation and maintenance manual.

The major areas of maintenance concern inspection and replacement. Three types of inspection must be practiced: hourly and daily, semiannual, and annual. Hourly and daily inspections require the operator to note the normal sounds of the pump and the bearing temperature and check for leaks in the staffing box. Semiannual procedures involve the inspection of the pumps, staffing box, packing, and alignment and the refilling of the lubrication oil and grease. Annual inspection requires that all parts of the pump assembly be thoroughly checked and cleaned and the connecting pipes be realigned with the pump.

Lastly, the pumping system should be completely overhauled whenever necessary, usually every 2–4 years.

Recordkeeping is also a very important issue in the operation and maintenance of the pumping system. Each individual pump should have its own complete record on file. This record should note the maintenance costs, the number and type of repairs, and the total operating hours. Experience has shown that photographs of the internal parts of the pump, taken during semiannual and annual inspection, prove to be very valuable.

BIBLIOGRAPHY

Alice-Chalmers, *Reference Data*, Alice Chalmer Pump, Inc., Cincinnati, Ohio, 1970.

Colt Industries, *Hydraulic Handbook*, Fairbanks Morse Pump, XXX, KS, 1975.

Compressed Air Magazine Company, *Cameron Hydraulic Data*, Ingersoll-Rand Co., XXXX, NJ, 1970.

Davis, C. V., *Handbook of Applied Hydraulics*, 3rd ed., McGraw-Hill, New York, 1969.

Hydraulic Institute, *Hydraulic Institute Standards*, 14th ed., Hydraulic Institute, Cleveland, 1983.

Karassik, I. J., et al., *Pump Handbook*, McGraw-Hill, New York, 1970.

Langteau, R. R., "Clarification Control in Pumps," *J. WPCF*, 38(4):585(April 1966).

Metcalf & Eddy, Inc., *Wastewater Engineering: Treatment, Disposal, Reuse*, 2nd ed., McGraw-Hill, New York, 1979.

4.8 OZONATION SYSTEM

Purpose The ozonation process recently gained popularity as a treatment alternative in the United States after the regulatory agencies placed restrictions on the disinfection by-products, especially the acceptable levels of trihalomethane. Ozonation is an established and proven disinfection alternative, as well as a preoxidant for the control of THM precursors. Ozonation also has the added benefits of (1) oxidation and volatization of organics, (2) control of algae and associatd taste- and odor-producing compounds, (3) destabilization (microflocculation) of certain types of turbidity, (4) removal of color-causing compounds, (5) oxidation of iron and manganese, (6) very short disinfection times, and (7) partial oxidation of organics for subsequent removal by microorganisms. This section discusses the design considerations for the ozonation system and presents some design issues with respect to ozone generation and the ozone contact tank.

Considerations Ozonation is considered to be a relatively new treatment process in the United States. However, the process has a history of almost 100 years of practical experience. Thus, basic knowledge should first be obtained from the extensive source of published articles and operational experiences prior to conducting design works. The second step, providing that there is both the time and budget, is to perform a pilot study: this will furnish the engineer with valuable design guides.

The design issues are (1) selection of a feed gas system, (2) preparation of the feed gas system, (3) selection of the ozone generator, (4) design of the ozone contact basin, and (5) destruction of off-gas ozone. The issue of safety will also be discussed briefly since ozone is both a toxic and flammable gas. The two most practical articles (published) on this subject so far are the *Design Manual on Municipal Wastewater Disinfection* (EPA/625/1-82/021) written by the EPA and "Ozonation of Drinking Water" by Domenic Grasso (*Ozone Science and Engineering*, Vol. 9, p.109, 1987).

Feed Gas Selection Ozone may be generated from air, oxygen-enriched air, or oxygen. The concentration of ozone produced by air is 1.5–2.5% by weight. The ozone concentration is increased to 3–5% if high-purity oxygen is processed by the same ozone generators.

Feed Gas Treatment Both air and pure oxygen must be treated prior to being fed to the ozone generator; this maximizes the ozone production rate and minimizes maintenance work on the ozonator. The pretreatment process includes the removal of dust, moisture, oil, and, in some instances, nitrogen gas: dust reduces the efficiency of ozone production; oil is capable of fouling the dielectric; and nitrogen has the ability to damage the ozone generator because of the production of nitric acid within the unit.

If the gas contains excessive moisture, the life span of the dielectric of the ozone generator becomes reduced and the power required to produce a specified level of ozone is increased. Thus, the removal of moisture is the most important of the gas pretreatments. This task is achieved by means of a dryer, which uses a combination of pressure, temperature, and desiccant.

The ozonation system for application to water treatment is generally composed of a precompressor with a 5 μm paper filter, main compressor, after-cooler (less than 95°F or 35°C), oil coalescer (if required), refrigerant dryer (optional), heat reactivated desiccant dryer with activated alumina and molecular sieves or silica gel, 1 μm filter, hygrometer, gas flow meter, and pressure-regulating valve.

The efficiency of ozone generation is a function of the feed gas treatment. The capability of the feed gas treatment may be maximized by reducing the moisture content of the feed gas to below 0.00596 lb of water per 1000 ft³ of air (0.14 g/m³), with a dewpoint of −40°F (−40°C). This dewpoint can only be attained if 99.71% of the mositure is removed. Yet, gas containing this moisture content is still considered to be in poor condition for the ozone generator. A dewpoint of −58°F (−50°C) is judged to be marginal for the generator. The desired feed gas humidity is achieved at a dewpoint of −76°F (−60°C); at this temperature the feed gas contains only 0.00051 lb of water per 1000 ft³ of air (0.01 g/m³) with 99.98% of the moisture removed.

The feed gas compressing process generally removes approximately 60% of the moisture content of the air; an additional 20% is removed by the refrigerant dryer and the desiccant dryer removes the remaining moisture so that a total of 99.98% is eliminated. Excessive moisture content in the feed gas (air) not only reduces ozone production and increases generator maintenance, but also damages the internal components of the generator. Since the pretreatment of the gas is vital to the efficient production of ozone, sufficient monitoring procedures and control devices should be implemented in the process train. Serious consideration should be given to duplicating critical elements such as the desiccant dryer. Table 4.8-1 lists the moisture content of air at various ambient temperatures.

The most popular air feed gas systems used in ozone generation systems are classified by their operating pressure. The system generally used is the low-pressure system. It operates at pressure of 10–40 psig (70–275 kPa). The high-pressure systems operate at pressures of 70–100 psig (480–690 kPa) and decrease the gas pressure prior to the ozone generator. These high-pressure systems are typically employed in small to medium sized applications.

As discussed, air feeding to the ozone generator is a complex, costly, and precisely controlled pretreatment process that is maintenance intensive. However, a great majority of the feed gas treatment process may be eliminated if high-purity oxygen such as liquified oxygen (LOX) is used to generate ozone. LOX is presently available at a very reasonable price ($3.00/ft³ or $0.40/lb). On-site generation of oxygen may also be achieved by means of a cryogenic or pressure-swing process.

However, these processes require highly complex mechanical units that may be too difficult for the personnel of an ordinary water treatment plant to operate and maintain. These units also create a negative visual impact and emit a certain degree of noise. Thus, design engineers should seriously evaluate the use of LOX and the oxygen feed ozone generation system for any size plant, especially for small to mid-sized treatment plants. This particular type of process offers several advantages for both the design engineer and

TABLE 4.8-1   Moisture Content of Air for Air Temperature from −80 to 40°C

Air Temperature		Moisture Weight lb H₂O/lb air	Weight Volume [a] lb/ft³	Moisture[b] Content lb H₂O per 1,000 ft³
(°C)	(°F)	lb H₂O/lb air	lb/ft³	
−80	−112	0.0000003168	0.07526	0.00002384
−75	−103	0.0000007713	0.07526	0.00005805
−70	−94	0.000001640	0.07526	0.0001234
−65	−85	0.000003342	0.07526	0.0002515
−60	−76	0.000006743	0.07526	0.0005074
−55	−67	0.00001311	0.07526	0.0009866
−50	−58	0.00002464	0.07526	0.001854
−45	−49	0.00004455	0.07526	0.003352
−40	−40	0.00007925	0.07526	0.005964
−35	−31	0.0001381	0.07526	0.010393
−30	−22	0.0002344	0.07526	0.017640
−25	−13	0.0003903	0.07526	0.029373
−20	−4	0.0006731	0.07526	0.050665
−15	5	0.001020	0.07526	0.07676
−10	14	0.001606	0.07526	0.1209
−5	23	0.002485	0.07526	0.1870
0	32	0.003788	0.07526	0.2851
5	41	0.005421	0.07526	0.4080
10	50	0.007658	0.07526	0.5763
15	59	0.01069	0.07526	0.8045
20	68	0.01475	0.07526	1.1100
25	77	0.02016	0.07526	1.5171
30	86	0.02731	0.07526	2.0552
35	95	0.03673	0.07526	2.7642
40	104	0.04911	0.07526	3.6959

[a]Air weight/volume is corrected to 68°F and 1 atm pressure. Air weight/volume = 1205 g/m³.
[b]g/m³ = lb/1000 ft³ × 16.012.
Source: Adapted from EPA Design Manual, EPA/625/1-86/02, October 1986.

the plant owners: the design engineer benefits from the simple overall design scheme, especially instrumentation and control, and a required floor space that is approximately half that of the air feed system. The owner benefits from the potentially significant savings in both capital and operation and maintenance costs.

Figure 4.8-1 is a schematic illustration of the once-through oxygen feed system, the oxygen feed with recycling system, and the air feed system.

Ozone Generator   Ozone is generated by passing a high voltage alternating current (6–20 kV) across a dielectric discharge gap through which oxygen-bearing gas is injected. There are currently three basic types of ozone generator for use in water treatment applications: (1) low-frequency units with variable voltage; (2) medium frequency units with constant voltage, variable voltage, or frequency control; and (3) high-frequency units. These units are tube type generators, in contrast to the plate type generators used exclusively in small ozone generators.

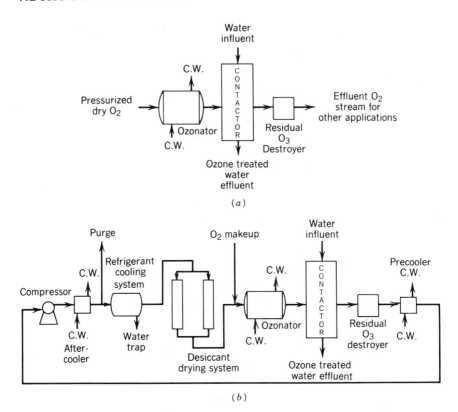

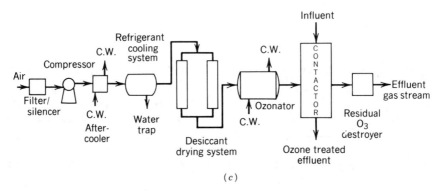

Figure 4.8-1 Comparison of feed gas systems. (a) Once-through oxygen feed system. (b) Oxygen feed with recycling system. (c) Air feed system.

The low-frequency units have a frequency of 50–60 Hz, single-phase power, and a peak voltage of 10–20 kV. The performance of these units is proved. The medium-frequency units operate at a frequency of 200–1000 Hz, are three phase, and have a peak voltage of 8–10 kV; these are relatively new but their performance is also proved effective. The newest ozone generators

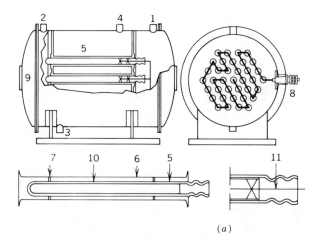

1. Air inlet.
2. Ozonized air outlet
3. Cooling water inlet.
4. Cooling water outlet.
5. Dielectric tube.
6. Silent discharge zone.
7. Tube support.
8. High–voltage terminal.
9. Port–hole.
10. Metal coating.
11. Contact brush.

(a)

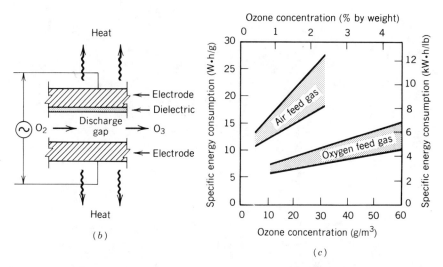

(b)

(c)

Figure 4.8-2 General concepts on ozone generation. (a) Horizontal tube, voltage-controlled, water-cooled ozonator. (b) Corona cell arrangement for ozone production. (c) Energy consumption versus ozone production for an ozone generator.

are the high-frequency units with 600–2000 Hz, three phases, and a peak voltage of 8–10 kV. However, they are manufactured by a very limited number of companies. Figure 4.8-2 provides a general concept of ozone-generating systems.

Much credence should not be given to the sales pitch of the manufacturers. The most practical ozone generators are either the low- or medium-frequency units, primarily because of their years of operational experience, reliability, and the large number of suppliers from which to choose. Units requiring

TABLE 4.8-2   Brief Comparison of Two Types of Ozone Generator

Unit Type	Degree of Electronic Complexity	Capital Cost	Operation and Maintenance Cost	Ozone Concentration (%)		Effect of Generator Cooling Temperature
				With Air	With Oxygen	
Low frequency	Lower	Lower	Higher (100% more power)	1–1.5	2–3	Moderate
Medium frequency	Higher	Higher	Lower (possibly higher)	2.5–3	5–6	Sensitive

lower voltages, including many medium-frequency units, generate less heat, thus minimizing the cooling water requirements and possibly increasing the life span of the dielectric. The medium-frequency units have a better power to production turn-down ratio than the low-frequency generators: the low-frequency units may achieve a production of 20% at 20% power if the air flow rate is reduced, whereas the medium-frequency units may achieve a 10% production rate at 10% power. Yet, the medium-frequency units contain more electronic components than the low; failure of any one of these components will cause the power supply to malfunction and terminate ozone production.

Selection between low- and medium-frequency ozone generators should therefore be based on the following considerations: (1) reliability and maintenance, (2) energy cost differential, (3) turn-down ratio, (4) cooling water temperature, (5) benefits attained by using oxygen-enriched feed gas, and (6) owner preference. Table 4.8-2 briefly summarizes the characteristics of the two types of ozone generator.

In the overall production of ozone, the electrical power is a critical element because the greater the power consumption the greater the heat generation. Gas flow also affects the generation of heat; a gas flow that is too low will cause the generator to heat up, whereas an excessively high gas flow rate will decrease ozone production.

Since 90–95% of the supplied power is converted to heat, the cooling system is a very important issue with respect to the generator. Fortunately, the manufacturers are highly competitive in their cooling designs.

The cooling water temperature should generally be 68°F (20°C) or less and approximately 0.75–1 gal (3–4 liters) of cooling water is required for each gram of ozone generated. Refer to Figure 4.8-3 for the various conditions that affect the rate of ozone production.

Figure 4.8-4 is an illustration of the bid prices of both low- and medium-frequency ozone generators in the United States after 1982. Figure 4.8-5 is a breakdown of the construction costs.

After the type of ozone generation unit is selected, the design engineer must establish the following key design issues:

- Size of the ozone generator
- Number of generators
- Cost of energy
- Availability and cost of oxygen
- Type of feed gas treatment system
- Reliability of each component
- Operation and maintenance costs
- Ozone Contactor design
- Destruction of the off-gas
- Use of catalysts such as UV and hydrogen peroxide

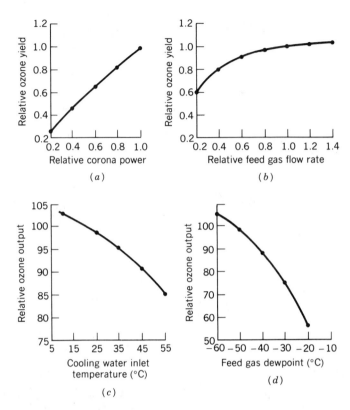

Figure 4.8-3 Various conditions that affect ozone production rate.
(a) Ozone yield versus corona power for a constant feed gas flow rate. (b) Ozone yield versus gas flow rate for constant corona power. (c) Effect of cooling water on ozone yield. (d) Effect of moisture on ozone yield. (Adapted from R. P. Rice and A. Hetzer, *Handbook of Ozone Technology and Applications*, Vol. 1, Ann Arbor Science Publishers, Ann Arbor, MI, 1982.)

The major manufacturers of ozone generators currently are Quantum (Emery), Welsback Ozone System, Griffin, Technics, PCI Ozone & Control, Asea Brown Boveri (ABB), Infilco Degremont Inc. (IDI), Trailigas, Schmidding-Werke (Megos) Mitsubishi, and Toshiba.

Ozone Contact Tank As in any other water treatment process unit, the effective mixing of ozone and process water is critical in maximizing the performance of the ozonation system. Due to the extremely low solubility of ozone in water, a special ozone contact tank must be designed. Table 4.8-3 lists the solubility of ozone in water. The contact tank is an effective means of transferring the ozone from the gas bubble to the bulk of the process water: the net ozone content in the gas bubbles is usually only 2–4%. This effective transfer of ozone to water is a critical step in the ozonation process.

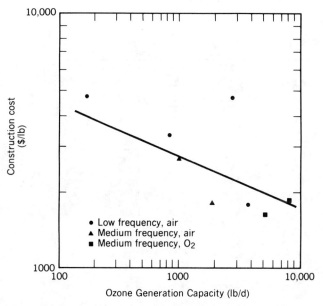

Complete Ozone System
(includes generation, building, and contactor)

- Low frequency, air
▲ Medium frequency, air
■ Medium frequency, O_2

Construction cost ($/lb)

Ozone Generation Capacity (lb/d)

Figure 4.8-4 Constraction cost of ozonation system. (J.M.M. Internal Seminar Handout, 1989.)

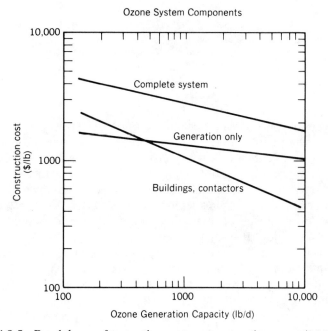

Ozone System Components

Complete system

Generation only

Buildings, contactors

Construction cost ($/lb)

Ozone Generation Capacity (lb/d)

Figure 4.8-5 Breakdown of ozonation system construction cost. (J.M.M. Internal Seminar Handout, 1989.)

TABLE 4.8-3   Solubility of Ozone in Water

Water Temperature (°C)	Henry's Constant Atm/Mole	Ozone Concentration mg/L	ppm-vol	Ozone Solubility mg/L
0	1,940	12.07	6,044	8.31
5	2,180	12.07	6,044	7.39
10	2,480	12.07	6,044	6.50
15	2,880	12.07	6,044	5.80
20	3,760	12.07	6,044	4.29
25	4,570	12.07	6,044	3.53
30	5,980	12.07	6,044	2.70
0	1,940	18.11	9,069	12.47
5	2,180	18.11	9,069	11.09
10	2,480	18.11	9,069	9.75
15	2,880	18.11	9,069	8.40
20	3,760	18.11	9,069	6.43
25	4,570	18.11	9,069	5.29
30	5,980	18.11	9,069	4.04
0	1,940	24.14	12,088	16.62
5	2,180	24.14	12,088	14.79
10	2,480	24.14	12,088	13.00
15	2,880	24.14	12,088	11.19
20	3,760	24.14	12,088	8.57
25	4,570	24.14	12,088	7.05
30	5,980	24.14	12,088	5.39
0	1,940	36.21	18,132	24.92
5	2,180	36.21	18,132	22.18
10	2,480	36.21	18,132	19.50
15	2,880	36.21	18,132	16.79
20	3,760	36.21	18,132	12.86
25	4,570	36.21	18,132	10.58
30	5,980	36.21	18,132	8.09

Note: The concentration of the ozone gas is determined at a standard temperature of 68°F (20°C) and a standard pressure of 1 atmosphere (101 kPa).

Source: Adapted from EPA Design Manual, EPA/625/1-86/02, October 1986.

Several types of contactor have been developed: (1) diffused bubbles (concurrent and countercurrent), (2) positive pressure injection (U-tube), (3) negative pressure (Venturi tube), (4) turbine mixer tank, and (5) packed tower. The countercurrent bubble contactor is considered to be the most efficient and cost effective of the alternatives and has been employed in the design of most ozonation systems (refer to Figure 4.8-6). Due to the rapid reaction kinetics of ozone, a contact time of only 3–10 min is considered to be a practical detention time for the contactor.

The ozone contact tanks have several common design features: the tanks must be completely enclosed; they are composed of concrete and are primarily located outside; and each tank has two cells and is capable of handling 50% of the maximum daily flow of the plant. Additionally, each cell should have a drain to allow periodic dewatering and cleaning; stainless steel hatches should be provided to facilitate inspection and maintenance; a slight negative pressure (2 in. of water) should be maintained in the tank to ensure that there

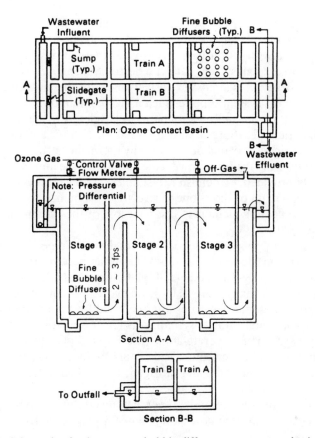

Figure 4.8-6 Schematic of a three-stage, bubble diffuser ozone contact basin. (Adapted from EPA Design Manual, EPA/625/1-86/02, Oct. 1986).

is no leakage of ozone above the water surface; all pipe (SS 304 or 316) connections should be welded or flanged inside the tank and a check valve should be installed in the gas feed line prior to the contactor to prevent the backflow of water into the gas line; the water depth in the tank should be 18–20 ft (5.5–6 m) to provide a minimum ozone transfer efficiency of 95%. Lastly, if the ozone concentration in the feed gas is near 10% (as in the Megos ozonator with oxygen feed), a different system consisting of a hydraulic eductor and in-line static mixer may be employed.

Off-Gas Destruction The ozone contact tank typically has a 90–95% transfer of ozone to the water, the remaining 5–10% results as off-gas. The ozone content of this off-gas must be reduced to levels below the OSHA and local Air Quality Management District (AQMD) standards prior to venting into the atmosphere.

According to OSHA, the maximum allowable ambient concentration of ozone is 0.0002 g/m^3 (0.1 ppm by volume) for an 8 h working day. The off-gas normally contains levels greater than 1 g/m^3 (500 ppm by volume). Three basic methods of destruction have been used: thermal destruction, thermal destruction with a catalyst, and catalytic destruction.

Thermal destruction is almost exclusively used for the air feed gas system; the oxgyen feed gas system supplies such a high content of oxygen in the off-gas that the use of thermal destruction is considered to be a fire hazard. Temperatures ranging from 570 to 660°F (300–350°C) for a period of 3 s are required to obtain 99% ozone destruction by thermal destruction. Due to the high temperature and the large power consumption of the unit, a heat recovery unit is usually provided. A demister, a stainless steel wire mesh, should also be provided upstream of the ozone destructor to minimize the accumulation of foam on the heating elements. The heat loss of the thermal destructor is approximately 30%.

The system of thermal destruction with catalyst consumes less energy than thermal destruction alone. Thus, this system has become more popular in recent years. The catalyst may be metal (platinum or palladium), metal oxides (aluminum oxide or manganese oxide), or hydroxides and peroxides. However, the hydroxides and peroxides are only used for pilot scale studies. The advantage of the metal catalyst system is its lower operating cost. The metal oxide catalyst system can operate at 85–120°F (30–50°C): this is much lower than the thermal unit. Yet, the disadvantage is its sensitivity to chlorine, chlorine derivatives, sulfides, and nitrogen; these chemicals may react with and destroy the catalyst.

Activated carbon adsorption has been employed, but the powdery carbon produced by the reactivation process is potentially explosive. Thus, this method has limited application.

In certain cases the off-gas is recycled back to the feed gas after conditioning or is injected into the filter waste backwash water, which is discharged to a holding lagoon. It should be noted that these alternatives are very rarely used because of the need to maintain an additional process and the lack of high reliability for the off-gas treatment.

Safety The issue of safety must be addressed because ozone is both a toxic gas and a fire hazard. Moreover, if the system uses oxygen as the feed gas, the situation becomes more dangerous. Although the ozonation system may be less hazardous than the chlorination system because it can be shut down if an ozone leak develops, it may also be more dangerous because the system must use high-voltage electrical power to generate the ozone.

The American Industrial Hygiene Association (AIHA), OSHA, and other associations recommend the following permissible levels of ozone in the air:

Workers will not be exposed to ozone concentrations in excess of a time weighted average of 0.2 mg/m^3 (0.1 ppm by volume) for eight hours or more per workday, and that no worker be exposed to a ceiling concentration of ozone in excess of 0.6 mg/m^3 (0.3 ppm by volume) for more than 10 minutes.

Local Air Quality Management Districts may have more stringent criteria that are part of the restrictions on smog.

A few other safety considerations include the installation of self-contained breathing apparatuses in the event of a severe ozone leak, the installation of eyewashes, and the availability of comprehensive safety and operation and maintenance manuals, which may be consulted during an emergency situation. Ideally, the ozone generation building and ozone contact tanks are isolated from the operations building for obvious safety reasons.

Design Criteria   The design criteria for the ozonation system are numerous. They range from the ozone generator, to the contact tank, to the destruction unit. The requirements for these and many more issues are presented below. Additionally, an example of a detailed design is presented.

Ozone dosage	1.5–3 mg/L (normal) depending on the purpose
Number of ozone generators	Minimum of two, and preferably three; one always acts as a standby
Ozone generator	
Minimum production	10–20% of rated capacity
Maximum production	75% of rated capacity
Cooling water temperature	Less than 75°F (24°C) at the inlet
Vessel construction	Pressure vessel (15 psig) constructed with 304 LSS or 315 LSS wit Hypalon or Teflon gaskets
Type of generator	Low frequency with variable voltage, medium frequency with frequency control, or another type, depending on the selection
Compressor	
Pressure	10–40 psig if heat reactive desiccant; 80–100 psig if pressure swing type desiccant
Number	Minimum of two; one acts as standby
Type	Liquid ring (<1000 lb/day ozone production), centrifugal (>1000 lb/day ozone production), piston type, oil-free, or oil lubricated with oil removal filters (<100 lb/day—small plant)
Filters	
Before compressor	Regular filter with silencer
Before dryer	3–5 μm
Before generator	0.3 μm
Pressure drop	0.5 psi when clean, 2 psi for coalescing filter
Moisture removal	
Refrigerant dryer	41°F (5°C) dewpoint (80% removal from air)
Desiccant dryer	−76°F (−60°C) dewpoint (99.98% removal from air)

Notes: Ambient air contains approximately 2.7 g of water per kg of air. A dewpoint of $-58°F$ ($-50°C$) is marginal for the desiccant dryer. Air with a dewpoint of $-76°F$ ($-60°C$) contains 6.7 mg of water per kg of air.

Desiccant regeneration cycle (heat reactivation type)	
Minimum cycle time	8 h
Design cycle time	16 h
Ozone contact tanks	
Number of tanks	Minimum of two
Transfer efficiency	Minimum of 95% if possible; may range from 90 to 95%
Detention time	5–15 min (usually less than 8 min)
Stage of contact	Normally two to three stages
Water depth	18–20 ft
Submergence of diffuser	16–18 ft
Freeboard	4–6 ft to allow the deposition of foam
Ozone diffuser	
Material	304 LSS, glass, ceramic, or Teflon
Bubble size	2–5 mm
Gas flow	0.5–4 cfm depending on the diffuser:
(each diffuser)	Rod diffuser (2.5 in. × 24 in. L): 4 cfm maximum
	Disk diffuser (7 in. diameter): 1.25 cfm maximum
	Disk diffuser (9 in. diameter): 1.8 cfm maximum
Headloss	Maximum of 0.5 psi
Permeability	8–20 cfm/ft^2
Porosity	35–45%
Residual ozone (ozonated water) monitor	
Type	UV ozone monitor (continuous monitoring) such as PCI Model HC or potassium iodide monitoring
Ozone destruction unit	
Type	Heat catalyst unit (most popular)
Temperature	80–100°F (27–38°C)
Catalyst	Metal (platinum) or metal oxides: these catalysts are proprietary items

Figure 4.8-7 is a schematic presentation of a typical ozonation system.

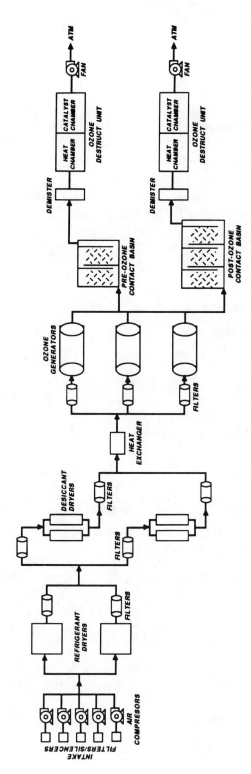

Figure 4.8-7 Schematic of ozone handling system.

Example Design Calculations

Example 1

Given The ozone concentration of the gas flowing from the ozone generator to the tank is customarily expressed as "percentage of ozone by weight," rather than the weight of ozone, in grams per cubic meter of gas. Assume that the ozone concentration is 20 g/m³.

Determine The percent weight of ozone for the air feed ozone generation system and the oxygen feed ozone generation system. Assume that the gases are all at 20°C and under 1 atmosphere (atm) of pressure (STP conditions).

Solution

$$\text{Gram molecular weight of air} = 29 \text{ g/mole}$$
$$\text{Gram molecular weight of oxygen} = 32 \text{ g/mole}$$
$$\text{Gram molecular weight of ozone} = 28 \text{ g/mole}$$

(i) For the air feed ozone generation system:

$$\text{Density of air} \quad \frac{MP}{RT} = \frac{29}{24.053} = 1.2056 \text{ g/L}$$
$$= 1205.6 \text{ g/m}^3 \quad \text{(see Appendix 10)}$$

$$\text{Density of ozone} \quad \frac{MP}{RT} = \frac{48}{24{,}053} = 1.9956 \text{ g/L}$$
$$= 1995.6 \text{ g/m}^3$$

$$\text{Volume of 20 g of ozone} \quad \frac{GRT}{MP} = \frac{20 \times 24.053}{48} = 10.02 \text{ L}$$
$$= 1\% \text{ of 1 m}^3$$

Since the volume of 20 g of ozone is 1% of 1 m³ of the gas, the weight of the gas is

$$1205.6 + [(1995.6 - 1205.6) \times 0.01] = 1213.5 \text{ g/m}^3$$

Thus, the percent weight is

$$(20 \div 1213.5) \times 100 = 1.648\%, \quad \text{say } 1.65\%$$

(ii) For the oxygen feed ozone generation system:

$$\text{Density of oxygen} \quad \frac{MP}{RT} = \frac{32}{24.053} = 1.3304 \text{ g/L}$$

$$= 1330.4 \text{ g/m}^3$$

Knowing that the volume of 20 g of ozone is 10% of the gas, the weight of the gas is

$$1330.4 + [(1995.6 - 1330.4) \times 0.01] = 1,337.05 \text{ g/m}^3$$

The percent weight is

$$(20 \div 1337.05) \times 100 = 1.496\%, \quad \text{say } 1.5\%$$

Example 2

Given The ozone concentration of the gas from an air feed ozone generator is measured by the potassium iodide wet-chemistry method as 25.6 mg/L.

Determine The percent weight of ozone at STP conditions.

Solution

$$25.6 \text{ mg/L} = 25.6 \text{ g/m}^3$$

$$\text{Weight of air} = 1205.6 \text{ g/m}^3 \quad \text{from Example 1}$$

$$\text{Weight of ozone} = 1995.6 \text{ g/m}^3 \quad \text{from Example 1}$$

$$\text{Volume of 25.6 mg/L of ozone} = \frac{GRT}{MP} = \frac{25.6 \times 24.053}{48} = 12.83 \text{ L}$$

$$= 1.28\% \text{ of the gas}$$

The weight of the gas is calculated to be

$$1205.6 + [(1995.6 - 1205.6) \times 0.0128] = 1215.7 \text{ g/m}^3$$

Thus, the percent weight of the ozone is

$$(25.6 \div 1215.7) \times 100 = 2.1\%$$

Example 3

Given An air feed ozone generator is operating at an air flow rate of 300 scfm; the standard conditions are 20°C or 68°F and 1 atm or 14.7 psi of pressure. The ozone concentration is measured to be 1.65% weight by the ozone monitor. The plant water flow rate is 20 mgd.

Determine

(i) The ozone production rate in #/day.
(ii) The ozone dosage in mg/L.

Solution

(i) The conversion factor for g/L to #/ft³ is 0.0753. The ozone production rate is calculated to be

$$\frac{1.65 \ \# \ O_3}{100 \ \# \ \text{Air}} \times 0.0753 \times (300 \times 1440) = 537 \ \#/\text{day}$$

(ii) The ozone dosage is

$$537 \div (8.34 \times 20) = 3.22 \ \text{mg/L}$$

Example 4

Design the ozonation system of a 20 mgd (average) water treatment plant using the design criteria presented in the "Design Criteria" section.

Step 1. Ozonation Production Rates

Maximum ozone production rate	$3 \times 8.34 \times 30$	$= 751$ ppd
Average ozone production rate	$3 \times 8.34 \times 20$	$= 500$ ppd
Minimum ozone production rate	$1.5 \times 8.34 \times 5$	$= 63$ ppd

Step 2. Type, Size, and Number of Ozone Generators A medium-frequency ozone generator is selected primarily because of its 10:1 turn-down ratio, lower power consumption, and its ability to enrich the feed air should the plant need to double the ozone output at a future date. Generally, any ozone generator should be operated for extended periods of time at approximately 75% of the rated capacity. The rated maximum daily production of ozone should be 751 ÷ 0.75 or 1000 ppd. Three generators, each capable of 335 lb/day, should therefore be provided. This type of generator can accommodate a turn-down ratio of 10:1 so that the minimum production rate is approximately 35 lb/day. Thus, a rate of 63 lb/day is easily achieved. Since two of the generators can satisfy the required 500 lb/day, under normal conditions, one unit should be left on standby.

Step 3. Size and Number of Air Compressors Since the maximum ozone concentration is 2.5% by weight for most air feed medium-frequency generators, the air flow at maximum conditions is

$$1000 \ \#/\text{day} \div \left(0.0753 \times 1440 \times \frac{2.5}{100}\right) = 369 \ \text{scfm} \quad \text{or} \quad 370 \ \text{scfm}$$

Add 60 scfm to this number for purge air flow to the dryers. Thus, the maximum air flow may be calculated as

$$370 + 60 = 430 \text{ scfm}$$

The minimum ozone concentration is approximately 1% by weight when this type of generator is employed. Thus, the minimum air flow rate is

$$75 \text{ #/day} \div \left(0.0753 \times 1440 \times \frac{1}{100}\right) = 69 \quad \text{or} \quad 70 \text{ scfm}$$

If 2.5% weight ozone is produced on a continuous basis, the minimum air flow rate would be

$$75 \text{ #/day} \div \left(0.0753 \times 1440 \times \frac{2.5}{100}\right) = 28 \text{ scfm}$$

This value is the smaller of the two. Add 22 scfm for purge air flow to the dryer. Therefore, the minimum air flow is

$$28 + 22 = 50 \text{ scfm}$$

Based on these calculations, the compressors may be either medium (35–50 psig) or low (10–30 psig) depending on the type of drying unit selected.

Provide:

One 50 scfm compressor
Two 100 scfm compressors
Two 150 scfm compressors

Step 4. Size and Number of Desiccant Dryers The design basis is as follows:

1. Operating pressure of 20–25 psig for low-pressure systems and 50 psig for medium-pressure systems.
2. Adjustable site elevation for the pressure.
3. Inlet air temperatures: 77°F (25°C) with after-cooler and 40°F (4.4°C) with refrigerated dryer.
4. A required maximum dewpoint temperature for the air flowing to the ozone generator of −76°F (−60°C).
5. Designed desiccant dryer cycle time of 16 h. A minimum drying time of 8 h.

Operation and Maintenance The ozonation system is both very sensitive and expensive when compared to other types of water treatment equipment. Plant operators must be very conscientious in maintaining optimum operational conditions for reasons of efficiency and safety.

The operation conditions of the ozonation system must be adjusted by the plant operator whenever the plant flow rate changes. This will provide the system with optimal operational conditions, thus facilitating maximum process efficiency and minimal operation and maintenance costs. Included in the optimization process is the determination of the optimum ozone dosage; a 0.5 mg/L reduction in dosage translates into a significant saving in power consumption. The regular performance of several other tasks also directly affects the efficiency of the ozonation system: checking the temperature of the cooling water, checking the dewpoint temperature at the desiccant dryer unit, checking and calibrating the ozone concentration monitors, checking the off-gas concentration, and adjusting the ozone dosage. The plant operation and maintenance manuals, as well as the manuals furnished by the equipment supplier, should describe these very important items in detail.

A common problem for many large metropolitan areas is air pollution due to smog. In these situations, the local AQMD sets the maximum allowable ozone concentration in the air. Thus, the efficiency of the off-gas destruction unit should be maintained at a level high enough to meet the stringent limits. Moreover, the exhausting air from the ozone contact tanks and the building housing the ozone generator must be monitored closely for excessive emission of ozone.

With regard to the safety aspects of the ozone system, strict adherence to the safety codes and regulations released by OSHA, the American National Standards Institute/American Society of Testing and Material (ANSI/ASTM), the American Conference of Government Industrial (ACGI), and the American Industrial Hygiene Association (AIHA) is absolutely necessary.

BIBLIOGRAPHY

Katzenelson, E., Kletter, B, & Shuval, H.I., "Inactivation Kinetics of Viruses and Bacteria in Water by Use of Ozone," *J. AWWA*, 66(12):725(December 1974).

LePage, W. L., "A Treatment Plant Operator Assesses Ozonation," *J. AWWA*, 77(8):41(August 1985).

Malorey, S. W., Sulffet, I. H., Bancroft, K., & Neukrug, H. M., "Ozone–GAC Following Conventional U.S. Drinking Water Treatment," *J. AWWA*, 77(8):66(August 1985).

Monk, R. D. F., Yoshimura, R. Y., Hoover, M. G., Lo, S. H., "Prepurchasing Ozone Equipment," *J. AWWA*, 77(8):49(August 19XX).

Rakness, K. L., "Ozone Disinfection," Chapter 6 of *EPA Manual—Municipal Wastewater Disinfection*, EPA/625/1-86/021 (October 1986).

Rice, R. G., "Uses of Ozone in Drinking Water Treatment," *J. AWWA*, 73(1):44(January 1981).

Rice, R. G., Netzer, A., *Handbook of Ozone Technology and Applications*, Vol. 1, Ann Arbor Science Publishers, Ann Arbor, MI, 1982.

Rice, R. G., Netzer, A., *Handbook of Ozone Technology and Applications*, Vol. 2, Butterworth Publishers, Boston, 1984.

CHAPTER 5

Design of Plant Components

5.1 FLOW MEASUREMENT

Purpose Flow measurement is the most important variable measured in a treatment plant. The three principal applications for flow measurement are in establishing the flow rate of the process and in determining the liquid chemical and gas flow rates of various subordinate processes. Differential pressure producers, direct discharge measurement, positive volume displacement measurement, and flow velocity–area measurement are the four basic types of measurement commonly used in water treatment plants.

Considerations Several questions must be considered in order to select the most appropriate type of flow measurement equipment.

1. Is liquid or gas flow being measured?
2. Is the flow occurring in a pipe or in an open channel?
3. What is the magnitude of the flow rate?
4. What is the range of flow variation?
5. Is the liquid being measured clean or does it contain suspended solids or air bubbles?
6. What is the accuracy requirement?
7. What is the allowable headloss by the flow meter?
8. Is the flow corrosive?
9. What type of flow meters are available to the region?
10. What type of postinstallation service is available to the area?

Various Flow Meters Flow meters commonly used in the field of water treatment are the Venturi type meter, orifice meter, propeller type meter, magnetic flow meter, ultrasonic flow meter, vortex meter, rotameter (variable-area meter), flumes, and weirs. Liquid chemical flow is almost exclusively measured by positive displacement pumps rather than flow meters, with the exception of the rotameter, which may be used in certain situations.

Differential Pressure Producer Type of Meter The differential pressure-producing types of meter currently on the market are the Venturi type, Dall tube, Hershel Venturi, Universal Venturi, and the Venturi inserts (Figure 5.1-1). The accuracy of these flow tubes is $\pm 1\%$ for a flow range of 10 to 1. However, since the square root extractor is used before the 4–20 mA signal is transmitted, the range of accurate flow measurement is restricted to no more than 5 to 1.

The headloss across the Venturi type flow meter is relatively small, ranging from 3 to 10% of the differential, depending on the ratio of the throat diameter to the inlet diameter (beta ratio). These flow tubes should be installed at an elevation that will continuously sustain the meter at full pipe flow and ideally will not allow negative pressure to form at the throat during maximum flow rate.

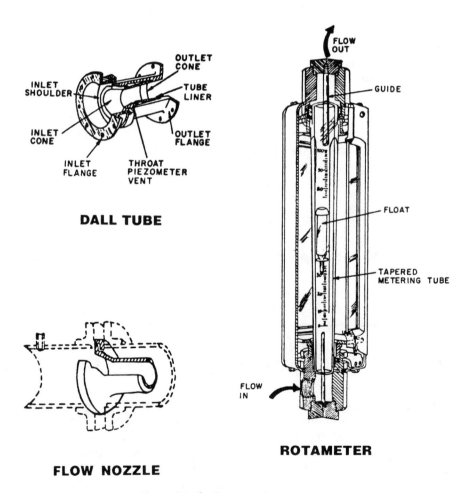

Figure 5.1-1 Common flow meters.

Orifice plates are another type of differential pressure producer. Both concentric and eccentric orifice plates are available. Their accuracy is $\pm 2\%$ for a flow range of 4 to 1. The orifice plate yields a high headloss ranging from 40 to 90% of the differential; these values depend on the ratio of the orifice diameter to the pipe diameter. This type of flow measurement equipment is suitable for measuring both gas and liquid flow.

A third type of differential producer is the flow tube (flow nozzle). The throat ratio determines the headloss of the flow tube; headloss ranges from 10 to 20% of the differential. However, its application is primarily in sewage works (Figure 5.1.2).

In order to assure uniform flow through the meter, a straight pipe with a length equivalent to at least 10 pipe diameters should ideally be provided upstream of all differential type flow meters. Yet, the meter manufacturer may only allow for a straight pipe that is five pipe diameters in length in conjunction with a flow straightener. Figure 5.1-2 presents the headloss characteristics of several Venturi type flow meters and their piping requirements.

Propeller and Turbine Meters   The Sparling, Rockwell, Badger, and Neptune meters have been on the market for decades. These types of meter have a magnetic coupling that senses the revolutions of the spinning rotor and sends out electronic pulses. The accuracy is 1–2% for a flow range of 10 to 1, provided that there is a uniform and straight pipe of at least five pipe diameters and straightening vanes ahead of the meter.

The minimum flow velocity should generally be 0.7 ft/s (0.2 m/s) and the maximum flow velocity must not exceed 6 ft/s (1.8 m/s). The propeller and turbine meters should be employed in situations where the water is relatively clean and noncorrosive and for liquids that are not too viscous since they are subject to wear, particularly in cases where the water contains suspended solids such as silt. Thus, the meter must be installed in a readily accessible location to facilitate easy maintenance.

Main-line meters are available in sizes up to 24 in. in diameter; the saddle type, up to 72 in.; and the intake meters, up to 78 in. in size.

Magnetic Flow Meter   The magnetic flow meters operate by having a conductor, such as water flowing in a magnetic field, create a voltage that is proportional to the velocity of relative motion. Fixed magnetic coils wrapped around the pipe produce a magnetic field that is perpendicular to the flow of liquid. The induced voltage is detected by two electrodes embedded in the pipe. The accuracy of this type of meter is $\pm 1\%$ for a flow velocity range of 3–30 ft/s (0.9–9 m/s) and within $\pm 2\%$ if the range is 1–3 ft/s (0.3–0.9 m/s). Yet, this type of direct discharge meter is relatively expensive and may require periodic calibration to correct drifting of the signal.

The magnetic flow meter may be used for measuring most any type of liquid, including the measurement of sludge flow. Sizes up to 48 in. are readily available and larger sizes may be custom manufactured. Since there is not

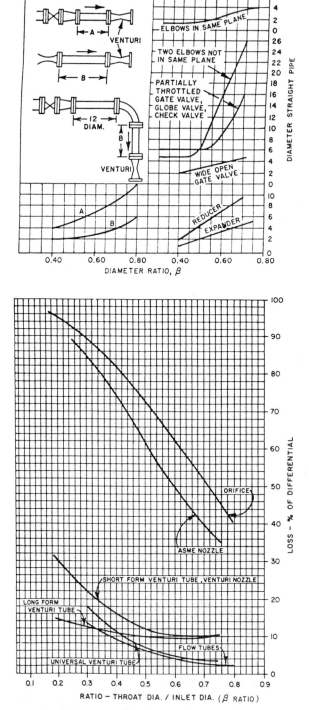

Figure 5.1-2 Piping requirement and headloss for Venturi type flow meters. (Adopted from BIF Catalog, 1968.)

flow restriction inside the meter, there is essentially no headloss within the unit. However, the size of the meter should be one size smaller than the size of the line so that the flow velocity requirements can be met. The two fundamental requirements are that a uniform and straight pipe of at least five pipe diameters be placed ahead of the meter and that the unit should receive a full pipe flow.

Ultrasonic Flow Meter   The flow measurement of the ultrasonic flow meters is based on the time required for an ultrasonic pulse to diagonally traverse a pipe or channel against the liquid flow. Application of this type of direct discharge meter should be to clean water so that it measures the transit time or time of flight and not the reflection of the pulse off particles or air bubbles (Doppler effect).

The flow velocity is proportional to the difference in the time of the two pulses. The accuracy is $\pm 1\%$ for a flow velocity ranging from 1 to 25 ft/s (0.3 to 7.6 m/s), but the meter reading is greatly affected by a change in the fluid composition. Thus, this type of meter may only be employed for water that is free of particles and air bubbles as much as possible. In the case of large pipes or a channel, more than one pair of sensors is required.

Vortex Meters   When a liquid flow hits a bluff object, eddies and vortices are formed; this is known as vortex shedding. The frequency at which the vortices are generated is proportional to the velocity of the liquid flow. The resulting pressure changes may be detected by probes mounted on the body of the bluff. The information is then relayed to an indicating recorder by a 4–20 mA signal.

The manufacturers claim that this type of meter has an accuracy of $\pm 1\%$ for a flow range of 12 to 1. The headloss is approximately two times the velocity head through the meter. The cost of the vortex meter is moderate. However, this type of direct discharge meter does not function well with viscous, dirty, abrasive, or corrosive fluids.

Rotameters (Variable Area Flow Meters)   Rotameters consist of a tapered glass tube containing a freely moving float. This glass tube is vertically mounted with the wide side up (Figure 5.1-1).

Rotameters may be used in measuring both gas flow and liquid flow rates. The flow rate of the particular gas or liquid is indicated when the float reaches a position of equilibrium between its weight and the upward force of the medium. The flow range for this type of meter is approximately 10 to 1 with an accuracy of $\pm 2\%$ of full scale. Yet, the accuracy is highly dependent on the skill of the individual reading the meter.

The rotameter may be applied in situations with very low flow rates, 0.1–140 gᵤh for water and 1–520 scfm for air, and the accuracy of the meter is not affected by the piping configuration immediately upstream of the rotameter. Basically, these meters are controlled manually and read or measured visually.

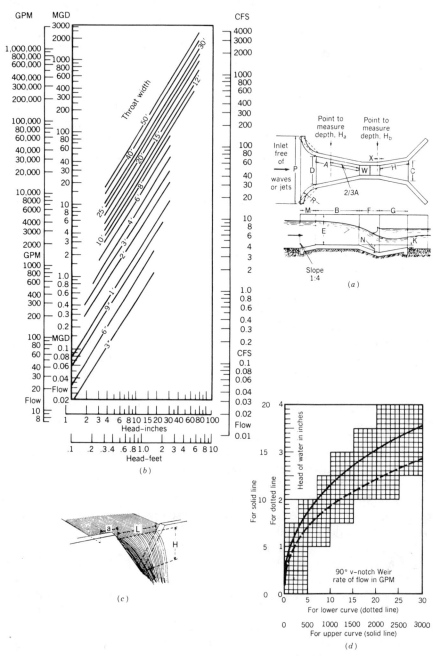

Figure 5.1-3 (a) Parshall flume. (b) Flow curves for Parshall flume. (c) V-notch weir. (d) Flow curves for 90° V-notch weir.

Flumes (Positive Volume Displacement Type) Flumes are used for open channel flow. The Parshall flume and Palmer–Bowles flume are the types predominantly used in wastewater treatment. These meters are used to measure small to very large flow rates (0.02 to 3000 mgd) and are applicable to both clean and dirty water. The accuracy of the flumes is ±5% over a range of 20 to 1, but the range may be expanded to 35 to 1 with some sacrifice in accuracy.

It is important that engineers provide a sufficient fall downstream of the flume to avoid the effects of the backwater because the level measurement is conducted at a specific point in the flume by either an ultrasonic level detector or a float located directly above the point. The flow rate may also be determined in a stilling well by a bubbler or float.

Weirs Several types of weirs—for example, the V-notch, rectangular, or trapezoidal—are currently available for use in the field of water treatment. The head above the crest of the weir is usually measured by a float, bubbler system, acoustic meter, hook gauge mounted above the weir, or staff board mounted on the side wall of the channel. The practical flow range of the weirs is approximately 30 to 1.

The rectangular and trapezoidal weirs have a power head–flow relationship of approximately 1.5, while the V-notch weirs have a power relation of 2.5. (see Appendix 14 for the formulas) Therefore, any inaccuracy in the measurement of the hydraulic head will greatly affect the accuracy of the flow measurement, especially in the case of the V-notch weirs. Yet, the V-notch weir plates may be used for a much wider flow range: an extreme case would be 3000 to 1.

Although the weirs are easy to manufacture and are cost effective, the main disadvantages are their high headloss and the tendency for settleable solids and large debris to collect on the upstream side of the weir.

Figure 5.1-3 illustrates flow curves for Parshall flume and 90° V notch weir. The equipment manufacturers can provide the engineer with detailed information on the various types of flow meter currently available and all engineers are advised to review the brochures prior to selection. Virtually all types of meter, with the exception of the rotameter, require an adequate length of uniform and straight conditions upstream of their location. Most of the flow measurement meters are capable of sending a standard 4–20 mA signal.

Example Design Calculation

Example 1

Given A water treatment plant has a daily maximum, average, and minimum design flow rate of 22, 15, and 4 mgd, respectively. However, this plant is projected to double its capacity in approximately 15 years.

Determine The appropriate size and elevation of the Venturi tube, which is connected to a plant inlet tank containing a water depth of 10 ft.

Notes: The selection guides for the Venturi type meter are as follows:

1. Select a meter size that will allow for an inlet flow velocity of 6–10 ft/s at the average daily flow rate.
2. Select a proper "beta ratio" for the Venturi tube so that a differential pressure of approximately 80 in. will be produced at the average daily flow.
3. Check the magnitude of the differential at the daily maximum and daily minimum design flows. The minimum differential should be at least 3 in. and the negative pressure should be no more than 2 ft at the throat of the Venturi (at the differential value) during the maximum flow rate.

Solution The plant influent pipe should be sized for a future flow rate of 30 mgd (daily average flow rate). Thus, the line size should be 36 in. in diameter ($v = 6.6$ fps). Yet, the flow velocity is only 3.3 fps at 15 mgd; this is the average daily flow rate of the initial stage. The Venturi tube selection table (Appendix 12) indicates that none of the 36 in. tubes will satisfy the criteria: a Type A throat (beta ratio A) of 36 in. is characterized by a 120 in. differential at 16 mgd, but the differential is 240 in. (20 ft) at 22 mgd; that is, the center line of the Venturi tube must be installed approximately 20 ft lower than the water level in the inlet tank in order to avoid producing a significant amount of negative pressure at the throat of the Venturi tube. The differential at the minimum design flow rate of 4 mgd is

$$120 \text{ in.} \div (16/4)^2 = 7.5 \text{ in.} > 3 \text{ in.} \quad \text{O.K.}$$

If the site condition allows this type of configuration for the plant inlet pipe, without significantly raising the cost, a Venturi tube with a Type A throat of 36 in. should be selected for the initial plant. When the plant capacity is doubled (at a future date), the Venturi may then be replaced with a Venturi tube with a Type C throat, which would produce a differential of 80 in. at a daily flow rate of 37 mgd, 160 in. (13.3 ft) at 45 mgd (maximum daily flow), and 5 in. at 8 mgd (minimum daily flow rate).

Alternatively, the pipe size may be reduced from 36 to 20 in. with a minimum pipe length of 30 ft, if the influent line cannot be installed over 20 ft below the water level of the tank due to the site condition or cost. Under these circumstances, a 20 in. Venturi tube ($v = 10.5$ fps at 15 mgd) should be selected: the Venturi tube selection table indicates that the Type D (20 in.) Venturi tube will give a differential of 80 in. at a flow rate of 16 mgd. The flow rate at 15 mgd is

$$80 \text{ in.} \div (16/15)^2 = 70.3 \text{ in.}$$

The differential at the daily maximum flow of 22 mgd is

$$80 \text{ in.} \times (22/16)^2 = (151 \text{ in.}) \ (12.5 \text{ ft})$$

and the differential at 4 mgd is 5 in. A minimum of 15 ft of uniform straight pipe and a straightener should be provided upstream of the Venturi tube.

When the plant capacity becomes doubled, replace the 20 in. line (containing the Venturi tube) with a 36 in. line and a 36 in. Type C Venturi tube. This alternative requires the center line elevation to be approximately 14 ft below the water level in the inlet tank; this is easily achieved in most cases.

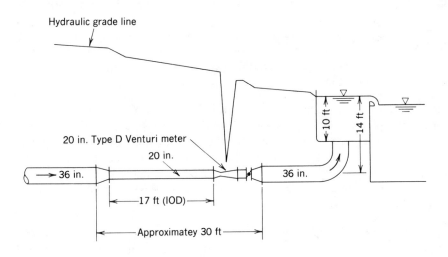

BIBLIOGRAPHY

Andrew, W. G., William, H. B., *Applied Instrumentation in the Process Industries*, Vols. 1 and 2, Gulf Publishing, Houston, 1980.

Babcock, R. H., "Instrumentation and Control in Water Supply and Wastewater Disposal," *Water Wastewater Eng.*, xx:xx (1968).

Department of Civil Engineering, "Instrumentation, Control, and Automation for Water Supply and Wastewater Treatment Systems," in *Proceedings of the 9th Sanitary Engineering Conference*, University of Illinois, 1967.

Knudsen, D. I., ed., *Process Instrumentation and Control System*, WPCF Manual of Practice No. OM-6, 1982.

5.2 LEVEL MEASUREMENT

Level measurement is an essential item in plant operations because the plant operator must be aware of the water level of all the process units, the levels of all chemical storage tanks and silos, and the pressure of water or compressed air lines—that is, the water level in the distribution mains and the utility lines.

There are several means of measuring liquid levels: via a float, pressure elements, bubbler system, or an ultrasonic system. Methods used in measuring dry, powdery materials include ultrasonic systems, photocell systems, rotary paddle switches, diaphragm units, and wire strain gauge systems. However, it is often very difficult to accurately measure the level of powdery materials held in storage bins due to the concavity of the surface or irregular surface pattern. The most accurate and reliable method of monitoring any dry or liquid material in storage bins or tanks is to use load cells to measure the total weight (tank plus content).

The rest of this section describes some features of the float system, pressure elements, bubbler tube system, and ultrasonic level detector. Other methods of level measurement are not commonly employed in water treatment and are therefore not discussed.

Float System In the past, the most common method of level measurement for open tanks or channels was the float-operated transmitter. This is a simple and reasonably accurate system. Nevertheless, the system is very unsightly, and the installation process is very time consuming and expensive due to the need for a stilling well and a collection of wires, wheels, and tackles. The float system also requires periodic maintenance to assure friction-free motion of the float and cable assembly.

Pressure Elements Level measurement via pressure elements is very common in the field of water treatment. This method determines liquid levels by measuring the pressure produced by the head or by measuring the level above the pressure measuring tap point. A pressure transducer connected to the pressure elements measures the water pressure at the base of the tank and directly reads the liquid level, if properly calibrated for the liquid.

The bourdon tube, bellow element, diaphragm element, and manometer are examples of pressure element type level measurers. The bourdon tube has both helical and spiral units and is suited for high-pressure (water depth included) measurements. The pressure range of the spiral units is commonly 0–15 and 0–200 psi (0–1 and 0–15 kg/cm²). The helical unit is typically used for measuring much higher pressures: 0–200, 0–400, and 0–6000 psi (0–15, 0–30, and 0–440 kg/cm²) or higher.

Bellow units measure pressures in the intermediate range. The units are commonly manufactured to measure water levels (depth) of 0–100 in. of water (0–2.5 m) and a pressure range of 0–5 and 0–30 psi (0–0.35 and 0–2 kg/cm²).

The diaphragm type pressure element is particularly suited for obtaining measurements within a small range in the low-pressure zone. The available units can cover a range of 0–10 and 0–20 in. (0–0.25 and 0–0.5 m), and 0–40 and 0–140 in. of water (0–1 and 0–3.6 m).

The use of manometers is limited to pilot studies or temporary use, when the prime meter is out of service.

Pressure elements often require the following conditions in order to yield accurate readings: zero suppression, zero elevation, or differential pressure for pressurized tanks (D/P cell).

Bubbler Tube System The bubbler system is used to measure the back pressure of the hydrostatic head. This is equivalent to the hydrostatic head of the liquid at the bottom of the bubbler tube. This type of level measurement is currently the most widely used method applied to open tanks.

The bubbler tube system has a tube placed inside a tank which runs from the top of the tank and opens approximately 3 in. from the bottom. During operation, compressed air is supplied to the submerged bubbler tube via a regulator (pressure regulator, differential pressure regulator) or a purge rotameter. The bubble purge rates are very low and as long as the bubbles escape periodically, the system is functioning properly. The purge rate is generally 0.5 ft^3 of free air per hour (14 L/h).

The source of the compressed air may be a small air compressor, a compressed air tank, or a nitrogen gas cylinder. If electrical power cannot be provided, for whatever reason, the use of a compressed air cylinder is the most economical method; a standard scuba diving compressed air tank can supply 1 month of air and is easily refilled.

The end of the bubbler tube must be set at a point of zero elevation or zero level measurement. For example, if the end of the bubbler tube is located 12 ft below the surface of the water, the bubbler tube would indicate a water level of 12 ft. If the water level drops by 5 ft, the measured depth would be 7 ft. This system detects the water depth (pressure) by means of an appropriately calibrated gauge or manometer.

The bubbler system has the advantages of (1) simple design, (2) easy accessibility and little concern over the corrosion of the pressure sensing device, and (3) the ability to be installed at the bottom of the tank, where accessibility is blocked or impossible. Figure 5.2-1 shows a bubbler system used in open tank level measurement.

Ultrasonic Level Detector The ultrasonic system is used to monitor either the water level in a tank or dry material stored in a storage bin open to the atmosphere. The height of the water/material is measured by means of an acoustic pulse; the ultrasonic transmitter and receiver units are located above the maximum level of the object.

The transmitter generates a pulse down through and perpendicular to the liquid level or surface of the dry material. The time elapsed between pulse

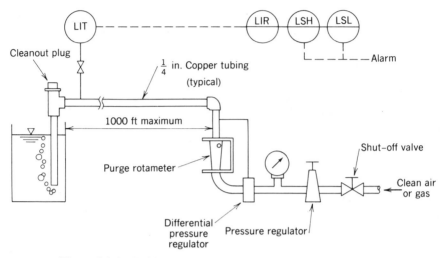

Figure 5.2-1 Bubbler tube system of tank level measurement.

generation and the detection of the reflected pulse energy is a function of the speed of sound in air. Since the speed of sound in air is dependent on temperature (0.18% per °C), the use of a temperature correction factor is necessary for accurate measurement.

Prior to selecting the level measuring device, the design engineer should be familiar with the following considerations: the required pressure range, required accuracy, equipment reliability, environmental conditions of the plant site, and operation and maintenance concerns.

BIBLIOGRAPHY

Andrew, W. G., Williams, H. B., *Applied Instrumentation in the Process Industries*, Vols. 1 and 2, Gulf Publishing, Houston, 1980.

Babcock, R. H., "Instrumentation and Control in Water Supply and Wastewater Disposal," *Water Wastewater Eng.*, xx:xx (1968).

Department of Civil Engineering, "Instrumentation, Control, and Automation for Water Supply and Wastewater Treatment Systems," in *Proceedings of the 9th Sanitary Engineering Conference*, University of Illinois, 1967.

Knudsen, D. I., ed., *Process Instrumentation and Control System*, WPCF Manual of Practice No. OM-6, 1982.

5.3 VALVE SELECTION

Purpose The flow of liquid or gas in pipes is regulated by valves, whereas gates are used to regulate the flow of water from reservoirs, tanks, or channels. Valves are used much more frequently because they are capable of performing a multitude of important functions. Their primary functions include shut-off, throttle, prevention of backflow, or a combination of these functions.

In order to be able to select an appropriate type of valve, all design engineers must have a basic knowledge of the various types of commonly used valves, the characteristics of the fluid that is to be regulated, and the hydraulic characteristics of the piping system in which the valve is to be installed. Valves are manufactured out of a wide variety of materials to meet a wide range of requirements and the engineer is responsible for specifying the appropriate valve type, as well as valve materials.

Engineers should also be aware that all large size (over 12 in.) valves, including gates, require adjustment after installation; without the necessary adjustments most valves will leak to some degree. Design engineers should therefore include the final valve adjustments and the associated costs in the valve specifications.

Considerations Items that must be considered during valve selection are the type of fluid or gas that is to be regulated, the temperature of the system, the flow range, the pressure of the system, valve function, valve location, the type of valve operator, and the reliability and cost of the valve. Each of these items is discussed.

Type of Fluid or Gas An important factor in selecting the appropriate type of valve is the characteristics of the liquid or gas. If the liquid or gas is corrosive in nature, then noncorrosive materials must be specified for all internal parts of the valve; Type 18-8 stainless steel is typically selected. Valves used in ozone gas lines must be composed of Type 316 stainless steel and Teflon seats, not rubber, must be used. If a chemical slurry is used in the treatment process, the valve must not have internal recesses, which may become filled with slurry solids. Lastly, if abrasive matter is present in the liquid, then the fluid passages must be composed of materials that are resistant to this type of erosion.

Temperature Although temperature is not a major issue when selecting the major valves of a water treatment process, it becomes a very important consideration when valves are used in conjunction with auxiliary equipment such as heating boilers and certain types of chemical feed system—that handle exothermic chemicals such as caustic soda and sulfuric acid. In these types of situation, valves ordinarily used in the water treatment process should not be employed because they do not perform well at operating temperatures above 150°F (65°C) due to thermal distortion, unless special metal parts are specified.

Flow Range Flow range is an important issue when selecting throttling valves. Control of the flow range is extremely important because most throttling valves have a limited range in which they yield satisfactory performance.

Unlike the throttling valve, flow range is not an important consideration for the multijet valves. These valves are designed and manufactured to service a wide range of pressures and flow control without cavitation.

For the purpose of simple shut-off, the flow range is usually not a design consideration. However, if the water velocity exceeds 35 ft/s (10.5 m/s), based on the valve port area, most valves are unsuitable for such service and the engineer must therefore specify special instructions for valve construction.

Pressure The maximum differential pressure across the valve and the normal and the extreme line pressures should be known prior to valve selection. Extremely high pressures often result from hydraulic surges in the piping system and selection of the valve operator should therefore be based on the maximum line pressure.

Valve Function Each type of valve has an optimal function. The valve function may be isolation of a line, drainage of a tank, prevention of backflow, reduction in pressure, or flow modulation. Thus, the design engineer should specify the most appropriate type of valve for each application.

Valve Location Valves may be located in a variety of locations: in a valve vault, a pipe gallery, in the wall at the entrance of a tank, at the exit of a pipeline, buried in the ground, or submerged in the water at all times. There are also differences in valve construction for buried and regular types of valve. Furthermore, valves installed in the tank wall require only one flange rather than a double flange. Engineers must select the proper type of valve and specify the conditions for valve installation.

Valve Operator The design engineer must specify whether a valve is to be operated manually or by power. If a manually operated valve is selected, the type of operator (i.e., a wheel or a square nut with key) and the orientation of both the operator and stem supports must be specified. Certain types, such as the flow modulating, automatic sequencing, and remote operated valve, necessitate the use of powered operators because manual power is insufficient.

Powered operators may be energized by means of electricity, compressed air, water, or oil. However, the first two options are the most commonly used in waterworks.

Strictly from the standpoint of reliability and maintenance, the use of air actuators is preferred during a power failure or during repair work because of its simplicity. However, the electrical motor offers better control over the modulating speed and is easier to incorporate into the instrumentation and control system.

Reliability and Cost The degree of valve reliability may play an important role in the control system, yet, may be independent of the application. If a high degree of reliability is essential to the treatment process, the cost of the valve is unimportant. However, the relative costs of the various sizes and types of valve should generally be compared for each application. A good cost comparison study will include the cost of the valve itself, as well as the projected maintenance costs and the cost of replacing equipment when necessary.

Types of Valve   There are five basic types of valve: (1) slide, (2) rotary, (3) globe, (4) swing, and (5) multijet.

The slide valve has a sliding disk that travels perpendicular to the flow direction. An example would be a gate valve.

Rotary valves operate by having a plug or disk move in a rotary fashion. The butterfly, ball, plug, and cone valves are classified as rotary type valves.

Globe valves are extensively used in ordinary home plumbing fixtures. They have disks or plugs that move parallel to the flow direction. The globe valves do not cavitate under most household usage conditions.

The swing valve is comprised of a swing check valve, which prevents reverse flow. The swing check valve is fundamentally a combination of a rotary and a globe valve.

The multijet valve, also known as a sleeve valve, is composed of two concentric pipes. Both the inner and outer pipes are covered with a multitude of small orifices. The inner pipe slides back and forth like a telescoping pipe, thus aligning or staggering the holes. An alternative type has perforations only in the outer sleeve. A third type of multijet valve has multiple orifices on the slide gate disk. The multijet valves became available in the early 1970s. They are used exclusively to reduce high pressure and to control flow rate without causing cavitation (under most conditions).

Table 5.3-1 summarizes the valve characteristics and the main valve applications commonly used in water treatment.

Flow Control Valves   The design engineer must carefully select the most appropriate type of valve for each application; the use of an improper type of valve may result in cavitation, corrosion, or erosion, and thereby degrade the structural integrity of the valve itself. Valve selection may be conducted in two steps: selection of the proper type of valve, followed by sizing.

Valve Selection   In order to select the most appropriate type of valve for a given application, the design engineer must evaluate the pressure drop characteristics and the working range of the valves. Figure 5.3-1 presents the examples of valve rangeability. Figure 5.3-2 shows the flow characteristics of certain types of valve when throttled.

Engineers should not confuse rangeability with turn-down. Rangeability is the ratio between the maximum controllable flow and the minimum controllable flow rates. The turn-down is a ratio of the normal maximum flow rate versus the minimum controllable flow rate.

TABLE 5.3-1 Characteristics and Applications of Common Valves

Type	Standard Size	Application	Rated Pressure (psi)	Remarks
Gate valves ANSI/AWWA C 509-87 for gate valves AWWA C 501-87 for sluice gate	2–48 in.	• On/off control but does not have a driptight shut-off • Not suitable for flow rate control by throttling	150	• This class includes wedge, double disk gate valves, and sluice gates • Debris deposits in the seats will cause valve closure problems • Full size openings allow pipe cleaning pigs to pass through • The valves often become inoperable when left in an open or closed position for extended periods of time • Double disk type should only be installed vertically • Slow opening and closing speeds minimize water hammer • Requires a large space for installation
Butterfly valves ANSI/AWWA C 504-87 "A" valves have a rated maximum line flow velocity of 8 fps; "B" valves, up to 16 fps	4–90 in.	• On/off control with driptight shut-off • Limited degree of flow control by throttling	75 and 150	• Easy to operate and at a low cost • Minimal maintenance requirements • Requires a small space for installation • Quick closing causes water hammer problem • Applicable to both air and liquid • Water should not contain any large debris • Does not allow the pipe cleaning pig to pass through
Ball valves ANSI/AWWA C 507-85	½–48 in.	• Driptight shut-off • Limited degree of flow control by throttling	Up to 300 available	• Full size openings allow pipe cleaning pigs to pass through • Very little head loss • The size of the valves is often necked down to $\frac{7}{10}$ of the size of the connecting line

Valve type	Size	Control		Features
Cone valves	6–24 in.	• On/off control with non-driptight closure • Limited degree of flow control by throttling	Up to 300 available	• Services clean water as well as slurry • Large size valves are heavy • Full size openings allow the pipe cleaning pigs to pass through • Operation of the valve begins to lift the plug off the seat before rotation of the cone • The valve operator is rather complicated and expensive • Should not be installed in an upside-down position • The port consists of a tapered plug or a full-area circular port.
Plug valves	$\frac{1}{2}$–36 in.	• On/off control with drip-tight closure • Often used as a flow control valve • Mild throttling is acceptable	120–170, depending on the size	• Plug valves are suitable for service of dirty water and slurry as well as clean water • Lubricated and eccentric plug valves are available • Lubricated type comes with cylindrical or tapered plugs with a diamond shaped port for throttling application • Lubricated types are easily operated, provided that they are frequently used; however, the plug type tends to "freeze" if is not operated for an extended period of time • Lubricated types require periodic lubrication with grease • Nonlubricated types are easier to operate and require less maintenance • Eccentric plug valves have rectangular ports whose area equals 80% of the pipe area • Eccentric valves may be installed in any position; however, the shaft must be kept horizontal and the plug must remain in the upper part of the valve body when the valve is opened

TABLE 5.3-1 (Continued)

Type	Standard Size	Rated Pressure (psi)	Application	Remarks
Globe valves	1–16 in.	150 but up to 300 available	• On/off control with a high pressure drop and drip-tight closure • Accurate throttling and flow control	• Disk globe, angle valve, Y-valve, needle valve, and altitude valves are in this class • Application to clean water only • The disk design is used for manual control and the needle valves are used for accurate and automatic control • Most all household valves are disk globe valves
Check valves ANSI/AWWA 506-78 and ANSI/AWWA 508-82	2–60 in.	150	• Prevents reversal of flow	• Lift, swing, and ball checks are the basic types of check valve • Internal spring-loaded types are called "silent" checks • Lift checks are often used with globe valves for both liquid and gas. Angle lift check valves are used for vertical line; horizontal lift types are used for horizontal lines • Swing checks are often used with gate valves and may be installed in either a horizontal or vertical position • Ball checks are used for air vent, vacuum valves, and chemical metering pumps—on both the suction and discharge sides • Headloss across the check valves is significant

Diaphragm valves	$\frac{1}{4}$–24 in.	Driptight shut-off for corrosive chemicals, slurry, and sewage	150 but up to 300 available	• The valves are used in lime slurry and PAC slurry lines • Intermittent operation may cause leakage • Poor flow control characteristics • Has a low cost but a high capacity
Multijet valves	8–72 in.	Flow control with a high pressure drop across the valve without cavitation	150	• Both in-line and submerged discharge types are available • Has a history of good performance; emits low noise and has minimal maintenance requirements • Limited number of valve manufacturers • These valves are expensive

Notes: (1) ANSI/AWWA C 540-87 standard for power-activated devices (operators) for valves and sluice gates.
(2) ANSI/AWWA C 550-81 standard for protective interior coating for valves and hydrants.

447

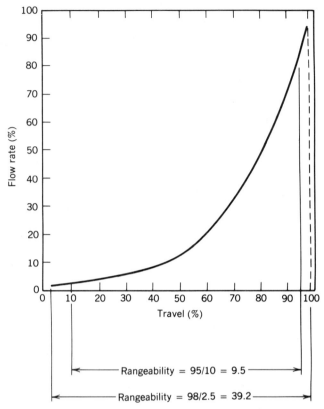

Figure 5.3-1 Rangeabilities for common valves. (From W. G. Andrew, *Applied Instrumentation in the Process Industries*, Gulf Publishing, Houston, 1980.)

If a valve is to be used in a situation requiring water pressure control, valves characterized by equal percentage flow, such as the ball and butterfly valves, should be selected for ordinary cases—where there is a normal pressure drop of at least 15% but less than 30%. If a higher pressure drop such as 50% is expected, a valve with linear characteristics (plug or multijet/sleeve valve) should be specified. In situations requiring a single valve to handle a pressure drop of over 50%, the sleeve valves perform effectively in most applications without cavitation.

For applications requiring the control of liquid levels, a valve with linear characteristics, such as the plug valve, is the most appropriate type in most cases.

Equal percentage valves are most appropriate under the following conditions: (1) a fast acting process; (2) in situations requiring high rangeability; (3) if the dynamics of the system are not well known; and (4) in the case of heat exchangers, where an increase in the production rate requires a correlative and greater increase in the volume of the heating or cooling medium.

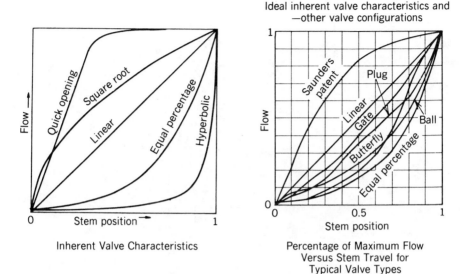

Ideal inherent valve characteristics and
—other valve configurations

Inherent Valve Characteristics

Percentage of Maximum Flow
Versus Stem Travel for
Typical Valve Types

Figure 5.3-2 Flow characteristics for common valves. (From W. G. Andrew, *Applied Instrumentation in the Process Industries*, Gulf Publishing, Houston, 1980.)

The linear valves should be applied (1) for slow acting processes and (2) when there is a system pressure drop of over 40% across the valve.

Quick opening valves should be used (1) for on/off control and (2) when maximum valve capacity must be obtained immediately. The quick opening valve exhibits linear characteristics for approximately one-quarter of its travel from shut-off. Beyond this point, the quick opening valve is only useful in an on/off function.

Valve Sizing The sizing of control valves involves the application of basic sizing formulas, the evaluation of the process fluid characteristics, and the use of basic data for the values. Some design engineers fail to take the size of the control valves into consideration and therefore end up with the wrong size valve. Nevertheless, system control by these valves is achieved to an acceptable degree for most of the time; this is possible because of the high rangeability of the control valves and the flexibility of the controller mode, and because the available pressure drops are often greater than the design points. However, optimum system control may only be achieved if the control valve is capable of handling both the minimum flow and a flow that is over 100% (120%) of the maximum flow rate.

Another factor that should be considered when sizing the control valves is the capacity factor (Cv). Cv is defined as the number of gallons of water (at room temperature, 60°F or 15.5°C) that will pass through a given flow restriction per minute at a pressure drop of 1 psi. The formula used to compute

TABLE 5.3-2 Capacity Comparison of Valve Body Types

Valve Type	$\frac{1}{2}$	$\frac{3}{4}$	1	1.5	2	3	4	6	8	10	12
					Cv Factor Comparison for Various Valve Types by Size (in.)						
Single-port globe	5.0	6.4	11.0	25.0	48.0	95	147	324	436	987	1180
Double-port globe			13.0	32.0	53.0	121	190	438	765	1320	1760
Split body		6.4	11.0	25.0	48.0	93	140	307			
Angle			16.5	38.2	68.7	152	293	492	766	1030	
Sauders	4.0	8.5	8.5	31.0	55.0	180	305	515	1130	1450	2350
Cage	6.5	14.0	21.4	38.0	67.2	150	235	460			
Butterfly (60°)			15.9	38.0	68.0	152	280	638	1163	1843	2679
Vee-ball				38.8	99.0	254	536	1040	1700	2690	3930

Cv is established by the Fluid Controls Institute, Inc. (FCI) and the formula for noncompressive fluids, such as water, is

$$Cv = Q(G/\Delta P)^{0.5} \quad \text{(volume base)}$$

where Cv = capacity factor in gpm,
Q = flow rate of the liquid in gpm,
G = specific gravity of the liquid (1.0 for water),
ΔP = differential pressure drop across the valve in psig.

$$Cv\phi = Cv/D^2$$

where Cvϕ = capacity factor through a 1 in. valve for a given valve angle with a pressure drop of 1 psig,
D = pipe diameter in inches.

Table 5.3-2 compares the capacity factor (gpm) of several types of valve. Most major valve manufacturers will provide the Cv values of their products.

The control valve is generally selected to open only 25–60% at the designed flow rate. The following general guides should be used in valve selection:

1. To select the appropriate size valve begin by computing the value of Cv at both maximum and minimum flow conditions.
2. Next, select a valve size that is capable of handling the maximum Cv, when the valve is 85–90% open, and the minimum Cv, when the valve is opened only 10–15%.

Cavitation Considerations   The formation and subsequent collapse of vapor-filled cavities resulting from the dynamic action in a liquid is defined as cavitation. The cavities may be bubbles, vapor-filled pockets, or a combination of the two. Dissolved gases are often liberated shortly before vaporization begins.

Cavitation causes noise and vibration and, if it is severe, both the disk and body of the valve will become pitted. Under these unfavorable conditions, cavitation may become so severe that the butterfly valve (flow modulator) must be replaced every few months.

The design engineer must make sure that the valve selected for flow control does not cavitate under critical conditions, that is, when the valve is at a high throttling position. The cavitation index (σ) is calculated from the following equation:

$$\sigma = \frac{H_d - H_v}{\Delta H + v^2/2g}$$

where σ = cavitation index,

$\quad\quad H_d$ = static head at the valve exit, in absolute feet,

$\quad\quad H_v$ = vapor pressure of the liquid,

$\quad\quad \Delta H$ = pressure drop across the valve, in feet,

$\quad\quad v$ = velocity through the pipe, which is directly connected to the valve, in ft/s.

Notes

1. The barometric pressures, expressed as feet of water column, at given ground elevations are as follows:

   33.9 at sea level

   32.8 at 1000 ft elevation

   31.6 at 2000 ft elevation

   30.5 at 3000 ft elevation

   29.4 at 4000 ft elevation

   28.3 at 5000 ft elevation

2. The absolute water vapor pressures, expressed as feet of water column, are as follows:

40°F or	4.4°C	0.28
50°F or	10°C	0.41
60°F or	15.6°C	0.59
70°F or	21.1°C	0.83
80°F or	26.7°C	1.18
90°F or	32.2°C	1.61
100°F or	32.8°C	2.21

The criteria for valve cavitation are as follows:

$\quad\quad \sigma > 2.5–3$ — In general, cavitation will not occur and no particular noise will develop.

$\quad 2.5 < \sigma > 1.5$ — There is a development of noise; "hammer blows" and a steady roar persists.

$\quad\quad \sigma < 1.5$ — Under these circumstances, the pipe and valve are apt to vibrate and cavitation develops. When σ reaches 0.5, cavitation will be fully developed.

Figure 5.3-3 illustrates a multijet (sleeve) valve. This type of valve is designed and used as a throttling valve in situations with a high pressure drop

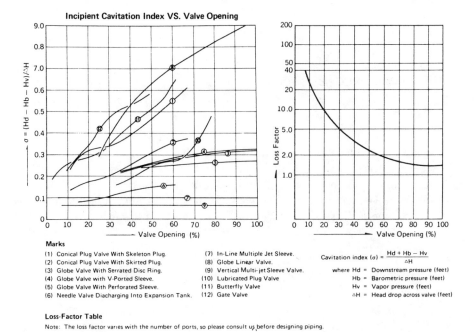

Marks

(1) Conical Plug Valve With Skeleton Plug.
(2) Conical Plug Valve With Skirted Plug.
(3) Globe Valve With Serrated Disc Ring.
(4) Globe Valve with V-Ported Sleeve.
(5) Globe Valve With Perforated Sleeve.
(6) Needle Valve Diacharging Into Expansion Tank.

(7) In-Line Multiple Jet Sleeve.
(8) Globe Linear Valve.
(9) Vertical Multi-jet Sleeve Valve.
(10) Lubricated Plug Valve
(11) Butterfly Valve
(12) Gate Valve

Cavitation index $(\sigma) = \dfrac{Hd + Hb - Hv}{\Delta H}$

where Hd = Downstream pressure (feet)
Hb = Barometric pressure (feet)
Hv = Vapor pressure (feet)
ΔH = Head drop across valve (feet)

Loss-Factor Table

Note: The loss factor varies with the number of ports, so please consult us before designing piping.

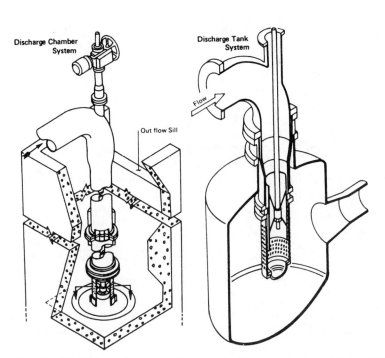

Figure 5.3-3 Multijet (sleeve) valve and cavitation index of values (Bailey Valve Brochure).

in the system. The multijet valve does not develop cavitation until the cavitation index approaches 0.2. This figure also compares the cavitation index of various types of multijet valve.

Example Design Calculation

Example 1

Given A 36 in. pipeline connects a water reservoir to a water treatment plant. The difference in static water level is 30 ft. A butterfly valve is selected to control the planned flow rate from 31.5 to 5 mgd. The pipeline elevation, with the Venturi flow tube and the flow control valve, is 15 ft below the water surface at the head of the plant. Assume that the headloss in the line is 14 ft at the maximum flow rate and 0.5 ft at the minimum flow rate. The plant elevation is 1000 ft above sea level and the ambient temperature is 80°F.

Determine The proper size flow control valve and evaluate the cavitation situation.

Solution

Step 1. Select a 36 in. butterfly valve, which is the same size as the pipeline. The capacity factor Cv at the maximum flow rate of 3.15 mgd is

$$Cv = Q \left(\frac{G}{\Delta P} \right)^{0.5}$$

$$= (31.5 \times 695) \left(\frac{1}{(45 - 14 - 15) \div 2.31} \right)^{0.5} = 8318$$

$$Cv\phi = Cv/D^2 = 8318 \div 36^2 = 6.4$$

Based on the graph in Figure 5.3-4, the disk angle of the valve is approximately 35°. The recommended disk angle of the butterfly valve, for the purpose of flow control, is generally 15–60° and 35° is therefore an appropriate valve.

The Cv at the minimum flow rate of 5 mgd is

$$Cv = (5 \times 695) \left(\frac{1}{(30 - 0.5) \div 2.31} \right)^{0.5} = 972$$

$$Cv\phi = Cv/D^2 = 972 \div 36 = 0.75$$

In this case the disk angle is approximately 9°, which is a very throttled position. This condition is unacceptable since it is less than the recommended angle of 15°.

Step 2. Since the rule of thumb is to size the flow control valve one size smaller than the size of the line, begin by selecting a 24 in. valve.

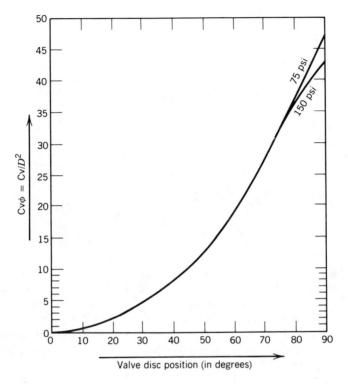

Figure 5.3-4 Unit capacity factor (Cv/D^2) versus approximate disk angle of butterfly valve.

The placement of a pipe size reducer and an increaser on the ends of the 24 in. valve will produce extra headloss. Although the head-loss from the reducer is negligible, the increaser will yield a head-loss of approximately 1 ft. The pressure drop across the valve is therefore $45 - 14 - 16 = 15$ ft or 6.5 psi.

$$Cv = (31.5 \times 695) \left(\frac{1}{6.5}\right)^{0.5} = 8587$$

$$Cv\phi = 1.49$$

The disk angle of the valve is 53°, which is within the accepted range.

The pressure drop across the valve, for the minimum flow rate, is practically the same as in the previous trial.

$$Cv\phi = 972 \div 24^2 = 1.7$$

The valve is open 16°. This valve is on the border line, but still within the proper range.

Now check the cavitation index (σ) when the 24 in. valve is in full throttle position. The hydraulic head at the valve exit is $15 - 1 = 14$ ft and the barometric pressure at an elevation of 1000 ft is 32.8 ft; thus,

$$H_d = 14 \text{ ft} + 32.8 \text{ ft} = 46.8 \text{ ft}$$

The water vapor pressure at 80°F is 1.18 ft; the flow velocity through the valve inlet is 2.46 pfs at 5 mgd; and the pressure drop across the valve is calculated to be $45 - 14 = 31$ ft.

$$\sigma = \frac{H_d - H_v}{\Delta H + v^2/2g} = \frac{46.8 - 1.18}{31 + [(2.46)^2 \div 64.4)]} = 1.47$$

This value is definitely in the cavitation range. The important decisionmaking factors are how often and how long this 5 mgd flow requirement occurs. If this low flow condition is infrequent and short in duration, then the 24 in. butterfly valve may be used with the knowledge that the cavitation condition exists and that the valve disk will cavitate every time. However, a plug valve or even a multijet valve should be considered if this low flow condition occurs frequently and if the duration of the 5 mgd flow is expected to last throughout most of the winter season.

BIBLIOGRAPHY

Andrew, W. G., Williams, H. B., *Applied Instrumentation in the Process Industries*, 2nd ed., Vols. 1 and 2, Gulf Publishing, Houston, 1980.

Benjes, H. H., et al., "How to Select Valves for Water and Wastewater," *Water Wastewater Eng.*, 5:62(May 1968).

Daneker, J. R., "Sizing Butterfly Valves," Parts 1 and 2, *Water Wastes Eng./Ind.*, O-10(July 1970), E-12(September 1970).

Edwards, J. A., "Valves, Pipes, and Fittings," *Pollution Eng. Mag.*, p. 22(December 1974).

Hutchinson, J. W., *ISA Handbook of Control Valve*, 2nd Edition, Instrument Society of America, North Carolina (1976).

5.4 PIPING SYSTEM

There are six basic piping systems in a water treatment plant: (1) raw water and finished water distribution mains; (2) plant yard piping that connects the unit processes; (3) plant utility, including the fire hydrant lines; (4) chemical lines; (5) sewer lines; and (6) miscellaneous pipings such as drainage and irrigation lines. The cost of the total plant piping system generally runs 5–7% of the total plant construction cost depending on the layout, site conditions, and winter temperatures. Careful attention must be paid to the design of the piping system because if the pipes in the main processes fail, the entire plant will be shut down for an indeterminate length of time.

The design considerations of the piping system are the function of the specifics of the system. However, all piping systems share a few common issues: the pipe strength must be able to resist internal pressure, handling, and earth and traffic loads; the pipe characteristics must enable the pipe to withstand corrosion and abrasion, and expansion and contraction of the pipeline (if the line is exposed to atmospheric conditions); engineers must select the appropriate pipe support, bedding, and backfill conditions; the design must account for the potential for pipe failure due to uneven subsidence in situations where the buried pipeline connects to a massive structure; and the composition of the pipe must not give rise to any adverse effects to the health of the consumers.

Raw Water Line Depending on the site specifics, the raw water line may be either a gravity flow or a pressure line. Factors that can adversely affect the functioning of the water line are the growth of clams and slime in the interior of the pipes; accumulation of sediment within the pipeline; air blockage at high points of the pipeline; presence of a vacuum condition within the pipeline since this can lead to the collapse of the pipeline; creation of a thrust force by the water flow, especially during flow surges; and the expansion and contraction of exposed pipeline resulting from changes in ambient temperatures.

Many case histories may be found to illustrate the problem concerning the growth of aquatic organisms in the raw water lines and how the carrying capacity of such pipelines can drastically be reduced. If there is a possibility that these conditions could arise, a temporary or permanent chlorine feed system may be incorporated into the piping system design, depending on the anticipated severity of the problem.

Since the majority of pipelines seldom have a constant pipeline grade from the intake to the plant, the air release and air vacuum valves should be installed at the high points of the pipeline and a blow-off assembly should be provided at the low points. If they must pass through areas with poor ground conditions, the pipelines must be extensively supported, for example, by the use of pile supports. Alternatively, the design may allow an adequate amount of flexibility to occur at the pipe joints.

Whenever a pipeline must cross a geological fault, the design engineer is faced with additional problems. The recommendations of geotechnical experts must be included in the design of the pipeline. Yet, there is one basic design rule: the design should allow the pipeline to undergo a sufficient degree of movement; therefore, flexible joints should be installed on either side of the fault and an articulated configuration (Z shaped pattern) should be used.

Figure 5.4-1 presents two types of flexible joint that can endure a certain degree of pipe displacement. The Dresser coupling provides for a deflection of approximately 6°, but the Flex-Tend assembly can tolerate a deflection of 15° and an expansion of 4 in.

Plant Yard Piping   The plant yard piping is a low-pressure system. The main design concerns are the flow velocity in the pipeline, allowable headloss, connections to the structures, physical strength of the pipes (in the case of deep installations), corrosion of the pipes, and cost of the system.

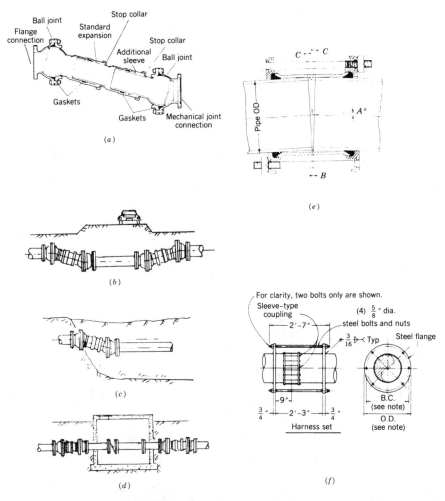

Figure 5.4-1   Flexible pipe joints and applications. (a) Flex-tend parts. (Courtesy of EBBA Iron Sales, Inc.) (b) Under roadway and earth fill. (c) Crossing unstable soil area. (d) Pipeline in and out of structure. (e) Dresser coupling. (f) Harness set. *Note*: B.C. (bolt circle) and O.D. (outside diameter) shall be sized to ensure 0.5 in. minimum clearance around coupling.

Flow velocity is an important consideration because of its effect on the fragile alum floc contained in the process water. If the flow is too fast, the alum floc will be destroyed; if the flow velocity is too slow, sludge will settle in the pipelines. The suggested flow velocity for each connecting pipeline (among treatment unit processes) is listed in Table 3.2-1 of Section 3.2.2, General Considerations.

Another important design issue is the pipe connections to the structures and tanks. Many pipe breakdowns occur within 6 ft of the structure wall due to differential settlement between the buried pipeline and structure. The solution is either to allow a rotating action to occur at the point where the pipe meets the wall through the use of a bell and spigot type of joint or to allow the pipeline to flex a couple of degrees of installing one or two pipe couplings in the pipeline adjacent to the wall. Nonetheless, sleeve type couplings have a limited degree of allowable deflection; if the angle of deflection is greater than 3–4°, water may begin to leak from the pipe. Figure 5.4-2 presents several examples of these designs. The Flex-Tend joint (Figure 5.4-1) is capable of tolerating three types of movement—rotation, deflection, and stretching or contracting—in one assembly, thus eliminating the need for a joint harness assembly.

The yard piping may have to be laid at a depth that is much deeper than the standard; the standard depth is dictated by the vehicle wheel loading. Additionally, the yard pipings are buried in twos or threes in a wide open trench. In both these cases, the thickness of the pipe is determined by the earth load and not by the internal water pressure; thus, the engineer must be careful when specifying the pipe thickness. Furthermore, the design engineer should specify the proper bedding conditions and details regarding backfill to avoid pipeline failure.

There are cases when large trench excavations are required, due to a maze of piping or because the pipes must be laid alongside large buried structures—to which they must connect. These types of yard piping conditions should be considered as wide trench or embankment conditions. Under these circumstances, the pipes may be protected from high external loading through the use of three basic methods: (1) reinforced concrete encasement, (2) reinforced concrete cradles, and (3) heavy walled pipe. Due to the generally congested nature of the yard pipings, the most common choice is the installation of heavy walled pipes that are specifically designed for embankment conditions.

Based on Marston's formula, the external loading, resulting from backfill on the pipe, under trench and embankment conditions are as follows:

$$\text{Trench condition:} \qquad W_d = C_d W\, B_d^2$$

where W_d = fill load in pounds per linear foot of pipe,
$\quad\ \ C_d$ = load coefficient (a function of H/B_d and K_μ); refer to Appendix 17,
$\quad\ \ W$ = unit weight of fill material in lb/ft^3,
$\quad\ \ B_d$ = width of the trench at the level of the top of the pipe in feet.

$$\text{Embankment conditions:} \quad W_c = C_c W B_c^2$$

where W_c = load per unit length of pipe,
      C_c = load coefficient (a function of H/B_c and $T_{sd}P$); refer to Appendix 18,
      W = unit weight of fill material,
      B_c = external diameter of the pipe in feet.

Pipe corrosion may occur internally and/or externally. The use of cement-lined cast iron and steel pipes provides adequate protection against internal corrosion since the process water of most water treatment plants is generally not highly corrosive. However, certain soil characteristics are very corrosive to the pipes. For this reason, the geotechnical study of the site must include soil resistivity measurements. If the soil is "hot," the engineer should obtain expert recommendations. In situations where corrosion poses a serious threat, the use of cathodic protection (as a backup to the protective coatings) should seriously be considered for steel and cast iron pipes.

Since the internal pipe pressure is relatively insignificant for the yard pipings, reinforced concrete pipes and steel cylinder concrete pipes should also be evaluated along with the cast iron or cement-lined and coated steel pipes. This recommendation considers the problem of corrosion, the need for flexibility at the pipe joints, the pressure imparted by the heavy earth, and the cost of the system.

Plant Utility and Water Distribution Lines These pipelines are high-pressure lines that must normally withstand pressure of 75 psi or more. The economic velocity of the distribution main is approximately 5 ft/s. However, a flow velocity of about 8 ft/s is quite common for utility lines carrying potable water, as well as utility water (nonpotable) and lines servicing fire hydrants.

There are four specific issues that greatly affect the design of the piping system: (1) extreme caution must be exercised when installing pipelines near existing sewer lines; (2) avoid cross connections with the chlorine injector actuation, line and other utility lines (not always potable water); (3) specify a joint harness when using a Dresser type pipe coupling so that the pipe will not slip out from the coupling; and (4) select pipe materials that will not degrade the quality of the potable water.

In most states, the Department of Health Services requires the water main and the sewer line to be physically separated. The basic rules set by the regulatory agencies are as follows. The horizontal separation between the water line and the drain or sewer line should be 10 ft. If the water main must cross a sewer line or a storm drain, the water line should be laid at an elevation which allows the bottom of the water main to be 18 in. above the top of the drain or sewer.

The regulatory agencies also dictate the type of cross connections which may be used between the potable and nonpotable water lines; this includes

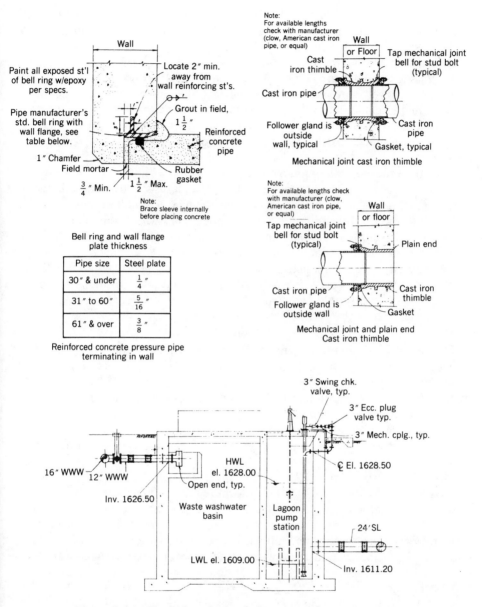

Figure 5.4-2 Typical methods of pipe connections to a structure.

steam condensate, the cooling water from the engine jacket or other heat exchanger, and the pipe connections to the chemical tanks or injectors. If a direct connection is absolutely necessary, an approved type of backflow prevention valve must be placed in the pipeline or an air gap of at least 6 in. must be provided in all connections to chemical storage tanks.

The installation of surge tanks is often required at the beginning of the distribution main whenever high service pumps are used to pump potable water into a lengthy distribution main. The surge tanks function to protect the pipeline and pumps.

Dresser couplings are commonly used in the high service pump discharge lines. Whenever sleeve type couplings, such as the Dresser coupling, are used in the high-pressure lines, the joint harness should be capable of resisting the unbalanced thrusts at the elbows or to tolerate the line pull caused by, among others, hydraulic thrusts. A mechanical type of coupling, such as the Victaulic coupling, does not require a harness to be installed; however, it does not allow the pipes to undergo much movement.

The recent EPA Drinking Water Quality Standards set stringent rules with respect to the allowable maximum concentration level of many items. Certain types of pipe can release undesirable components, such as pipe material or coating material, into the potable water. Of major concern are asbestos cement pipes and pipes coated with bituminous materials. Although lead pipes are primarily used in servicing residential homes, engineers should be aware that small lead pipes pose a threat to the health and safety of the public. All new pipe designs should avoid specifying the aforementioned types of pipe when establishing the potable water supply line.

Chemical Feed Lines   The main concerns surrounding the chemical feed lines are corrosion, abrasion, formation of scale, and heat resistance. When the plastic pipes are exposed to the sun, they can potentially be deteriorated by the ultraviolet rays and through weathering. Moreover, these pipes should have a proper type of pipe support in order to minimize pipe sagging between the supports, especially when plastic pipes are used.

The selection of an appropriate type of pipe material depends on the type and solution strength of each chemical that will flow through the pipe; refer to Table 4.1-1 in Section 4.1 and Appendix 13.

Any chemical that is fed in slurry form requires a minimum flow velocity of 5 ft/s and if the chemical feed line is exposed to the atmosphere, the pipes must be insulated from freezing and direct sunlight; certain chemical solutions tend to crystallize when the ambient temperature is well above 32°F (0°C). When employing scale-forming chemical solutions, the pipe design must have a minimum number of sharp turns and should provide clean-outs and a water flushing line.

If the acid, alkali, oxidizing, and organic compound chemical lines are installed in a pipe chase, double-walled chemical feed pipes must be used, as dictated by local OSHA regulations or other codes. This practice will minimize the potential for contamination of soil and fire or explosion in the event that a pipe ruptures. Similarly, from a safety standpoint, chlorine should be fed to the application point as chlorine gas (under negative pressure), rather than as a strong solution under pressure, unless the feed line is short and double-walled pipe is used.

Miscellaneous Pipings Included in the category of miscellaneous pipings are the sewer, drain, storm drain, and irrigation pipelines. If these pipelines are designed to enter any building, the uniform plumbing codes must be adhered to strictly. Should these pipelines be laid outside, the following materials may be used, depending on the conditions of the site: cast iron, salt-glazed vitrified clay, concrete, or PVC. With respect to sewer lines, the standard sewer line design should be followed when establishing details such as the size, flow velocity, minimum slopes, and access hole requirement.

Piping Materials The most commonly used materials for water supply are cast iron, steel, concrete, and PVC. In the past, asbestos cement pipes were extensively employed; however, recent health concerns over the detrimental effects of the asbestos fibers have caused the EPA to ban their use (June 1989).

Cast iron pipes have a certain resistance to corrosion. The standard sizes range from 2 to 24 in. in diameter. Most of the larger cast iron pipes are cement lined. The Hazen–William C-value may be as high as 145 for new pipes larger than 8 in. in diameter. The disadvantages of this type of pipe are its slightly brittle nature, its heaviness, and its rather high cost—particularly in the case of large pipes. Yet, ductile cast iron pipes may be manufactured as large as 78 in. (2 m) in diameter due to their ductile nature. However, the length of the large sized pipes must be limited to reduce the weight of the pipes and this is achieved by using a large number of pipe joints. Mechanical joints are generally used in these types of circumstances, rather than the bell and spigot, and the flange joints normally associated with the cast iron pipes (ANSI/AWWA C 110 to C 151-86 for cast iron and ductile iron pipes and fitting).

Steel pipes are commonly used when large and relatively small pipelines are required. In most cases, both the interior and exterior of these pipes are lined with cement for protection against corrosion.

Both longitudinal and spiral welded steel pipes are available for use. The joints between the pipes are usually constructed by welding or through the use of a bell and spigot with rubber gasket, sealed flanges, sleeve couplings, or mechanical couplings.

The advantages of the steel pipe include thinner walls, lighter weight, and, in general, lower cost when large diameter pipes are specified. Yet, this type of pipe also has several disadvantages: its inability to carry high external loads, its tendency to collapse when subjected to negative pressures, and the additional expense of providing corrosion protection measures such as cathodic protection (ANSI/AWWA C 200 to C 213-85, steel water pipe standards).

Concrete pipes are available in various forms: precast, reinforced concrete, prestressed with steel bars, and steel cylinder pipes. These pipes have low maintenance costs, are resistant to corrosion under normal conditions, and are capable of withstanding external loads. However, the concrete pipes also have some undesirable features, such as a short pipe length due to the weight, free lime being leached from the concrete, the tendency to crack under pressure and cause leaking, the permeability of the concrete, and corrosion in the presence of strong acids (ASTM C 14, C 76, C 361 and AWWA C 300 to C 303, specifications for concrete to modified concrete pipes).

Plastic pipes are primarily used as chemical solution, drain, sewer, and irrigation lines. The standard specifications for plastic and glass fiber reinforced pipes may be found in ANSI/AWWA C 900 to 950-81 and the thickness design may be found in the ANSI/AWWA standards piping handbook or the reference data and/or catalogs provided by the pipe manufacturer. The recommended pipe trench dimensions and details concerning pipe support, bedding, compact fill around the pipes, and backfill may also be found in the aforementioned publications and various other publications.

BIBLIOGRAPHY

American Concrete Pipe Association, *Concrete Pipe Design Manual*, Seventh printing, American Concrete Pipe Association, Vienna, Va., 1987.

AWWA, *PVC Pipe—Design and Installation*, M-23, AWWA, Denver, 1980.

AWWA, *Steel Pipe—A Guide for Design and Installation*, M-11, AWWA, Denver, 1985.

Bealey, M. and Lemons, J. D., eds., *Concrete Pipe and the Soil–Structure System*, American Society for Testing & Material, STP 630, Philadelphia, 1977.

IAPMO, *Uniform Plumbing Code*, 18th ed., International Association of Plumbing & Mechanical Officials, Los Angeles, Ca., 1988.

King, R. C., Crocker, S., *Piping Handbook*, 5th ed., McGraw-Hill, New York, 1973.

5.5 STANDBY POWER SUPPLY

All water treatment plants should be operated in a reliable and cost effective manner. Power outages will undoubtedly occur for a number of reasons, including damage to the power generation system or power supply system as the result of accidents such as earthquakes, fires, floods, tornadoes, hurricanes, lightening, or even sabotage. Certain equipment, those of the monitoring and control systems, requires a continuous power supply. This may be generated by a standby emergency power supply unit.

The standby unit also offers a monetary benefit for water treatment plants using high service pumping stations with large electric motors. Variations in water demand affect pumping requirements, which in turn affect the power demand. Since the rate structure or amount the utility charges for power is based in part on anticipated peak demand, a standby power unit may be used to reduce the base rate by peak shaving.

Power Outage   It is important to investigate the reliability of the local power supply system over a period of at least 10 years. Issues that should be studied prior to the design work are the frequency of the power outages and the duration of each episode. The type and scope of the standby power supply system are dependent on several factors: (1) the frequency of the power outages, (2) duration of the outages, (3) amount of essential equipment, (4) total power load required by the essential equipment, (5) available types of fuel for the power generator, (6) cost of the fuels, and (7) availability of alternative power supplies from other substations in the vicinity of the plant.

Water treatment plants that use raw water pumps will have a different size standby power supply system than plants receiving raw water from the source via gravity. The standby power system is installed in substantially greater numbers for the latter type of treatment plant. This is because the raw water keeps flowing into the plant regardless of whether the power is on or off. Thus, the treatment process units must be capable of processing the water based on standby power. Although the plant flow rate can be reduced and power consumption cut down, the plant flow may not be allowed to be reduced if the water source is shared by more than two or three users, even in the event of a power failure. Such a reduction would result in a significant increase in pressure or a massive overflow at the transmission line. In most cases, the allocated daily flow cannot be changed without giving advanced 24 h notice to the agency controlling the source.

For any given situation, the minimum capacity of the standby unit should be capable of providing a power supply which will operate 25% of the designed total capacity of the plant. Yet, specific local conditions, as well as the characteristics of the water supply system, can also dictate the minimum supply quantity.

The following is a list of equipment whose functions must essentially be maintained during a power outage:

- All equipment associated with the instrumentation system and the major units of the control system.
- The computer system: all data stored in the RAM of the system will be lost during the time between the loss of power and the start-up of the standby generator; thus, it is crucial to have an uninterrupted power supply.
- The emergency lighting of all site structures.
- Electrically operated valves.
- Major chemical feeders such as coagulant and disinfectant feeders.
- Chlorinator booster pumps, leak detectors, and emergency fans.
- The flash mixer.
- Twenty-five to 50% of the mechanical flocculators.
- Sampling pumps.

- Sump pumps.
- A designated number of raw water and treated water pumps; less than 50% of the units (for each type) must generally be functioning.
- Certain laboratory equipment: refrigerators, incubators, compressors, vacuum pumps, room lights, and other essential analyzers.
- The hot water circulating pumps of the heating system.

Depending on the duration of the power outage, other major process units, such as the filter backwash equipment, may or may not require the services of a standby power unit.

If a reliable power supply system can be obtained from other electrical substations in the vicinity, the treatment plant may opt to use a dual power supply system rather than a standby generator.

Peak Power Shaving   The economic evaluation of the various peak shaving alternatives is influenced by the rate structures of the utility companies. The four basic treatment plant operational modes used to achieve peak shaving are:

1. *Load Shading.* Significant load shading may be accomplished by reducing the load through the implementation of a computer or microprocessor-based controller.
2. *Reduction of the Peak Utility Demand.* This method involves the transfer of part of the power load from the utility to the engine generator supply during periods of high electricity demand (peak).
3. *Paralleling with Utility.* In this scenario, the power generator is connected directly to the normal distribution system. Energy required during peak power consumption is supplied directly to the system and the amount of furnished power is determined by the engine governer control.
4. *Prime Power Cogeneration.* A prime power system is a system in which the on-site generator is the prime source of power. Only certain dedicated loads may be connected to the on-site generator and the utility may be used as a standby source through the use of an automatic transfer switch.

Natural gas or diesel engine generators are generally used to supply the standby power necessary to maintain the essential equipment during power outages; under these circumstances, the second peak shaving method is normally considered.

The start-up and shut-down functions of the engine generator should be fully automatic. However, a time delay of approximately 60 s should be provided, before the generator reaches the proper speed, in order to differentiate the extremely short-term power outages, resulting from lightening and other causes, from the long-term power failures.

CHAPTER 6

Elements of Detailed Designs

6.1 PLANT HYDRAULICS

One of the major design efforts of a water treatment plant is the in-plant hydraulics. Since each water treatment unit process is functionally dependent on hydraulic factors, a hydraulic analysis must be conducted during the design phase because it plays a crucial role in assuring overall plant function and the efficiency of each treatment process.

Considerations

Obtaining an even flow distribution to parallel process units is an extremely important hydraulic consideration. It is rather amazing to find that a number of existing treatment plants have problems with uneven flow distribution: among sedimentation basins, between the treatment process modules, and even among the filters. This has resulted in the presence of significant variations in the flow distribution. Differences as high as 47% have been observed among the process units. A sample computation for sizing the inlet channel may be found in section 3.2.3, Example Design Calculations, Example 1.

Another common hydraulic problem is the occurrence of severe flow short-circuiting in the tanks of process units such as flocculation and sedimentation, as well as the clearwell and the disinfection contact tank. Flow short-circuiting in sedimentation tanks has been studied extensively. These studies show that only 20–30% of the mean detention time (tank volume divided by the flow rate) may be considered to be the actual flow through time. Figure 3.2.5-3 presents some examples of flow short-circuiting resulting from density flow and improper design of the inlet diffuser wall.

Many college engineering courses currently teach that selection of the pipings connecting two separate tanks should be based on the most economical size that is capable of carrying a particular flow rate from point A to point B using computer analysis. However, practical and operational water treatment experience indicates that the proper pipe size is determined by two factors: the required flow velocity, as dictated by the process, and the extreme operational conditions of the plant. For instance, all pipes or channels designed to transport floc to the sedimentation tank must be sized to neither break the fragile floc nor allow floc settling. If the pipe is used to carry sludge, a minimum flow velocity of 2 ft/s (0.6 m/s) must be maintained to prevent the sludge from settling.

Another hydraulic consideration is the maximum and minimum plant flow rate. Should one side of a module be taken off line, the remaining side is required to carry nearly twice the designed flow rate and an extremely high flow will occur. Thus, all pipes and channels should generally be sized at 150% of the designed flow rate (maximum daily flow rate). If this scheme is followed, the output of the treatment plant may easily be increased by 50% should the regulatory agencies decide to revise the current conservatively set filtration rates and allow an increase of 50% (i.e., 4–6 gpm/ft^2) at some future date. The raw water main and the main distribution lines are generally sized to handle future plant flow rates, in anticipation of plant capacity expansion after construction of the original plant.

The required available headloss across an ordinary conventional water treatment plant averages approximately 16.5 ft (5 m). However, an advanced treatment plant with preozonation, postozonation, and GAC adsorption processes requires an available head of approximately 25 ft (7.5 m): available head is defined as the difference between the water surface elevation in the tank at the head of the plant, where the plant influent flows in, and the high water level in the clearwell at the designed plant flow rate.

The flow hydraulics through granular media beds, including filter beds, have been studied extensively and may be found in various textbooks and technical journals. However, it is important to note that most inexperienced design engineers tend to spend many hours calculating the headloss through a filter bed at various filtration rates and water temperatures, determining the results up to two decimal points. Yet, this whole process is quite unnecessary; the engineer actually only needs to calculate a conservative approximate figure, which may be assumed from similar existing filter beds or pilot study data. Furthermore, design engineers should realize that the headloss across the filter is a function of many factors that come into play after the medium is in place: changes in the medium gradation after numerous backwashings, changes in medium size due to attrition loss and the loss of the fine grains, formation of mud balls within the bed, and changes in the depth of the filter bed resulting from medium loss.

Generally, the filtration cycle requires a filter design with over 8 ft (2.4 m) of available headloss across the bed. Consequently, a difference of ±3 in. (75 mm) in the initial headloss does not have a significant impact on the filtration process from an operational standpoint. The same analogy may be applied to computations on backwash hydraulics and filter bed expansion.

The calculation of headloss through the filter underdrain orifices is another example where practical experience saves the expenditure of a lot of unnecessary time and effort. The engineer should not spend time evaluating the orifice coefficient when performing hydraulic calculations because most filter underdrains are covered with a gravel bed. Consequently, over 30% of the orifices do not have a clean orifice area. Thus, the most conservative coefficient (0.65) should be used for the calculations.

The hydraulics of water troughs, gullets, channels, conduits, orifices, weirs, pipes, and fitting are not discussed in this book because they may be found in a number of other textbooks.

Example Design Calculation

Example 1

Given A 10 mgd (0.44 m³/s) water treatment plant has two flocculation/ sedimentation tanks in parallel and four filters. A 20 in. Venturi type flow meter and a 20 in. butterfly valve are located just upstream of the head of the plant.

Each flocculation tank has two 14 in. butterfly valves (as the inlet) and each of the two baffle walls (one per tank) produces a 0.1 ft headloss across the wall.

Each sedimentation tank is equipped with two launders that are 35 ft long with a total of 550 V-notches (90°) to decant the settled water.

Each filter has an inlet weir that is 4 ft in width and a 14 in. inlet butterfly valve as shown in Figure 6.1-1. The required water elevation in the filters is elevation 100.

Determine The hydraulic grade line between the filter and the inlet to the 20 in. Venturi flow meter at a flow rate of 10 mgd.

Solution

Step 1. Determine the water level control points in the process train. In this case, the filter inlet weir and the V-notched weirs in the sedimentation tanks are the level control points. Both weirs should be set at an elevation that will allow a free-fall condition (not submerged) to be maintained, resulting in a uniform flow rate at each weir.

Step 2. Compute the headloss or required head over the weirs between the level control points. Use a computation sheet (tabulation form as shown here) so that the values may easily be checked. Refer to Appendix 14 for the proper K-factors and the basic hydraulic formulas that are applicable to this problem.

(i) Between the Filters and the Settled Water Conduit.

Item	Q	Diameter	v	$v^2/2g$	K	Friction Loss		L.O.H.	Water Elevation
						L	Δh		
Water level in the filter cell									100
14 in. BV	2.5 mgd	14 in.	3.62	0.2	0.35	—	—	0.07 ft	100.07
Inlet hole	2.5 mgd	14 in.	3.62	0.2	1.6	—	—	0.32 ft	100.39
					$(C = 0.8)$				
Inlet weir*	2.5 mgd = 3.97 cfs = 3.33 × 4 $H^{1.5}$, where H = 0.44 ft							0.44 ft	101.44
Filter inlet channel								Negligible	101.44
20 in. pipe exit	10 mgd	20 in.	7.09	0.78	1.0	—	—	0.78 ft	102.22
Pipeline	10 mgd	20 in.	—	—	—	200 ft	0.9 ft/100 ft $(C = 120)$	1.8 ft	104.02
Pipe inlet	10 mgd	20 in.	7.09	0.78	0.5	—	—	0.39 ft	104.41
Water level in the settled water conduit								Negligible	104.41

*Note: Set the crest of the weir at 101.00 to ensure a free discharge condition.

(ii) *Between the Settled Water Conduit and the Head of the Plant.* The proper size of each launder in the sedimentation tank can quickly be determined by using Figure 3.2.6-15. Since each launder carries 2.5 mgd (1740 gpm), the width should be 1.5 ft and the height should be specified as 1.67 ft. Set the bottom elevation of all launders at an elevation of 104.75. Each V-notch on the launders should have a free discharge flow. Thus, the elevation of the V-notch should be set to give 106.50 water elevation in the tank. Since there is a total of 1100 V-notched weirs, the flow rate of each 90° V-notched weir is

$$(10 \times 695) \div 1100 = 6.32 \text{ gpm} \quad \text{or} \quad 0.0141 \text{ cfs}$$

Since $0.0141 \text{ cfs} = 2.54H^{1.5}$, $H = 0.125$ ft or 1.5 in.

Item	Q	Diameter	v	$v^2/2g$	K	L.O.H.	Water Elevation
Water level in the settled water conduit							104.41
Water level over the V-notched weirs* on the launders						0.125 ft	106.50
Sedimentation tank						Negligible	106.50
Diffuser wall (sedimentation tank inlet)						0.1 ft (given)	106.60
Baffle wall (flocculation tank)						0.1 ft (given)	106.70
Inlet BV	2.5 mgd	14 in.	3.62	0.2	0.3	0.06 ft	106.76
Inlet hole	2.5 mgd	14 in.	3.62	0.2	1.6	0.32 ft	107.08
Inlet channel						Negligible	107.08

Note: The crest of the V-notch is set at an elevation of 106.375.

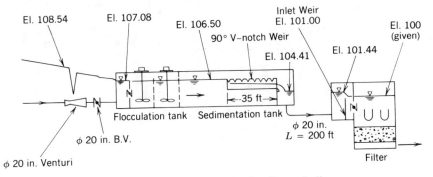

Figure 6.1-1 Process train and hydraulic grade line.

(iii) *Inlet of the Venturi Meter to the Inlet Channel to the Flocculation Tanks.*

Item	Q	Diameter	v	$v^2/2q$	K	Friction Loss		L.O.H.	Water Elevation
						L	Δh		
Water level in the inlet cell									107.08
20 in. Pipe exit	10 mgd	20 in.	7.1 fps	0.78 ft	1.0	—	—	0.78 ft	107.86
20 in. BV (full open)	10 mgd	20 in.	7.1 fps	0.78 ft	0.25	—	—	0.20 ft	108.06
20″ Venturi†	10 mgd	20 in.	—	—	—	—	—	0.34 ft	108.40
Pipeline	10 mgd	20 in.	—	—	—	15 ft	0.9 ft/100 ft ($C = 120$)	0.14	108.54

†*Note:* The Venturi differential is 80 in. (6.67 ft) at 10 mgd and the headloss across the Venturi is approximately 5% of the differential.

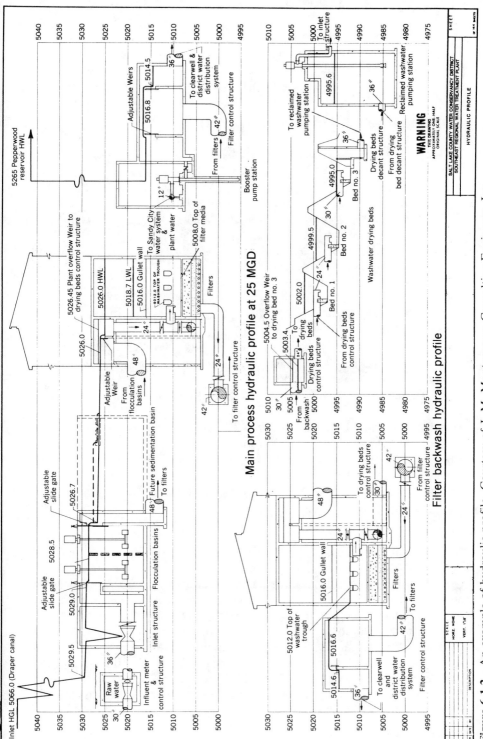

Figure 6.1-2 An example of hydraulic profile. Courtesy of J. M. Montgomery Consulting Engineers, Inc.

473

Refer to the earlier chapters for examples of hydraulic calculations on baffled walls in basins, density flow and flow stability (Froude number) of sedimentation tanks, and the basic hydraulics of filter structures. An example of the hydraulic grade line across a plant is illustrated in Figure 6.1-2.

BIBLIOGRAPHY

Benefield, L. D., et al., *Treatment Plant Hydraulics for Environmental Engineers*, Prentice-Hall, Englewood Cliffs, NJ, 1984.

Chao, J. L., Trussell, R. R., "Hydraulic Design of Flow Distribution Channels," *ASCE J. Environ. Eng. Div.*, 106(2):321(April 1980).

Davis, C. V., *Handbook of Applied Hydraulics*, 3rd ed., McGraw-Hill, New York, 1970.

Hudson, J. E., *Water Clarification Processes—Practical Design and Evaluations*, Van Nostrand Reinhold, New York, 1981.

Huisman, L., *Sedimentation and Floatation*, Delft University of Technology, Delft, The Netherlands, 1973.

Hydraulic Institute, *Engineering Data Book*, Hydraulic Institute, Cleveland, Ohio, 1979.

Karassick, I. J., et al., *Pump Handbook*, McGraw-Hill, New York, 1976.

King, H. W., Krutzsch, W. C., Fraser, W. H. & Messina, J. P., *Handbook of Hydraulics*, McGraw-Hill, New York, 1976.

Miller, D. S., *Internal Flow Systems*, Vol. 5 in the BHRA Fluid Engineering Series, BHRA Fluid Engineering, London, 1978.

Montgomery, J. M., Consulting Engineers, *Water Treatment—Principles and Design*, Wiley, New York, 1985.

Shaw, G. V., Loomis, A. W., *Cameron Hydraulic Data*, Compressed Air Magazine Co., Ingersoll-Rand, Phillipsburg, NJ, 1965.

Yao, K. M., "Hydraulic Control for Flow Distribution," *ASCE Proc. Sanitary Engr. Div.*, SA 2:275(April 1972).

6.2 SPECIFICATIONS

Clear and precise specifications containing drawings are a mandatory part of any modern day project. Vague or erroneous specifications result in inferior quality products, in addition to a large number of change orders and delays in the project. Moreover, they are a potential source of lawsuits. General guidance on the various types of engineering specifications are available from several agencies, including the Construction Specification Institute, American Consulting Engineers Council, American Society of Civil Engineers, National Society of Professional Engineers, American Institute of Architects, and the General Services Administration.

Project engineers should write the specifications in a manner that is both clear and convenient for the contractors since the specifications are written for the contractors who must furnish all the materials and equipment, build the structures, perform the civil works, provide and assemble the components, and finish the electrical works. For instance, a system such as ozonation or chlorination contains many subsystems; thus, the system specifications should include all the components and subsystems under one heading. This frees the contractor from having to search through several other divisions and sections to find the specifications for each subsystem. If the specifications are not organized in this manner, the contractor will waste precious time when formulating a bidding price for the project and may even overlook some components and subsystems.

Lastly, project engineers must be aware that the specifications of certain projects must be tailored to meet the particular needs of the system. Although the master and standard specifications certainly save time, minor adjustments are always required in most treatment plant constructions.

Contents of Specifications   The specifications are generally comprised of two major parts: the administrative requirements and the detailed requirements of the work to be performed. Some major items that must be considered by the engineer are the size of the project, the duration of construction, and the nature of the project—whether it is a privately or publicly financed project.

The outline of the specifications should contain the following items:

- Bid advertisement (invitation to bid)
- Information to bidders
- Form of the proposal
- Form of the contract agreement
- Bond forms
- Contract revisions
- Standard specifications
- Master specifications
- Special provisions/conditions
- General provisions
- Technical specifications

Bid Advertisement   It is standard practice for public agencies to publicly advertise all public works projects. Sealed bids are received through advertising on various news media for a required period of time. After the bids are opened, publicly read aloud, tabulated, and evaluated, the lowest bidder is selected.

Information to Bidders   It is customary to issue a set of drawings and specifications to all prospective bidders for a reasonable charge.

Form of the Proposal   In most cases, the proposals are accompanied by a guaranty in the form of a certified check or surety bond. This practice ensures that the successful bidder will enter into a contract; the guaranty is returned when the bid is awarded. However, if the lowest bidder fails to execute the contract, the certified check is forfeited as liquidated damages and the obligations of the surety (under the bond) is enforced as compensation to the owner for the cost of awarding the contract to the next lowest bidder or for the additional cost of readvertising.

If the project is financed by private owners, it is not necessary to obtain competitive bids and negotiations are often conducted with selected contractors.

Types of Contract   For both public and privately financed works, the two basic types of contract are the unit price and lump sum. Negotiated and special contracts are also possible.

UNIT PRICE   The unit price contract should be used when it is impossible to define the exact limits of the various work items encompassed in the contract and/or bid documents. For payment purposes, the work is broken down into major segments with respect to the type of work and trade involved. Each segment is designated as a payment item (estimated quantities) and are listed in the proposal. Bidders are requested to submit a written bid for each unit. For example, a specific number of cubic yards of concrete will be bid at a certain unit price per cubic yard.

The total bid is the sum (in dollars) of all items listed in the proposal and is the basis by which all received bids are compared in order to determine the lowest bidder. Payment to the contractor is determined by the actual quantity measured for each item incorporated into the work at the contract unit price.

LUMP SUM CONTRACT   The lump sum contract is selected when it is possible to accurately delineate, through drawings, the limits of the work encompassed in the contract. With this type of contract, the bidder makes quantity take-offs and presents them as the basis for the bid. However, it is essential that the bidder be provided with drawings and specifications that are extremely detailed with respect to all features and requirements of the work. Payment is made in accordance with the lump sum bid to cover all work and services required by the drawings and specifications.

NEGOTIATED CONTRACT   Private work contracts often take the form of negotiated contracts. Several different payment methods are available: lump sum or unit price, fixed fee plus reimbursable costs (with a ceiling price), fixed fee plus reimbursable costs (no ceiling), reimbursable costs plus a percentage of the cost, and a construction–management contract. Furthermore, incentives such as a bonus for economic construction and/or completion ahead of schedule may be added. Conversely, the contractor is penalized for inefficiency and/or failure to meet the completion date.

SPECIAL CONTRACT   The use of a special contract may be applicable in specific circumstances such as the procurement and installation of highly specialized equipment. In this case, the owner simply invites proposals from a select group of prequalified contractors without advertising the project. However, contracting procedures are generally followed for contracts procured through public bidding.

Contract Revisions   There are four basic types of contract revision: addenda, stipulation, change order, and supplementary agreement.

ADDENDA   Revisions to contract documents, made before bids are received from contractors, are classified as addenda. The revisions are usually changes to the contract drawings and specifications resulting from errors, omissions, or the need for clarification. Bidders should be notified of any changes in the bid opening dates by means of an addendum.

All addenda should be delivered sufficiently in advance of the bid opening date to all persons who have picked up the bid documents so that they may adjust their proposals. In turn, the bidders must acknowledge receipt of all addenda; otherwise, their bids will not be accepted.

STIPULATION   Stipulations are issued to the bidder when the contract is executed. They contain modifications of the contract terms as proposed by the owner.

CHANGE ORDER   This is a written order to the contractor authorizing a change in work from the original drawings and specifications. It usually increases or decreases the quantity of work to be performed. Thus, a payment adjustment should be made accordingly. The change order must be approved by the owner and signed by the engineer.

Good design work will have a total adjusted payment of only 2–3% of the total contract price resulting from change orders. If the total cost of the changes exceeds a specified percentage (such as 20%) of the total contract price, a supplementary agreement, acceptable to both the owner and the contractor, should be executed before to proceeding with the affected work.

SUPPLEMENTARY AGREEMENT   This written agreement encompasses all modifying work that is beyond the scope and terms of the contract or work changes that are within the scope of the contract yet exceed a set percentage of the original contract price. The supplementary agreement must be signed by both the owner and the contractor.

Standard Specifications The standard specifications are often preestablished by public agencies in order to ensure uniformity in administrative procedures and in the quality of the facilities. These criteria list the specific requirements for all materials and work quality. The specifications are periodically updated or amended and published.

Although in some instances the standard specifications may be used without any changes, they generally require minor modifications and additions. The assembled modifications and additions are referred to as supplementary specifications, special provisions, or special conditions.

Master Specifications Many consulting engineering firms provide master specifications that are used for projects owned by either private or local agencies. These specifications are written in-house and pertain to general items.

When the master specifications are applied to a project, the engineer should delete those requirements that are irrelevant to the particular project. This practice helps to reduce the time required to write the contract specifications, as well as serving as a check list, thereby minimizing errors and omissions.

General Provisions The general provisions define the rights and responsibilities of both the owner and the contractor to the construction contract, surety, and requirements, as well as delineating the legal relationships and the authority of the engineer.

The general provisions are usually comprised of the following sections:

- Definitions and abbreviations
- Bidding requirements
- Contract and subcontract procedures
- Scope of the work
- Control of the work
- Legal and public relations
- Damage claims
- Worker's compensation insurance
- Comprehensive general liability of the contractor
- Laws, ordinances, and regulations
- Responsibility for work
- Explosives
- Sanitary provisions
- Public safety and convenience
- Property damage
- Public utilities
- Prosecution and progress

- Time of completion
- Suspension of work
- Unavoidable delays
- Annulment and default of contract
- Liquidated damages
- Time extensions
- Measurement and payment
- Payment
- Termination of contractor responsibility
- Guaranty against defective work

These sections are often referred to as "boiler plates."

Technical Specifications   The common types of technical specifications pertain to materials and work quality, material procurement, and performance. Each of these is discussed.

MATERIALS AND WORK QUALITY SPECIFICATIONS   The material and work quality specifications are commonly included in most construction projects. They cover general and specific conditions related to the performance of the work, material requirements, construction details, quantity measurements, and the criteria used for determining payment.

MATERIAL PROCUREMENT SPECIFICATIONS   These are usually written for special situations such as accelerated construction schedules, for projects requiring the use of materials that have a long lead time, and for projects involving several general construction contracts. This type of technical specification ensures that the materials are available to the job site on time. Furthermore, it ensures uniformity in the materials.

The technical specifications should clearly delineate all elements concerning materials, work quality, and the mode and time of delivery. Examples are filter media (i.e., anthracite and filter sand), filter underdrains, special valves, and certain precast concrete items.

PERFORMANCE SPECIFICATIONS   The performance specifications are divided into two categories. One type specifies the water treatment unit process or an entire water treatment plant. These generally include a basic site plan, the maximum required hydraulic loading rate for each unit process, a guaranteed water quality limit (required) from each process for the product water, the time of completion, and other miscellaneous basic requirements. This type of performance specification is commonly employed in most European and Asian countries.

The second type of performance specification concerns the prepurchasing of equipment by the owner prior to signing a construction contract; this is often practiced in the United States. Certain types of equipment must be prepurchased due to their long lead time or because the owner is determined to obtain a particular type of equipment which may not be purchased through competitive bidding procedures. The performance specifications should therefore define the quality, function, and all requirements pertaining to the water treatment equipment, in addition to the required tests, affidavits, and other supporting evidence of compliance. It is vital that the test conditions and procedures be clearly defined in order to prevent conflict and/or future lawsuits.

Construction Specification Institute (CSI) The purpose of the CSI is to organize and improve construction specifications. It was established by architects, engineers, and certain manufacturers.

The CSI recommends that companies use the CSI format for all water treatment projects. Their format consists of 16 divisions, with several sections per division pertaining to specific construction activities. The CSI format is widely used in the construction of buildings and public works and for product specifications released by manufacturers.

The CSI also publishes a *Manual of Practice*, which contains a great deal of information on writing specifications. Instruction documents and project manuals such as *Specification Language*, *Specification Writing and Production*, *Performance Specifications*, *Methods of Specifying*, and *Bidding Requirements* are valuable reference items for all project engineers.

BIBLIOGRAPHY

Goldbloom, J., White, J. J., "Specifications," in Section 3 of Standard Handbook for Civil Engineers, 3rd ed., McGraw-Hill, New York, 1983.

White, J. J., "Specifications," in Section 3 of *Standard Handbook for Civil Engineers*, 3rd ed., McGraw-Hill, New York, 1968.

6.3 FINAL COST ESTIMATE

Cost estimation for a project is generally required in three stages: feasibility, preliminary design, and the design completion phase. Each of these phases would be expected to have a respective level of accuracy of ± 45, ± 30, and $\pm 7\%$.

Cost estimates conducted during the early stages of a project are often used to prepare funding for the project budget, through the use of cost estimation curves and by adjusting the cost to a geographical area, as described in Section 2.3, Preliminary Engineering Study. This section discusses the final cost estimate for bid evaluation.

The final cost estimate is not merely a simple summary of the anticipated costs of various items based on quantity take-offs, extensive cost data, and statistics. Many other factors are an integral part of the final cost estimate: the complexity of the project, site and climatic conditions, job site location, judgment of the cost estimator with respect to economic trends, dynamism of the construction industry, and extent to which the project attracts contractors. Professional cost estimators generally require 2–5 work-hours per sheet of construction drawings to complete an estimate.

If the project engineer is required to make the final construction cost estimate, he/she should be aware of three weaknesses characteristic of most engineers: (1) the tendency to devote too much time to the design aspects and not enough to cost estimation: (2) the tendency to focus on direct costs and to neglect the necessary indirect costs incurred by the contractor; and (3) the omission or insufficient allotment of funds to cover specific business costs—risks, escalation, and legal requirements—incurred by the contractor.

Development of Strategy and Data   The first step in preparing a construction cost estimate is to develop the strategy and data used in the estimating process. One method involves dividing this task into four basic components: site conditions, classification of the work and quantity take-offs, construction program, and unit cost development.

Site Conditions   The topographical, geological, cultural, climatic, and socioeconomic conditions of the site all directly affect project cost. Site conditions such as the topography, nature of the soil, physical characteristics of the soil, underground water table, construction costs, the source of the imported materials, and the available disposal sites (for the excess soil) directly relate to the total design and construction costs. Furthermore, items such as site access, working conditions, productivity of the local labor, and cost of materials and labor (in the vicinity of the site area) should not be overlooked.

If the project is in a foreign country, the cost estimator should review all applicable regulations with respect to taxes, import duties, port and freight charges, currency conversion rates, local inflation rate, and limitations on money transferred between countries. Other factors such as specific laws or regulations governing construction by foreign contractors and the employment of local labor must also be evaluated.

Classification of Work and Quantity Take-offs   Once the design plans and specifications are completed, the engineer should divide the project into a series of specific bid items. Each of these items should then be broken down into the most logical and practical individual categories of work so that payments may be proportioned as the work progresses.

Since each bid item generally includes a number of separate and distinct construction operations or classes of work, a separate cost estimate must be prepared for each item. The quantity take-offs are further divided to define the scope or size of each construction operation based on the site condition and the engineer's judgment. The quantities (numerical figures) are then tabulated using the appropriate units: feet, yards, cubic yards, or days. Be aware, however, that certain projects benefit from having the items expressed as a lump sum.

Construction Program　Having executed the classification of work and quantity take-offs, the cost estimator must now develop a realistic construction program. Items that must be tabulated are the required type and level of laborers and professionals, required type and number of construction equipment, and the frequency of use and duration of need (anticipated) for such equipment and the crews to operate them. The cost estimator should establish a preliminary but fairly accurate overall construction progress schedule, including sequential and concurrent construction operations. The schedule should provide a reasonable time for mobilization and preparation at the beginning of the project, in addition to demobilization and completion of the work.

Unit Cost Development　In anticipating the direct costs of a project, the cost estimator must develop the unit cost for (1) direct labor, (2) equipment operation, (3) consumable materials, and (4) nonconsumable materials.

The unit cost for direct labor is determined by establishing a schedule and dividing the project into various classifications such as crafts, skills, and labor. The availability, quality, and rates for the various trade classifications, in addition to the hourly rates of payment for regular work hours, must be designated for each classification. Various taxes and fringe benefits should be taken into account when determining the hourly rates.

Direct equipment operating costs include the cost of fuel, oil, tires, all equipment parts, and labor—operators, maintenance, and repair personnel.

Consumable materials are defined as items that are used during the construction operation; they are not part of the permanent installation. This category includes items such as small tools and accessories. Depending on the type of construction, the unit cost for consumable materials may be expressed as cost per hour of use or as a percentage of the direct labor cost.

Nonconsumable goods are defined as materials and equipment that are installed on a permanent basis. A schedule of these items should be prepared early in the estimating work because a considerable amount of time will be spent on obtaining quotes from all suppliers and manufacturers, including transportation or freight charges.

Direct Costs　Direct costs encompass direct labor, the operation of construction equipment, consumable and nonconsumable materials, and subcontracted services. If the project contains several construction operations, these costs must be estimated separately for each operation and the costs should then be totaled to yield the cost for each bid item. The direct costs of each bid item are subsequently summarized.

TABLE 6.3-1 An Example of Detailed Cost Estimates for a Water Treatment Plant Construction Project

UTAH VALLEY WATER PURIFICATION PLANT
DETAILED CONSTRUCTION COST ESTIMATE SUMMARY

MARCH 2, 1977

PROJECTED BID DATE: MAY 18, 1977

Description	Estimate ($)	Adjusted Estimate a ($)
Earthwork, Grading, Paving	381,699	343,529
Yard Piping	1,225,249	1,102,724
Landscaping & Irrigation	68,779	68,779
Operations Building	844,730	824,938
Pumping Station	722,961	702,206
Meter Vault No. 1	297,612	286,484
Meter Vault No. 2	83,497	77,865
Valve Vault No. 1	61,927	57,665
Valve Vault No. 2	71,639	66,597
Valve Vault No. 3	219,241	203,525
Valve Vault No. 4	47,871	45,312
Filters	1,735,277	1,628,577
Filter Waste Washwater Reclamation Basin	328,034	321,342
Filter Waste Washwater Basin	169,746	157,571
Flocculation Basins	374,815	349,080
Chemical Storage & Activated Carbon Tanks	226,435	218,832
Treated Water Reservoir	465,491	426,906
Retaining Wall	76,328	68,695
L. P. Gas Tanks	14,330	12,897
Testing and Disinfecting	20,000	20,000
Fencing and Gates	25,675	23,107
Overflow Structure to Salt Lake Aqueduct	43,848	39,463
Sludge Supernatant Pumping Station	16,620	15,412
Move On and Off	150,000	150,000
Subtotal	7,671,824	7,211,506
Contractor's Overhead & Profit	1,534,703	1,442,301
Subtotal	9,206,527	8,653,807
Electrical, Instrumentation, Telemetry	1,341,189	1,341,189
Total	10,547,716	9,994,996

[a] Adjusted downward by 10% to account for lower Salt Lake City area costs of earthwork, paving, fencing, yard piping, concrete, reinforcement steel, miscellaneous metalwork, structural steel, waterstops and certain other construction materials and associated labor.

Indirect Costs   The principal indirect cost items are administrative and overhead labor; transportation and equipment overhead; general office expenses; staff expenses; expenses incurred in establishing an on-site project office, in drawing up the construction plan and shop drawings; expenses associated with the operation and maintenance of the facilities; payroll taxes and burden; equipment mobilization and demobilization; equipment ownership; insurance and bond; taxes; and licenses and fees.

Items such as depreciation, interest on the investment, major repair and storage costs, overhead, insurance, and taxes are categorized as equipment ownership costs. If the equipment is rented, rental charges will be substituted for the ownership costs.

Estimate Summary   The estimate summary should be comprised of the direct and indirect costs, make-up, and the cost escalation adjustment. The make-up is defined as the allowance granted to the contractor for profit and contingencies. The time schedule, working conditions, uncertain ground and climatic conditions, in addition to the economic situation, all affect the amount of the make-up; the amount is generally expressed as a percentage of the total cost and is approximately 20% for many water treatment construction projects. The percentage will be higher for small sized projects and vice versa.

The final cost estimate described above is a general approach that is applicable to almost any type of construction project, including the construction of water treatment plants. Table 6.3-1 is an example of a cost estimate summary for a water treatment plant construction project.

Whether it is regarded as an art, science, or equal parts of good data, good judgment, and good fortune, cost estimators are increasingly relying on computers to help prepare the final cost estimate; since computers are more accurate and efficient, they reduce project costs. There are approximately 170 different types of computerized cost estimating software currently available on the market. The popular packages are distributed by Contractors Management System (CMS) of San Diego, California; Management Computer Controls (MCC) of Memphis, Tennessee; Timberline Software Corporation and Bidtek, Inc., of Wilsonville, Oregon; and G-2, Inc., of Boise, Idaho.

BIBLIOGRAPHY

Dickson, R. D., in *Water Treatment Plant Design—Estimating Water Supply System Costs*, R. L. Sank, ed., Ann Arbor Science Publishers, Ann Arbor, MI, 1978.

Guttman, D. L., et al., *Computer Cost Models for Potable Water Treatment Plants*, EPA-600/2-78-181, EPA (September 1978).

Hopper, T. W., "Estimating Construction Costs," *J. Consulting Eng.*, p. 94 (August 1968).

Montgomery, J. M., Consulting Engineers, Inc., *Water Treatment Principle and Design*, Wiley, New York, 1985.

6.4 SUPPLEMENTAL STUDIES

The designing of a water treatment plant requires the combined talents of experts from various disciplines. In many cases, the entire plant design cannot be handled by one design firm since some areas are simply beyond the capability of the firm or because it is more cost effective to subcontract them out. These areas include the geotechnical/geological study, cold weather design, corrosion study, acoustic study, and surveying.

Geotechnical/Geological Study The geotechnical/geological study and recommendations may be obtained from reputable local geotechnical consultants for a reasonable fee. A local firm or a large firm with a local office is generally preferred simply because it is assumed to be the most familiar with the local conditions and history of events. These firms usually provide services, such as test boring of several or more strategic plant site locations and laboratory analysis of test samples, and furnish a report on their findings. Additionally, the firm should provide recommendations on the plant design, especially with respect to the structural design, the design of the foundation, and the construction aspects of the civil engineering work.

The subcontracting firms are expected to supply the following information: in situ moisture, density, a laboratory maximum density and optimum moisture curve, direct shear tests, moisture sensitivity test, consolidation tests, resistivity, pH, and an analysis of the soluble sulfate and chloride content of the soil.

The design recommendations should also address: (1) seismic design parameters; (2) structural design parameters such as the type and size of foundations, the various anticipated earth pressures for subterranean walls, and the design of slabs on various grades; (3) design data on earthwork, such as cut, fill, and compaction requirements, and the type of equipment required for excavation and compaction; (4) corrosivity of the soil, complemented by recommendations on the major underground pipeline design and the required type of cement for concrete structures set in the ground; and (5) a prediction on the groundwater level during plant construction and after construction is completed.

Cold Weather Design If the water treatment plant is located in a region with extremely cold winter weather conditions, the plant design requires many special considerations. For instance, all major unit processes should be either housed or buried in the ground (sedimentation tanks and the clearwell) to prevent freezing of the process water. Certain items in the building and construction design should be specifically outlined, especially the roofing system and the vapor barriers: these should be designed to reduce condensation; otherwise, the doors will freeze in place. Heat traces should be conducted for all outdoor equipment and chemical feed pipes; these items should also be insulated. Drain pipes, especially filter waste wash and sludge pipes, should be protected from freezing and be capable of draining tanks or lagoon that may be frozen solid. The exit end of these pipes must be located at the bottom of the tank, beneath the anticipated thickness of surface ice. Furthermore, the capacity of the tank should be large enough to accommodate a required volume of water or sludge below the ice caps.

Another important design consideration concerns the pipelines. All pipes transporting liquid must be buried below the local frost depth so that the piping system functions on a continuous basis and in order to avoid pipes ruptures—which result from the freezing of water.

The design and construction of buildings and treatment plants in cold weather regions are unique tasks requiring very special knowledge. Thus, projects located in arctic regions should employ project engineers who are licensed as cold weather design specialists (in the engineering field). The engineer, as well as the architect, mechanical engineer, and other project staff members, should be either registered as cold weather design specialists or hired/selected only if qualified to design and construct structures for cold weather regions.

Corrosion Studies   Corrosion is a common concern in the field of water treatment. Corrosion of pipes due to the carrying water is rarely a serious problem because all steel and cast iron pipes, with the exception of pipes less than 4 in. in diameter, are now cement lined. Moreover, in the United States, the pH of the filtered water is now commonly adjusted prior to distribution to the consumer, thereby significantly reducing internal corrosion in the main distribution lines. Nevertheless, certain sites, primarily marshy areas, can have very corrosive soils. Under these circumstances, the issue of corrosion must be given special consideration during the design of the yard pipings and concrete structures. There are also a few cases where iron and steel pipes (buried) severely corroded due to stray currents emanating from an electric railroad track or as the result of galvanic action along the pipelines. A corrosion expert should be involved in the project if pipe corrosion is anticipated to be a major problem.

Acoustic Studies   If a water treatment plant is located in a quiet residential area, the noise generated by certain water treatment equipment may be a significant issue. Noise emanating from air compressors, blowers, high service pumps, and engine-driven power generators are loud and disturbing to both plant operators and neighboring residents. OSHA recommends that the maximum noise level be maintained below 80 decibels (dB) for the health and safety of the operators. Yet, this criterion is generally much more stringent with respect to neighboring communities. The design engineer should consult an acoustic control specialist for recommendations on how to properly solve the noise problem in a cost effective manner.

Surveying Although most engineering firms are capable of surveying, an outside firm of surveying specialists should be used if it is the more cost effective alternative. Whenever the job site is remote or does not have easy access, or if it is difficult to conduct surveying using regular surveying techniques, an aerial surveying specialist should be chosen.

CHAPTER 7

Specific Water Treatment Processes

7.1 LIME–SODA ASH SOFTENING

7.1.1 General Discussion

The main purpose of water softening by means of lime–soda ash or lime softening is to reduce the levels of calcium and magnesium in the process water to a total hardness of approximately 80 mg/L (as $CaCO_3$) as well as reducing the magnesium hardness to approximately 40 mg/L (as $CaCO_3$). Over 1000 municipal water softening plants currently operate in the United States, with the majority being located in the Midwest and Florida.

Prior to the mid-20th century, the primary problem of the domestic consumer was the high consumption of soap by hard water. However, due to the widespread use of synthetic detergents, this is no longer a major issue today. Aside from the more immediate consumer concerns, the additional benefits of lime softening are quite substantial, including the removal of heavy metals, metallic elements, and organic compounds, as well as effectively killing bacteria, viruses, and algae. Lime softening also improves water quality with respect to pipe corrosion, boiler feed, and cooling waters. The spontaneous removal of elements such as iron, manganese, lead, mercury, and chromium is particularly attractive to purveyors of potable water, while the reduction of silica and total dissolved solids is beneficial to industries.

Water hardness is defined as the amount of divalent metallic cations in the water and is expressed in mg/L as $CaCO_3$. The major divalent metallic cations that contribute to water hardness are Ca, Mg, Sr, Fe, and Mn. However, in most cases, the principal contributors to hardness are calcium and magnesium.

A few major European countries are adopting their own scales of water hardness and this confusing situation is compounded by the fact that many industries in the United States still use the old expression of hardness: grains per gallon. Table 7.1.1-1 presents the relationship between the expressions of hardness.

Engineers should note that hard water is not currently known to adversely affect human health in any significant manner. Quite a few studies performed in the 1970s in the United States, Canada, United Kingdom, and Japan have demonstrated an inverse correlation between the incidence of cardiovascular

TABLE 7.1.1-1   **Relationships Among Hardness Units**

United States	mg/L as $CaCO_3$, which is 0.02 meq/L
	1 grain of $CaCO_3$/gal $= 17.1$ mg/L as $CaCO_3$
Germany	Equivalent to 10 mg/L of lime expressed as CaO
	1 German degree $= 17.86$ mg/L as $CaCO_3$
France	1 French degree $= 10$ mg/L as $CaCO_3$

The level of water hardness is generally classified as follows:

0–75 mg/L	Soft water
75–150 mg/L	Moderately hard water
150–300 mg/L	Hard water
Over 300 mg/L	Very hard water

disease and the hardness level of drinking water. Yet, conflicting studies have also been reported.

Lime softening, as well as other processes such as ion exchange, electrodialysis, reverse osmosis, distillation, and freezing may be employed to soften hard water. However, only the process of lime softening is discussed in this chapter. Yet, engineers should be aware that there are a few technical articles that demonstrate that use of magnesium carbonate, in conjunction with lime softening, improves the calcium carbonate sludge characteristics.

7.1.2   Basic Chemical Reactions

This section presents the basic chemical reactions of lime and lime–soda ash softening, caustic soda softening, and water stabilization of softened water.

Lime and Lime–Soda Ash Softening

$$CaO + H_2O \rightarrow Ca(OH)_2 \tag{1}$$

$$CO_2 + Ca(OH)_2 \rightarrow CaCO_3 + H_2O \tag{2}$$

$$Ca(HCO_3)_2 + Ca(OH)_2 \rightarrow 2\,CaCO_3 + 2H_2O \qquad pH \geq 9.5 \tag{3}$$

$$Mg(HCO_3)_2 + Ca(OH)_2 \rightarrow CaCO_3 + MgCO_3 + 2\,H_2O \qquad pH \leq 9.5 \tag{4}$$

$$MgCO_3 + Ca(OH)_2 \rightarrow Mg(OH)_2 + CaCO_3{}^* \qquad pH \geq 11 \tag{5}$$

$\qquad$ *plus $Ca(OH)_2$ due to an excess lime dosage

$$MgSO_4 + Ca(OH)_2 \rightarrow Mg(OH)_2 + CaSO_4 \tag{6}$$

$$CaSO_4 + Na_2CO_3 \rightarrow CaCO_3 + Na_2CO_3 \tag{7}$$

$$CaCl_2 + Na_2CO_3 \rightarrow CaCO_3 + 2\,NaCl \tag{8}$$

Caustic Soda Softening

$$CO_2 + 2\,NaOH \rightarrow Na_2CO_3 + H_2O \tag{9}$$

$$Ca(HCO_3)_2 + 2\,NaOH \rightarrow CaCO_3 + 2\,Na_2CO_3 + 2\,H_2O \tag{10}$$

$$Mg(HCO_3)_2 + 4\,NaOH \rightarrow Mg(OH)_2 + 2\,Na_2CO_3 + 2H_2O \tag{11}$$

$$MgSO_4 + 2\,NaOH \rightarrow Mg(OH)_2 + Na_2SO_4 \tag{12}$$

$$CaSO_4 + 2\,NaOH \rightarrow Ca(OH)_2 + Na_2SO_4 \tag{13}$$

$$CaCl_2 + 2\,NaOH \rightarrow Ca(OH)_2 + 2\,NaCl \tag{14}$$

Note: (13) and (14) need to add CO_2 to form $CaCO_3$ precipitate.

Water Stabilization of Softened Water

$$Ca(OH)_2 + 3\,CO_2 \rightarrow CaCO_3 + Ca(HCO_3)_2 + H_2O \qquad pH = 8.8 \tag{15}$$

$$CaCO_3 + CO_2 + H_2O \rightarrow Ca(HCO_3)_2 \qquad pH \leq 8.3 \tag{16}$$

$$2\,CaCO_3 + H_2SO_4 \rightarrow Ca(HCO_3)_2 + CaSO_4 \qquad pH \leq 8.3 \tag{17}$$

$$2\,CaCO_3 + 2\,HCl \rightarrow Ca(HCO_3)_2 + CaCl_2 \qquad pH \leq 8.3 \tag{18}$$

7.1.3 Softening Process Alternatives

There are five alternatives to the softening process. These depend on the type of hardness, degree of softening, operational convenience, degree to which the production of lime sludge must be reduced, and desired savings in chemical cost. The alternatives include (1) particle softening, (2) excess lime softening, (3) lime–soda ash softening, (4) caustic soda softening, and (5) softening with both lime and caustic soda.

Alternative 1: Partial Lime Softening This process only removes calcium carbonate. Lime (alone) is fed to the process water to raise the pH to approximately 9.5 (Equation 3). The softened water is then stabilized by recarbonation (Equations 15 and 16) or through the addition of acid (Equations 17 and 18) to reduce the pH to approximately 8.8, thus minimizing the occurrence of heavy scaling within the pipes and valves.

Alternative 2: Excess Lime Softening This alternative process is used in removing the carbonate form of calcium and magnesium. An excess amount of lime is applied to the process water at a minimum pH of 10.6, preferably 11–11.3 (Equations 3, 4, and 5) since magnesium hydroxide will not adequately form unless the pH of the water is raised above 11. Following the formation of magnesium hydroxide, the process water should be recarbonated to reduce the pH to 8.7–8.8. This step will convert the hydroxide form to carbonate or bicarbonate because the water has a calcium hydroxide concentration of 40–50 mg/L at pH 11 and the water is unstable.

At pH 8.8, approximately 10% of the alkalinity is carbonate and 90% is in the bicarbonate form. Calcium bicarbonate is dissolved in the water, yet, calcium carbonate precipitates out and must be removed by means of sedimentation or filtration in order to minimize its effect of hardening the softened water. The sedimentation process generally follows the recarbonation process because the filtration process receives a rather excessive amount of calcium carbonate precipitate.

When clean water such as underground water is softened by means of this process, a portion of the raw water is blended to the softened water (called split treatment). The carbon dioxide and bicarbonates (in the raw water) react with the calcium hydroxide in the softened water to yield the calcium carbonate precipitate.

Alternative 3: Lime–Soda Ash Softening This alternative should be used to remove both carbonate and noncarbonate hardness. Noncarbonate hardness is typically produced by calcium sulfate and calcium chloride; soda ash must be used to remove these compounds (Equations 6, 7, and 8).

The pH of the softening water should be a minimum of 10.6; therefore, recarbonation or acidification is required to stabilize the water prior to filtration.

Alternative 4: Caustic Soda Softening Caustic soda may be used as a substitute for lime in Alternatives 1 through 3. The advantages of caustic soda are its decreased sludge production, the elimination of problems associated with chemical dust, and the option of utilizing simpler storage and feed systems. However, caustic soda also has several major disadvantages: the cost of caustic soda is eight to ten times higher than lime; it is a potential health hazard to operators should there be a massive leak; and freezing problems occur for 50% solutions at 55°F (13°C).

Alternative 5: Softening with Both Lime and Caustic Soda This last alternative is a variation of Alternative 4; it is used to reduce the overall chemical cost and to reduce the capital cost of the lime feed system. The chemical reaction involved in this alternative is as follows:

$$2Ca(HCO_3)_2 + Ca(OH)_2 + 2NaOH \rightarrow 3CaCO_3 + Na_2CO_3 + 4H_2O$$

7.1.4 Overall Softening Treatment Process

All five water softening alternatives basically have the same overall water treatment process as the conventional complete water process: flash mixing, flocculation, sedimentation, recarbonation, second sedimentation, filtration, and chlorination. If the softened water is stabilized through acid rather than carbon dioxide, the recarbonation and second sedimentation processes are deleted. However, the addition of acid increases either the sulfate or chloride concentration, whereas the use of carbon dioxide does not.

The normal lime softening process can manage occasional turbidity spikes of up to 500–1000 NTU because the large dosage of lime (for softening) also acts as a coagulant. Nevertheless, turbidity removal and filter performance will both improve if metal coagulants such as sodium aluminate or ferric sulfate are fed in conjunction with lime. If the turbidity frequently exceeds 1000 NTU, engineers should consider installing an additional clarification process (alum flocculation and sedimentation) upstream of the regular lime softening process train.

Design engineers should note that recycling lime sludge to the head of the softening process train always improves the efficiency of the softening process; recycling gives a seeding effect in the formation of calcium carbonate floc. The solids contact and sludge recirculation types of reactor clarifier are frequently used in lime softening plants precisely because of this feature and are proved to be quite effective in this capacity.

If the carbon dioxide content of the raw water exceeds 10 mg/L, an aeration process such as a coke tray aerator should be considered for use in gas stripping. This scheme reduces the dosage of lime or caustic soda required for the removal of carbon dioxide (Equations 2 and 9), as well as reducing the volume of sludge production.

7.1.5 Design Criteria

The design criteria for flash mixing, flocculation, sedimentation, recarbonation, filtration, chlorination, and sludge handling and disposal (for lime–soda ash softening) are presented below.

Flash Mixing

Mixing energy (G)	$700-1000 \text{ s}^{-1}$
Mixing time	$2-5$ s

Flocculation

Mixing energy (G)	$5-50 \text{ s}^{-1}$
Mixing time	$30-40$ minutes
Flow velocity through ports	$0.5-1.2$ fps
	$(0.15-0.36$ m/s$)$

Note: Adequate mixing time is very important because of the slow reaction rate.

Sedimentation

Surface loading	0.75–1 gpm/ft^2, rectangular tank (1.9–2.5 m/h) 1–1.75 gpm/ft^2, sludge blanket type (2.5–4.4 m/h)
Detention time	Minimum of 2 h, rectangular tank Minimum of 1 h, sludge blanket type
Weir loading	15 gpm/ft, rectangular tank (11 m^3/m·h) 20 gpm/ft, sludge blanket type (15 m^3/m·h)
Total water loss	< 3% due to sludge withdrawal

Recarbonation

Carbon dioxide diffusion tank	3 min minimum contact time
Recarbonation tank	20 min minimum detention time
Water depth	Minimum of 12 ft
pH of water after recarbonation	8.7–8.8

Note: The use of liquid carbon dioxide has become more commonly practiced for recarbonation.

Filtration (Dual Media Bed)

Filtration rate	4–6 gpm/ft^2 (10–15 m/h)
Anthracite coal	
Depth	20 in. (0.5 m)
E.S.	1.0–1.1 mm
U.C.	<1.5
S.G.	1.67–1.7
Sand	
Depth	10 in. (0.25 m)
E.S.	0.55–0.6 mm
U.C.	<1.5
S.G.	>2.63
Backwash rate	20 gpm/ft^2 average (50 m/h)

Notes

1. A surface washing system must be installed because it plays an essential role in maintaining the top portion of the filter bed free of mud balls and mud pans; these form as the result of the cementing action of the calcium carbonate precipitates.

TABLE 7.1.5-1 pH versus Hydroxide Alkalinity

pH	9.7	10	10.2	10.4	10.5	10.6	10.8	11.0
OH (mg/L as CaCO$_3$	2–3	4–5	8–9	12–13	14–16	17–20	26–30	41–50

Source: Adapted from *The Nalco Water Handbook*, F. N. Kemmer, ed., McGraw-Hill, New York, 1979.

2. The design criteria for the surface wash system are identical to those used in designing regular filters (refer to Section 3.2.6, Granular Medium Filtration).

3. Based on the bed expansion, the use of filter backwash (alone) is not an effective method.

Chlorination Even though the Ct rule by EPA does not give any credit, the softening process is always characterized by a good disinfecting capability; the level of THM formation due to subsequent chlorination is usually also quite low.

The required chlorine dosage is defined as the minimum amount required to maintain a chlorine level of 0.5 mg/L throughout the distribution system to fight postcontamination. This restriction is imposed because under high pH conditions residual chlorine is in the hypochlorite form. The disinfecting

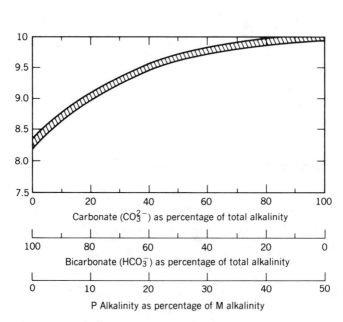

Figure 7.1.5-1 Relationships between pH and carbonate and bicarbonate ions. (Adapted from *Nalco Water Handbook*, F.N. Kemmer, ed., McGraw-Hill, New York, 1979.)

power of this ion is much less effective than hypochlorous acid. Under a pH range of 8.6–8.7, approximately 90% of the chlorine residual is in the hypochlorite form. Thus, the chlorine disinfecting power increases with a decrease in pH after recarbonation or acid neutralization.

Sludge Handling and Disposal

Sludge production rate	2–3.5 lb of solids for each pound of hardness removed
Sludge dewatering	Sludge lagoon type of drying bed or a mechanical dewatering system (refer to Section 4.3)

Note: Lime recovery by lime sludge recalcification has been practiced to some degree. However, due to recent laws against air pollution, most urban areas severely limit the use of this process.

Table 7.1.5-1 lists the approximate levels of hydroxide alkalinity at various pH levels (of the water). Figure 7.1.5-1 illustrates the relationship between carbonate and bicarbonate ions at various pH levels which are common to the recarbonation process.

7.1.6 Operational Considerations

The lime–soda ash softening process is based on a stoichiometric chemical reaction between the elements of hardness and the softening chemicals. A properly trained operational staff and a basic chemical laboratory are therefore essential in ensuring successful water softening.

One of the most important operational parameters is proper chemical dosage. The molecular weights of the major items involved in the softening process are listed in Table 7.1.6-1.

Chemical Dosages When softening water containing carbonate hardness, the following general formula will determine the lime dosage (the chemical feed details are discussed in Section 4.1):

$$\text{Lime (CaO in mg/L)} = (A + B + C) \times D/E$$

where A = CO_2 as CO_2 (mg/L) $\times$ 56/44,
  B = HCO_3 alkalinity as $CaCO_3$ (mg/L) $\times$ 56/100,
  C = Mg as Mg (mg/L) $\times$ 56/24.3,
  D = excess lime required to raise pH to 11 to remove magnesium; it is generally 10–20% of A + B + C: thus, D = 1.1–1.2,
  E = purity of the quick lime (0.88–0.95); therefore, E = 0.88–0.95.

TABLE 7.1.6-1 Molecular Weights of Major Items

CaO	$Ca(OH)_2$	Na_2CO_3	$CaCO_3$	NaOH	CO_2	Ca	Mg
56	74	106	100	40	44	40	24.3

For slaked lime $(Ca(OH)_2)$,

$$\text{Slaked lime } (Ca(OH)_2) = (A' + B' + C') \times D/E$$

where A $'$ = CO_2 as CO_2 (mg/L) $\times$ 74/44,
   B$'$ = HCO_3 alkalinity as $CaCO_3$ $\times$ 74/100,
   C$'$ = Mg as Mg (mg/L) $\times$ 74/24.3,
   D = excess lime: D = 1.1–1.2,
   E = purity of the lime: generally, E = 0.93–0.95.

The soda ash dosage for noncarbonate hardness is determined as follows:

$$\text{Soda ash } (Na_2CO_3 \text{ in mg/L}) = (NH - L) \times 106/100$$

where NH = noncarbonate hardness as $CaCO_3$ (mg/L),
   L = noncarbonate hardness remaining in the softened water (mg/L).

Since commercially available soda ash is nearly 100% pure, a purity correction factor is usually not necessary.

If caustic soda is employed instead of lime, in the same manner as described above, refer to Equations 9–14 and use a molecular weight of 40 (for NaOH) rather than 56 (for CaO). However, the practicality of this alternative is questionable because the unit cost of caustic soda is approximately 10 times higher than lime.

The carbon dioxide dosage for recarbonation is

$$CO_2 \text{ (mg/L)} = \{[HA + (CA \times R)] \times 44/100\} \times F$$

where HA = hydroxide alkalinity as $CaCO_3$ in mg/L,
   CA = carbonate alkalinity as $CaCO_3$ in mg/L,
   R = the ratio used to convert CO_3 to HCO_3 (generally ranges from 0 to 0.2),
   F = the excess CO_2 dosage factor: usually F = 1.2.

The pH of the recarbonated water must be approximately 8.8 in order to facilitate the formation of HCO_3 and $CaCO_3$; these compounds are vital in removing excess calcium hardness by means of settling and filtration.

The treatment plant operator should remember the basic relationships of hydroxide, carbonate, and bicarbonate alkalinities:

1. When the phenolphthalein alkalinity (PA) is zero, the water does not contain hydroxide or carbonate alkalinities. Methyl orange alkalinity (MA) is the bicarbonate form.
2. If PA < 0.5MA, there is no hydroxide alkalinity. The carbonate alkalinity is 2PA and the bicarbonate alkalinity is MA − 2PA.
3. When PA = 0.5MA, both the hydroxide and bicarbonate alkalinities are zero, and MA is all carbonate form.
4. When PA > 0.5MA, the bicarbonate alkalinity is zero. The hydroxide alkalinity is 2PA − MA and the carbonate alkalinity is 2(MA − PA).
5. When PA = MA, both the carbonate and bicarbonate alkalinities are zero and the hydroxide alkalinity is equal to either PA or MA.

It is also worth remembering that the pH for phenolphthalein endpoint titration is 8.3.

Sludge Handling and Disposal   This issue also has great bearing on the operations of the softening plant. Section 4.3 presents a detailed discussion of this topic.

7.1.7   Example Calculation

Example 1

Given   The raw water contains the following: 250 mg/L of bicarbonate hardness, 30 mg/L of magnesium, 75 mg/L of noncarbonate hardness, and 15 mg/L of carbon dioxide. Assume that the softened water delivered to the consumer has a pH of 8.7 and a total hardness of 85 mg/L, of which 35 mg/L is carbonate hardness.

Determine   The required chemical dosages, using commercially available chemicals, based on a plant flow rate of 15 mgd.

Solution

$$\text{Dosage of quick lime (mg/L)} = (A + B + C) \times D/E$$

where A = 15 × 56/44 = 19 mg/L,
    B = (250 − 35) × 56/100 = 120 mg/L,
    C = 30 × 56/24.3 = 69 mg/L,
    D = 1.15 (15% excess lime),
    E = 0.88 (88% purity is assumed).

TABLE 7.1.7-1   Changes in Water Characteristics During the Softening Process (Assumed)

Characteristic	Raw Water	Softened Water	Recarbonated Water
CO_2	15	0	0
pH	7.5	11	8.7
Mg	30	3	3
Total hardness	325	110	85
Noncarbonate hardness	75	35	35
Total alkalinity (MA)	250	75	50
P alkalinity (PA)	0	59	15
HCO_3 alkalinity	250	0	20
CO_3 alkalinity	0	32	30
OH alkalinity	0	43	0

Notes: (1) MA represents methyl orange alkalinity.
(2) PA is phenolphthalein alkalinity.
(3) Magnesium cannot be completely removed by this process. Therefore, a residual of 3 mg/L is assumed to be present in the softened water.

Thus,

$$CaO \ (mg/L) = (19 + 120 + 69) \times (1.15/0.88) = 272 \ mg/L$$

$$Dosage \ of \ soda \ ash \ (mg/L) = (NH - L) \times 106/100$$

where NH = 75 − 35 = 40 mg/L,
   L = 35 mg/L.

Thus,

$$Na_2CO_3 = 40 \times (106/100) = 42.4 \ mg/L, \quad say \ 43 \ mg/L$$

It is assumed that the water qualities of the softened water and recarbonated water are the same as those listed in Table 7.1.7-1.

Recarbonation is achieved by means of liquid carbon dioxide, since it is nearly 100% pure, is easily handled, is easily fed, and has low maintenance and capital costs when compared to flue gas: flue gas is attained by burning oil or natural gas and only yields approximately 12% CO_2.

The carbon dioxide dosage is determined from Table 7.1.7-1. Since a hydroxide alkalinity of 43 mg/L and a carbonate alkalinity of 2 mg/L must be converted to bicarbonate alkalinity,

$$CO_2 \ (mg/L) = \{[HA + (CA \times R)] \times (44/100)\} \times F$$

where HA = 43 mg/L
   CA × R = 2 mg/L or 32 × 0.063 = 2

Thus,

$$CO_2 = \{(43 + 2) \times 0.44\} \times 1.2 = 24 \text{ mg/L}$$

Summary of Chemical Dosages

Lime (CaO) dosage = $272 \times 8.34 \times 15 = 34{,}027$ lb/day

Soda ash dosage = $43 \times 8.34 \times 15 = 5{,}380$ lb/day

Carbon dioxide dosage = $24 \times 8.34 \times 15 = 3{,}002$ lb/day

BIBLIOGRAPHY

AWWA/ASCE, *Water Treatment Plant Design*, AWWA, New York, 1969.

Cox, C. R., "Operation and Control of Water Treatment Processes," WHO, Geneva, 1964.

Craun, G. F., McCabe, L. J., "Problems Associated with Metals in Drinking Water," *J. AWWA*, 67(11):593(November 1975).

Graveland, A., VonDijk, J. G., Moel, P. J. & Oomen, J. H. C. M., "Developments in Water Softening by Means of Pellet Reactors," *J. AWWA*, 75(12):619(December 1983).

Hess, J. S., "Lime and Caustic Soda Softening at Fremont, Ohio," *J. AWWA*, 60(8):980(August 1968).

Judkins, J. F., Wynn, R. H., "Crystal-Speed Conditioning of Lime-Softening Sludge," *J. AWWA*, 64(5):306(May 1972).

Kemmer, F. N., ed., *The Nalco Water Handbook*, Nalco Chemical Co. McGraw-Hill, New York, 1979.

National Research Council, *Drinking Water and Health—Water Hardness and Health*, National Academy of Science, Washington, DC, 1977, p. 439.

National Research Council, *Drinking Water and Health—Water Hardness and Health*, Vol. 3, National Academy of Science, Washington, DC, 1980, p. 21.

Reh, C. W., in *Water Treatment Plant Design—Lime-Soda Softening Processes*, R. L. Sanks, ed., Ann Arbor Science Publishers, Ann Arbor, MI, 1982.

Riehl, M. L., *Hoover's Water Supply and Treatment*, 10th ed., National Lime Association, Washington, DC, 1970.

Schroeder, J. A., "Municipal Drinking Water and Cardiovascular Death Rates," *J. Am. Med. Res.*, 195(2):125(January 1966).

Singhal, A. K., "Conventional Lime–Soda Ash Softening vs. Split Treatment," *J. AWWA*, 69(3):158(March 1977).

Snoeyink, V. L., Jenkins, D., *Water Chemistry*, Wiley, New York, 1980.

Thompson, C. G., Singley, J. E., & Black, A. P., "Magnesium Carbonate—A Recycled Coagulant," parts I and II, *J. AWWA*, 64(1):11(January 1972) and 64(2):93(February 1972).

Van der Veen, C., Graveland, A., "Central Softening by Crystallization in a Fluidized-Bed Process," *J. AWWA*, 80(6):51(June 1988).

Winton, E. F., McCabe, L. J., "Studies Relating to Water Mineralization and Health," *J. AWWA*, 62(1):26(January 1970).

Wood, F. O., "Selecting a Softening Process," *J. AWWA*, 64(12):820(December 1972).

Wyness, D. K., "The Helical Flow Reactor Clarifier," *J. AWWA*, 71(10):580(October 1979).

7.2 IRON AND MANGANESE REMOVAL

Iron and manganese are minerals commonly found in soil, where they exist in the form of oxides, namely, insoluble ferric oxide and manganese oxide (very insoluble). However, when water contains carbon dioxide, or in an acidic water, ferric iron is reduced to the ferrous form under anaerobic conditions; this ion *is* soluble in water. Under the same conditions, the dioxide form of manganese is reduced from a valence of 4 to a valence of 2 and is also soluble in water. Consequently, a significant amount of iron and manganese are often found in deep well water and the hypolimnion water of stratified lakes and reservoirs. Iron and manganese may occasionally be found in groundwater containing hydrogen sulfide. Organically bound iron and manganese may also be found in groundwater containing humic acid.

The iron and manganese concentrations of surface water seldom exceed 1 mg/L. The iron content of groundwater may reach levels up to 10 mg/L under conditions of low alkalinity (less than 50 mg/L) and up to 2 mg/L of manganese.

The presence of iron and manganese in potable water is not known to cause health problems. These minerals are primarily associated with aesthetic factors, such as unpleasant taste and odor, and domestic problems, such as staining of laundry and fixtures, and can potentially be a serious problem for many industries—for example, staining of the product and/or formation of mineral deposits. Another potential problem is the growth of iron bacteria within the distribution main.

The Secondary Drinking Water Standards (EPA) recommends MCL levels of 0.3 mg/L iron and 0.05 mg/L manganese.

7.2.1 Alternative Methods

Several alternatives exist for the removal of iron and manganese in the field of water treatment. The basic methods are oxidation followed by clarification and filtration, ion exchange, stabilization by means of a sequestering agent, or lime softening. Each of these schemes is discussed. A case history is also presented at the end of the section.

Oxidation Oxidation may be achieved through aeration, chlorination, chlorine dioxide, potassium permanganate, or ozonation.

Aeration   Aeration can be used to fulfill two purposes: transfer of oxygen to water and the removal of volatile gases.

$$4Fe\,(HCO_3)_2 + O_2 + 2H_2O \rightarrow 4Fe\,(OH)_3 + 8CO_2$$

$$2MnSO_4 + 2Ca(OH)_2 + O_2 \rightarrow 2MnO_2 + 2CaSO_4 + 2H_2O$$

The air diffusion type of aeration uses air diffusers to supply air to the process water at a depth of 12–15 ft (3 to 4.5 m). The air to water volume ratio is 0.75:1.0. This type of unit has an average oxygen transfer efficiency of only 5–10% and is therefore not considered to be effective.

The coke tray type of contact aerator is a more effective method of oxidation. This aerator consists of a series of trays that are each 12–18 in. deep (0.3–0.45 m). The bottom of the trays are perforated and coke, crushed stone, limestone, or plastic medium is placed in each tray as the contact material; the loading rate is 15–20 gpm/ft^2 (37–50 m/h).

Theoretically, 1 mg of oxygen oxidizes 7 mg of divalent iron and 3.4 mg of divalent manganese. Organically bound iron cannot be oxidized through aeration.

The rate of oxidation is a function of the pH of the process water: the higher the pH, the better the result. In order to complete the process of oxidation within 15 min, the pH of the water should be higher than 7.5, preferably 8. Manganese oxidizes very slowly and is not effectively oxidized at pH values below 9.5. In fact, it would take over 1 h to oxidize manganese at pH 9.5.

Alum flocculation and sedimentation are generally required after aeration. These processes are then followed by filtration, especially in cases where the iron concentration of the raw water exceeds 5 mg/L.

Chlorination   Chlorine is often used in the oxidation of divalent iron and manganese because, unlike aeration, it has a faster oxidation rate and is capable of oxidizing organically bound iron.

$$2Fe(HCO_3)_2 + Ca(HCO_3)_2 + Cl_2 \rightarrow 2Fe(OH)_3 + CaCl_2 + 6CO_2$$

$$Mn(HCO_3)_2 + Ca(HCO_3)_2 + Cl_2 \rightarrow MnO_2 + CaCl_2 + 4CO_2 + 2H_2O$$

The above equations show that 1 mg/L of chlorine oxidizes 1.58 mg/L of iron and that 0.78 mg/L of manganese is oxidized by 1 mg/L of chlorine.

The rates of the oxidation reactions are pH dependent. In order for iron to be oxidized within 15–30 min, the pH of the process water should be approximately 8.0–8.3, preferably 8.5. Under these same conditions, manganese will be oxidized within 2–3 h.

When the raw water contains ammonia, chloramines are formed upon addition of chlorine. Consequently, the oxidation rate is reduced. If THM formation is discouraged, chlorination cannot be employed as an oxidant.

Generally, the standard process train for iron and manganese removal is alum flocculation, followed by clarification and filtration; the practice of prechlorination leaves a minimum level of 0.5 mg/L of available free chlorine. This process train quickly coats the filter medium grains with a layer of manganese dioxide and ferric hydroxides or ferric oxides. Once the filter grains have been coated, the filter bed is capable of effectively removing manganese and iron by means of continuous prechlorination.

Chlorine Dioxide   Chlorine dioxide is a strong oxidant that effectively oxidizes organically complexed iron and manganese. The pH level necessary for oxidation is a very important factor in the reaction rate and should be a minimum of 7.0.

$$Fe(HCO_3)_2 + NaHCO_3 + ClO_2 \rightarrow Fe(OH)_3 + NaClO_2 + 3CO_2$$

$$Mn(HCO_3)_2 + 2NaHCO_3 + 2ClO_2 \rightarrow MnO_2 + 2NaClO_2 + 4CO_2 + 2H_2O$$

Theoretically, 1 mg/L of chlorine dioxide will oxidize 0.83 mg/L of iron and the same amount of chlorine dioxide will oxidize 0.41 mg/L of manganese.

Engineers must remember that many regulatory agencies are limiting the use of chlorine dioxide due to the potential toxicity of the chlorite and chlorate ions formed during the production of chlorine dioxide. Another important issue is the operational cost. The chlorine dioxide process is approximately twice as expensive as chlorination.

Potassium Permanganate   Potassium permanganate is a strong oxidant. The reaction time is therefore fast in a wide range of pH. The oxidation time varies from as short as 5 min to 10 min for both iron and manganese, provided that the pH is over 7.0.

$$3Fe(HCO_3)_2 + KMnO_4 + 7H_2O \rightarrow 3Fe(OH)_3 + MnO_2 + KHCO_3 + 5H_2CO_3$$

$$3Mn(HCO_3)_2 + 2KMnO_4 + 2H_2O \rightarrow 5MnO_2 + 2KHCO_3 + 4H_2CO_3$$

The above reactions indicate that 1 mg/L of potassium permanganate oxidizes 1.06 mg/L of iron and 0.52 mg/L of manganese. If both iron and manganese are present in the raw water, the oxidation process may be more cost effective if chlorine is added prior to feeding potassium permanganate; the chlorine oxidizes iron rather easily and consequently leaves the more expensive potassium permanganate to oxidize the manganese.

In the field of water treatment, it is well recognized that the addition of potassium permanganate to the head of a conventional complete treatment process helps to improve flocculation and sedimentation. This phenomenon is due to the seeding effect of the oxidized iron and manganese particles. Under these conditions, the medium grains in the upper portion of the filter bed soon become coated with manganese oxide and ferric hydroxide in the same manner as the prechlorination process.

Ozonation The use of ozone for the sole purpose of oxidizing ferrous iron and manganous manganese is very unlikely due to the availability of other cost effective methods. Yet, ozonation has been practiced successfully in Europe for the removal of both iron and manganese by means of a conventional complete treatment process coupled with pre-ozonation. The experience has shown that excessive ozone oxidation of manganese forms permanganate and the water in the clearwell becomes pink in color.

$$2Fe(HCO_3)_2 + O_3 + 2H_2O \rightarrow 2Fe(OH)_3 + O_2 + 4CO_2 + H_2O$$

$$Mn(HCO_3)_2 + O_3 + 2H_2O \rightarrow MnO_2 + O_2 + 2CO_2 + 3H_2O$$

Theoretically, 1 mg/L of ozone will oxidize 2.3 mg/L of iron and 1.5 mg/L of manganese.

Ion Exchange (Zeolite Process) When the raw water contains less than 0.5 mg/L of iron and manganese, either a cation or hydrogen cation type of ion exchange unit may be employed to remove these elements; these units are also capable of removing hardness.

Yet, if the processing water contains any amount of dissolved oxygen, there is always the danger of fouling the ion exchange material and clogging the exchanger bed. Thus, it is vital that the process be run in the absence of oxygen to prevent the elements from oxidizing.

The ion exchange process is generally limited to the processing of industrial water.

Manganese Zeolite Filtration Green sand impregnated with manganese (glauconite) is commonly known as zeolite. Each grain of the sand is coated with iron and manganese oxides. This type of filter medium is a form of ion exchange commonly used in industries.

In this particular process, potassium permanganate is fed to the filter influent on a continuous basis, oxidizing soluble iron and manganese, and continuously regenerating the filter medium. It is very important that a proper dosage of potassium permanganate be applied to the process water, since an excessive dosage yields an effluent that is pink in color, while an inadequate dosage results in the leakage of manganese into the filter effluent.

In the case of groundwater treatment, zeolite is placed in the pressure filter cells, rather than the gravity filters, in order to preserve the discharge pressure of the well pump. This scheme eliminates the problem of repumping and allows one filter to be backwashed by using the pressured filtered water from the remaining filters.

The filtration rate of this process usually ranges from 3 to 4 gpm/ft^2 (7–10 m/h). Manganese zeolite filtration is generally limited to water containing a maximum level of 1 mg/L of iron or manganese. However, it has been reported that zeolite has the capacity to remove 0.09 lb/ft^3 (1.5 kg/m^3) of either iron or manganese and to regenerate 0.18 lb/ft^3 (2.9 kg/m^3) of filter medium by potassium permanganate.

Sequestering Process The purpose of the sequestering process is to hold both iron and manganese in solution by means of sequestering chemicals. This process is generally applicable to water containing less than 2 mg/L of iron and manganese; both elements must be present in their bicarbonate form. The sequestering agents include compounds such as sodium silicate, trisodium phosphate, hexametaphosphate, and zinc orthophosphate.

The proper dosage of the sequestering chemicals should be calculated on the basis of the manufacturer's suggested dosage. Approximately 2 mg/L of hexametaphosphate may be used for 1 mg/L or iron. However, polyphosphate dosages are limited to less than 10 mg/L due to its tendency to stimulate biological growth in the distribution system. In this case, an adequate amount of residual chlorine must be present in the process water so that the growth of the microorganisms can be controlled.

Engineers should note that the sequestering process is only effective in cold water: the sequestering agents lose their dispersing properties if the water is either highly heated or boiled. For this reason, this process is seldom employed in treating medium to large sized domestic water supply systems. Yet, small water systems are often financially incapable of implementing oxidation and filtration or lime softening processes and the much simpler and less costly (capital) sequestering process may therefore be the appropriate choice for treating their well water.

Lime Softening As described in Section 7.1, iron and manganese are effectively removed by the softening process provided that the pH of the process water is above 9.5.

Based on the pH and solubility relationships, 83% of the iron precipitates at a pH of 8.4, but 92 and 100% of the iron precipitates at the respective pH values of 8.8 and 9.6. Manganese is more difficult to remove: no manganese will precipitate out at a pH of 8.8; however, a pH of 9.4 and 9.8 will remove 98 and 100% of the manganese, respectively.

7.2.2 Case History

A water treatment plant located in the northern part of the United States treats shallow groundwater containing approximately 5–5.5 mg/L of ferrous ion and 0.6 mg/L of manganese. The water has a total hardness of 200 mg/L and a pH of approximately 7.1. A sludge blanket type of reactor clarifier and standard dual media filters are used to treat this water.

The basic treatment process is partial softening with lime. This scheme automatically reduces the iron level of the filtered water to nearly 0.1 mg/L and also reduces the manganese level to below 0.05 mg/L.

The plant conducted an experimental study to evaluate the feasibility of reducing the required lime dosage, as well as the feasibility of reducing the sludge production rate based on the oxidation of ferrous iron through the application of an appropriate amount of chlorine. The results were very in-

TABLE 7.2.2-1   Main Results of the Iron Removal Test

Lime dosage—range (mg/L)	45–55	55–65	65–75	75–85	85–95
pH of the settled water	7.8–7.9	7.9–8.2	8.3–8.6	8.7–8.9	8.9–9.1
Turbidity of the filtered water	0.7–2.2	0.1–1.2	0.05–0.2	0.05–0.12	0.05–0.15
Iron (mg/L)					
Settled water	3.9–4.1	3.1–3.9	2.6–3.1	2.1–2.5	2.0–2.5
Filtered water	0.5–0.8	0.1–0.5	0.01–0.2	0.01–0.05	0.01–0.05
Filter run length (h)	10–16	12–22	26–30	24–34	24–30

Notes: (1) Approximately 5 mg/L of chlorine were fed with lime for all cases.
  (2) The hardness of the filtered water increased in proportion to the lime dosages for all cases where the pH was less than 8.5. However, a 5–10 mg/L reduction in hardness only occurred in cases where the pH exceeded 8.5.
  (3) The water temperature ranged from 6 to 7°C.
  (4) The filtration rate was approximately 2 gpm/ft^2.

teresting and proved to be educational for those individuals designing iron removal systems under similar conditions. Table 7.2.2-1 presents some of the results.

These results indicate that the chlorinated water must have a pH of at least 8.2 if the treatment process is expected to produce a filtered water containing less than 0.2 mg/L iron within a reasonable period of time: a filter run length

TABLE 7.2.2-2   Jar Test Results for the Polymer Evaluation Tests

Beaker Number	Cl_2 Dosage	$KMnO_4$ Dosage	Polymer Dosage	Floc Size	Settled Water	
					Turbidity	Iron
1	0	5	0	Small	5.8	4.5
2	0	5	0.2	Medium to large	3.5	2.3
3	2.5	2.5	0.2	Large	2.7	2.1
4	5	0	0	Small	7.0	4.7
5	5	0	0.2	Medium to large	3	1.6
6	5	0	0.2	Large	2	1.4

Notes: (1) Iron content of the raw water was 5 mg/L.
  (2) The jar test was conducted at $G \times t = 3 \times 10^4$, with a total mixing time of 10 min and a settling time of 20 min.
  (3) Polymer (Nalco 7763) was added approximately 1 min after the oxidant was applied to the beakers, with the exception of beaker numbers 5 and 6.
  (4) The polymer was added 2.5 min after chlorine application for beaker number 5 and 5 min after the addition of chlorine for beaker number 6.
  (5) The pH was adjusted to 8.2 by NaOH prior to the addition of oxidant.
  (6) The temperature of the raw water was 5.6°C.
  (7) The chemical dosages are all in mg/L. The unit of iron in water is also mg/L.

of 24 h. The increase in filtered water hardness at pH values less than 8.5, after the addition of lime, is the obvious result of the conversion of most of the applied lime (calcium hydroxide) to calcium bicarbonate; the calcium bicarbonate is completely soluble in water at lower pH conditions.

Bench scale studies were also conducted at the plant in order to evaluate the effectiveness of anionic polymer for improving the settleability of iron hydroxide prior to filtration. A portion of the test results are listed in Table 7.2.2-2.

The results listed in Table 7.2.2-2 demonstrate two major issues. The first is an obvious improvement in both the iron hydroxide characteristics and iron removal (via sedimentation) through the addition of anionic polymer in concentrations as low as 0.2 mg/L. The second issue is the time lag between the addition of oxidant and polymer. With respect to the polymer, a time delay of 5 min significantly improves the removal of both turbidity and iron by means of sedimentation.

BIBLIOGRAPHY

Adams, R. B., "Manganese Removal by Oxidation with Potassium Permanganate," *J. AWWA*, 52(2):219(February 1960).

AWWA, *Water Treatment Plant Design*, AWWA, New York, 1969.

Boatby, J. R., "Optimum Mn Removal and Washwater Recovery at Filtration Plant in Brazil," *J. AWWA*, 80(12):71(December 1984).

Conneley, E. J., "Removal of Iron and Manganese," *J. AWWA*, 50(5):697(May 1958).

Culp/Wesner/Culp, *Handbook of Public Water System*, Van Nostrand Reinhold, New York, 1986.

Knocke, W. R., Hamon, J. R. & Thompson, C. P., "Soluble Mn Removal on Oxide-Coated Filter Media," *J. AWWA*, 80(12):65(December 1988).

Knocke, W. R., Hoehn, R. C., & Sinsabaugh, R. L., "Using Alternative Oxidants to Remove Dissolved Manganese from Water Laden with Organics," *J. AWWA*, 79(3):75(March 1987).

Montgomery, J. M., Consulting Engineers, *Water Treatment—Principles and Design*, Wiley, New York, 1985.

"Research Needs for the Treatment of Iron and Mn," Committee Report, *J. AWWA*, 79(9):119(September 1987).

Robinson, R. B., Ronk, S. K., "The Treatability of Mn by Sodium Silicate and Chlorine," *J. AWWA*, 79(11):64(November 1987).

Sanks, R. L., *Water Treatment Plant Design*, Ann Arbor Science Publishers, Ann Arbor, MI, 1978.

Victoreen, H. T., "Controlling Corrosion by Controlling Bacteria Growth," *J. AWWA*, 76(3):87(March 1984).

Viraraghavan, T., Winchester, E. L. & Landine, R. C., "Removing Mn from Water at Frederiction, N.B., Canada," *J. AWWA*, 79(8):43(August 1987).

Welch, W. A., "Potassium Permanganate in Water Treatment," *J. AWWA*, 55(6):735(June 1963).

Willey, B. F., Jennings, H. & Welch, W. A., "Iron Removal with Potassium Permanganate," *J. AWWA*, 55(6):729(June 1963).

Wong, J. M., "Chlorination Filtration for Fe and Mn Removal," *J. AWWA*, 76(1):76(January 1984).

7.3 TASTE AND ODOR CONTROL

In the water supply industry, it is not unusual for water purveyors to receive customer complaints regarding the objectionable taste and odor of the potable water. These complaints are more frequent when the source is surface water rather than groundwater, because surface water is more likely to be affected by microorganisms and organic substances, as well as manufactured wastes. Some consumers also consider the odor of the residual chlorine in tap water to be offensive.

The most frequent causes of taste and odor in the water supply are algae and *Actinomycetes* (during certain seasons). The second is decaying vegetation. Taste and odor problems are occasionally the result of hydrogen sulfide, agricultural runoff, industrial chemical spills, illegally discharged industrial chemicals, and pollution by sewage. The current (1989) drinking water quality standards set the Threshold Odor Number (TON) to less than 3. It is generally difficult to control taste and odor problems once they have developed. The plant design should therefore provide the operator with a certain amount of flexibility, that is, several available options. It is very important to allow the use of alternative chemicals, as well as supplying alternative feed points, so that the location and type of chemical can be changed in accordance with changes in raw water quality. Moreover, the intake design should have provisions that allow the intake to withdraw water from selected depths in the lake or reservoir: this is an essential design factor.

7.3.1 Major Taste- and Odor-Producing Substances

The detection and intensity of taste and odor are very subjective issues; some individuals are very sensitive, while others have a very high level of tolerance. One method that has been developed in recent years is the Flavor Profile Analysis (FPA). The FPA is designed to reduce/eliminate the subjective nature of this problem by using a panel of people who are trained to analyze odors associated with processed water.

Most odor problems are the result of algae or *Actinomycetes*. Table 7.3.1-1 lists the major odor-producing compounds and the responsible species of algae. Among the typical odor-producing compounds, geosmin and 2-methlyisoborneal (MIB) often impart objectionable odor at very low concentra-

TABLE 7.3.1-1   Taste- and Odor-Producing Algae and Characteristics of the Taste and Odor

Algal Genus	Algal Group	Odor When Algae Are: Moderate	Odor When Algae Are: Abundant	Taste	Tongue Sensation
Anabaena	Blue-green	Grassy, nasturtium, musty	Septic	—	—
Anacystis	Blue-green	Grassy	Septic	Sweet	—
Aphanizomenon	Blue-green	Grassy, nasturtium, musty	Septic	Sweet	Dry
Asterionella	Diatom	Geranium, spicy	Fishy	—	—
Ceratium	Flagellate	Fishy	Septic	Bitter	—
Dinobryon	Flagellate	Violet	Fishy	—	Slick
Oscillatoria	Blue-green	Grassy	Musty, spicy	—	—
Scenedesmus	Green	—	Grassy	—	—
Spirogyra	Green	—	Grassy	—	—
Synura	Flagellate	Cucumber, muskmelon, spicy	Fishy	Bitter	Dry, metallic slick
Tabellaria	Diatom	Geranium	Fishy	—	—
Ulothrix	Green	—	Grassy	—	—
Volvox	Flagellate	Fishy	Fishy	—	—

SOURCE: Adapted from Palmer (1962).

Compound	Structure	Associated Organisms
Methylisoborneol (MIB)	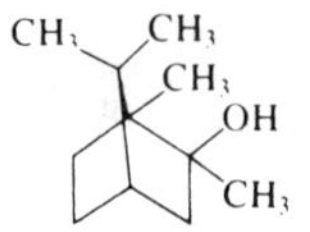	*Actinomycetes* *Oscillatoria curviceps* *Oscillatoria tenuis*
Geosmin	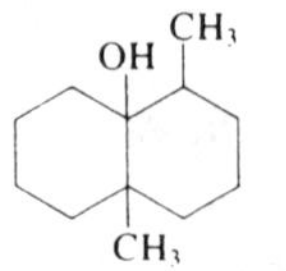	*Actinomycetes* *Sympioca muscoum* *Oscillatoria tenuis* *Oscillatoria simplicissima* *Anabaena scheremetievi*
Mucidone	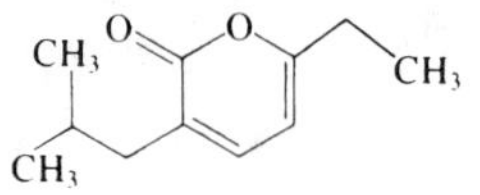	*Actinomycetes*
Isobutyl mercaptan	$CH_3{>}CHCH_2{-}SH$ (CH₃)	*Microcystis flos-aquae*
N-Butyl mercaptan	$CH_3(CH_2)_3{-}SH$	*Microcystis flos-aquae* *Oscillatoria chalybea*
Isopropyl mercaptan	$CH_3{>}CHSH$ (CH₃)	*Microcystis flos-aquae*
Dimethyl disulfide	$CH_3{-}S{-}S{-}CH_3$	*Microcystis flos-aquae* *Oscillatoria chalybea*
Dimethyl sulfide	$CH_3{-}S{-}CH_3$	*Oscillatoria chalybea* *Anabaena*
Methyl mercaptan	CH_3SH	*Microcystis flos-aquae* *Oscillatoria chalybea*

Figure 7.3.1-1 Odor-producing algae and chemical structures of odors. (Adapted from J. M. Montgomery, Consulting Engineers, *Water Treatment—Principles and Design*, Wiley, New York, 1985.)

TABLE 7.3.1-2 Troublesome Algae and Required Dosages of Control Chemicals

	Organism	Trouble	Copper Sulfate (mg/L)	Chlorine (mg/L)
Algae				
Diatoms	*Asterionella, Synedra, Tabellaria*	Odor: aromatic to fishy	0.1–0.5	0.5–1.0
	Fragillaria, Navicula	Turbidity	0.1–0.3	—
Grass-green	*Melosira*	Turbidity	0.2	2.0
	Eudorina,[a] *Pandorina*[a]	Odor: fishy	2–10	—
	Volvox[a]	Odor: fishy	0.25	0.3–1.0
	Chara, Cladophora	Turbidity, scum	0.1–0.5	—
	Coelastrum, Spirogyra	Turbidity, scum	0.1–0.3	1.0–1.5
Blue-green	*Anabaena, Aphanizomenon*	Odor: moldy, grassy, vile	0.1–0.5	0.5–1.0
	Clathrocystis, Coelosphaerium	Odor: grassy, vile	0.1–0.3	0.5–1.0
	Oscillatoria	Turbidity	0.2–0.5	1.1
Golden or yellow-brown	*Cryptomonas*[b]	Odor: aromatic	0.2–0.5	—
	Dinobryon	Odor: aromatic to fishy	0.2	0.3–1.0
	Mallomonas	Odor: aromatic	0.2–0.5	—
	Synura	Taste: cucumber	0.1–0.3	0.3–1.0
	Uroglenopsis	Odor: fishy. Taste: oily	0.1–0.2	0.3–1.0
Dinoflagellates	*Ceratium*	Odor: fishy, vile	0.2–0.3	0.3–1.0
	Glenodinium	Odor: fishy	0.2–0.5	—
	Peridinium	Odor: fishy	0.5–2.0	—
Filamentous bacteria	*Beggiatoa* (sulfur)	Odor: decayed. Pipe growths	5.0	—
	Crenothrix (iron)	Odor: decayed. Pipe growths	0.3–0.5	0.5
Crustacea	*Cyclops*	c	—	1.0–3.0
	Daphnia	c	2.0	1.0–3.0
Miscellaneous	*Chironomus* (bloodworm)	c	—	15–50
	Craspedacusta (jellyfish)	c	0.3	—

[a] These organisms are classified also as flagellate protozoa.
[b] Classification uncertain.
c These organisms are individually visible and cause consumer complaints.

Source: Adapted from J. M. Montgomery, Consulting Engineers.

tions (20 ng/L or 0.02 μg/L). Therefore, the acceptable level of these compounds is considered to be approximately 5 ng/L.

The taste of the potable water can generally be attributed to the presence of small amounts of metals, such as iron and copper, and phenolic compounds. For example, 0.05–0.1 mg/L of dissolved iron, 2–5 mg/L of divalent copper, and 4–9 mg/L of dissolved zinc are considered to be taste thresholds. In the case of phenol, the threshold level is 1 mg/L, but 0.002 g/L for chlorophenol (chlorophenol is formed by the chlorination of phenol-bearing water). Generally, tastes and odors are objectionable to the consumer. However, under normal conditions, they are not a threat to the health of the consumer, provided that the water treatment plants are properly operated.

Refer to Figure 7.3.1-1 for a list of the taste- and odor-producing algae and their characteristic features. Table 7.3.1-2 presents the troublesome algae and the recommended copper sulfate and chlorine dosages that should be used in their control.

7.3.2 Control Measures

Three basic taste and odor control measures are used in the treatment of potable water: (1) prevention at the source, (2) removal at the treatment plant, and (3) control within the distribution system. The fundamentals of each of these measures are discussed.

Preventative Measures   Methods used in prevention at the source include reservoir mixing, aquatic plant control, and water pollution control.

Reservoir Mixing   Most lake and reservoirs stratify during warm seasons, frequently causing the bottom layer (hypolimnion) to be oxygen deficient. Under such anaerobic conditions, both hydrogen sulfide and the degradation of organic substances result in the emission of objectionable odors.

Experience gained during the last half of the 20th century has shown that artificial mixing of lakes by means of mechanical mixers, submersible pumps, or air-lift pumps significantly improves water quality, thereby reducing the burden on the water treatment plants in controlling taste and odor problems, as well as iron and manganese. The design engineer should therefore evaluate the feasibility of installing a reservoir mixing device as a preventative measure for taste and odor.

Aquatic Plant Control   The control of marginal aquatic plants is a very effective method in preserving the (good) water quality of a reservoir. However, this task is not easily achieved when the lake or reservoir is large.

One method of controlling marginal plants, including algae, is to drastically change the water level several times a year. When the water level is lowered by 4–5 ft, the roots of the marginal plants become exposed to sun and therefore dry up.

The most common method of controlling algae growth is the use of copper sulfate. The required dosage varies with the type of algae and ranges from 0.05 to 0.8 mg/L. Sensitive species such as blue-green algae—which frequently cause intense taste and odor—may be treated with lower dosages. However, green algae, which occasionally causes taste and odor problems, require much higher dosages. Yet, dosages higher than 0.8 mg/L are not recommended since they are lethal to most types of fish.

Potassium permanganate is an alternative to copper sulfate. Its effective dosage ranges from 0.4 to 4 m/L. Nonetheless, factors such as low solubility (5 g/100 mL at 20°C), higher cost ($2000/ton), and the development of a purple color generally discourage engineers and plant operators from using potassium permanganate.

Other methods that have been tried in the past include PAC—to create a "blackout" condition—and the application of chlorine compounds. Carbon treatment is usually limited to small reservoirs and is a temporary measure because the carbon settles out rather quickly unless a certain degree of agitation is provided in the reservoir. The effective dosage of PAC ranges from 2 to 10 mg/L. Chlorine compounds such as calcium hypochlorite and sodium hypochlorite may be used to control algae with 0.2–1 mg/L of residual free chlorine. Yet, their effectiveness is quickly lost with exposure to sunlight and water-borne organic materials.

Groundwater Source Management Groundwater is frequently contaminated by pollutants such as gasoline, industrial solvents, and a wide range of volatile organics. The prevention of groundwater contamination has just begun in recent years, but this issue will most likely continue being a source of engineering concern for several more years, since today's problems are generally the result of practices and politics that have been ongoing for the past 10–50 years.

Reservoir Management In regions that experience the four seasons, most lakes and reservoirs undergo thermal stratification. Under these conditions, a stagnation zone (hypolimnion) is formed and the water below 25 ft (7.5 m) is nearly stagnant during both the warm and cold months. Since the stagnation zone has little to no dissolved oxygen, the sulfates and nitrates serve as a source of oxygen for biochemical oxidations by anaerobic bacteria and the sulfate ion is reduced to sulfide ion to form hydrogen sulfide. The level of hydrogen sulfide formed in this manner may be as high as 5–6 mg/L. However, the well water in some areas may have hydrogen sulfide levels as high as 20–30 mg/L. Conversely, the water of impoundments overturns twice a year—spring overturning and autumn overturning—due to the change in water temperatures. This phenomenon brings the hydrogen sulfide, as well as other odor-producing compounds and unfavorable elements, to the surface.

Positive and effective control of hydrogen sulfide formation may be attained by mixing the entire body of lake water, thus preventing the formation of a stagnation zone within the reservoir. This artificial destratification control method has proved to be effective for most reservoirs. Mixing may be achieved through the use of mechanical mixers, pumps, or diffused air pipes.

The energy requirement for artificial destratification is amazingly low. For example, a case history showed that only one 30 hp compressor—operated four times during the summer season for periods of 4–6 days—was sufficient to destratify a lake having 142 acres (57.5 hectares) of surface and an average depth of 23 ft (7 m). The watershed management is also a preventive measure but this is not discussed here.

Removal at the Treatment Plant   If the preventative measures are ineffective or impractical, the problem of taste and odor must be controlled at the treatment plant. The three basic measures are aeration, oxidation, and adsorption.

Aeration   Aeration is a practical solution in controlling taste and odor problems caused by volatile compounds such as hydrogen sulfide. It is generally not the best method in controlling geosmin and MIB, the most frequent taste- and odor-producing compounds. Examples of the aeration process include the diffused, mechanical, nozzle spraying, multiple tray cascading, and the packed tower type.

Oxidation   In most cases, oxidation is the most practical and effective method of controlling taste and odor problems. Oxidation can be potassium permanganate oxidation, ozonation, ozone and hydrogen peroxide oxidation, chlorine dioxide, or chlorination.

The most frequently used oxidizing chemical is potassium permanganate. The dosage ranges from 0.5 to 5 mg/L; however, taste and odor are commonly reduced to an acceptable level of 3 TON through the use of 0.6–1.2 mg/L.

Ozonation is also an effective method of oxidation. This process actually changes the characteristics of the odor and flavor, in addition to reducing the level of the odor-producing compounds. The effective dosage of ozone ranges from 1 to 5 mg/L, with 2 mg/L being the average effective dosage. With this method, the key to successful taste and odor control is to provide alternative chemical feed points and an alternative chemical feed system—in case it becomes necessary to use an alternative oxidant.

Several pilot studies conducted in recent years indicate that the concurrent use of ozone and hydrogen peroxide may or may not be superior to the use of ozone alone. Thus, bench test or pilot studies should be conducted to evaluate the effectiveness of this method.

Chlorine dioxide has been used effectively in the past. However, its use is now discouraged by both the EPA and local branches of the Department of Health Services due to the potential harmful effects on human health. Similarly, chlorination is also an effective method of taste and odor control but its use must be carefully evaluated due to the possible formation of THM and chlorophenol (chlorophenol is produced when phenol is oxidized in water).

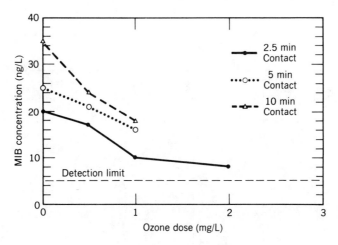

Figure 7.3.2-1   Effect of ozone contact time on MIB removal.

Generally, if any taste and odor problems are anticipated, the water treatment plant must have a potassium permanganate and a chlorine feed system. The design of these chemical feed systems is covered in Section 4.1. It is also essential to conduct bench scale studies to help select the optimum type and dosage of oxidant prior to plant design.

Figure 7.3.2-1 illustrates the relation between MIB concentration and ozone dosage with a contact time of 2.5, 5, and 10 min. Figure 7.3.2-2 presents the changes in flavor at various ozone dosages and contact times.

Adsorption   The two basic and effective adsorption processes in removing taste and odor are the addition of PAC to the process water and the use of GAC adsorption beds (to filter the water). The use of PAC adsorption is the most appropriate in situations where moderate taste and odor problems are infrequent: mid to small sized plants.

The two basic types of PAC storage and feed systems are dry carbon storage and dry feeding. The rule of thumb dictates that if the hourly feed rate of PAC is less than 150 lb, the PAC dry feeder should be used in combination with a solution mixing tank or a vortex mixer. If the hourly feed rate exceeds 150 lb and there is a need for frequent PAC addition, the use of a slurry feed system should be considered. The PAC dosage varies from 1 to 50 mg/L. However, based on past experience, 25 mg/L should be considered to be the maximum allowable dosage, regardless of the type of feeding (dry or slurry). Refer to Section 4.1 for the design considerations of the PAC feed system.

Whenever a moderate to severe chronic taste and odor problem exists, the use of GAC adsorption beds should be considered. The GAC adsorption is similar to ordinary filters; however, the empty bed contact time (EBCT) is a very important design issue. The EBCT values typically range from 3 to 10 min for the purpose of taste and odor removal; the filtration rate of the GAC

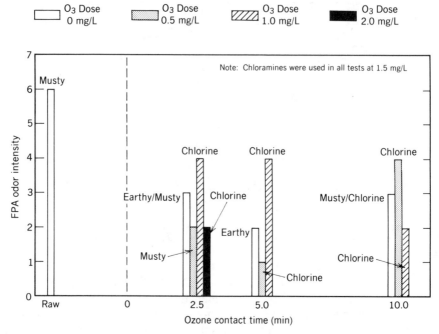

Figure 7.3.2-2   Changes of flavor and odor intensity by ozonation.

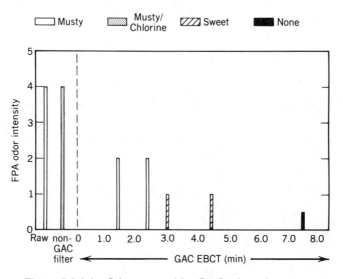

Figure 7.3.2-3   Odor removal by GAC adsorption process.

adsorption beds ranges from 3 to 6 gpm/ft^2 (7.5–15 m/h). The standard size of the medium is approximately the same as that of rapid sand filter beds and the depth of the layer is approximately twice the size. Operational experience gained from European facilities indicates that the GAC beds need to be regenerated every 4–5 years. Further details on the GAC adsorption process may be found in Section 7.7.

Figure 7.3.2-3 illustrates the effect of the EBCT on the removal of several different types of odor based on a pilot study. It is very important to run a bench scale or pilot test so that the appropriate type and size of GAC may be selected, and in order to find the optimum EBCT for dominant taste and odor compounds.

7.3.3 Hydrogen Sulfide Removal

The removal of hydrogen sulfide is discussed separately from the removal of other odor-producing compounds because of its unique characteristics. In natural water, under normal conditions, the sulfur species exists in five stable forms: namely, HSO_4^-, SO_4^{2-}, S^0, H_2S, and HS^-. Other species such as thiosulfate, polysulfide, and polythionate also exist in the natural water. However, they are not thermodynamically stable.

The following relationship exists among the sulfur species. In waters with a pH of 8 or below, a normal pH range for natural waters, H_2S and HS^- are the dominant form of sulfur. At pH 7, approximately 80% of the sulfur is in the form of hydrogen sulfide and almost 100% at pH 6. However, at pH levels of 8 and above, the reduced sulfur exists in the water as HS^- and SO_4^{2-} ions and the amount of free H_2S is very small and insignificant.

The most common method of removing hydrogen sulfide from water is oxidation. Adsorption by activated carbon is also an effective method but rather expensive. The oxidation process used in the field of water treatment includes aeration, chlorination, ozonation, and potassium permanganate oxidation.

Aeration There are three basic types of aeration: spraying, diffused air, and multiple tray. The aeration process is both practical and satisfactory if the level of total sulfides (expressed as H_2S) is less than 3–4 mg/L. It is essential to adjust the pH of the water to below 7, preferably 6, prior to aeration. Under the proper pH conditions, aeration may liberate CO_2 and other volatile organic substances. The effective air to water flow rate ratio should be in the range of 80–120:1 and the hydraulic loading for the tray type aerator—the type most commonly used in well water treatment—is 7–15 gpm/ft^2. For diffused air aeration, the power requirement for the blower is approximately 1 kW per each million gallons of water treated.

$$H_2S + \tfrac{1}{2}O_2 \rightarrow H_2O + S^0$$

Chlorination Chlorination is often used to oxidize hydrogen sulfide in well water. The reaction is as follows:

$$4 \, Cl_2 + H_2S + 4 \, H_2O \rightarrow H_2SO_4 + 8 \, HCl$$

Therefore, 8.3 parts of chlorine are required to oxidize one part of hydrogen sulfide. Factors that affect this reaction include pH, temperature, and reaction time. The oxidation rate sharply increase the pH to 6.5–7.0.

Ozonation The ozonation process is seldom used to oxidize hydrogen sulfide. The efficiency of this process is strongly affected by the level of pH; the pH of the water should be maintained at 7 or less. The reaction equation is

$$H_2S + 4 \, O_3 \rightarrow H_2SO_4 + 4 \, O_2$$

The ratio of ozone to hydrogen sulfide is 5.7:1.

Potassium Permanganate Potassium permanganate is a common oxidant that is often used in the water treatment process.

$$4 \, KMnO_4 + 3 \, H_2S \rightarrow S^0 + 2 \, K_2SO_4 + 3 \, MnO + MnO_2 + 3 \, H_2O$$

The above equation illustrates that 6.2 parts of potassium permanganate are required to oxidize one part of hydrogen sulfide. Once again, pH control plays an important role in maintaining the pH in the range of 6.5–7.

Elemental sulfur exists in water in a virtually insoluble form and thus may be removed by filtration. However, sulfur may also exist in water in colloidal form, as a milky-white emulsion. In this form, the sulfur may not be removed satisfactorily by ordinary filtration processes and the odor problem will remain. The use of sodium pyrosulfite or sulfur oxide has been recommended by Monscvitz in treating this colloidal form of sulfur.

7.3.4   Control in the Distribution System

Taste and odor problems may also occur in the distribution system, primarily as the result of corrosion of pipe material and/or the growth of iron bacteria such as *Crenothrix* and *Leptothrix* within the water main. Additionally, in cases where the sulfate content of the water is high and the water is allowed to stand in dead-end mains, taste and odor problems may be compounded by sulfate reducing bacteria. These problems can only be rectified if design engineers minimize the use of dead-end mains and provide blow-off and clean-out assemblies at strategic points in the distribution system. Furthermore, maintenance crews must keep the water lines clean by periodically flushing out deposits and microorganisms that accumulate within them. In cases where the interior of the pipe main is badly corroded, the main must be cleaned by the "pig" and relined with new cement.

Water treatment plants can also minimize these taste and odor problems by maintaining an adequate level of residual chlorine to combat bacteria in the distribution system and by taking anticorrosion measures—either adjusting the pH of the water based on the saturation index or through the addition of a corrosion inhibitor to the finished water.

BIBLIOGRAPHY

Ansleme, C. I. H., et al., "Effects of Ozonation on Taste and Odor," *J. AWWA*, 80(10):45(October 1988).

AWWA, "Research on Taste and Odors," Committee Report, *J. AWWA*, 62(1):59(January 1970).

AWWA, *Water Treatment Plant Design*, AWWA, New York, 1969.

Dougherty, J. D., and Morris, R. L., "Studies on the Removal of Actinomycetes Musty Tastes and Odors in Water Supply," *J. AWWA*, 59(10):1320(October 1967).

Krasner, S. W., McGuire, M. J., and Ferguson, V. B., "Taste and Odors: The Flavor Profile Method," *J. AWWA*, 77(3):34(March 1985).

Lalezary, S., Pirbazari, M., McGuire, M. J., and Krasner, S. W., "Air Stripping of Taste and Odor Compounds from Water," *J. AWWA*, 76(3):83(March 1984).

Lalezary, S., Pirbazari, M., and McGuire, M. J., "Oxidation of Five Earthy-Musty Taste and Odor Compounds," *J. AWWA*, 78(3):2(March 1986).

Meadows, M. D., "Reservoir Management," *J. AWWA*, 79(2):26(February 1987).

Means, E. G., "An Early Warning System for Taste and Odor Control," *J. AWWA*, 78(3):77(March 1986).

Medsker, L. L., Jenkins, D., and Thomas, J. F., "Odorous Compounds in Natural Waters," *Env. Sci. Technol.*, 2(6):461(June 1968).

Monscvitz, J. T., and Ainsworth, L. D., "Treatment for Hydrogen Polysulfide," *J. AWWA*, 66(9):537(September 1974).

Montgomery, J. M., Consulting Engineers, *Water Treatment—Principles and Design*, Wiley, New York, 1985.

Nelson, M. K., "Sulfide Odor Control," *J. WPCF*, p. 1285(October 1963).

Palmer, C. M., "Algae in Water Supplies," U.S. Department of Health, Education and Welfare, Water Supply, 1962.

Powell, S. T., "Removal of Hydrogen Sulfide from Well Water," *J. AWWA*, 40(12):1277(December 1948).

Raman, R. K., "Controlling Algae in Water Supply Impoundments," *J. AWWA*, 77(8):41(August 1985).

Schiller, B., "Vacuum Degasification of Water for Taste and Odor Control," *J. AWWA*, 47(2):124(February 1955).

Silvey, J. K., "Growth and Odor—Production Studies," *J. AWWA*, 64(1):35(January 1972).

Suttet, I. H., "An Evaluation of Activated Carbon for Drinking Water Treatment: A National Academy of Science Report," *J. AWWA*, 72(1):41(January 1980).

Vajdic, A. H., "Gamma-Ray Treatment of Taste and Odors," *J. AWWA*, 63(7):459(July 1971).

Walker, G. S., Lee, F. P., and Aieta, E. M., "Chlorine Dioxide for Taste and Odor Control," *J. AWWA*, 78(3):84(March 1986).

7.4 CONTROL OF THMs AND VOCs

Trihalomethanes (THMs) are by-products of chlorine disinfection. Synthetic volatile organic compounds (VOCs) are industrial wastes frequently found in many groundwaters. Both substances are categorized by the EPA as a threat to public health since most of them are suspected of being carcinogenic compounds. Since THMs and VOCs are organic contaminants that predominantly cause problems in the field of water supply, among others, the control methods are briefly discussed.

7.4.1 THMs

THMs are formed by chlorination of naturally occurring organic precursors by free residual chlorine. The organic precursors are generally humic substances and fluvic acids. Chloramines are also capable of producing THMs but at such low levels that this process is usually considered to be a non-THM-forming process.

In 1979, the EPA established a maximum contaminant level (MCL) of 0.1 mg/L for the total trihalomethane (TTHM) concentration. TTHM is defined as the sum of trichloromethane ($CHCl_3$), tribromomethane ($CHBr_3$), bromodichloromethane ($CHBrCl_2$), and dibromochloromethane ($CHBr_2Cl$).

In order to be able to establish THM control methods and to comply with the THM regulation, the characteristics of THM formation under different conditions must be understood. There are four basic definitions for THM: instantaneous THM, terminal THM, THM formation potential, and maximum total THM potential.

Instantaneous THM   The instantaneous THM is the THM concentration at the moment of sampling. Compliance with the MCL of THM is determined by this value.

Terminal THM   The THM concentration at the farthest end of the distribution main is known as the terminal THM. If a bench scale test is used to evaluate the terminal THM, the test should simulate the actual chlorine contact time in the treatment plant, plus the anticipated detention time in the distribution system. Furthermore, the temperature and pH of the sample water should be identical to the actual water treatment and supply conditions.

THM Formation Potential   The THM formation potential is determined by subtracting the instantaneous THM concentration from the terminal THM.

Maximum Total THM Potential   This value represents the maximum THM concentration level under the most favorable conditions for THM formation. The test condition, as defined by the EPA, requires either a period of 7 days at a water temperature of 25°C with a minimum of 0.2 mg/L of free chlorine residual or an initial chlorine concentration of 5 mg/L with a pH of 9–9.5 for a period of 7 days
   The rate of THM formation and the terminal THM concentration depend on five major factors: the amount of organic precursors, level of free residual chlorine, water temperature, water pH, bromide concentration in the water, and chlorine contact time.

Precursors   If organic precursors are not present in the water, there will be no THM formation. Thus, the concentration and type of precursor have direct impact on THM formation. The purpose of the maximum total THM potential test is to measure the quantity of organic precursors available to react with free residual chlorine.

Level of Free Chlorine Residual   Since THM is a by-product of chlorination, the amount of chlorine residual directly influences the level of THM formation. Therefore, the initial mixing conditions and the design of the contact tank also affect the rate of THM formation.

Water Temperature   Water temperature directly influences the rate of THM formation: a higher temperature induces a faster reaction rate and therefore results in a higher rate of formation. In many cases, THM levels during the summer months, when the water temperature is 25°C, is nearly twice as great as THM levels during the winter season (water temperature is generally 4°C). Thus, no special THM control measures are necessary during the winter months. Although water temperature is a major factor in the rate of THM formation, the type and concentration of organic precursors also greatly affect this phenomenon.

pH of the Water   The rate of THM formation and terminal THM are greatly affected by the pH of the water. The higher the pH, the faster the reaction rate and the higher the THM levels. This phenomenon is believed to be due to pH-induced changes in the functional groups of the precursor molecules.

Bromide Concentration   The rate of reaction between bromide and the precursor molecules is faster than that of chlorine and the precursors: the level of THM increases within a shorter period of time. If bromide is present in the process water, the concentration of terminal THM is higher than the chlorine compounds.

The mechanism involved in this reaction is that the bromide ion is oxidized by free residual chlorine to form bromine, which is capable of quickly reacting with precursors to form THMs. In coastal regions the bromide ions are often supplied to surface water, as well as groundwater, from seawater. Yet, bromide ions may also be present in water obtained from many other sources, in regions that were below sea level during prehistoric times.

Chlorine Contact Time The THM level is a function of chlorine contact time: the longer the contact time, the higher the level of THM concentration. This phenomenon is particularly distinct during the first several hours but generally tapers off after 2–3 days.

7.4.2 THM Control Measures

THM control measures should be implemented in two basic types of situation: the design of a new water treatment plant and during the modification of an existing plant (due to high THM levels). The THM control strategies are the same for both cases and the basic steps are as follows.

Step 1. Look for alternative local water supplies such as groundwater. If alternative supplies do not exist or are not economically feasible, the second step should be evaluated.

Step 2. This step contains several options, which are modifications of the conventional treatment technique.

   (i) Improve flocculation and sedimentation to increase the removal of organic precursors (enhanced coagulation).
   (ii) Change the points of chlorination to shorten the contact time.
   (iii) Substitute potassium permanganate for prechlorination and use small amounts of chlorine dioxide as pre-oxidant.
   (iv) Use chloramines as an alternative disinfectant. This option should also include the addition of ammonia 10–20 min after chlorine application and observation of the THM formation level.
   (v) Add PAC to remove precursors and THM (seasonal use).

If any of these options appears feasible, after bench scale test evaluation, a pilot scale test (or even an actual plant) should be conducted to confirm the effectiveness of the option and to establish the anticipated costs.

Should any of the second step options fail, a more drastic and expensive treatment technique should then be evaluated (step 3).

Step 3. The alternatives in this step of the THM control strategies are as follows:

   (i) Use pre-ozonation to oxidize and remove organic precursors.
   (ii) Strip THMs through the use of an aeration tower.
   (iii) Provide a GAC adsorption bed to remove THMs.

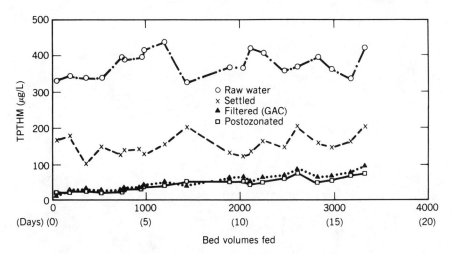

Figure 7.4.2-1 Total potential trihalomethane (TPTHM) reduction through pilot plant processes (Adapted from The Predesign Report for NBR Water Treatment Plant by J. M. Montgomery, Consulting Engineers.)

(iv) Utilize a membrane process to remove THMs.

(v) Change to a water supply source containing a minimal level of precursors.

Use of ozone and the GAC adsorption bed have been reported to be effective means of controlling THM levels since the mid-1970s; technical articles concerning this subject may be found elsewhere. Figure 7.4.2-1 is an example of the effective control of THMs using a combination of GAC adsorption bed and postozonation. In this case, the THMs were removed by GAC, and postozonation was necessary to combat the significant leakage of microorganisms from the GAC bed: the GAC bed is literally a nest of microorganisms due to the abundant supply of nutrients.

7.4.3 Operational Issues

The plant operator must run the treatment plant in a manner that produces a processed water that is in compliance with the MCL of the THM standard. Refer to Figure 7.4.3-1 for the appropriate steps. The first of these steps is the quarterly sampling requirement. The quarters end on March 31, June 30, September 30, and December 31. The THM levels of these samples must be submitted to the regulatory agency 30 days after receipt of the analysis. The sampling requirements and the conditions for reduced sampling frequency are different for surface water and groundwater supplies.

Quarterly Sampling for Surface Water Supplies Each treatment plant is required to take and analyze a minimum of four samples per quarter. The samples must be collected from the distribution system within 24 h of each

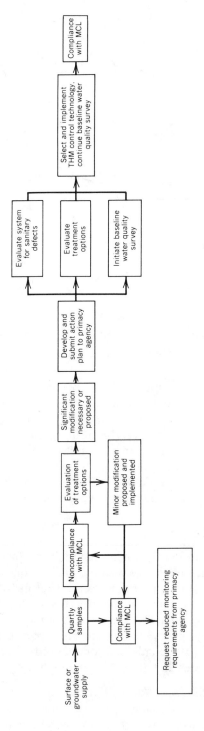

Figure 7.4.3-1 Steps required to comply with MCL of the THM regulation.

other and at least 25% must be collected from a location that reflects the maximum resident time within the system. The remaining samples are collected from representative locations in the distribution system based on the individual served, the water source, and the treatment methods used.

Compliance with the MCL of THM standard is based on a running average of the four quarterly averages. If the instantaneous TTHM concentration never exceeds 0.1 mg/L after 1 year of compliance monitoring, a request to decrease the monitoring frequency can be made to the primary agency. If the agency agrees to a reduced sampling frequency of one sample per quarter, this sample should be collected at a point (in the distribution system) reflecting the maximum resident time.

Quarterly Sampling for Groundwater Supplies The quality of groundwater is generally more consistent and has a lower concentration of precursors than surface water. For these reasons, the monitoring requirement for THMs are less stringent than for surface water supplies.

Distribution systems that use groundwater as their sole source of water may have a monitoring frequency as low as one maximum total THM potential test per year per aquifer. This sample must be collected from a location in the distribution system with the longest residence time. However, the primary agency will only agree to this reduced frequency if the utility can substantiate that the maximum total THM potential is less than 0.1 mg/L during periods of maximum total THM potential formation (high temperatures and pH) and/ or during periods of highest total organic compound (TOC) concentrations.

7.4.4 VOCs

It has been estimated that approximately 20% of all groundwater supplies in the United States are contaminated by synthetic volatile organic compounds (VOCs). This constitutes a real threat to public health since approximately 80% of the public water supplies depend on groundwater and 95% of rural areas use groundwater to satisfy domestic water needs.

Volatile organic chemicals are widely employed in industrial, agricultural, commercial, and domestic activities. Yet, it is believed that the majority of VOC contamination (of groundwater) results from improper disposal, leaks, or spills of industrial wastes. In some instances, VOCs may enter the potable water as a by-product of chlorination: certain joint compounds used in reservoir membrane liners and certain types of plastic pipe or plastic pipe liner react with chlorine to yield carbon tetrachloride.

Once the VOCs enter an aquifer, they can be transported over great distances due to their low affinity for soil, the lack of opportunity to evaporate, and the minimal chance for biodegradation.

Recent federal and state studies reveal that the VOC most frequently found in the highest concentrations is trichloroethylene (TCE); tetrachloroethylene (PCE) ranks second. The most common VOCs and their sources are briefly discussed.

Trichloroethylene (TCE) TCE is a nonflammable liquid with an odor resembling chloroform. Its primary application is as a degreasing solvent in the metal industry. Some household products, such as spot removers, rug cleaners, dry cleaning liquid, and air fresheners, also contain TCE.

Tetrachloroethylene (PCE) Tetrachloroethylene is also known as perchlorethylene and is therefore commonly abbreviated as PCE to distinguish it from trichloroethylene (TCE). PCE is a colorless, nonflammable liquid with an etherlike odor. It is primarily used as a commercial dry cleaning solvent. Other usages vary from rug and upholstery cleaner; stain, lipstick, and rust remover; degreaser (metal industry); textile scouring; to vegetable fumigant.

Carbon Tetrachloride (Tetrachloromethane) This VOC is a colorless, nonflammable, heavy liquid used as a fire extinguisher; in the cleaning of cloth; as a solvent for oil or fat; and as a fumigant. Its application in consumer goods was banned in 1970 and it ceased being used as an aerosol propellant in 1978.

1,1,1-Trichloroethane (Methylchloroform) This compound is a nonflammable liquid that is virtually insoluble in water. It is primarily used for cold type metal cleaning. Other applications include the cleaning of plastic molds and drains, leather tanning, degreasing of septic tanks, and as a shoe polish.

1,2-Dichloroethane This chemical is an intermediary in the manufacturing of vinyl chloride monomers and chlorinated solvents such as TCE and PCE. It is used as a solvent in cleaning textiles, PVC processing equipment, and so on. It is also used in the manufacturing of paint, adhesives, and other similar substances.

Vinyl Chloride (Chlorethylene) This compound is a colorless, flammable gas that is soluble in alcohol, ether, carbon tetrachloride, and benzene; it is also slightly soluble in water. Vinyl chloride is used primarily in the manufacturing of plastic and PVC resins. Since plastic and PVC products are widely consumed in the modern world, vinyl chloride is used extensively by many industries.

Benzene (Benzol) Benzene is a colorless and highly flammable liquid, possessing a characteristic odor, which is fairly soluble in water. Originally derived from coal, benzene has been used in the manufacturing of a wide variety of goods: medicinal chemicals, dyes, artificial leather, linoleum, oil cloth, varnishes, lacquers, and numerous other organic compounds and goods.

7.4.5 VOC Control Measures

The VOC control measures can be classified into two basic schemes: management and treatment. Each of these is discussed.

Management Control Measures   The management alternatives are (1) finding a new source, (2) blending contaminated water with water of a good quality, (3) containment of pollutants, and (4) elimination of the pollution source.

The first alternative is generally not feasible since it requires a significant expenditure of capital or is simply not practical. The blending alternative may be attractive to some water purveyors: those who own multiple wells or own both groundwater and surface water supplies. Yet, this second alternative is limited by the lack of system flexibility, insufficient dilution capability, and lack of quality water sources. The third and fourth alternatives are theoretically good but are very difficult to implement because there is no easy way of locating the source of contamination, predicting the migration of pollutants, or stopping the migration.

Treatment Control Measures   The treatment control strategies are (1) stripping by aeration, (2) removal by adsorption, and (3) a combination of aeration and adsorption.

Air-Stripping Method   The process of aeration transfers volatile substances from water to air or from air to water. The former process is defined as air-stripping.

Air-stripping has been used to strip carbon dioxide, hydrogen sulfide, and certain taste- and odor-producing compounds from water. It has also been used successfully in the removal of VOCs. However, air-stripping has certain limitations: the process is temperature dependent and is therefore not applicable to treatment plants located in cold climates; its exhaust pollutes the air; and a high air to water ratio is required to effectively remove VOCs. Other minor drawbacks include the potential degradation of water quality by airborne particulates; the oxidation of the reduced form of metallic inorganics, thereby promoting floc formation; the increased potential for corrosion-related problems due to the increase in dissolved oxygen; and increased biological growth within the unit.

Three basic types of air-stripping method have been tested and applied to actual treatment plants: (1) diffused air, (2) spray, and (3) packed tower. According to Hess (1983) (Figure 7.4.5-1), if over 90% of the VOCs must be removed, the packed tower type of air-stripping is the only method capable of meeting this requirement.

When designing air-stripping units, five basic design considerations (minimum) must be addressed:

1. Characteristics of the VOCs to be removed.
2. Temperature of the water and air.
3. Air to water ratio.
4. Contact time.
5. Required surface area for mass transfer.

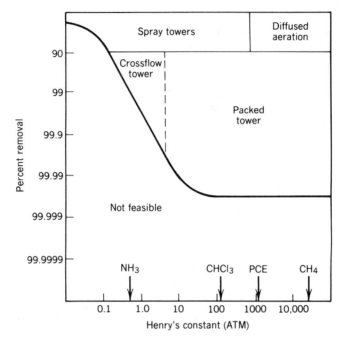

Figure 7.4.5-1 Selection of a feasible aeration process for VOCs. (Adapted from A. F. Hess et al., AWWA Research Foundation Report, 1983.)

Generally, an air to water ratio (cfm/cfm) ranging from 20:1 to 25:1 is required to obtain over 90% removal of both TCE and PCE. Based on the results of several pilot scale studies, this air to water ratio must be at least 20:1 for THM removal. The basic formula for designing a packed tower was developed by chemical engineers prior to the mid-20th century. However, modifications of this basic design (for special water treatment applications) have been published by several people in the water treatment industry within the past 10 years; these articles are listed in the Bibliography at the end of this section.

Adsorption Methods   The three basic methods of adsorption used in the removal of VOCs are PAC, synthetic resin adsorption, and GAC. PAC adsorption is not very promising in the removal of VOCs due to the nature of the medium: Compounds with high molecular weights tend to adsorb more readily to PAC; but VOCs are low molecular weight compounds that do not readily adsorb to PAC unless very high dosages are used.

Synthetic resins such as Amersort XE-340, manufactured by Rohm and Hass, were designed to remove nonpolar organic compounds with low molecular weights. This alternative has had some promising results but is not yet commercially available.

Currently, the most effective adsorption process for VOCs is the GAC bed. Results obtained from pilot studies indicate that this type of bed effectively removes PEC and other VOCs. The prime factors that affect the efficiency of adsorption are solubility and affinity. Thus, a crucial consideration for design engineers is the potential desorption of VOCs and THM from the GAC bed. The desorption process can result from competitive adsorption and from the reversal of the adsorption process due to a decreased influent concentration. Since the adsorption rate is a function of water temperature, higher temperatures improve the adsorption rate; pH does not affect the adsorption rate of VOCs.

Combination of Aeration and Adsorption   This alternative is an effective method of VOC removal. Aeration reduces the loading of the contaminants to the GAC bed, thereby decreasing the frequency of regeneration for the GAC bed.

BIBLIOGRAPHY

Aieta, E. M., Regan, K. M., Lang, I. S., McReynolds, L., and Kang, J., "Advanced Oxidation Processes for Treating Groundwater Contaminated with TCE and PCE: Pilot Scale Evaluation," *J. AWWA*, 80(5):64(May 1988).

AWWA, *Organic Chemical Contaminants in Ground Water: Transport and Removal*, AWWA Seminar Proceedings (June 1981). No. 20156.

AWWA Research Foundation, *Occurrence and Removal of VOC from Drinking Water*, Cooperative Research Report (1983).

AWWA Research Foundation, *Advanced Oxidation Process for Control of Off-gas Emissions from VOC Stripping*, AWWA Research Fund/AWWA, (October 1989).

Bilello, L. J., and Singley, J. E., "Removal of Trihalomethanes by Packed-Column and Diffused Aeration," *J. AWWA*, 78(2):62(February 1986).

Crittenden, J. C., Hand, D. W., Arora, H., and Lykins, B. W. Jr., "Design Considerations for GAC Treatment of Organic Chemicals," *J. AWWA*, 79(1):74(January 1987).

Crittenden, J. C., Cortright, R. D., Rick, B., Tang, S., and Perram, D., "Using GAC to Remove VOC's from Air Stripper Off-Gas," *J. AWWA*, 80(5):73(May 1988).

Glaze, W. H., and Kang, J., "Advanced Oxidation Process for Treating Groundwater Contaminated with TCE and PCE: Laboratory Studies," *J. AWWA*, 80(5):57(May 1988).

Hand, D. W., "Design and Economic Evaluation of a Treatment of VOC's from a Contaminated Groundwater," *J. AWWA*, 78(9):87(September 1986).

Hand, D. W., Crittenden, J. C., Gehin, J. L., and Lykins, B. W. Jr., "Design and Evaluation of an Air-Stripping Tower for Removing VOCs from Groundwater," *J. AWWA*, 78(9):87(September 1986).

Hand, D. W., Crittenden, J. C., and Thacker, W. E., "Simplified Model for Design of Fixed-Bed Adsorption Systems," *J. Environ. Eng.*, 110(2):440(April 1984).

Hess, A. F., Dyksen, J. E., and Dunn, H. J., "Control Strategy—Aeration Treatment Technique"—Occurrence and Removal of VOCs from Drinking Water. A Report by AWWA Research Foundation, Denver, CO, 1983.

Kavanaugh, M. C., and Trussell, R. R., "Design of Aeration Towers to Strip Volatile Contaminants from Drinking Water," *J. AWWA*, 72(12):684(December 1980).

Lykins, B. W. Jr., Clark, R. M., and Adams, J. Q., "GAC for Controlling THM's," *J. AWWA*, 80(5):85(May 1988).

Perry, R. H., and Green, D. W., *Chemical Engineering Handbook*, 6th ed., McGraw-Hill, New York, 1986.

Ram, N. M., Christman, R. F., and Cantor, K. P., *Significance and Treatment of VOC in Water Supplies*, Lewis Publishers, Chelsea, MI, 1990.

Singley, J. E., *Trace Organics Removal by Air Stripping*, AWWA Research Foundation, Denver, CO, 1980.

Snoeyink, V., "Adsorption as a Treatment Process for Organic Contaminant Removal from Groundwater," in *Proceedings of AWWA Seminar*, 1981.

Symons, J. M., *Treatment Techniques for Controlling THM's in Drinking Water*, EPA 600/2-81-156 (September 1981).

Umphries, M. D., Tate, C. H., Kavanaugh, M. C., and Trussell, R. R., "A Study of Trihalomethane Removal by Packed-Tower Aeration," *J. AWWA*, 75(8):414(August 1983).

7.5 FLUORIDATION AND FLUORIDE REMOVAL

During the early 20th century, studies confirmed that drinking water containing fluoride ion concentrations in excess of 1.5 mg/L caused dental fluorosis (mottled enamel of human teeth) and that the severity of mottling increased with an increase in the fluoride level. Other studies found that dental caries were less prevalent in people with mottled enamel and concluded that it was desirable for the public water supply to have a fluoride ion concentration of approximately 1 mg/L to promote optimal dental health.

The fluoride concentration of potable water should be maintained at an optimal level based on the figure set by the local health department. Consequently, design engineers must design either a fluoride feed system to maintain an optimal level of fluoride or a process that removes fluoride when the water supply contains excessive levels of fluoride.

7.5.1 Fluoridation

General Considerations   As previously mentioned, the recommended level of fluoride in potable water is 1 mg/L. The Department of Health Service suggests that this figure be an annual average at water temperatures near 15°C; the recommended fluoride concentration falls to 0.8 mg/L during hot months (summer) and rises to 1.2 mg/L during cold seasons. The reasoning for the adjusted levels is that the average water intake by consumers varies with the seasons—more water is consumed during warm seasons than the winter months.

Fluoride is not found in the free state; it always exits in combination with other elements. All fluoride compounds dissociate to yield fluoride ion when added to the process water.

The most frequently used fluoride compounds are sodium fluoride (NaF), sodium silicofluoride (Na_2SiF_6), and hydrofluosilicic acid (H_2SiF_6). These three compounds are chosen because of their availability, reasonable cost,

TABLE 7.5.1-1 Fluoride Compounds and Feed System

Item	Na_2SiF_6	NaF	H_2SiF_6
Commercially available form	Dry	Dry	Liquid
Commercial purity (%)	98–99	90–98	23–24
Molecular weight (MW)	188.05	42.0	144.08
Specific gravity	—	—	1.220 at 17.5°C (10.17 lb/gal)
Percentage of fluoride in MW	60.7	45.25	79.2
Solubility (g/100 mL at 25°C)	0.762	4.05	Infinite
Freezing point of the solution	0°C	0°C	−22°C
Shipping containers	100 lb bags or 125–400 lb drums or in bulk form		55 gal drums or in bulk form
Approximate cost ($/lb)	0.25	0.45	0.3
Feed system components	• Dry chemical storage • Water softener if total hardness of tap water exceeds 75 mg/L as $CaCO_3$ • Two solution tanks for a batch system • A scale to monitor daily dosage • A reliable metering pump • A dust collector for dry feeder or solution preparation area		• Liquid storage tanks such as steel tank with Neoprene lining • A day tank with a scale to monitor daily usage • A reliable metering pump

and safety. The solubility of the first two compounds is very low and the maximum solution concentration is 4% for sodium fluoride and only 0.76% for sodium silicofluoride at 25°C. Fluosilicic acid is available in liquid form and its solubility is therefore infinite. However, the concentration of commercially available solutions are generally only 23%. Refer to Table 7.5.1-1 for a summary of the chemical characteristics and the major components of the feed system.

Both sodium fluoride and sodium silicofluoride are available in powder form or as fine crystals. Thus, problems associated with dust must be addressed during preparation of the solution. Plant operators must wear the proper type of protective gear and use an exhaust fan when handling the storage bags because the dust is harmful to human health. Furthermore, if the hardness of the dissolving water exceeds 75 mg/L (as $CaCO_3$), it must be softened prior to preparing the fluoride solution or the fluoride ion will react with calcium and magnesium: the resulting precipitate is capable of clogging the feeders and the piping system.

Fluosilicic acid is much easier to handle than the other two compounds but is a very corrosive chemical (strong acid). Moreover, if the process water has low alkalinity, the application of a very small amount of this chemical, to produce a dosage of 1 mg/L, can significantly depress the pH of the water.

The design of these chemical feed systems should follow the guidelines presented in Section 4.1.

Example Design Calculation

Example 1

Given A 20 mgd water treatment plant in Oklahoma requires a new fluoridation system. The water supply consistently has a natural fluoride concentration of 0.3–0.4 mg/L. Hydrofluosilicic acid will be used for fluoridation because of its easy handling and safety (to the operator).

Determine

(i) The required type and capacity of the feeder.
(ii) The capacity of the day tank.
(iii) The capacity of the storage tank.
(iv) The application point.

Solution

(i) Since the commercially available product is a 23% solution and the percentage of fluorine in the fluosilicic acid is 79.2%, based on the molecular weight derived from the chemical formula, the fluorine content ($\%$) of the product is

$$79.2 \times 0.23 = 18.2\%$$

Maximum dosage as fluoride ion $= (1 - 0.3)$

$$\times\ 8.34 \times 20 = 116.8 \text{ ppd}$$

Minimum dosage as fluoride ion $= (1 - 0.4)$

$$\times\ 8.34 \times (20 \times 0.25) = 25 \text{ ppd}$$

Knowing that the specific gravity of the solution is 1.220 at 17.5°C, the weight of the solution is

$$8.34 \times 1.22 = 10.17 \text{ lb/gal}$$

Thus, the actual chemical dosages in gallons per hour are

Maximum dosage of 23% solution $= 116.8 \text{ ppd} \div (10.7 \times 0.182)$

$$= 63 \text{ gpd} \quad \text{or} \quad 2.6 \text{ gph}$$

Minimum dosage of 23% solution $= 25 \text{ ppd} \div (10.17 \times 0.182)$

$$= 13.5 \text{ gpd} \quad \text{or} \quad 0.56 \text{ gph}$$

The required capacity for the metering pump is therefore 3 gal/h. Since the turn-down ratio of any metering pump is 10:1, the minimum dosage is 0.3 gph.

The type of pump that should be specified is a diaphragm–plunger combination pump, because it is best for handling very corrosive solutions. The pump should have "Hastelloy C" stainless steel parts, a polyethylene or Neoprene diaphragm, and PVC pipings.

Two identical pumps are provided, with one acting as a standby. The metering pumps are preferably paced to the plant flow rate.

(ii) The maximum amount of chemical used per day is 63 gal. Therefore, provide a day tank with a minimum capacity of 80 gal/day. This tank should be set on a weighing scale to facilitate the monitoring of daily use.

(iii) Bulk shipments of hydrofluosilicic acid are more cost effective and are safer: operators do not have to frequently handle the 55 gal drums. Since the capacity of a tank (trailer) is 4000 gal, a storage tank with a 5000 gal capacity should be installed. This storage tank should be fabricated from steel, for protection from tornadoes and potential vandalism, and lined with Neoprene rubber to protect the steel from corrosion. Based on the average dosage, the 4000 gal supply provides approximately 100 days worth of storage.

The tank is installed outside for reasons of safety. A heating system is not provided since the ambient temperatures in this area do not fall below $-22°C$ ($-8°F$), which is the freezing point of this chemical.

(iv) The application point for fluosilicic acid is at the inlet to the clearwell. A diffuser with a static mixer is provided for good mixing. This location is chosen because application of fluoride, as well as lime (for pH control), at the head of the plant results in the loss of some of the added fluoride due to alum flocculation. Refer to Section 7.5.2 on fluoride removal.

7.5.2 Fluoride Removal

The two alternative methods for removing excess fluoride from the water supply are chemical precipitation and the process of ion exchange.

Chemical Precipitation The fluoride content of the process water can be reduced to an acceptable level by means of either conventional alum flocculation and sedimentation, provided that a large alum dosage (200–300 mg/L) is used, or a lime softening process. Results obtained from bench scale tests indicate that fluoride concentrations up to 3.5 mg/L (raw water) may be reduced to 1 mg/L by employing alum dosages up to 350 mg/L. Lime softening also effectively reduces fluoride to a safe level by producing calcium fluoride precipitates.

Ion Exchange Process There are presently only three types of ion exchange material which have practical application in defluoridation: (1) tricalcium phosphate, including bone char or bone meal; (2) activated alumina; and (3) ion exchange resins. The effectiveness of these materials has been proved by actual plant scale tests and/or pilot scale tests.

Bone Char and Bone Meal These two materials are very similar in that they are both composed of processed bone material. Their only difference is the slight variation in the manufacturing process and products.

An example of defluoridation by ion exchange using bone char is as follows:

$$Ca(PO_4)_6 \cdot CaCO_3 + 2F^- \rightarrow Ca(PO_4)_6 \cdot CaF_2 + CO_3^{2-}$$

Bone char, expressed as $Ca_3(PO_4)_2$, can be regenerated through the application of 1% caustic soda.

Activated Alumina (Al_2O_3) Fluoride removal by means of activated alumina has been demonstrated by an actual plant (Gila Bend, Arizona) and several pilot scale studies. In each of these cases, granular activated alumina (filter sand size)—approximately 5 ft deep—was used to reduce fluoride from an initial level of 3 mg/L to less than 0.5 mg/L using a filtration rate of 5–7 gpm/ft^2 (12.5–17.5 m/h). These studies also demonstrated that it was necessary to adjust the pH of the raw water to 5.5 prior to filtration in order to maximize the efficiency of fluoride removal. Regeneration of the bed was achieved through the addition of 1% caustic soda solution.

Ion Exchange Resin Process Fluoride can be removed from the process water through the use of ordinary cation exchange and anion exchange resin columns arranged in series. The first column exchanges its polarized hydrogen with sodium to form the equivalent acid:

$$2NaF + H_2R^+ \rightarrow H_2F_2 + Na_2R^+$$

The hydrogen fluoride is subsequently removed by the second column by the following reaction:

$$2R^- + H_2F_2 \rightarrow 2RHF$$

Design engineers should note that one problem does exist for all three types of ion exchange: the disposal of the concentrated fluoride. These processes cannot be implemented unless proper disposal methods are used or an appropriate site is found.

As a final note, it must be stressed that both bench scale and pilot scale tests must be conducted prior to the design of the fluoride removal process. These procedures are of extreme importance because defluoridation is significantly affected by the presence of other ions in the raw water for these ions compete with the fluoride removal reaction.

BIBLIOGRAPHY

AWWA, *Water Fluoridation—Principle and Practice*, AWWA Manual No. M4, AWWA, Denver, 1977.

Harmon, J. A., and Kalichman, S. G., "Defluoridation of Drinking Water in Southern California," *J. AWWA*, 57(2):245(February 1965).

Kreft, P. H., "Removal of Fluoride by Actived Alumina," Masters Thesis, Loyola University, Los Angeles, 1986.

Maier, F. J., "Defluoridation of Municipal Water Supplies," *J. AWWA*, 45(8):879(August 1953).

Raboski, J. B., and Miller, J. P., "Fluoride Removal by Lime Precipitation and Alum and Polyelectrolyte Coagulation," in *Proceedings of the 29th Purdue Industrial Waste Conference*, Purdue University, Lafayette, IN, 1974, p. 669.

Savinelli, E. A., "Defluoridation of Water with Activated Alumina," *J. AWWA*, 50(1):33(January 1958).

Sawyer, C. N., *Chemistry for Sanitary Engineers*, McGraw-Hill, New York, 1960.

Smith, H. R., and Smith, L. C., "Bone Contact Removes Fluoride," *Water Works Eng.*, 90(5):600(1937).

Williams, R. B., and Culp, G. L., *Handbook of Public Water Systems*, Van Nostrand Reinhold, New York, 1986.

Wu, Y. C., and Nitya, A., "Water Defluoridation with Activated Alumina," *J. Environ. Eng. Div. ASCE*, 105(EE2):359(1979).

7.6 CORROSION CONTROL

Businesses in the industry of water treatment and supply are hampered by the serious problem of corrosion (iron pipes and metals) and degradation of concrete. Corrosion protection is therefore a major design consideration for design engineers, while painting, coating, proper maintenance of the cathodic protection system (for metal structures), and the continuous feeding of corrosion inhibitors to the distribution system are the major tasks of the plant operation personnel.

Corrosion is far from being a simple phenomenon. It is a combination of many factors: electrochemical, physical, chemical, biological, and metallurgical. Since corrosion and corrosion control are subjects which on their own could fill a book, only a brief discussion is presented here.

7.6.1 Corrosion of Metals

The electrode potential is the potential difference between a metal and a solution of its ions. It is expressed by the Nernst equation:

$$E = \frac{RT}{nF} \ln \frac{P}{P'} = \frac{RT}{nF} \ln \frac{C}{C'}$$

where n = valency of the metal ions,
  T = absolute temperature,
  R = gas constant,
  P = pressure at which the metal dissolves
  P' = osmotic pressure of the solution,
  C = a constant,
  C' = activity of the metal ions in the solution,
  F = Faraday number.

The electrode potential is standard for a standard solution of its ions and is expressed as E_0, thus simplifying the Nernst equation to give the general equation defining the electrode potential of a metal at 25°C:

$$E = E_0 + \frac{0.058}{n} \log C'$$

In the case of iron, this equation becomes

$$E = -0.44 + 0.029 \log Fe^{2+}$$

Table 7.6.1-1 presents the oxidation reactions and potentials for the major types of common metal at 25°C. Metals with positive electrode potentials are classified as noble metals (cathodic); those with negative electrode potential are base metals (anodic).

The corrosion of iron is caused primarily by electrochemical cell action. The presence or absence of oxygen alters the rate of corrosion. Other factors that influence the corrosion of iron are the total mineral concentration, the presence of dissolved gases, the temperature of the water, the surface condition of the iron, the flow velocity across its surface (pipe flow), and biological action.

General Corrosion of Metals   General corrosion of metals is induced by galvanic action. A good example is the case where two dissimilar metals are connected and exposed to water: one metal becomes cathodic while the other anodic, thus setting up a galvanic cell. For instance, when a steel and copper pipe are connected (which is often the case in most household plumbing), the steel pipe becomes the anode. Since metal loss occurs at the anode, the steel pipe corrodes. As shown in Table 7.6.1-1, there is a 0.96 V difference in electrode potential between steel and copper, and the corrosion is therefore

TABLE 7.6.1-1   Oxidation Potentials Against Normal Hydrogen Electrode at 25°C

Metal	Electrode Reactions	Equilibrium Potential (V)
Magnesium	$Mg = Mg^{2+} + 2e^-$	-2.34
Beryllium	$Be = Be^{2+} + 2e^-$	-1.70
Aluminum	$Al = Al^{3+} + 3e^-$	-1.67
Manganese	$Mn = Mn^{2+} + 2e^-$	-1.05
Zinc	$Zn = Zn^{2+} + 2e^-$	-0.762
Chrome	$Cr = Cr^{3+} + 3e^-$	-0.71
Iron	$Fe = Fe^{2+} + 2e^-$	-0.440
Nickel	$Ni = Ni^{2+} + 2e^-$	-0.250
Lead	$Pb = Pb^{2+} + 2e^-$	-0.126
Hydrogen	$H_2 = 2H^+ + 2e^-$	-0.000 by convention
Copper	$Cu = Cu^{2+} + 2e^-$	$+0.345$
Copper	$Cu = Cu^+ + e^-$	$+0.522$
Silver	$Ag = Ag^- + e^-$	$+0.800$
Platinum	$Pt = Pt^{2+} + 2e^-$	$+1.2$ approx.
Gold	$Au = Au^{3+} + 3e^-$	$+1.42$
Gold	$Au = Au^+ + e^-$	$+1.68$

Source: Adapted from Degremont, *Water Treatment Handbook*, 5th ed., Halsted Press, New York, 1979.

severe. The corrosion rate of metals that are close to one another in the table is less than that of metals that are widely separated.

Just as in the case where dissimilar metals are exposed to water of a uniform concentration, a galvanic current is also produced when a single metal is exposed to water of varying concentrations (ionic strengths). The portion of the metal located in the more concentrated solution becomes the anode and therefore corrodes. This process is known as concentration cell corrosion.

As long as a treatment plant is characterized by poor design and/or has deposits of debris on the metal surface, concentration cell corrosion may occur at any plant site. These factors create a localized concentration of a specific element such as oxygen, sulfate, or chloride, which is significantly different from other portions of the water. For instance, the oxygen concentration beneath the deposits of debris is less than the surrounding water; thus, this portion of the metal becomes anodic and acts as the starting point for corrosion. This type of corrosion mechanism is called an oxygen differential cell. Coexisting with most oxygen cells are the corresponding sulfate and chloride concentration cells. These coexist because chloride and sulfate ions penetrate the deposit and become concentrated.

The corrosion rate is influenced by the surface area ratio of cathodic and anodic metals: the larger the cathode area the higher the corrosion rate and vice versa. For example, a large anode will produce a general corrosion condition, whereas a small anode yields a severe pitting type of corrosion.

As previously mentioned, the surface condition of the metal greatly influences the conditions for corrosion. When a metal surface has deposits of debris or when alum floc or lime particles create nonaerated zones (after precipitation) beneath the deposits, corrosion begins in these anodic regions. Furthermore, if bacteria are growing under the deposits, they create an oxygen-deficient (reducing) condition, which intensifies the corrosion condition. Corrosion of stainless steel often occurs under this type of condition.

Anaerobic microbes also play a role in corrosion. The presence of sulfate reducing bacteria within the deposit creates a very aggressive condition, which further accelerates metal loss. Some anaerobic bacteria such as iron bacteria (i.e., *Lepthothrix*) actually consume ferrous iron and thereby increase the rate of corrosion.

Secondary Corrosion Factors   Secondary factors that influence corrosion are (1) the concentration of dissolved salts in the water, (2) the level of dissolved gases (in water), (3) water temperature, (4) stress corrosion, (5) fatigue, and (6) impingement attacks. Each of these is briefly described.

Dissolved Solids   The effect of the dissolved solids content on water corrosivity is a complex issue. Both the species and the concentration of the ions are important factors. Some species, such as carbonate and bicarbonate, reduce corrosion, whereas chloride, sulfate, bromide, and nitrate ions markedly accelerate corrosion. In regard to concentration, less than 1 mg/L of copper ion can significantly increase the rate of corrosion: copper ions are deposited on the metal and act as an anode, causing a severe pitting type of corrosion to occur. Household aluminum utensils have been known to corrode as the result of copper ions.

Dissolved Gases   Dissolved gases such as free carbon dioxide, oxygen, hydrogen sulfide, and ammonia strongly influence the corrosion of metals, especially iron and copper.

Water Temperature   Higher water temperatures accelerate the rate of corrosion by increasing the rate of the cathodic reaction. The chemical reaction rate generally doubles for every 8°C (15°F) increase in temperature.

Stress Corrosion   Metals that are under tensile or bending stress, due to external loads or internal stress locked into the metal during fabrication, will exhibit stress corrosion when subjected to a corrosive environment.

Fatigue Corrosion   Fatigue corrosion occurs when a metal undergoes repeated stress in a corrosive environment.

Impingement Attacks   Cavitation, erosion, abrasion by water-borne grit and other solids, and erosion of the protective coating on metal surfaces are examples of impingement attacks.

7.6.2   Corrosion Control for Metals

Five basic measures exist for corrosion control: (1) the selection of construction materials, coatings/linings, and insulation; (2) application of corrosion inhibitors; (3) installation of cathodic protection systems; (4) installation of sacrificial anodes; and (5) pH control by alkali chemicals.

Design Efforts   When corrosion problems are expected for a treatment plant, the design engineer may consider using corrosion resistant materials such as stainless steel, concrete, plastic, or copper as a substitute for carbon steel. If these materials cannot be implemented, special coatings and/or linings should be applied to the metal to separate it from the corrosive water or environment. Paints, plastic sheet, rubber, and cement are often used in this capacity.

Galvanic corrosion resulting from the connection of dissimilar metals must be avoided by insulation. Nonconductive materials such as plastic or synthetic rubber (Figure 7.6.2-1) should be inserted between the metals.

Application of Corrosion Inhibitors   Despite the best design efforts, most water treatment systems also include a system that applies chemical corrosion inhibitors to the process water. The inhibitors control corrosion by stopping the anodic and/or cathodic reaction. Chromate, orthophosphate, nitrite, and

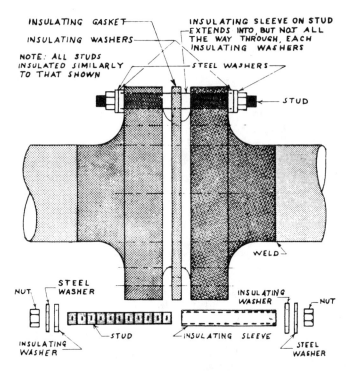

Figure 7.6.2-1 Suggested insulation flange assembly. (Adapted from A. W. Peabody, *Control of Pipeline Corrosion*, NACE, Houston, 1967.)

silicate are anodic inhibitors; calcium carbonate, polyphosphate, and zinc are typical cathodic inhibitors. There are some chemicals that act as both anodic and cathodic inhibitors: mercaptobenzothiazole, benzotriazole, and tolytriazole. These chemicals form an organic film of amines on the surface of the metal, thereby inhibiting corrosion (similar to painting).

The corrosion inhibitors most commonly used in the field of water treatment are polyphosphate (cathodic inhibitor requiring a dosage of approximately 2 mg/L), zinc orthophosphate (anodic inhibitor requiring a dosage of 1 mg/L or less), and sodium silicate (anodic inhibitor requiring a dosage of 8–10 mg/L as silica).

Cathodic Protection Cathodic protection is used to lower the electrode potential of metals and is used to protect buried pipes or equipment exposed to seawater. Cathodic protection is generally effective in applying an electrical potential of 0.85–1 V.

Sacrificial Anodes Sacrificial anodes reduce galvanic cell type corrosion for steel or iron structures. They are composed of zinc or magnesium and become anodic under corrosive conditions, thereby supplying electrons to cathodic surfaces.

pH Control by Alkali Chemicals The most common method of corrosion control for municipal water distribution systems is pH control of water. This method provides a slightly positive ($+0.2$) Langelier Index or a Satability Index (Ryznar Index) of 6 or below.

The chemical generally fed to the process water is lime, since it provides calcium and alkalinity, in addition to raising the pH. However, lime particles tend to settle out in the distribution main, which creates a condition conducive to concentration cell type corrosion. Thus, caustic soda is often employed as a substitute for lime, due to its ease of handling and because it eliminates the precipitation of lime particles.

$$\text{Langelier Index (LI)} = \text{pH} - \text{pH}_s$$

where pH = pH of the sample water,
  pH_s = saturation pH.

$$\text{Stability Index (SI)} = 2\text{pH}_s - \text{pH}$$

Figure 7.6.2-2 and Appendix 16 are graphical determinations of pH and LI. SI can be determined by simple calculation after LI is obtained.

Table 7.6.2-1 illustrates the relation between corrosion characteristics and LI and SI values.

As a means of corrosion control, some treatment plants feed lime ahead of filtration (pH $\geq$ 8) in the belief that this scheme will eliminate the after-precipitation of undissolved lime in the distribution system. In actuality, this practice increases the finished water concentration of aluminum; the alum floc in the filter influent is dissolved by the lime due to the solubility characteristics of the floc (refer to Figure 3.2.3-1) and the level of residual aluminum is often 0.3 mg/L or more.

Plant operators should realize that an adjustment in pH based on LI and SI does not mean that corrosion does not occur. Indeed, in many cases, iron structures became corroded, although the water had a slightly positive Langelier Index. This situation often occurs with soft water. The best way to monitor metal corrosion and to optimize corrosion inhibitor dosage or pH control by alkali chemicals is to conduct a continuous corrosion coupon test.

7.6.3 Corrosion Coupon Test

A number of tools have been developed to measure corrosion by indirect means. One method is the corrosion coupon test: Figure 7.6.3-1 presents a typical installation of the corrosion coupon test assembly. The major steps of this test are as follows:

1. The pretest weights of the metal specimens are recorded prior to placing the metals into the system for a period of 30–100 days.

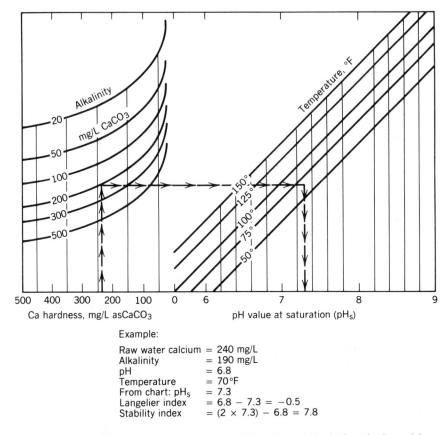

Example:

Raw water calcium	= 240 mg/L
Alkalinity	= 190 mg/L
pH	= 6.8
Temperature	= 70°F
From chart: pH_s	= 7.3
Langelier index	= 6.8 − 7.3 = −0.5
Stability index	= (2 × 7.3) − 6.8 = 7.8

Figure 7.6.2-2 Graphical determination of pH_s and Langelier index. (Adapted from *The Nalco Water Handbook*, F. N. Kemmer, ed., McGraw-Hill, New York, 1979.)

2. Following removal, the specimens are cleaned, reweighed, and evaluated. The loss of weight and type of corrosion (general, pitting, etc.) are observed and recorded.

3. The corrosion rate is expressed as metal loss in mils per year (mpy), a mil being 0.001 in. One mpy is equivalent to 0.025 mm/year.

TABLE 7.6.2-1 Corrosion Characteristics Versus LI and SI

	LI	SI
Medium to heavy scaling	+0.5 to +1.0	4–5
Slight scale formation	+0.2 to +0.3	5–6
Equilibrium	0	6–7
Slightly corrosive	−0.2 to −0.3	7–7.5
Medium to heavy corrosion	−0.5 to −1.0	7.5–8.5

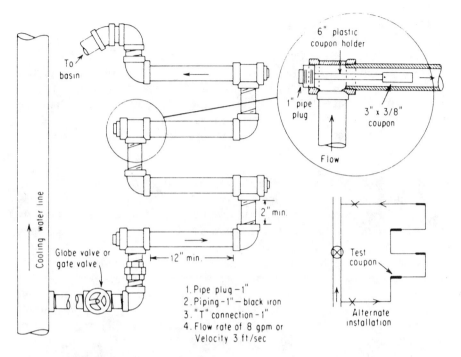

Figure 7.6.3-1 Coupon test assembly. (Adapted from *The Nalco Water Handbook*, F. N. Kemmer, ed., McGraw-Hill, New York, 1979.)

The corrosion coupon test is occasionally run for visual observation only.

The results of the coupon test can vary widely, depending on the condition of the coupon: whether the metal was sandblasted versus polished, the location of the coupon within the test system, the length of exposure (duration of the test), the type of metallurgy, and the quality of the water.

7.6.4 Corrosion (Degradation) of Concrete

Today, almost all concrete structures are composed of steel bar reinforced concrete. Theoretically, the steel bars do not corrode due to the positive potential in the concrete, that is, until the covering concrete (approximately 2 in. thick) degrades. The electrode potential of the steel bars (in concrete) is approximately $+0.1$ V and is referred to as the normal hydrogen electrode for concrete at a pH of approximately 11.6. Yet, in reality, corrosion of the reinforcing steel bars is an important concern for engineers since degradation of concrete or the formation of cracks in the concrete can be attributed to three causes: (1) physical causes, (2) chemical attacks, and (3) bacterial actions.

Physical Causes The degradation of concrete will be accelerated if the concrete is excessively permeable. This characteristic results from the use of the improper size of gravel; the wrong mixing proportion of cement, sand, and gravel; and poor concrete casting techniques. If the concrete is poorly constructed—that is, has cracks and cavities, an insufficient thickness of concrete over the steel bars, or badly constructed expansion joints that leak water—these defects will contribute to its degradation. The last of the physical causes is erosion resulting from heavy vehicle traffic, scouring by sand or grit, or a high velocity of water flow (over 15 fps).

Chemical Attacks Carbon dioxide, strong acids, ammonia, sulfate, and strong alkali chemicals all attack concrete. Of the five, carbon dioxide has the least effect because the high lime residue of the cement simply converts the carbon dioxide into calcium carbonate. However, if a concrete surface is continuously exposed to water containing over 15 mg/L of CO_2, its surface will begin to degrade. In or near large cities, concrete corrosion due to exhaust gas from industries and cars has become a serious problem.

Strong acids such as sulfuric, hydrochloric, nitric, and phosphoric, as well as organic acids, are highly destructive to concrete. The only effective method of protecting concrete from the action of these acids is to line or coat it with a suitable type of material, such as plastic membrane, rubber, or paint.

Ammonia degrades concrete by replacing calcium (in concrete) with ammonia. This loss of calcium from cement causes the concrete to corrode rapidly.

In recent years, many water treatment plants began feeding ammonia (in addition to chlorine) to the process water. The hydrochloric acid produced by chlorine and ammonia is capable of degrading concrete, provided that the water is soft and has very little alkalinity, although its dosage is so small that it will not quickly attack concrete except at ammonia feed point.

Certain regions have water supplies containing a high level of sulfates. Since sulfates are capable of reacting with aluminates—an ingredient of cement—and cause the original volume to swell by a factor of 2–2.5, they may be considered to be the primary cause of concrete deterioration in water treatment structures.

A number of existing plants, especially those located in regions with a high sulfate content in the water, exhibit severe degradation of their concrete surfaces, especially in areas continuously exposed to cascading water and flowing water: concrete troughs in sedimentation basins and filters and the side walls of channels and basins. Engineers are therefore strongly advised to consider a dense and high-strength concrete with Type 5 cement when designing plants for these regions (sulfate content over 100 mg/L). Table 7.6.4-1 lists the available types of cement and their major ingredients.

Concrete can also be deteriorated by strong alkali chemicals such as caustic soda, potassium hydroxide, and soda ash because the aluminum oxides in the cement (see Table 7.6.4-1) tend to become soluble when in contact with water

TABLE 7.6.4-1   Major Ingredients of Portland Cement

Composition (%)	Type 1 General Purpose	Type 2 Modified	Type 3 High Early Strength	Type 4 Low Heat	Type 5 Sulfate Resisting
$3CaOSiO_2$	53	47	58	26	38
$2CaOSiO_2$	24	32	16	54	43
$3CaAl_2O_3$	8	3	8	2	4
$4CaOAl_2O_3Fe_2O_3$	8	12	8	12	8
Total	93	94	90	94	93

with a high pH; the strong alkali chemicals react with the sodium molecule in the molecular structure. Thus, engineers are advised either to select cement with a low aluminum content (Type 2 or Type 4) or to coat the concrete surface with an appropriate type of paint. If general purpose cement is specified, a PVC membrane should be used, provided that the liquid pH exceeds 12.

Bacterial Action   The third cause of concrete degradation is bacterial action. A typical example is crown corrosion of concrete sewer pipes. In this situation, the concrete is deteriorated by sulfuric acid formed by anaerobic bacteria, which consume water-borne sulfate. This phenomenon rarely occurs in water treatment plants.

7.6.5   Special Construction/Design Considerations

The construction supervising engineer must give special consideration to the contact between steel reinforcing bars and iron pipes passing through concrete walls. Absolutely no direct contact should exist between these two elements, since contact will produce galvanic cell corrosion in the iron pipe.

The same type of problem will occur whenever steel members are imbedded in the reinforced concrete structure. If the end of a steel member, which is either exposed to water or buried in the ground, must be welded to reinforcing steel bars (in concrete), it must be well coated or cathodically protected from galvanic cell corrosion. Design engineers should be very careful when designing these types of contacts and are advised to make notations on the drawings.

Additional design considerations include the use of insulation flanges in cases where dissimilar metal pipes must be connected and in situations where the pipeline exits a concrete wall to enter the ground or a water tank (see Figure 7.6.2-1). Furthermore, stray current electrolysis type of corrosion may occur in areas near direct current (DC) transit systems. It is also very important to specify a dense and high-strength concrete (compressive strength of over 5000 psi) reducing the water/cement ratio to less than 0.45. Recent case histories have shown a clear superiority of the high-strength reinforced concrete structures against both corrosion and erosion. It is also known that an over use of form release agents have an obvious effect on the rate of concrete corrosion.

For further information, refer to *Control of Pipeline Corrosion* (Peabody, 1967) released by the National Association of Pipe Corrosion (NAPC) Engineers as listed in the Bibliography.

BIBLIOGRAPHY

Degremont, *Water Treatment Handbook*, 5th ed., Halsted Press, New York, 1979.

Fair, G. M., Geyer, J. C., and Morris, J. C., *Water Supply and Wastewater Disposal*, Wiley, New York, 1954.

Good, R. J., Moor, G. C., and Wojciak, C. J., "Coping with Deterioration of Concrete Surfaces in Adelaide's Early Water Filtration Plants," A Report by Engineering and Water Supply Department, Adelaide, South Australia.

Hatch, G. B., and Rice, O., "Influence of Water Composition on the Corrosion of Steel," *J. AWWA*, 51(6):719(June 1959).

Kemmer, F. N., ed., *The Nalco Water Handbook*, Nalco Chemical Co. McGraw-Hill, New York, 1979.

Larson, T. E., "Chemical Control of Corrosion," *J. AWWA*, 58(3):354(March 1966).

Montgomery, J. M., Consulting Engineers, *Water Treatment Principle and Design*, Wiley, New York, 1985.

Peabody, A. W., *Control of Pipeline Corrosion*, National Association of Corrosion Engineers, Houston, 1967.

Pisigan, Jr., R. A., and Singley, J. E., "Effects of Water Quality Parameters on Corrosion of Galvanized Steel," *J. AWWA*, 77(11):76(November 1985).

Pisigan, Jr., R. A., and Singley, J. B., "Influence of Buffer Capacity, Chlorine Residual, and Flow Rate on Corrosion of Mild Steel and Copper," *J. AWWA*, 79(2):62(February 1987).

Reiber, S. H., Ferguson, J. F., and Benjamin, M. M., "Corrosion Monitoring and Control in the Pacific Northwest, *J. AWWA*, 79(2):71(February 1987).

Stone, A., Spyridakis, D., Benjamin, H., Ferguson, J., Reiber, S., and Osterhus, S., "The Effects of Short Term Changes in Water Quality on Copper and Zinc Corrosion Rates," *J. AWWA*, 79(2):75(February 1987).

Strum, W., "Corrosion Studies," *Public Works*, p. 78(1957).

Strum, W., "Investigation on the Corrosive Behavior of Waters," *J. Sanit. Eng. Div. ASCE*, 86:27(1960).

7.7 GRANULAR ACTIVATED CARBON ADSORPTION

7.7.1 Purpose

Granular activated carbon (GAC) has recently been used as a substitute for granular filter medium or as an additional process in the conventional treatment process, for the removal of organic compounds, including disinfection byproducts, those producing taste and odor, pesticides, and other synthetic organic compounds. GAC has been proved to be effective in removing these various organic compounds and even organic mercury.

The GAC filter bed is not only capable of functioning in the same manner as ordinary filters—removal of suspended matter but also removes organic compounds. However, the Europeans use GAC adsorption beds for the removal of taste and odor, other trace organic compounds, and total organic carbon, by installing the GAC contactors downstream of ordinary filters (in most cases) preceeded with pre-ozonation process.

In recent years, the GAC adsorption process has suddenly gained a great deal of attention in the United States as the result of the 1986 Amendments to the Safe Drinking Water Act. These amendments set the maximum contaminant levels (MCLs) for over 50 organic compounds, of which 26 are synthetic organic compounds.

The water treatment industry in the United States has recently begun to take large steps in improving the aesthetic quality of drinking water, by reducing color, objectionable tastes, and odors, in addition to making it safer. The GAC adsorption process is certainly capable of meeting these objectives. Yet, the operation and maintenance costs are significantly higher, primarily due to the need to regenerate the GAC.

7.7.2 Characteristics of GAC

GAC may be manufactured from a variety of materials: wood, nut shells, coal, peat, or petroleum residues. However, GAC used in water treatment is usually produced from bituminous or lignite coal through slow furnace heating under anaerobic conditions. The absence of oxygen ensures that the coal does not burn but transformed into a porous carbon material.

This product is then activated by exposure to a mixture of steam and air at a temperature of 1500°F. The activation process oxidizes the surface of the carbon pores, thereby allowing the surface to attract and hold organic compounds.

According to the *Calgon Bulletin*, the physical properties of GAC are as follows:

Total surface area of GAC	$890-900$ m^2/g
Bulk density of wetter GAC	30 lb/ft^3
Particle density	$1.4-1.5$ g/cm^3
Effective size	$0.8-0.9$ mm for Filtrasorb 100
	$0.55-0.65$ mm for Filtrasorb 200
Uniformity coefficient	1.9 for Filtrasorb 100
	1.7 for Filtrasorb 200
Iodine number	$850-875$
Abrasion number	$70-86$
Moisture as packed	$0.5-2\%$

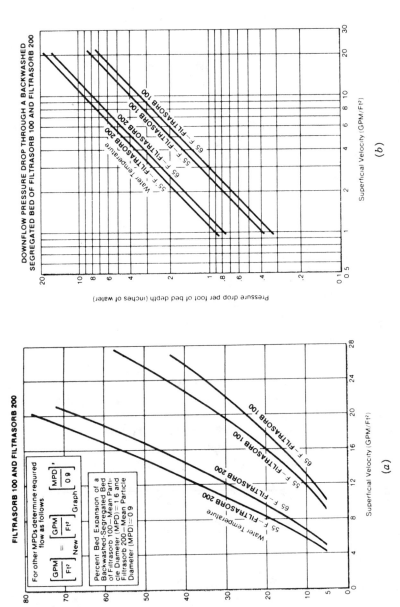

Figure 7.7.2-1 Hydraulic characteristics of typical GAC medium beds. (a) Backwash rate versus GAC bed expansion rate. (b) Headloss versus backwash rates. (Adapted from *Calgon Bulletin 20-60*, 1976.)

Figure 7.7.2-1a illustrates the hydraulic characteristics of a GAC bed. The optimum backwash rate is approximately 30–40% of the bed expansion rate due to the light specific gravity of the material. The GAC beds must be backwashed periodically to remove suspended matter and microorganisms, even though the mass transfer zone is completely destroyed by backwashing.

Figure 7.7.2-1b depicts the headloss across a clean GAC bed at various filtration rates.

7.7.3 Design Considerations

Design engineers are advised to conduct the following sequence of activities whenever the GAC contactors are being considered as part of a treatment process train:

1. Intensively survey all available literature on the subject.
2. Collect all available water analysis data on the subject water. Identify the viable options for removing the organic compounds.
3. Procure a fresh sample of water and conduct a complete water analysis centered on the organic compounds.
4. Reevaluate the preliminary options and establish both the carbon usage rate and the pilot testing conditions.
5. Conduct a GAC pilot study.
6. Establish the design parameters of the process after analyzing the pilot test data.

The major design considerations of the GAC contractors are:

1. The designed flow rate and available headloss for this process.
2. The nature and amount of all organic compounds in the water supply.
3. The MCLs of the regulated organic compounds.
4. The overall treatment process train and each unit process.
5. The type of GAC.
6. The size of the GAC.
7. The filtration rate (superficial velocity).
8. The empty bed contact time (EBCT).
9. The type of contact unit.
10. The location of the GAC contactor in the process train.
11. The regeneration frequency and the method to do it.

Designed Flow Rate The designed flow rate should allow for the present required flow rate as well as for future expansion of the facility (if needed).

Nature and Amount of All Organic Compounds Both the physical and chemical properties of the organic compounds strongly affect the type and the level of pretreatment prior to the GAC adsorption bed. Alum coagulation and flocculation with clarifier are effective schemes for high molecular weight organic compounds; the majority of highly volatile, low molecular weight compounds may possibly be removed by air-stripping upstream of the GAC contact beds.

MCL of Regulated Organic Compounds Refer to Section 2.3, Preliminary Engineering Study, for a detailed discussion.

Overall Treatment Process Train and Each Unit Process Refer to Section 2.3, especially Figure 2.3.2-1b.

Type of GAC The pore size of the GAC is an important factor in the removal of organic substances. For example, GAC with large pore volumes may be effective in removing water-borne compounds with high molecular weights. Moreover, activated carbon adsorbs inorganic ions in a preferential order:

$$H^+ > Al^{3+} > Ca^{2+} > Li^+ > Na^+ \quad \text{for cations}$$

$$NO_3^- > Cl^- \quad \text{for anions}$$

Engineers are therefore advised to consult the literature provided by manufacturers prior to selecting the GAC medium.

Size of GAC The size of the GAC strongly influences both the depth of the bed and the headloss across the bed given a specific flow rate. Coarse GAC has definite advantages for application to deep bed gravity contactors under high filtration rates. However, smaller sized GAC is characterized by a higher rate of adsorption, given an equal level of activity with shallower bed, since it has a greater active surface area.

Filtration Rate Generally, the most common filtration rate is 2–5 gpm/ft^2 (5–12.5 m/h), although higher rates are also used.

EBCT A practical EBCT is one that ranges from 7.5 to 30 min. The shorter EBCT may be used when removing synthetic organic compounds (SOCs) since these are easily adsorbed, but the longer contact time must be used in removing the more soluble and less adsorbable organic compounds.

Figure 7.7.3-1 illustrates the EBCT versus the THM breakthrough at EBCT values of 5, 7.5, and 15 min. The graph shows that the EBCT should be a minimum of 15 min, and possibly 30 min, given a reasonable regeneration cycle (a 6 month to 2 year cycle). It is very interesting to see that increasing the EBCT from 7.5 to 15 min significantly increases the carbon life. Similar

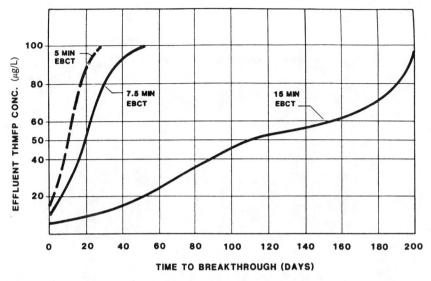

Figure 7.7.3-1 EBCT versus THM breakthrough against time. (Adapted from X. Westerhoff, in *AWWA Proceedings for GAC*, 1987.)

results have been obtained from a number of pilot tests, including the test presented in an EPA publication entitled *Treatment Technique for Controlling THM's in Drinking Water* (1981).

All these design parameters (1 through 8) may be evaluated by a GAC pilot study, provided that the project has both the time and the budget to do so. The pilot study should test a variety of pretreatment options such as ozonation, aeration, alum coagulation, sedimentation, and filtration. Since the actual raw water usually contains a unique combination of suspended solids, organic compounds, and inorganic compounds indigenous to the region, the only way to properly establish most of the design criteria is to conduct a pilot scale study on the plant site.

Type of Contact Unit The configuration of the GAC bed may be upflow, downflow in parallel trains, or downflow in series trains. The upflow configuration is not suitable for application in a potable water treatment process unless the ordinary filter is located downstream of the upflow GAC bed. Otherwise, the GAC grains will be present in the finished water due to the escape of fine GAC particles during the later stages of the filtration cycle: the result of bed expansion due to the development of headloss across the bed.

Location of the GAC Contactor in the Process Train The GAC contact bed can be placed upstream of the flocculation process, before the ordinary filters, or downstream of the regular filters, depending on the nature of the substance

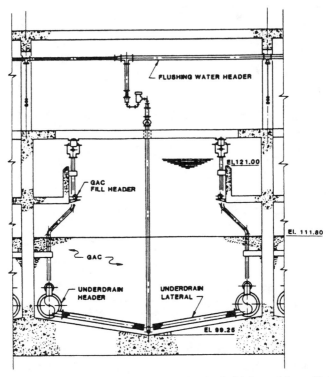

Figure 7.7.3-2 Cross section of a GAC adsorption bed. (Adapted from X. Westerhoff, in *AWWA Proceedings for GAC*, 1987.)

to be removed and the result of the pilot study. The GAC contact bed is occasionally used as both a filtration and adsorption bed. Figure 7.7.3-2 is a cross section of a GAC adsorption bed designed for the city of Cincinnati.

7.7.4 GAC Regeneration

GAC is a very expensive product ($1800/ton); thus the saturated medium cannot be discarded like spent PAC. Fortunately, GAC is fairly easy to regenerate. The need to regenerate the GAC ranges from 6 months to 5 years, depending on the amount and type of organic contaminants.

The two basic alternatives for GAC regeneration are off-site regeneration and on-site regeneration. The on-site alternative is generally not cost effective unless the carbon exhaustion rate is very large (over 5 tons/day). Moreover, the additional air pollution from the furnace off-gas can pose a great problem in large metropolitan areas, which already have poor quality air.

The off-site regeneration area can be any site outside urban areas where air pollution is not an issue. In most cases, the GAC manufacturer is capable of regenerating the spent GAC at its facility on a contractual basis.

The three methods used in GAC regeneration are: regeneration by steam, thermal regeneration, and chemical regeneration. Since only an outline of these methods is presented, refer to the *Carbon Adsorption Handbook* edited by Cheremisinoff et al. (1978) (see Bibliography) and/or the manufacturer for further details.

Steam Regeneration This method is limited to carbon that has been saturated with a limited type and level of volatile compounds. For further details, refer to the *Carbon Adsorption Handbook.*

Thermal Regeneration This is a pyrolytic process that burns off adsorbed compounds. The dewatered GAC is heated to approximately 1500°F (815°C) under controlled conditions; steam is used to prevent the carbon from igniting. The GAC is then quenched with water for reuse. Approximately 10–15% of the original GAC is generally lost during this entire process.

The types of furnace commonly used in thermal regeneration of GAC are multiple-hearth, fluidized bed, and rotary kiln. All these types are very expensive investments.

Both spent and regenerated GAC is transferred in slurry form by means of hydraulic eductors. The slurry is transported through stainless steel piping for most of the distance; stainless steel screw conveyers are used to feed the spent GAC to the furnace. Figure 7.7.4-1 is an example of the off-gas handling system.

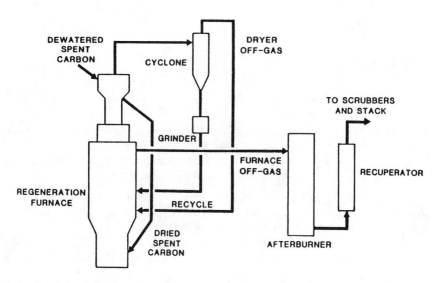

Figure 7.7.4-1 An off-gas control system. (Adapted from X. Westerhoff, in *AWWA Proceedings for GAC*, 1987.)

Chemical Regeneration This method uses a solvent to extract adsorbed compounds at a temperature of approximately 210°F (100°C) by means of pH. The process was developed by the French firm Degremont and its application is very limited at present.

7.7.5 Capital and Operational Costs

Capital cost of GAC adsorption system can be estimated based on a total volume of GAC medium required for the system. The capital cost includes GAC medium, reactor cells, pipings, pumping, flow control, electrical, and concrete works. The estimated capital cost of the system is shown below;

Volume of GAC in ft³	2500	5000	10000	25000	50000	100000
Cost in $ million	0.48	0.84	1.40	3.90	7.80	13.0

The cost shown above does not include GAC medium storage and spent GAC reactivation system because they are site specific. In general, a total cost is 2.5 to 3.5 times higher than the cost shown above depend upon the size of the facility and other factors if the storage and the reactivation facilities are included.

Operational cost includes labor, fuel, power, electrical and control, virgin GAC makeup, miscellaneous indirects, and maintenance cost. The operational cost is usually estimated based on the GAC exhaustion rate.

GAC exhaustion rate (lb/day)		1000	2500	5000	10000	25000	50000	100000
	On site regeneration	2.1	1.3	0.9	0.72	0.65	0.61	0.58
Cost	Off site regeneration	2.7	2.65	2.6	2.6	2.6	2.6	2.6
($/lb)	Throwaway	4.5	4.45	4.43	4.4	4.4	4.38	4.35

BIBLIOGRAPHY

Adams, J. Q., and Clark, R. M., "Cost Estimates for GAC Treatment Systems," *J. AWWA*, 81(1):34(January 1989).

AWWA, *AWWA Seminar Proceedings—GAC Installations: Conception to Operation*, Annual Conference in Kansas City—1987, AWWA, Denver, 1987.

Bablon, G. P., "Developing a Sand-GAC Filter to Achieve High Rate Biological Filtration in France," *J. AWWA*, 80(12):47(December 1988).

CALGON Corp., *Calgon Bulletin*, #20-60, Calgon Corp., Pittsburgh, PA, 1976.

Cheremisinoff, P. N., and Ellerbusch, F., editors, *Carbon Adsorption Handbook*, Ann Arbor Science Publishers, Ann Arbor, MI, 1978.

Crittenden, J. C., Hand, D. W., Arora, H., and Lycins, B. W. Jr., "Design Considerations for GAC Treatment of Organic Chemicals," *J. AWWA*, 77(1):132(January 1987).

Culp, R. L., and Clark, R. H., "Granular Activated Carbon Installations," *J. AWWA*, 75(8):398(August 1983).

EPA, *Treatment Techniques for Controlling THM's in Drinking Water*, EPA-600/2-81-156 (September 1981).

Grease, S. L., Snoeyink, V. L., and Lee, R. G., "GAC Filter Adsorbers," *J. AWWA*, 79(12):64(December 1987).

Hand, D. W., Crittenden, J. C., Arora, H., Miller, J. M., and Lykins, B. W., "Designing Fixed-Bed Adsorbers to Remove Mixtures of Organics, *J. AWWA*, 81(1):67(January 1989).

Hassler, J. W., *Purification with Activated Carbon*, Chemical Publishing, New York, 1974.

Hyde, R. A., Hill, D. G., Zabel, T. F., and Burke, T., "GAC as Sand Replacement in Rapid Gravity Filters," *J. AWWA*, 79(12):33(December 1987).

Lykins, B. W., Clark, R. M., and Adams, J. Q., "Granular Activated Carbon for Controlling THMs," *J. AWWA*, 80(5):85(May 1988).

Quinn, J. E., and Snoeyink, V. L., "Removal of Total Organic Halogen by GAC Adsorbers," *J. AWWA*, 72(8):483(August 1980).

Roberts, P. V., and Summers, R. S., "Performance of GAC for Total Organic Carbon Removal," *J. AWWA*, 74(2):113(February 1982).

Suffett, I. H., *Activated Carbon Adsorption of Organics from the Aqueous Phase*, Vols. 1 and 2, Ann Arbor Science Publishers, Ann Arbor, MI, 1980.

Suffett, J. J., "An Evaluation of Activated Carbon for Drinking Water Treatment: A National Academy of Science Report," *J. AWWA*, 72(1):41(January 1980).

Symons, J. J., "A History of the Attempted Federal Regulation Requiring GAC Adsorption of Water Treatment," *J. AWWA*, 76(8):34(August 1984).

Weber Jr., W. J., and Jocellah, A. M., "Removing Humic Substances by Chemical Treatment and Adsorption," *J. AWWA*, 77(1):132(April 1985).

Wiesner, M. R., Rook, J. J., and Fiessinger, F., "Optimizing the Placement of GAC Filtration Units," *J. AWWA*, 79(12):39(December 1987).

7.8 ION EXCHANGE AND MEMBRANE SEPARATION PROCESSES

The ion exchange and membrane separation processes are not ordinarily used in municipal water treatment practices. These processes are the products of highly specialized chemical engineering. Consequently, water treatment engineers not involved in the detailed design of these processes—the design, fabrication, and operational aspects—are advised to seek the assistance of specialized equipment manufacturers and resin suppliers.

Provided that the project has both the budget and the time, water treatment engineers may run a pilot test on one of the processes to establish the design and operational parameters (for a specific water supply). Although engineers should understand the basic concept and the important characteristics of these processes, in order to write performance specifications and possibly P&ID diagrams, it is very important that engineers obtain assistance and advice from equipment manufacturers since they possess a great deal more knowledge and experience.

This section discusses some fundamental features of the ion exchange process and four membrane processes: microfiltration, ultrafiltration, electrodialysis, and reverse osmosis.

7.8.1 Ion Exchange

The process of ion exchange removes undesirable ions from raw water by exchanging them with a number of desirable ions; the desirable ions are stored in a solid material (ion exchanger). Since the ion exchanger has a limited capacity for exchange, it must be washed with a regenerating solution in order to regain its function. This cycle generally entails backwashing, regeneration, rinsing, and service.

The most primitive ion exchangers are natural soil and sodium aluminosilicates. The first synthetic ion exchanger was composed of processed natural greensand (Zeolite). However, insoluble synthetic plastic (resin) is almost exclusively used today. Most commercial resins are composed of synthetic plastic materials such as copolymers of styrene and divinyl benzene and are expected to survive 5–10 years of continuous use.

The resins are classified as either cation exchangers—that is, resins with negatively charged sites that take up positively charged ions—or anion exchangers. The resin has a porous and permeable structure; thus, the whole resin particle participates in the exchange process.

Resins are typically spherical in shape and range from 20 to 40 mesh (0.8–0.4 mm). Cation resins generally have an exchange capacity of approximately 2 milliequivalents per dry gram (meq/g). The typical strong base anion resins have a 1.3 meq/g exchange capacity.

In the field of water treatment, the ion exchange process is used for water softening, the removal of fluoride or nitrate, and the production of demineralized water for laboratory use. The resinous ion exchange process is the only process, with the exception of evaporation, that has the potential to completely remove minerals from water, thus yielding ultrahigh purity water.

Characteristics Whether ion exchange is performed in fixed beds, continuous loops, pulsed beds, or by countercurrent extraction, it is still a batchwise process with respect to exhaustion and regeneration of the medium. Resins have a discrete exchange capacity and must be regenerated when exhausted. The cost of the chemicals per unit volume of water treated is almost directly proportional to the amount of minerals removed.

The reaction for the removal of water hardness is

$$\left.\begin{matrix} Ca^{2+} \\ Mg^{2+} \end{matrix}\right\} + Na_2R \rightarrow \left.\begin{matrix} Ca \\ Mg \end{matrix}\right\} R + 2Na^+$$

where R is the exchange resin.

A more comprehensive expression of ion exchange, with a hydrogen cycle (H-form exchange), used in treating a typical water containing a variety of ions is as follows:

$$\left.\begin{array}{c} Ca^{2+} \\ Mg^{2+} \\ Fe^{2+} \\ 2Na^+ \\ 2NH^+ \end{array}\right\} + H_2R \rightarrow \left.\begin{array}{c} Ca \\ Mg \\ Fe \\ Na_2 \\ (NH_4)_2 \end{array}\right\} R + 2H^+$$

A typical anion exchange resin, with a chloride cycle, used to remove nitrate ions is

$$NO_3^- + ClR \rightarrow NO_3R + Cl^-$$

Ion exchange resins are characterized by selectivity; the preferential removal of certain ions. Selectivity primarily varies with the type of resin, ionic strength of the solution, relative amounts of different ions, and water temperature. This selectivity is a very important characteristic in the regeneration process. For example, in the sodium cycle the exchanger has a preference for calcium ion over sodium at concentrations of 1000 mg/L or below. However, at 100,000 mg/L the exchanger prefers sodium ions over calcium; thus, it can be regenerated with brine.

The order of ion selectivity (in ordinary water) is listed below.

For cations: $Fe^{3+} > Al^{3+} > Pb^{2+} > Ba^{2+} > Sr^{2+} > Cd^{2+} > Zn^{2+}$

$> Cu^{2+} > Fe^{2+} > Mn^{2+} > Ca^{2+} > Mg^{+2} > K^+ > NH_4^+ > H^+ > Li^+$

For anions: $CrO_4^{2-} > SO_4^{-2} > SO_3^{2-} > HPO_4^{2-} > CNS^- > CNO^-$

$> NO_3^- > NO_2^- > Br^- > Cl^- > CN^- > HCO_3^- > HSiO_3^- > OH^- > F^-$

The ion exchange units can be used to deionize water by simply allowing the water to flow continuously through a cationic resin column (hydrogen cycle) and a strong base anionic resin column (hydroxide cycle).

$$\left.\begin{array}{l} Ca(HCO_3)_2 \\ Mg(HCO_3)_2 \\ CaSO_4 \\ MgSO_4 \\ CaCl_2 \\ MgCl_2 \\ NaCl \\ SiO_2{\cdot}H_2O \end{array}\right\} + HR \rightarrow \left.\begin{array}{l} CO_2 \\ H_2O \\ H_2SO_4 \\ HCl \\ SiO_2{\cdot}H_2O \end{array}\right\} \rightarrow ROH \rightarrow H_2O$$

$$\left.\begin{array}{c} Ca \\ Mg \\ Na \end{array}\right\} R \qquad\qquad R \left\{\begin{array}{l} SO_4 \\ Cl \\ HSiO_2 \end{array}\right.$$

The important issues that must be addressed when designing an ion exchange process are reduction of chemical costs, reduction of waste products, and improvement of the processed water. However, it is vital that the water be pretreated by means of a proper process such as alum flocculation and sedimentation (possibly with filtration) prior to ion exchange (reverse osmosis or lime softening) to reduce the ionic loading to the exchanger system and to remove particulates, such as iron, manganese, and organic substances, which may foul the resins.

Tables 7.8.1-1 and 7.8.1-2 can help in selecting the most appropriate type of ion exchange process and alternative.

The major U.S. suppliers of resin are Dow Chemical, Rohm & Haas, and Zeolite. Many water treatment equipment manufacturers are also capable of producing ion exchange process units.

7.8.2 Membrane Separation Process

In the field of water treatment, in-depth filtration with granular medium is a regular filtration system that is capable of removing particles that are approximately 0.1 µm or above in size, provided that chemical coagulation pretreatment is performed. In response to the EPA's recent emphasis on safe drinking water, engineers have begun to reevaluate all the best available technology (BAT), including the membrane separation process. This process has distinct advantages over the conventional treatment process: particulates can be removed with little or no coagulant chemicals; the process requires little or no disinfection prior to filtration; it requires less space than conventional filters; it is easily integrated into the instrumentation and control system; and its operation potentially requires fewer personnel.

The membrane separation process uses semipermeable membranes to separate impurities from water. Semipermeable membranes are selectively permeable to water and certain solutes and play an extremely important role in nature: nutrients and waste products are transferred across the semipermeable membrane of cells. Similarly, the membrane separation process used in the field of public water supply uses a driving force to pass the water through the membrane, leaving the impurities behind as a concentrate. The degree of purification is dictated by the type of membrane, the type and level of driving force, and the characteristics of the water (Table 7.8.2-1). However, the process does have several associated problems: membrane fouling, disposal of the concentrate, and the operation and maintenance costs.

Due to conflicting reports and the rare application of the membrane separation process to ordinary water treatment practice, there is some confusion in regard to the relative merits and application of the ion exchange and membrane techniques, especially with respect to electrodialysis and reverse osmosis. Table 7.8.2-2 briefly compares the characteristics and applications of these processes, while Table 7.8.2-3 indicates the size ranges of suspended matter in raw water and selected separation processes.

TABLE 7.8.1-1 Basic Types of Demineralizer and Resins Used

Scheme	Removal		
	Cations	Strong anions	Weak anions
(1) Two bed — SCR, WBR	By SCR	By WBR	None
(2) Two bed — ① SCR or ② WCR │ ① SBR or ② MB	1. By SCR 2. By WCR + SCR in MB	◄─── By SBR ───► ◄─── By SBR in MB ───►	
(3) Mixed bed — MB	By SCR	◄─── By SBR ───►	
(4) Three bed — SCR, WBR, SBR	By SCR	By WBR	By SBR
(5) Three bed — WCR, WBR, MB	By WCR + SCR in MB	By WBR	By SBR in MB
(6) Two bed with DG — ① SCR or ② WCR │ DG │ ① SBR or ② MB	1. By SCR 2. By WCR + SCR in MB	By SBR By SBR in MB	By DG + SBR By DG + SBR in MB
(7) ① SCR or ② WCR │ DG │ WBR │ ① SBR or ② MB	1. By SCR 2. By WCR + SCR in MB	By WBR By WBR	By DG + SBR By DG + SBR in MB

SCR — strong cation resin; WCR — weak cation resin
SBR — strong base resin; WBR — weak base resin
DG — degasifier; MB — mixed bed

Source: Adapted from *The Nalco Water Handbook*, F. N. Kemmer, ed., McGraw-Hill, New York, 1979.

TABLE 7.8.1-2   Ion Exchange Dimineralization Unit Selection Chart

Flow rate, gpm	Amount of impurity to be removed, mg/l			Resins required			Units to be used			
	FMA	CO₂	SiO₂	C	WB	SB		C	DG	A	MB
1. Any	Any	None	None	x	x	—	(1)	x	—	x	—
							(2)	—	—	—	x
2. Any	Any	Any	None	x	x	—	(1)	x	x	x	—
							(2)	—	x	—	x
3. 0–20	Any	Any	Any	x	—	x	(1)	x	—	x	—
							(2)	—	—	—	x
4. 20–50	Any	0–50	Any	x	—	x	(1)	x	—	x	—
							(2)	—	—	—	x
5. Over 100	0–100	Over 100	Any	x	—	x	(1)	x	x	x	—
							(2)	x	x	—	x
6. Over 100	Over 200	Over 100	Any	x	x	x	(1)	x	x	x	—
							(2)	x	x	x	x

NOTES: 1. C = cation, A = anion, DG = degasifier, MB = mixed bed, WB = weak base, SB = strong base, FMA = free mineral acidity (SO₄ + Cl + NO₃).
2. Numbers in parentheses: (1) multi-bed plant, (2) mixed bed for 1 micromho effluent.
3. Intermediate ranges (50–100 gpm flow, 50–100 mg/l CO₂, 100–200 mg/l FMA) require careful evaluation for best balance in capital and operating costs.

Source: Adapted from *The Nalco Water Handbook*, F. N. Kemmer, ed., McGraw-Hill, New York, 1979.

Microfiltration (MF)   The purpose of microfiltration is to remove rather large particles that are over 0.5 μm in diameter. One application of MF is to the field of sanitary engineering: Millipore filters are used to isolate coliform bacteria from water to collect suspended solids from water samples. Another application is in the removal of airborne particulates for air-feed type ozone generators.

TABLE 7.8.2-1   RO Membrane Characteristics (Based on 90% Rejection)

	Type membrane		
	Triacetate hollow fibers	Polyamide hollow fibers	Cellulose acetate spiral wound
Flux at 400 psi, gpd/sq ft	1.5	1.0	15–18
Back pressure, psi	75	50	0
pH range	4–7.5	4–11	4–6.5
Maximum temperature, °F	86	85	85
Cl₂, maximum mg/l	1.0	0.1	1.0
Bio-resistance	Good	Excellent	Fair
Backflushing	Ineffective	Ineffective	Effective

Source: Adapted from *The Nalco Water Handbook*, F. N. Kemmer, ed., McGraw-Hill, New York, 1979.

TABLE 7.8.2-2 Comparison Between Membrane Separation and Ion Exchange Process

Item	Economically Treatable Water	Pore Size of Membrane or Resin	Driving Force	Removal Objects	Notes
Microfiltration (MF)	Filtered	0.1–2 μm 0.45 μm is often used	Pressure >10 psig (>0.7 kg/cm²)	Insoluble particlelike bacteria, for example, but only a few colloidal matters	A batch process; mostly analytical and industrial uses; removes particles over 0.5 μm in diameter
Ultrafiltration (UF)	Filtered	0.002–0.1 μm 0.01 μm is often used	Pressure >20 psig (>1.4 kg/cm²)	Molecular size compounds, including all microorganisms	Mostly used in medical and industrial fields; has clogging problems if pretreatment is not provided; a batch process
Electrodialysis (ED)	Filtered 500–8000 mg/L salts (TDS)	<1 nm (<10 Å)	DC electric current 0.27–0.36 kW/lb salts	Ionized salt ions	Continuous process; incapable of producing fully demineralized water
Reverse osmosis (RO)	Filtered 100–36,000 m/L salts (TDS)	<1 nm (<10 Å)	Pressure >200 psig (>14 kg/cm)	Ionized salt ions and all colloidal matter	Continuous process; removes 90–95% of inorganic salts and 95–99% of organic matter
Ion exchanger (IX)	Settled or filtered 50–1000 mg/L salts (TDS)	<1 nm (<10 Å)	Pressure <7 psig (<5 kg/cm²)	Ionized ions: cationic and/or anionic	Batch process; possible to have complete removal of mineral ions

TABLE 7.8.2-3 Size Ranges of Suspended Material in Water and Corresponding Membrane Processes

Source: J. G. Jacangelo et al., "Assessing Hollow-Fiber Ultrafiltration for Particle Removal," *J. AWWA*, 81(11):68 (1989).

The MF membranes may be manufactured from a variety of materials, including cellulose acetate. The pore size of these membranes is generally larger than 0.1 μm and a pore size of 0.45 μm is frequently employed. Thus, removal of all colloidal matter (generally 0.01–1 μm in size) and dissolved materials cannot be expected. Yet, the membrane is capable of swelling and changing its characteristics, thereby allowing it to retain certain types of colloidal and soluble substances.

Whenever the MF membrane becomes clogged, the filtration rate drops below a practical level, thereby necessitating their replacement. Consequently, the water should be pretreated and have most of its suspended solids removed prior to the membrane filtration process. For example, if ordinary tap water is applied, the microfilter membrane will become clogged (and must be replaced) after filtering 250 gallons per square foot (1 L/cm²) of membrane surface area.

Ultrafiltration (UF)   Ultrafiltration uses membranes with pore sizes that are significantly smaller than 0.1 μm and generally require a driving pressure ranging from 30 to 90 psi (210–620 kPa or 2–6 kg/cm²). UF is capable of removing colloids, bacteria, viruses, and high molecular weight organic compounds. The membranes are consequently susceptible to clogging. However, certain types of UF membrane can be backwashed.

The UF membranes are usually composed of two layers: an extremely thin (1–5 μm) and fine porous skin over a more thick (25–50 μm) and porous substrate. The flow rate of the filtration process is reduced by the suspended matter and by the concentration of polarization, resulting in a localized increase of rejected impurities. This increase in the density and viscosity of the solution at the surface of the membrane causes a reduction in the flux (flow) rate: the water production rate is approximately 50 gal/day per square foot (2 m³/day·m²) of membrane surface area and the energy requirement ranges from 15 to 45 kW·h per 1000 gal (4–12 kW·h/m³) of the product water.

The UF membranes may be flat with a filter press type of support, flat and spirally wound, tubular with a supporting cylinder (inside), or a tubular (hollow) fiber with an external diameter of 0.5–1.5 mm, that does not have a supporting member (the raw water flows inside the tubes). This last configuration requires prefiltering through membrane filters with a pore size of 10–20 μm in order to prevent the membranes from becoming quickly clogged.

The UF process has been used in the medical field for the separation of blood plasma and also has a number of industrial applications: concentration of macromolecular suspensions (i.e., polyvinyl alcohol), recovery of certain substances from wastes, and concentration of emulsions.

Electrodialysis (ED)   The semipermeable dialysis membrane is impermeable to water but permeable to either cations or anions under different driving forces. The driving force may be a concentration gradient (simple dialysis), pressure (piezodialysis), or an electrical potential (electrodialysis). The most commonly used membranes are composed of hydrated cellophane and the selectivity of a given membrane is largely a function of its pore size. Although dialysis is rarely used in the fields of water treatment and water reclamation, a few industries, such as the textile industry, use this process to reclaim caustic soda from mercerizing baths.

The electrical potential is the driving force in the process of ED. Membranes having anion and cation exchange properties are alternately stacked in a press with a narrow water passage between them: the channels are made as narrow as possible to minimize electrical consumption. When direct current is applied to the electrodes (located on either side of the stacked membrane), the anions migrate to the anode and vice versa. Since the cation exchange membrane only permits cations to pass through, and only anions for the anion exchange membrane, alternating concentration and dilutions are created in the various (alternating) compartments of the membrane stack.

Unlike ion exchange, ED is a continuous process where the amount of salt removed from the water is proportional to the applied electrical current. However, application of an excessive amount of current results in electrolysis, forming hydrogen and oxygen gas, and thereby reduces overall efficiency.

The efficiency of ED may also be reduced by concentration polarization and by the formation of scale and deposits on the membrane surface; these phenomena increase the electrical current requirement. For example, calcium carbonate and magnesium hydroxide scales will develop at the anion exchange membrane due to the high pH resulting from the hydroxide ions. Although improvements in the stack design, as well as in the membranes, have helped to reduce both polarization and scale formation, they have not eliminated these problems.

If the raw water contains nonionized molecules, such as organic compounds and colloids, they remain partially demineralized in the water. It is therefore necessary to pretreat the raw water to remove turbidity, organic compounds, and iron, prior to being applied to the ED process; otherwise, they will foul the membranes.

The power consumption of the ED process is approximately 0.27–0.35 kW·h per pound of salt removed (0.6–0.8 kW·h/kg).

The ED process has three major limitations: (1) it is incapable of producing highly demineralized water; (2) the cost of the process rapidly increases with the salinity of the feed water, with the maximum salinity limit for economical treatment being 8000 mg/L; and (3) the number of equipment manufacturers is very limited: Ionix, Inc., Aqua-Chem Inc., and Dow Chemical USA are the only major supplier in the United States.

Reverse Osmosis (RO) When pure water flows through a semipermeable membrane (separating the two waters) into water with a high salt content, the water level in the salted water compartment rises until the osmotic pressures on either side of the membrane are equalized: this is osmosis. In reverse osmosis, an osmotic pressure greater than that of the salted water is applied

TABLE 7.8.2-4 Typical Performance of RO Process

Demineralization (Softening) by RO Process*

	Constituent (mg/L)	
	Raw	Finished
Hardness, as $CaCO_3$	380	20
Alkalinity, as $CaCO_3$	215	16
Total electrolyte, as $CaCO_3$	445	29
Silica, as SiO_2	25	3
pH	7.2	6.0
CO_2	25	25

Performance of a B-10 Module in Treating Seawater†

Operating conditions:

P = 56 bar	Conversion factor Y = 25%	Acidification H_2SO_4 = 810 psig

	Feed Water (mg/L)	Purified Water (mg/L)	Salt Passage (%)
Ca^{2+}	329	1.5	0.5
Mg^{2+}	1 031	3	0.3
Na^+	9 419	77	0.8
K^+	355	3	0.8
Sr^{2+}	13	0.2	—
SO_4^{2-}	2 200	1	0.05
PO_4^{3-}	9	0	—
HCO_3^-	68	26	38
Cl^-	15 825	120	38
F^-	0	0	—
NO_3^-	3.4	0	—
SiO_2	0	0	—
Total salinity	32 680	185	0.6
Conductivity (μS/cm)	37 250	390	1
pH	6.3	6.3	—

*Adapted from *The Nalco Handbook*, F. N. Kemmer, ed., McGraw-Hill, New York, 1979.
†Adapted from Degremont, *Water Treatment Handbook*, 5th ed., Halsted Press, New York, 1979.

to the compartment containing the salted water. This then forces pure water to flow back across the membrane. The required pressure depends on the difference in salt concentration but is often greater than 300 psig (2070 kPa or 21 kg/cm²). The rule of thumb is to apply approximately 10 psig (70 kPa) for each 1000 mg/L difference in TDS.

The RO process may be applied in cases where the water has a salinity (TDS) that is less than 500 mg/L to produce a 25 mg/L TDS product water using a small amount of energy (compared to the EO process). RO is also capable of demineralizing seawater (36,000 mg/L TDS) to water containing approximately 300 mg/L TDS (Table 7.8.2-4).

The RO process is occasionally called hyperfiltration since a high pressure is required to drive the system. The major differences between RO and ultrafiltration, which also applies a fairly high pressure, are the magnitude of the applied pressure and the size of the substances that can be removed. Ultrafiltration is essentially incapable of removing solutes of low to intermediate molecular weight. Reverse osmosis can remove ionic species with low molecular weight, as well as all types of colloids, bacteria, and viruses.

The ratio between the amount of demineralized water and the amount of raw water fed to the system is defined as the conversion factor and is expressed as a percentage.

Two types of RO membrane are currently being marketed: cellulose acetate and aromatic polyamide. Membranes composed of cellulose acetate will yield a high flow rate per unit of surface area. They are available in tube shape, as spirally wound flat sheets, and hollow fibers. Conversely, polyamide membranes have a low rate of flow and are marketed as hollow fiber to provide the maximum surface area per unit volume. The advantage of the polyamide membrane is its excellent resistance to chemical degradation and biological attacks, thereby providing it with a longer service life (3–5 years) than the cellulose acetate membranes (2–3 years); refer to Table 7.8.2-1.

Since RO is a continuous process, the membranes must be replaced every 3–5 years. However, unlike ED, there are a number of membrane and RO

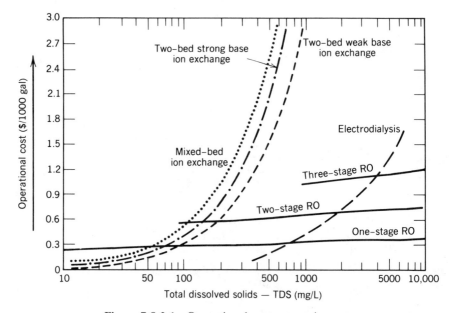

Figure 7.8.2-1 Operational cost comparisons.

process unit manufacturers to choose from. The major suppliers of the RO process in the United States are: Crane Co., Culligan International, DOW Chemical USA, DuPont Inc., Envirogenics System Co., Hydranautics, Ionics, Inc., LA Water Treatment, Osmonic, Inc., Permutit Co., and Infilco Degremont, Inc. Moreover, the operational costs associated with demineralization are lowest with RO when compared with ED and ion exchange; it also is uniquely characterized by a constant cost production factor.

Figure 7.8.2-1 is an example of the demineralization costs associated with ion exchange, electrodialysis, and the process of reverse osmosis excluding the capital amortization cost.

During the late 1980s, the "nanofiltration" process (see Table 7.8.2-3) was first applied to the public water supply for a small French village. Nanofiltration uses a membrane similar to ultrafiltration and the driving force for this membrane is less than 20 psig (1.4 kg/cm^2). It is reported that this process is successfully treating a surface water with an average turbidity of 6 NTU, and a turbidity over 100 NTU for a short period of time, without the use of coagulant, thus providing almost 12 h filtration cycles and a conversion factor of approximately 90%.

At this early stage, the application of the nanofiltration membrane process to the public water supply appears to be promising. However, application of this process to a large scale public water supply system may not occur until after the year 2000.

BIBLIOGRAPHY

Applegate, L. E., "Post Treatment of Reverse Osmosis Product Waters," *J. AWWA*, 78(5):59(May 1986).

Bersillon, J. L., Anselme, C., Mallevialle, J., and Fiessinger, F., "Ultrafiltration Applied to Drinking Water Treatment: Case of a Small System," AWWA Annual Convention in Los Angeles, 1989.

Degremont, *Water Treatment Handbook*, 5th ed., Halsted Press, New York, 1979.

Fair, G. M., Geyer, J. C., and Morris, J. C., *Water Supply and Wastewater Disposal*, Wiley, New York, 1954.

Jacangelo, J. G., Aiela, E. M., Carns, K. E., Cummings, E. W., and Mallevialle, J., "Assessing Hollow-Fiber Ultrafiltration for Particulate Removal," *J. AWWA*, 81(11):68(November 1989).

Lauch, R. P, and Guter, G. A., "Ion Exchange for Removal of Nitrate from Well Water," *J. AWWA*, 78(5):83(May 1986).

Martin, H. J., and Miller, G. R., "A Zero Discharge Stream Electric Power Generating Station," *J. AWWA*, 78(5):52(May 1986).

Montgomery, J. M., Consulting Engineers, *Water Treatment—Principle and Design*, Wiley, New York, 1985.

Nalco Chemical Co., Kemmer, F. N., ed., *The Nalco Water Handbook*, McGraw-Hill, New York, 1979.

TABLE 7.9-1 Effective Processes for the Removal of Inorganic Contaminants

Contaminant	Chemical Coagulation and Sedimentation		Ion Exchange	Reverse Osmosis	Remarks
Arsenic Ar^{3+} and As^{5+}	$Fe_2(SO_4)_3$ Alum Lime	(pH 6–8) (pH 6–7) (pH 10.5–11.3)	—	—	IX process (Na and H cycles) are not effective
Barium Ba^{2+}	Lime	(pH 10–11)	IX (Na cycle) is effective	—	Alum and ferric coagulant are not effective
Cadmium Cd^{2+}	Lime $Fe_2(SO_4)_3$	(pH 8.5–11.3) (pH 8–9)	—	—	Alum coagulation is not effective
Chromium Cr^{3+} and Cr^{6+}	Lime $Fe_2(SO_4)_3$ Alum	(pH 10–11.3) (pH 7–9) (pH 7–8.5)	—	—	Cr^{6+} can be removed by $FeSO_4$ (pH 7–7.9) but not by $Fe_2(SO_4)_3$
Fluoride F^-	Lime	(pH 9.5–11.3)	IX by activated alumina is good	—	Bone char is also good
Lead Pb	$Fe_2(SO_4)_3$ Alum Lime	(pH 7–9) (pH 6–9) (pH 9.5–11.3)	—	—	
Inorganic Hg	$Fe_2(SO_4)_3$ Lime	(pH 7–8) (pH >11)	—	—	Both GAC and PAC (>50 mg/L) are effective for mercury
Organic Hg	—	—	—	—	Both strong base IX and weak base IX are good
Nitrate N^-	—	—	IX resin process is effective	—	
Selenium Se^{4+} and Se^{6+}	—	—	Cation IX is good	RO is effective	Coagulation by metal coagulants is not very good
Silver Ag	$Fe_2(SO_4)_3$ Alum Lime	(pH 6–9) (pH 6–8) (pH >10)	—	—	
Asbestos fibers	$Fe_2(SO_4)_3$ Alum	(pH 7–9) (pH 6–7.5)	—	—	Filtration should be followed by coagulation and sedimentation

Reinhard, M. R., Goodman, N. L., McCarty, P. L., and Argo, D. G., "Removing Trace Organics by RO Using Cellulose Acetate and Polyamide Membranes," *J. AWWA*, 78(4):163(April 1986).

Sinisgalli, P. D., and McNutt, J. L., "Industrial Use of Reverse Osmosis," *J. AWWA*, 78(5):47(May 1986).

7.9 INORGANICS REMOVAL PROCESSES

In 1985 the EPA released its recommendations on the maximum contaminant level (MCL) for 10 inorganic contaminants: arsenic, barium, cadmium, chromium, fluoride, lead, mercury, nitrate, selenium, and silver. This section briefly covers the applicable and tested processes used to remove these contaminants to levels below the allowable level.

In 1978, the EPA published a manual that presented the recommended treatment techniques for meeting the interim primary drinking water standards; these are listed in Table 7.9-1. Since the EPA did not present effective methods for removing asbestos fibers, it has been added to the table.

BIBLIOGRAPHY

EPA, *Manual of Drinking Water Treatment Techniques for Meeting the Interim Primary Drinking Water Regulations*, EPA-600/8-77-005, U.S. EPA, Cincinnati, OH, 1978.

CHAPTER 8

Management of Procurement and Construction Phases

The design phase of a project is completed once the final plans and specifications of the project are presented to the owner and the required regulatory agencies. As part of the design phase, the design engineer should compile a complete set of project files, including all design calculations and correspondence, since these documents may prove to be essential for the project in the later stages and particularly in the case of future lawsuits.

Upon completing the design phase, the engineer must begin managing the procurement and construction phases. The sequence of events are (1) bid advertising, (2) bid opening, (3) bid evaluation and award of contract, and (4) the construction phase. There are generally three principal parties involved in the construction phase: the owner, the designer, and the bidder.

8.1 PROCUREMENT PHASE (BID PHASE)

As previously stated, the procurement phase is composed of bid advertising, bid opening, and bid evaluation and award of contract.

8.1.1 Bid Advertising

During bid advertising the project engineer and the resident engineer (inspector) should review the contract documents together to ensure that the constructability of the structures and important field considerations have been incorporated into the plant design and specifications. Since the period between bid advertising and bid opening is generally a minimum of 6–8 weeks (unless the project is very small)—to allow the prospective contractors to perform the necessary activities—the engineers have time to carefully review the documents. If any omissions or errors are found during this period, the project engineer should issue an addendum to the plans or specifications to all bidders a minimum of 10 days before bid opening. The project manager should also look ahead and meet with the resident engineer to develop an agenda and a list of key subjects that must be discussed with the selected contractor at the preconstruction meeting, which is held with the successful bidder.

During the bidding phase, the project and resident engineers should perform the items listed below.

1. Arrange for bid advertising and select a bid opening date. Log the issuance of bid documents and release the project data to construction trade papers.
2. Keep an accurate log of all sets of contract documents issued: for example, name of the firm, address of each bidder, name of the individual acquiring the documents for the firm, the date of pickup, and deposit fee received.
3. If addenda are issued to the bidding documents, a copy of the addenda should be mailed to each holder of a set of documents by certified mail with a "return receipt requested."
4. If the demand for the documents exceeds the prepared supply, exact reproductions of the original documents must be supplied to the requesting firms.

For obvious reasons, it is vital that all bidders receive identical sets of bidding documents and that the same information be provided to all bidders whenever they request clarification by phone or by mail.

In order to protect the owner against financial loss, which may occur because the bidder later refuses or is unable to sign the contract after submitting a bid, the requirement for a bid bond (security) should be included in the bidding documents. The bid bond should be 5–10% of the contractor's bid price and may be in the form of a certified or cashier's check and, occasionally, a negotiable security. The bid securities are returned to the bidders after all bids are opened and the contract is awarded.

All public works contracts require the selected contractor to submit performance and payment (labor and materials) bonds once the contract is awarded. The performance bond acts as a surety to cover costs associated with the completion of the contract, should the contractor fail to execute the contractual requirements. Under the payment bond, if the contractor becomes insolvent, the surety guarantees payment of all legitimate bills (labor or the acquisition of material) resulting from performance of the contract.

8.1.2 Bid Opening

The bid opening procedure generally follows the sequence shown below:

1. Receipt of sealed bids at the designated time and place.
2. Confirmation that all bids are responsive.
3. Acceptance and logging of each bidder's name and the bid amount of all responsive bids.
4. Summary of all line-item prices for all unit-price bids.

After all the bids have been opened, the nonresponsive or informal bids are rejected. The project manager then identifies the "apparent" low bidder, which after careful bid review and evaluation is subject to confirmation.

8.1.3 Bid Evaluation and Award of Contract

Following the bid opening session, all bids are tabulated and each bidder is checked for financial responsibility, licensing, integrity, and reputation. The remaining tasks for the project and resident engineers are:

1. Analyze the bids and compare them with the engineers' estimate.
2. Make recommendations in regard to awarding the contract.
3. Prepare notification of award of contract and procure bid bonds from the three lowest bidders.
4. Remove copies of payment and performance bonds and certificate of insurance from the file prepared on the selected contractor and have the contractor submit the payment and performance bond with a copy of the actual insurance policy.
5. Have the owner sign the contract and return all other bid bonds.

8.2 CONSTRUCTION PHASE

There are several types of contractual form for a construction project. Fisk (1982) describes the following types as the most commonly used.

1. Traditional architect/engineer (A/E) contract.
2. Design/construction manager (D/CM) contract.
3. Professional construction manager (PCM) contract.
4. Design–build (turnkey) contract.

8.2.1 Traditional Architect/Engineer Contract

The traditional A/E contract limits the engineering services provided during construction to occasional visits to the construction site and a few other functions, such as review and approval of shop drawings, interpretation of plans and specifications during construction, evaluation of claims by contractors, and review of payment to the contractor.

8.2.2 Design/Construction Manager Contract

The D/CM contract is similar to the A/E contract except that the project manager of the A/E firm is responsible to the owner during both the design and construction phase. The project manager must satisfy all project needs including scheduling, cost control, quality control, prepurchasing, and coordination of all work. The responsibility of the firm does not end until the owner's final acceptance of the project.

8.2.3 Professional Construction Manager Contract

Under the PCM contract, the professional construction management firm only acts as a representative for the owner during the entire period of the project; the firm is not involved in the design or the construction of the treatment plant. Under this type of contract, the construction manager operates as an organization, not as an individual, and is responsible for the total project time, cost and quality control, and supervision and has control over the functions of the A/E firm and the contractor.

8.2.4 Design–Build Contract

The design–build contract is commonly called "turnkey" construction. The owner contracts with a single firm that plans, designs, and constructs the entire project with its own in-house capabilities. Furthermore, the turnkey contractor often offers the owner financial incentives.

The major advantages of turnkey construction are (1) the elimination of claims by the contractor against the owner for errors or omissions in the plans and specifications and (2) the ability to design construction as the design of each independent phase of the project is completed without having to wait for completion of the overall project design, which is often referred to as "fast track." However, the disadvantage of turnkey contracts is that there is no competitive bidding process for plant construction. Thus, turnkey construction contracts are illegal in the United States for all publicly funded construction projects: although it is commonly used to build water treatment plants in the rest of the world and is applicable to any project in the United States provided that the project solely uses private funds.

If a design–build contract is chosen, the firm should be selected on the basis of individual expertise, previous designing experience in the particular type of project, financial status, and the reputation of the firm. Although it appears that this concept is beneficial and saves a great deal of cost to the owner, case histories have proved that this is not necessarily true: without competitive bidding procedures, the construction costs tend to be higher and the quality of the design and specifications is not necessarily superior.

8.3 CONSTRUCTION MANAGEMENT

Construction involves the combined skills of engineering, science, and organization, in addition to the use of educated guesses and calculated risks. Numerous and complex details must be carefully coordinated between various parties: the owner, engineers, architects, general and specialty contractors, manufacturers, material suppliers, equipment distributors, governmental agencies, utilities, labor, and so on.

Construction contractors are private enterprises that engage in various types of construction projects, including water supply. The prime objectives of the contractor are (1) to complete the construction project to the satisfaction of the owner, within the set budget, and on schedule, and (2) to make a profit. Since a contractor's goal is to complete the project as quickly and as economically as possible, there is always the possibility of poor quality work on the project.

Engineers, on the other hand, tend to believe that if their design is correct and the plans and specifications are carefully prepared, the field construction will proceed without any problems. However, many case histories have proved this idea to be far from true and engineers therefore need to ensure that all work and construction practices are equal to or are in excess of the standards dictated in the construction contract documents. Because of these facts, it is not surprising that owners are reluctant to place the responsibility for overseeing the quality of the work on the project solely in the hands of the contractor.

The construction management program is a vital tool in completing the construction phase of a project in a satisfactory manner because it meets the requirements of both the owner and the engineers. As a matter of fact, many federal construction contracts require that the resident inspector, who is employed by the contractor, meet the contractor quality control requirements.

Ideally, construction management is handled by the firm that designed and prepared the plans and specifications, since it is intimately familiar with the plant site and design details and therefore capable of effectively administering the construction phase of providing quality control for the owner, of securing the public safety, and of maintaining the professional reputation of both the design firm and the contractor.

Construction administration and construction quality control should be characterized by continuous on-site inspection by one or more experienced and competent engineers who are technically qualified. These inspectors are often known as resident engineers.

The resident engineer is responsible for ensuring that all the design details shown on the plans and specifications and on the shop drawings approved by the design engineer are constructed in strict accordance with the requirements. The resident engineer must also ensure that the work quality and construction practices are equal to or better than the standards dictated in the construction documents.

One important restriction is that both the resident project representative and the inspector are not authorized to change the plans and specifications nor are they allowed to implement their own interpretations of the documents. If errors or problems associated with construction practice arise during construction, they should immediately be brought to the attention of the project manager for further clarification.

Whenever there are obvious errors or if the changes vastly improve the situation or save time and material without sacrificing quality, these changes should be implemented by having the owner sign a formal "change order" (consent). The change order generally involves an adjustment in the construction fee.

The construction operation is a dynamic process, requiring maximum efficiency, speed, and economy for the contractor. Therefore, the resident engineer(s) should make quick, yet appropriate, decisions on all minor technical problems in order not to delay the operation of the contractor; resident engineers have the authority to approve materials and work.

As in the other fields, computerized project management programs are available for tasks such as scheduling, budgeting, and the plotting of job progress. Use of this type of software is recommended because the time required to perform these tasks is accelerated to a short period of time.

8.4 RESPONSIBILITY OF RESIDENT ENGINEERS

The function of the resident engineer is discussed in Section 8.3. However, the major responsibilities are as follows:

1. Be thoroughly familiar with the plans and specifications.
2. When the work does not meet the requirements, immediately notify the contractor of the construction site and record the infraction and date of notification in the daily log. If the contractor chooses to ignore the notice, immediately advise the project manager of the inaction.
3. Write a daily report that includes the daily activities, the items inspected, any instructions given, any agreements made with the construction team, the weather conditions, and other pertinent items. The daily report and diary/log book have legal importance in the event of contract disputes.
4. Perform all on-site testing specified in the contract documents. The tests should be performed in a expeditious and careful manner. All test failures must be reported to the contractor without delay.
5. Perform all inspections in a timely manner.
6. Discover unacceptable work in its early stages and report it to the contractor as soon as possible.
7. All problems beyond the capability of the resident engineer should be reported to the project manager of the engineering firm.
8. Do not be forced into making hasty decisions; carefully evaluate the situation and the possible consequence of the decision.

9. Be prepared to stand behind any decision made in regard to the contractor's work.
10. Be concerned with the safety of the public and all persons associated with the construction project and attempt to be objective in all situations (control emotions).

8.5 PROTESTS AND CLAIMS

Protests are defined as disputes arising from the issuance of a contract change order by the engineer against the objection of the contractor. When the engineering firm receives a letter of protest from the contractor, it should be examined carefully before acknowledging its receipt, to assure that the basic requirements of the specifications are included in the letter. The project manager and the owner should then review the merits of the protest and issue a letter advising the contractor of his rights under the contract to file a legal claim.

One of the most important items that is considered before presenting any legal claim is documentation. When negotiations between the owner and the contractor fail, the data documented by the resident engineer(s) and the contractor become valuable in the litigation or arbitration procedures.

8.6 PROJECT CLOSEOUT

The main activities during project closeout are final inspection, acceptance of the plant for the owner, and final payment to the contractor. However, many important issues still remain: review of the record ("as built") drawings, transfer of the guarantee, obtaining the operation and maintenance instructions for the equipment, return of bonds, composing the plant operation and maintenance manual, and compiling the punch list items. The punch list is a detailed check-off list that enumerates all items still requiring completion or correction before the work will be accepted and a certificate of completion is issued.

In some cases, the project is accepted as being "substantially complete" where only minor items remain to be completed or the "beneficial use" of the facility is accepted by the owner. Beneficial use (occupancy) is defined as the owner's use of the facility prior to its full completion. The date of beneficial use is the beginning of the 1 year guarantee period and the total responsibility of proper maintenance and operation of the facility is transferred to the owner.

Even if the certificate of substantial completion is filed with the punch list, the owner will not release the retained funds until all remaining deficiencies are corrected and a waiver of liens is completed. Final payment signifies the release of all retainers held during the project; this generally amounts to 5–10% of the total project cost.

Liquidated Damages   If the contractor fails to complete the project by the specified date, he is financially liable to the owner for a pre-agreed sum for each day beyond the original finish date. This amount of money represents the opportunity cost of the owner resulting from such delays and in the case of water treatment plants, this is usually the loss of income from the sale of potable water to customers.

Clean-up   The contractor is responsible for cleaning up the construction site at the end of the job. The work will not be accepted until all temporary utilities, construction access roads, security fences (for material storage), field offices, stockpiles, surplus materials, and scraps are removed from the plant site.

Site clean-up is the obligation of the contractor and must be conducted at his own expense. The contractor should not be permitted to bury any waste or rubbish on the plant site unless the owner issues written approval.

8.7   MAJOR CLOSEOUT ACTIVITIES

The major closeout activities are as follows:

1. Closeout inspections as presented in the punch list.
2. The preparation of clear and concise documentation of the closeout inspection on each item listed on the punch list.
3. A partial reduction of the field office inspection staff.
4. Final inspection with qualified engineers; a final punch list must be developed for any outstanding deficiencies requiring correction.
5. Preparation of the record drawing with the contractor.
6. Obtain the following items from the contractor before accepting the project.
   a. Guarantees.
   b. Certification of inspection.
   c. Operating manuals and instructions for all equipment.
   d. Keying schedules.
   e. Spare parts that are specified to be received.
   f. Record drawings.
   g. Bonds (guarantee bonds, maintenance bonds, etc.).
   h. Certificate of inspection and compliance issued by local agencies.
   i. Waivers of liens.
   j. Consent of surety for final payment.

7. Receive the request for final progress payment from the contractor.
8. Check that all remaining work requiring completion is properly fixed.
9. Obtain signatures from the A/E project manager, the contractor, and the owner on the certificate of completion. File this certificate with the office of the County Recorder.
10. Notify the owner that the project is complete and ready for operation.
11. Arrange for final payment and release of the retainer by the owner.
12. Clean up and close the field office.

8.8 CONSTRUCTION PHASE CHECKLIST

The project manager is responsible for the following major tasks during the construction phase:

1. Approve the cost breakdown documents submitted by the low bidder.
2. Identify and list the contractor's key personnel and the phone numbers at which they may be contacted.
3. Finalize the quality control and inspection plans.
4. Hold preconstruction meetings with the owner and with the contractor and owner.
5. Arrange for surveys, test labs (i.e. determine concrete strength), and mill and factory inspections.
6. Arrange for the construction documentation tools (photos, diary, field log).
7. Issue a notice to proceed.
8. Review the mobilization and construction schedule as submitted by the contractor.
9. Provide the necessary construction management service for on-site quality assurance of materials and work; the scope of this service depends on the type of construction contract.
10. Prepare daily construction reports and a diary/field log. Transmit these to the project manager.
11. Prepare and distribute monthly reports.
12. Review and process shop drawings with the transmittals.
13. Receive, log, and transmit all submittals from the contractor.
14. Review the record drawings and update them on a monthly basis.
15. Confirm the amount of each payment.
16. Process all progress payments.
17. Process the relocation of retainer fees.
18. Estimate the proposed changes and the extra work required by these changes.

19. Prepare, log, and process all change orders.
20. Document all extra work and work order protests.
21. Process and document time delays and requests for time extensions.
22. Assist in negotiations related to disputed work.
23. Administer the owner's acceptance of "beneficial use."
24. Compile a punch list and arrange for final inspection and job acceptance.
25. Prepare a notice for project completion, final payment, and final billing. Advise the owner that a 1 year guarantee bond is required from the contractor prior to occupation.
26. Finalize the record ("as built") drawings.
27. Prepare a construction contract summary report and file this in the construction management project document file.
28. Inspect the project prior to the end of the warranty; prepare a report on the findings and make it part of the project management file.
29. Evaluate the performance of the construction management team.
30. File all pertinent documents for permanent record.

BIBLIOGRAPHY

Fisk, E. R., *Construction Project Administration*, 2nd ed., Wiley, New York, 1982.

Montgomery, J. M., Consulting Engineers, *Project Control Manual*, J. M. Montgomery Consulting Engineers, Pasadena, CA, 1976.

O'Brien, J. J., *CPM in Construction Management*, 3rd ed., McGraw-Hill, New York, 1983.

Project Management Institute, *Project Management Seminar/Symposium—Innovation*, Project Management Institute, Drexel Hill, PA, 1984.

Project Management Institute, *Research Goals with Project Management*, Project Management Institute, Drexel Hill, PA, 1987.

CHAPTER 9

Operations and Maintenance Manual and Operator Training

Since water treatment plants generally require a period of 2 years to be constructed, adequate time exists for the preparation of the operation and maintenance manual. This manual must be provided to the plant personnel at least 1–2 months before plant start-up. Training sessions for the plant operators must begin during the final month of plant construction, when a substantial portion of the plant is completed. Thus, the owner should either hire or assign the key positions of the operational staff by this date. All new members of the plant operating staff—especially the superintendent, chief operator, and maintenance supervisor—are advised to frequently visit the construction site, beginning in the early stages of construction, so that they are familiar with the location, elevation, and method of installation of all pipes, weirs, valves, and other items which are not visible after plant construction is completed. Most of these structures are either buried in the ground or embedded in the concrete walls and slabs once the plant is finished.

Today, all municipal and public works officials, as well as directors of private water purveyors, are aware of the new and stringent drinking water quality standards set by the EPA and the states, and the requirements imposed on the water treatment process. In the midst of all these complex interactions of needs and concerns are the public officials and the directors of the private sector, who have identified the needs of their communities for improved and more sophisticated water treatment and control. Furthermore, the public is demanding safe drinking water. Thus, millions of dollars have been funded for the construction of technologically sophisticated, complex, and expensive water treatment plants and the modification of existing plants to meet the new requirements. Yet, use of high-quality equipment and computer-based instrumentation and control does not guarantee the continuous production of safe water: the plant must be operated by qualified and well trained personnel. Thus, the need for a well organized and practical plant operation and maintenance manual, as well as systematic and highly skilled on-site training, is clearly obvious. Unfortunate disasters such as Three-Mile Island clearly demonstrate this need.

9.1 OPERATION AND MAINTENANCE MANUAL (O&M MANUAL)

Too many O&M manuals are relegated to occupying shelf space on the book-shelves of water treatment plants. They are never used as a reference because they are either abstracts of textbooks or "boiler plates" that have little practical value. Although the EPA issued guidelines for composing the O&M manual in 1974, the key to creating a useful manual is to be practical and specific. The O&M manual should have practical features such as discussions of the basic principles used in the plant; talk-through presentations of the process layouts; detailed descriptions of each process component; step-by-step operating instructions for each process and process component; discussions of the variables that arise during process control and process status evaluation; descriptions and catalog cuts of all major equipment; tabulated troubleshooting guides listing the probable causes and solutions for problems commonly encountered in the field; and emergency response procedures.

The format of the O&M manual can be any style; however, the manual should at least include the following subjects:

1. Technical literature (equipment).
2. Process operating manual.
3. Preventative operation procedures.
4. Standard operating procedures.
5. Maintenance servicing schedules.
6. Emergency response procedures.
7. Policy formulation procedures.

The process operating manual listed above (point 2) is the heart of the O&M manual. This section should include the following items:

- Process description
- Design criteria
- Operating procedure
- Troubleshooting guides
- Emergency response program
- Servicing and maintenance requirements

Photographs of the control systems and the major equipment of each process train and a narrative instruction—which provides information on how to start-up, shut-down, and troubleshoot respective pieces of equipment—have been found to be very effective in educating plant personnel. Use of such documents (videocassettes) provides even newcomers to the plant with a visual direction on how to operate and maintain any system. The use of schematic illustrations is always helpful in creating a better understanding of the electrical, electronic control, and extensive valving schemes.

One of the most popular and effective O&M manuals is that prepared by television and automobile manufacturers. Their concise manuals are almost always only a few pages long: a summary of the key items supported by a detailed description of each function. The plant O&M manual can also be

TABLE 9.1-1   Operations Checklist (Example)

OPERATIONS CHECKLIST

Date:_____

Operator:_____

Item Description	Equipment No.	Status

Source: J. M. Montgomery, Consulting Engineers.

arranged in a similar manner. This type of format allows the plant operator to obtain key information from a small pocket guide that summarizes the key subjects. If this concept is adopted, the O&M manual may be organized as shown below.

Main O&M Manual

1. Purpose of the treatment process
2. Process descriptions
3. Component descriptions
4. Detailed operating procedures
5. Performance evaluation and process control
6. Troubleshooting guide
7. Maintenance and servicing schedule
8. Emergency response program

Operator Pocket Guide

Flow schematics
Concise operating procedures
Troubleshooting guides
Sample problems, including emergency response

Maintenance Pocket Guide

Maintenance procedures
Servicing schedules
Schematics of major processes and control loops

The O&M manual should additionally contain the recommended types of forms used for records, logs, and reports. Today, data logging by means of computers is almost the standard method of producing the plant operation log and report. Therefore, an additional operational log sheet is not necessary. However, it is recommended that a few additional forms be distributed to the operators so that they can familiarize themselves with each piece of equipment, its location, and its physical condition (e.g., the need for painting). Over-reliance on the computerized system will make the plant operators ill prepared for any emergencies resulting from failure of the computer. Tables 9.1-1, 9.1-2, and 9.1-3 are some examples of this type of form.

9.2 OPERATOR TRAINING BEFORE PLANT START-UP

There are two key requirements for successful plant start-up: (1) good operation and proper maintenance and (2) familiarity with all plant facilities and a working knowledge of all water treatment processes. The goal of the

TABLE 9.1-2   Equipment Data Form

EQUIPMENT DATA FORM

Plant No._____ Equipment Description_____

Equipment No._____ Component of_____

Supplier Address

Manufacturer_____   _____
Model No._____   _____
Serial No._____   _____
Installation Date_____   _____
Component Cost_____   _____

Recommended Spare Parts

Capacity_____   _____
Speed_____   _____
TDH_____   _____
Frame_____   _____
Type _____   _____

Comments_____
_____
_____
_____

DRIVE MOTOR

Supplier Address

Manufacturer_____   _____
Model No._____   _____
Serial No._____   _____
Installation Date_____   _____
Component Cost_____   _____

Horsepower _____ DC/AC _____   Type _____
Frame_____ Ins. Class_____   °C Rise_____
Speed_____ Phase_____   Enclosure _____
Amperes _____ Cycles _____   Condition _____
Voltage _____ S.F._____
Comments_____
_____
_____

Source: J. M. Montgomery, Consulting Engineers.

operator training sessions (prior to start-up) is to educate the plant operators in these two essential items.

The training sessions should last a minimum of three working days with the O&M manual as the textbook. The project engineer and the heads of the mechanical, electrical, instrumentation, and control disciplines are assigned as the instructors for each area of expertise. Representatives of the major equipment manufacturers should also be invited to participate in the training

TABLE 9.1-3   Problem Identification Form

PROBLEM IDENTIFICATION FORM

O b s e r v e r	**PROBLEM STATEMENT** Observer: _____ Date: _____ Subject: _____ Description of Problem:(Attach sketches or additional sheets if needed.)_____ _____ _____ _____ _____ _____ Suggested Solution _____ _____ _____ _____ Submit one copy each to: _____ _____ _____ Plant Superintendent Project Coordinator Sup. San. Engineer
P r o j e c t **M a n a g e r**	RESOLUTION OF PROBLEM (Decision /Corrective Actions) _____ _____ _____ _____ _____ _____ DISTRIBUTION Sup. San. Engineer _____ Project Coordinator Plant Superintendent (JMM/OCI) Plant Superintendent _____

Source: J. M. Montgomery, Consulting Engineers.

sessions. Full use of audiovisual aids, such as videotapes and color slides, and a chalkboard is highly recommended.

BIBLIOGRAPHY

EPA, *Considerations for Preparation of Operation and Maintenance Manuals*, EPA-430/9-74-001, U.S. EPA, Washington DC, 1974.

CHAPTER 10

Plant Start-up and Follow-up Services

10.1 START-UP

Even if the plant is constructed exactly according to the plans and specifications, plant operators must always debug a few items during plant start-up. Moreover, certain processes, such as ordinary granular medium filters and sludge blanket type reactor-clarifiers, require a period of maturation in order to function properly.

Plant start-up is a hectic time for plant operators. Many equipment and process control units require adjustment. Therefore, serious problems and complications are most likely to arise during this period. The plant owner is strongly advised to contract the services of a qualified professional consulting firm so that problems often associated with plant and process start-up can be avoided, thereby quickly placing the plant into service with fewer complications; the firm may also troubleshoot and optimize process efficiency. The most appropriate choice would be to hire the engineering firm that conducted the design, as well as construction management, simply because it is the most familiar with the plant, provided that it is qualified and experienced in doing so.

Optimization of coagulant dosage, adjustment of mixing conditions for flocculation, optimization of filter wash conditions, ensuring the proper pump operation sequence, and careful analysis and control of the finished water quality prior to delivery to the customer are items that are effectively executed by the consulting firm.

10.2 FEEDBACK

Feedback from the owner (client) with respect to the quality of the design, the performance of the engineers, and the performance and efficiency of the facility (water treatment plant) is absolutely necessary for maintaining quality engineering services. Most owners are reluctant to voice their dissatisfaction or their constructive opinions, unless outrageous mistakes have been made. Instead, the unhappy client never awards another job to the engineering firm and the engineers have no idea as to what caused the dissatisfaction. Thus, from both a business and engineering standpoint, it is imperative that engineers ask the client "How have we been doing?" and "How can we improve our service?" (feedback). It is a cold fact that the performance of the engineers and the firm will neither improve nor progress unless proper feedback is obtained from the client.

Generally, once all the debugging has been performed and the water treatment plant has been operating in a satisfactory manner, the engineers who were assigned to the project begin work on the next project. Seldom do engineers have a chance to view the operational plant, unless they are asked by the client to remedy significant problems. However, these problems are often minor in nature: problems with mechanical equipment, problems associated with the electrical or control systems, or problems with the civil works or architectural items. Engineers from these various disciplines are capable of handling these types of problem; thus, the project engineer and the project manager are generally not involved or even supplied with the details of the problem. This is especially true for busy consulting engineers. Consequently, some engineers continue making the same mistakes over and over. For this reason, a project engineer should visit each plant that he/she has designed at least once a year for the first 5 years. This allows the engineer to (potentially) obtain feedback from the plant operators. Furthermore, the engineer should compose a memorandum describing the findings during the plant visitation, including the owner's criticisms and comments (Table 10.2-1). The memo should be filed in the project file and should also be disseminated among the key engineers of the project team; this allows every key engineer to benefit from the findings.

When an error of a serious nature is discovered, some project engineers find it very difficult to disclose this information to their boss and other engineers and the error tends to be covered up. This is not in the best interest of the engineer or the firm. All grave errors should immediately be reported to both the chief engineer and the legal department of the firm. Numerous cases have needlessly ended up in court, costing the engineering firm 10 or even 100 times more than if the problem had quickly been remedied; the latter case usually costs less than $50,000. Any delay in corrective action not only aggravates and angers the client but can potentially lead to a lawsuit.

10.3 OPERATOR TRAINING SERVICES

Both the EPA and public officials have recognized the importance of operator training for wastewater treatment plants in improving plant effectiveness and, more importantly, in cutting operation and maintenance costs: the EPA requires that all publicly funded plants provide operator training services. Due to the requirements imposed by both the states and the EPA on water treatment plants, many administrators have begun to feel the same need for operators of public facilities including water treatment plant. This concept is

TABLE 10.2-1 Project Feedback Report Form (Example)

JAMES M. MONTGOMERY,
CONSULTING ENGINEERS, INC.   **PROJECT FEEDBACK REPORT**

FROM: _____

TO: _____   Division of Human Resources and Technology

PROJECT	_____	CHECK APPROPRIATE SPACE
JOB NO.	_____	____ General Design Problem
CLIENT	_____	____ Specification Problem
OBSERVER	_____	____ Standard Detail
DATE INITIATED	_____	____ Special Detail
SPEC. SECT. & PAGE	_____	____ Special Legal Problem
DRAWING SHEET NO.	_____	____ Other

CASE HISTORY OR PROBLEM: (Attach sketches or additional information)

_____
_____
_____
_____
_____
_____
_____
_____
_____

SUGGESTED IMPROVEMENT:

_____
_____
_____
_____
_____
_____
_____
_____

ACTION TAKEN IN THIS CASE:

_____
_____
_____
_____
_____
_____
_____
_____

not a new idea: operator training has been used for a number of years to help plant operators to obtain their operator certification (required by most states).

The operator training program has been offered as a state sponsored short school program to help operators pass the certification examination, as a correspondence course, as in-plant practical technical assistance, and as in-plant training. Of these programs, the in-plant training is undoubtedly the most comprehensive and effective, since it is a combination of "hands-on" and classroom instruction conducted at the treatment plant. This type of program provides operators with a sound understanding of the water treatment process and plant management, as well as useful information with lasting value. Several well qualified consulting firms and professional organizations offer this type of training service.

The operator training program generally teaches plant operators how to analyze plant processes and provides instruction in mathematics, chemistry, and biology. Additional topics include housekeeping, safety equipment inspection techniques, and emergency procedures.

In addition to the operator program, several other types of specific instruction are available in categories such as management, maintenance, and laboratory procedures. Management training is intended for the plant management staff and focuses on human resources development, quality control, quality assurance, process analysis, and writing skills. Maintenance training provides instruction in equipment performance standards, electrical and control system wiring diagrams, the scheduling of preventative and corrective maintenance, maintenance management programming, mechanical and electrical troubleshooting, and emergency services.

The laboratory training program is designed to provide detailed and functional training for plant operators in standard testing for routine operations and on-site specific process control, including tasks such as jar testing, core sampling of filter bed, and filter medium analysis. Laboratory training should not be limited to plant operators; the laboratory staff should be provided with on-site instruction, as well as supplemental training at a state certified laboratory in *Giardia* and virus monitoring; the use of gas chromatography/mass spectrometer equipment for the analysis of organic pollutants; adsorption and emission spectroscopy, inductively coupled plasma, and other techniques used in heavy metal analysis; and techniques used in conducting asbestos fiber counts with the transmission electron microscope. Since the tracer test, for evaluating the tank detention time, has become an important issue with respect to disinfection criteria, it is essential that plant personnel be trained to perform the test and be allowed to practice the technique.

10.4 PLANT OPERATION AND MANAGEMENT SERVICES

Before any type of training program can be implemented, the owner must hire the water treatment plant personnel. The entire plant operation and management are often conducted by a private professional organization. The primary reasons for this arrangement are cost effective plant operation, better process control, and the steady production of safe drinking water by a well educated and well trained (licensed) professional staff.

The cost effectiveness and effective performance provided by qualified private professional firms are both well known and proved in many sectors of society. Although all major water treatment plants in Paris are operated and managed by a private professional firm (those in London have also considered this option), this concept has not yet gained popularity in the United States. However, all public administrators and managers of private water purveyors should seriously begin evaluating the merits of this concept. Consulting firms and private professional organizations should also be ready to offer this type of service in the near future.

One of the newest approaches to effective plant management is the comprehensive maintenance program using a computerized maintenance management system. One type, the Computer Assisted Management System for Environmental Operations (CAMEO), developed by Envirotech Operating Services, generates data on issues such as the equipment age and the number of breakdowns. The master report also provides information on equipment cost and repair. By using this type of system, an organized maintenance program, including preventative maintenance work and prioritized work orders, can easily be established. The effective use of thermographic analysis, vibration detection, and analysis systems such as the "Snapshot" system can identify potential problems in major equipment before those problems become serious, thus avoiding costly and major repairs. These types of assistance make the overall management of the water treatment plant more efficient and economical and are provided by a few professional service firms.

Another special type of service which may become essential for water purveyors is that of a complete laboratory, which performs tasks such as sample collection, analysis, data evaluation, and report preparation. As a result of the new EPA regulations, based on the Safe Drinking Water Act, very specific requirements are imposed on all public water systems in regard to sampling: the analysis of over 100 and over 180 regulated contaminants will be required by the early 1990s and the year 2000, respectively, and the report will have to follow a specific format. It is obvious that most small to mid-sized public water supply agencies are presently incapable of handling the analysis of all these contaminants due to the need for special sophisticated analyzers and well qualified scientists/chemists. In the author's opinion, one of the best solutions is either to have the state use the services of certified laboratories or to have each water purveyor enter a service contract with a certified private laboratory that will handle these requirements. Several certified laboratories presently operate in major metropolitan regions and they are capable of meeting these new needs for a reasonable fee.

Abbreviations

AASHTO	American Association of the State Highway Transportation Officials
ANSI	American National Standards Institute
ASCE	American Society of Civil Engineers
ASTM	American Society for Testing and Materials
ASU	Areal Standard Units
AWWA	American Water Works Association
BIF	BIF-Unit of General Signal Corporation
cfh	cubic foot per hour
cfm	cubic foot per minute
cfs	cubic foot per second
CMS	Construction Management Service
COD	chemical oxygen demand
CPM	critical path method
CPU	central processing unit
CRT	cathode ray tube
CSI	Construction Specification Institute
$C \times t$ or CT	product of concentration (mg/L) and time (min)
Cv	capacity factor for valves
DBP	disinfection by-product
DO	dissolved oxygen
ED	electrodialysis
EIS	environmental impact study
ENR	*Engineering News Record*
EPA	Environmental Protection Agency of the United States
Fr	Froude number
ft	foot, feet
GAC	granular activated carbon
G	velocity gradient or mixing intensity (s^{-1})
$G \times t$	product of G and time (s)
g	acceleration of gravity
gph	gallons per hour
gpm	gallons per minute

gpm/ft^2 or gpm/sf	gallons per minute per square foot
HPC	heterotrophic plate count
HVAC	heating, ventilation, and air conditioning
IEEE	Institute of Electrical and Electronic Engineers
ISA	Instrument Society of America
ISPM	Integrated System of Project Management
kW	kilowatt
kW·h	kilowatt-hour
LI	Langelier index
LOX	liquid oxygen
lb	pound
max	maximum
MCL	maximum contaminant level
Min	minimum
min	minutes
mg/L	milligrams per liter
mgd	million gallons per day
MIB	2-methylisoborneol
m^3/s	cubic meters per second
μm	micrometer (10^{-6} m)
μg/L	micrograms per liter
mL	milliliter
MPN	most probable number
NAS	National Academy of Sciences
NEMA	National Electrical Manufacturers Association
ng	nanogram
NIPDWR	National Interim Primary Drinking Water Regulation
nm	nanometer
NPDES	National Pollutant Discharge Elimination System
NPDWR	National Primary Drinking Water Regulations
OSHA	Occupational Safety and Health Administration
O&M	operation and maintenance
PAC	powdered activated carbon
P&ID	process and instrumentation diagram
PC	programmable controller
PCE	trichloroethylene (perchloroethylene)
PCM	process control module
pc/L	picocuries per liter
PD	process diagram
POADMA	polydiallyl dimethyl ammonium
PERT	program evaluation and review technique
PLC	programmable logic controller
PVC	polyvinylchloride
ppd	pounds per day
ppm	parts per million

R_e	Reynolds number
RFP	request for proposal
RFQ	request for qualification
RO	reverse osmosis
RTU	remote terminal unit
SCADA	supervisory control and data acquisition
SI	international system of units
SOCs	synthetic organic chemicals
SMCLs	secondary maximum contaminant levels
SWTR	surface water treatment rule
SDWA	Safe Drinking Water Act
TCE	trichloroethylene
TDS	total dissolved solids
TDH	total dynamic head
THMs	trihalomethanes
TON	threshold odor number
UBC	Uniform Building Code
VOCs	volatile organic compounds

The SI (International System) and Alphabet Table

Systems of Basic Units

		Systems		
Designation	Dimensions	English (FPS)	Metric (MKS)	International (SI)
Length	(L)	foot (ft)	meter (m)	meter (m)
Mass	(M)	pound (lb)	kilogram (kg)	kilogram (kg)
Time	(t)	second (sec)	second (s)	second (s)
Electric current	(A)	ampere (A)	ampere (A)	ampere (A)
Temperature	(T)	degree Fahrenheit (°F)	degree Celsius (°C)	degree Kelvin (°K)
Luminous intensity	(l)	candela (cd)	candela (cd)	candela (cd)

Decimal Multiples and Fractions of SI Units

Factor	Prefix	Symbol	Factor	Prefix	Symbol
10^1	deka	D (da)	10^{-1}	deci	d
10^2	hecto	h	10^{-2}	centi	c
10^3	kilo	k	10^{-3}	milli	m
10^6	mega	M	10^{-6}	micro	μ
10^9	giga	G	10^{-9}	nano	n
10^{12}	tera	T	10^{-12}	pico	p
10^{15}	femta	F	10^{-15}	femto	f
10^{18}	atta	A	10^{-18}	atto	a

Systems of Derived Units, Geometry, Mass

Designation	Dimensions	FPS	MKS	SI
Area	$(L)^2$	ft^2	m^2	m^2
Static moment of area	$(L)^3$	ft^3	m^3	m^3
Moment of inertia of area	$(L)^4$	ft^4	m^4	m^4
Product of inertia of area	$(L)^4$	ft^4	m^4	m^4
Polar moment of inertia of area	$(L)^4$	ft^4	m^4	m^4
Volume	$(L)^3$	ft^3	m^3	m^3
Static moment of volume	$(L)^4$	ft^4	m^4	m^4
Moment of inertia of volume	$(L)^5$	ft^5	m^5	m^5
Product of inertia of volume	$(L)^5$	ft^5	m^5	m^5
Polar moment of inertia of volume	$(L^5$	ft^5	m^5	m^5
Mass^a $= M = W/g$	(M)	lb	kg	kg
Static moment of mass	$(M)(L)$	lb-ft	kg-m	kg-m
Moment of inertia of mass	$(M)(L)^2$	lb-ft^2	kg-m^2	kg-m^2
Product of inertia of mass	$(M)(L)^2$	lb-ft^2	kg-m^2	kg-m^2
Polar moment of inertia of mass	$(M)(L)^2$	lb-ft^2	kg-m^2	kg-m^2
Curvature of a curve	$(L)^{-1}$	1/ft	1/m	1/m
Torsion of a curve	$(L)^{-1}$	1/ft	1/m	1/m
Plane angle^b	(R)	rad	rad	rad
Solid angle^b	(S)	sr	sr	sr

^{a}In the English system (FPS) and in the metric system (MKS), the mass M is a derived unit, defined as the weight W divided by the acceleration due to gravity g. lb = pound mass, kg = kilogram mass.
^{b}The unit of plane angle called *radian* (rad) and the unit of solid angle called *steradian* (sr) are supplemental units.

Alphabet Table

Greek Letter	Greek Name	English Equivalent	Russian Letter	English Equivalent
Α α	Alpha	(a)	А а	(ä)
Β β	Beta	(b)	Б б	(b)
Γ γ	Gamma	(g)	В в	(v)
			Г г	(g)
Δ δ	Delta	(d)	Д д	(d)
Ε ε	Epsilon	(e)	Е е	(ye)
Ζ ζ	Zeta	(z)	Ж ж	(zh)
			З з	(z)
Η η	Eta	(ā)	И и	(i, ē)
Θ θ	Theta	(th)	Й й	(ē) 7
Ι ι	Iota	(ē)	К к	(k)
			Л л	(l)
Κ κ	Kappa	(k)	М м	(m)
Λ λ	Lambda	(l)	Н н	(n)
			О о	(ô, o)
Μ μ	Mu	(m)	П п	(p)
Ν ν	Nu	(n)	Р р	(r)
Ξ ξ	Xi	(ks)	С с	(s)
			Т т	(t)
Ο ο	Omicron	(o)	У у	($\overline{oo}$)
Π π	Pi	(p)	Ф ф	(f)
Ρ ρ	Rho	(r)	Х х	(kh)
			Ц ц	(ts)
Σ σ ς	Sigma	(s) 6	Ч ч	(ch)
Τ τ	Tau	(t)	Ш ш	(sh)
Υ υ	Upsilon	(ü, $\overline{oo}$)	Щ щ	(shch)
			Ъ ъ	8
Φ φ	Phi	(f)	Ы ы	(ë)
Χ χ	Chi	(H)	Ь ь	9
			Э э	(e)
Ψ ψ	Psi	(ps)	Ю ю	(ū)
Ω ω	Omega	(ō)	Я я	(ya)

Quantities and SI Units

Unit Type	Quantity	Unit	Symbol	Expressions in Terms of Other Units
Base units	Length	meter	m	
	Mass	kilogram	kg	
	Time	second	s	
	Electric current	ampere	A	
	Thermodynamic temperature	kelvin	K	
	Amount of substance	mole	mol	
	Luminous intensity	candela	cd	
Supplementary units	Plane angle	radian	rad	
	Solid angle	steradian	sr	
Derived units having special names	Frequency	hertz	Hz	s^{-1}
	Force	newton	N	$kg \cdot m/s^2$
	Pressure, Stress	pascal	Pa	N/m^2
	Energy, Work, Quantity of heat	joule	J	$N \cdot m$
	Power	watt	W	J/s
	Electric charge	coulomb	C	$A \cdot s$
	Electric potential Potential difference Electromotive force	volt	V	W/A
	Electric resistance	ohm	Ω	V/A
	Electric conductance	siemens	S	A/V
	Electric capacitance	farad	F	C/V
	Magnetic flux	weber	Wb	$V \cdot s$
	Inductance	henry	H	Wb/A
	Magnetic flux density	tesla	T	Wb/m^2
	Luminous flux	lumen	lm	$cd \cdot sr$
	Illuminance	lux	lx	lm/m^2
Some other derived units	Area	square meter		m^2
	Volume	cubic meter		m^3
	Velocity—angular	radian per second		rad/s
	Velocity—linear	meter per second		m/s
	Acceleration—angular	radian per second squared		rad/s^2
	Acceleration—linear	meter per second squared		m/s^2
	Density (mass per unit volume)	kilogram per cubic meter		kg/m^3
	Moment of force and torque	newton meter		$N \cdot m$

Quantities and SI Units (Continued

Unit Type	Quantity	Unit	Symbol	Expressions in Terms of Other Units
Some other derived units (cont'd)	Viscosity—dynamic	pascal second		Pa·s
	Thermal conductivity	watt per meter kelvin		W/m·K
	Thermal flux density or irradiance	watt per square meter		W/m^2
	Thermal capacity or entropy	joule per kelvin		J/K
	Permeability	henry per meter		H/m
	Permittivity	farad per meter		F/m
	Magnetic field strength	ampere per meter		A/m
	Luminance	candela per square meter		cd/m^2
	Molar entropy	joule per mole kelvin		J/mol·K

Metric (SI) Conversion Factors*

PHYSICAL QUANTITY	CUSTOMARY ("British" or "Imperial")	METRIC†† spelled out	METRIC†† symbolic	RECIPROCAL ***
DISTANCE	1 foot	0.3048 meter	0.3048 m	3.281
	1 yard	0.9144 meter	0.9144 m	1.094
	1 mile	1.609 kilometer	1.609 km	0.6215
AREA	1 square inch	6.452 square centimeter	6.452 cm²	0.155
	1 square foot	0.0929 square meter	0.0929 m²	10.76
		or 929 square centimeters	929 cm²	0.001076
	1 square yard	0.836 square meter	0.836 m²	1.196
	1 acre	4,047 square meters	4,047 m²	0.000247
		or 0.4047 hectare	0.4047 h	2.47
	1 square mile	2.590 square kilometers	2.590 km²	0.386
		or 259.0 hectares	259.0 h	0.00386
VOLUME	1 cubic inch	16.39 cubic centimeters	16.39 cm³	0.0610
	1 pint (liquid)	473.2 cubic centimeters	473.2 cm³	0.002113
	1 quart	946.4 cubic centimeters	946.4 cm³	0.001057
		or 0.9464 liter	0.946 l	1.057
	1 gallon	3.785 liters	3.785 l	0.2642
	1 cubic foot	0.0283 cubic meter	0.0283 m³	35.3
	1 cubic yard	0.765 cubic meter	0.765 m³	1.308
	1 acre-foot	0.1233 hectare-meter	0.1233 h·m	8.11
		or 1.233 megaliter	1.233 Ml	0.811
		or 1,233 cubic meters	1,233 m³	0.000811
TIME	1 minute	60 seconds	60 s	0.01667
	1 hour	3.6 kiloseconds	3.6 ks	0.2778
	1 day	86.4 kiloseconds	86.4 ks	0.01157
		or 0.0864 megasecond	0.0864 Ms	11.57
	1 week	0.6048 megasecond	0.6048 Ms	1.6534
	1 dozen days	1.0368 megaseconds	1.0368 Ms	0.96451
	1 month (30 days)	2.592 megaseconds	2.592 Ms	0.3858
	1 year (365¼ days)	31.56 megaseconds	31.56 Ms	0.03169
VELOCITY	1 foot per hour	84.6 micrometer per second	84.6 µm/s	0.01182
	1 foot per minute	5.080 millimeters per second	5.080 mm/s	0.1969
	1 foot per second	0.3048 meter per second	0.3048 m/s	3.281
	1 mile per hour	0.4470 meter per second	0.4470 m/s	2.237
FLOW	1 gallon per minute	63.08 milliliter per second	63.08 ml/s	0.01585
	1 cubic foot per minute	0.4719 liter per second	0.4719 l/s	2.119
	1 cubic foot per second	28.32 liters per second	28.32 l/s	0.0353
	1 million gallons per day	43.812 liters per second	43.812 l/s	0.022824
		or 0.043812 kiloliter per second	0.043812 kl/s	22.824
MASS	1 ounce mass	28.35 grams	28.35 g	0.03527
	1 pound mass	453.6 grams	453.6 g	0.002205
		or 0.4536 kilogram	0.4536 kg	2.205
	1 ton mass	0.907 megagram	0.907 Mg	1.102
	(short, 2000 pounds mass)	or 0.907 metric ton	0.907 t	1.102
		or 0.907 tonne	0.907 t	1.102
FORCE	1 ounce force	0.2780 newton	0.2780 N	3.597
	1 pound force	4.448 newtons	4.448 N	0.2248
	1 ton force (2,000 pounds force)	8.897 kilonewtons	8.897 kN	0.11240
PRESSURE	1 pound per square foot	47.88 newtons per square meter	47.88 N/m²	0.02089
	1 pound per square inch	6.895 kilonewtons per square meter	6.895 kN/m²	0.11240
	1 millimeter of mercury	133.3 newtons per square meter	133.3 N/m²	0.0075
	1 foot of water	2.989 kilonewtons per square meter	2.989 kN/m²	0.3346
	1 atmosphere	0.1013 meganewton per square meter	0.1013 MN/m²	9.87
ENERGY	1 watt-second	1.000 joule	1.000 J	1.000
	1 foot-poundforce	1.356 joule	1.356 J	0.7375
	1 BTU	1.055 kilojoule	1.055 kJ	0.948
	1 watt-hour	3.60 kilojoules	3.60 kJ	0.2778
	1 horsepower-hour	2.684 megajoules	2.684 MJ	0.3726
	1 kilowatt-hour	3.60 megajoules	3.60 MJ	0.2778
POWER	1 horsepower	745.7 watts	745.7 W	0.001341
		or 0.7457 kilowatt	0.7457 kW	1.341
	1 joule per second	1.000 watt	1.000 W	1.000
	1 BTU per hour	0.293 joule per second	0.293 J/s	3.41
TEMPERATURE	1 degree Fahrenheit	5/9 degree Celsius for each Fahrenheit degree above or below 32 °F	$(5/9 \times (T_F - 32))$ °C	1.8 degree Fahrenheit for each Celsius degree, plus 32 °F

*From J. J. Tuma, *Handbook of Physical Calculations*, McGraw-Hill, New York, 1976.

Appendix 3 (*Continued*)

SPECIAL COMPOUND UNITS			
BTU per cubic foot	37.30 joules per liter	37.30 J/l	0.0268
BTU per pound of mass	2.328 joules per gram	2.328 J/g	0.4296
BTU per square foot per hour	3.158 joules per square meter	3.158 J/m²	0.3167
1 pound per thousand cubic feet per day	185.4 grams per cubic centimeter per second	185.4 g/cm³s	0.00539
	or 0.1854 kilogram per cubic meter per second	0.1854 kg/m³s	5.39
1 cubic foot per minute per thousand cubic feet	16.67 microliters per liter per second	16.67 μl/l s	0.0600
	or 16.67 x 10⁻⁶ per second*	16.67 x 10⁻⁶ s⁻¹	60.000
1 cubic foot per second per acre	69.98 liter per second per hectare	69.98 l/s h	0.01429
	or 6.998 micrometers per second†	6.998 μm/s	0.1429
1 gallon per acre	9.35 milliliters per hectare	9.35 ml/h	0.1070
	or 9.35 centimeters**	9.35 cm**	0.1070
1 gallon per day per linear foot	143.7 microliters per second per meter	143.7 μl/s m	0.00696
	or 0.1437 square millimeters per second	0.1437 mm²/s	6.96
1 gallon per day per square foot	0.4715 milliliter per square meter per second	0.4715 ml/m²s	2.12
	or 0.4715 micrometers per second	0.4715 μm/s	2.12
1 pound mass per acre	1.122 kilogram per hectare	1.122 kg/h	0.892
	or 0.1122 gram per square meter	0.1122 g/m²	8.92
1 pound mass per cubic foot	16.02 grams per liter	16.02 g/l	0.0624
1 pound mass per linear foot	1.488 kilogram per meter	1.488 kg/m	0.672
1 pound per horsepower-hour	0.1690 milligram per joule	0.1690 mg/J	5.917

*i.e., 16.67 parts in 1,000,000, per second
†i.e., 6.998 micrometers depth change per second
**i.e., 9.35 centimeters depth

***Multiply quantity known in metric units by this number to get British equivalent.
††Multiply quantity known in British units by this number to get metric equivalent.

Temperature Conversions*

*From the California Water Pollution Control Association, *The Bulletin*, Vol. 12, No. 2, October 1975.

Equivalent Temperature Readings for Fahrenheit and Centigrade Scales

$$F° = \tfrac{9}{5} C° + 32°$$

Equivalent Temperature Readings for Fahrenheit and Centigrade Scales

(Continued)

$$C° = \tfrac{5}{9}(F° - 32°)$$

Fahrenheit deg	Centigrade deg	Fahrenheit deg	Centigrade deg
−459.4	−273	−21	−29.4
−436	−260	−20.2	−29
−418	−250	−20	−28.9
−400	−240	−19.	−28.3
−382	−230	−18.4	−28
−364	−220	−18.	−27.8
−346	−210	−17.6	−27.7
−328	−200	−16.	−26.7
−310	−190	−15.	−26.1
−292	−180	−14.8	−26
−274	−170	−14.	−25.6
−256	−160	−13.	−25
−238	−150	−12.2	−24.4
−220	−140	−11.2	−24
−202	−130	−10.	−23.9
−184	−120	−9.4	−23.3
−166	−110	−9.	−23
−148	−100	−8.	−22.2
−139	−95	−7.6	−22
−130	−90	−7.	−21.7
−121	−85	−6.	−21
−112	−80	−5.	−20.6
−103	−75	−4.	−20
−94	−70	−3.	−19.4
−85	−65	−2.2	−18.9
−76	−60	−2.	−18.3
−67	−55	−1.4	−18
−58	−50	+1.	−17.8
−49	−45	2.	−17.2
−40.	−40	3.	−16.7
−39.4	−39.4	4.	−16.1
−38.2	−38.9	5.	−15.6
−38.	−38.3	6.8	−15
−36.4	−37.8	7.	−14.4
−35.	−37.2	8.	−14
−34.	−36.7	8.6	−13.3
−33.8	−36.1	9.	−13
−32.	−35.6	10.4	−12.8
−31.	−35	11.	−12.2
−30.	−34.4	11.2	−12
−29.2	−33.9	12.2	−11.7
−28.	−33.3	14.	−11.1
−27.	−32.8	14.8	−10.6
−25.6	−32.2	15.	−10
−24.	−31.9	15.8	−9.4
−23.8	−31.1	16.	−8.9
−23.	−30.	17.	−8.3

Fahrenheit deg	Centigrade deg
17.6	−8.
18.	−7.8
19.4	−7.2
20.	−7.
21.	−6.1
21.2	−5.6
22.	−5.
23.	−4.4
24.8	−4.
25.	−3.9
26.	−3.3
26.6	−2.8
27.	−2.2
28.4	−1.7
29.	−1.1
30.2	−0.6
31.	0.
32.	+0.6
33.8	1.
35.	1.7
35.6	2.2
36.	2.8
37.4	3.3
38.	3.9
39.2	4.4
41.	5.
42.	5.6
42.8	6.1
44.	6.7
44.6	7.2
46.	7.8
46.4	8.3
48.	8.9
48.2	9.4
49.	10.
50.	10.6
51.8	11.1
52.	11.7
53.	12.2
53.6	12.8
54.	13.
55.4	

Fahrenheit deg	Centigrade deg	Fahrenheit deg	Centigrade deg	Fahrenheit deg	Centigrade deg	Fahrenheit deg	Centigrade deg
56.	13.3	95.	35.	134.	56.7	173.	78.3
57.	13.9	96.	35.6	134.6	57.2	174.	78.9
57.2	14.4	96.8	36.1	135.	57.8	174.2	79.4
58.	15.	97.	36.7	136.	58.3	175.	79.
59.	15.6	98.	37.2	136.4	58.9	176.	80.6
60.8	16.1	98.6	37.8	137.	59.4	177.	81.1
61.	16.7	99.	38.2	138.2	60.	177.8	81.7
62.6	17.2	100.4	38.8	140.	60.6	178.	82.
63.	17.8	101.	38.9	141.	61.	179.	82.8
64.4	18.3	102.	39.4	141.8	61.7	179.6	83.3
65.	18.9	102.2	40.	142.	62.2	180.	83.9
66.	19.4	103.	40.6	143.6	62.8	181.	84.4
66.2	20.	104.	41.1	144.	63.	181.4	85.
67.	20.6	105.8	41.7	145.	63.3	182.	85.6
68.	21.1	106.	42.2	145.4	63.9	183.	86.1
69.8	21.7	107.	42.8	146.	64.4	183.2	86.7
70.	22.2	108.	43.3	147.	65.	184.	87.2
71.6	22.8	109.	43.9	147.2	65.6	185.	87.8
72.	23.3	109.4	44.	148.	66.1	186.8	88.3
73.4	23.9	110.	44.4	149.	66.7	187.	88.9
74.	24.4	111.	45.	150.8	67.2	188.	89.4
75.2	25.	112.	45.6	151.	67.8	188.6	89.
76.	25.6	113.	46.1	152.	68.3	190.	90.6
77.	26.1	114.8	46.7	152.6	68.9	190.4	91.
78.8	26.7	115.	47.	153.	69.4	191.	91.7
79.	27.2	116.6	47.8	154.4	70.	192.	92.2
80.6	27.8	117.	48.3	155.	70.6	192.2	92.8
81.	28.3	118.4	48.9	156.	71.1	193.	93.3
82.4	28.9	119.	49.	156.2	71.7	194.	93.9
83.	29.4	120.2	49.4	157.	72.2	195.8	94.
84.2	30.	121.	50.	158.	72.8	196.	94.4
85.	30.6	122.	50.6	159.	73.3	197.	95.6
86.	31.1	123.8	51.1	159.8	73.9	197.6	96.1
87.	31.7	124.	51.7	160.	74.4	198.	96.7
87.8	32.2	125.6	52.2	161.6	75.	199.	97.2
88.	32.8	126.	52.8	162.	75.6	199.4	97.8
89.	33.3	127.	53.3	163.	76.1	200.	98.3
89.6	33.9	127.4	53.9	163.4	76.7	201.2	98.9
90.	34.4	128.	54.4	164.	77.2	202.	99.
91.		129.2	55.	165.2	77.8	203.	
91.4		130.	55.6	166.		204.8	
92.		131.	56.1	167.		205.	
93.2		132.		168.8		206.	
94.		132.8		169.		206.6	
		133.		170.6		207.	
				171.		208.	
				172.		208.4	
						209.	
						210.2	

Fahrenheit deg	Centigrade deg
211.	99.4
212.	100.
213.	100.6
213.8	101.
214.	101.1
215.	101.7
215.6	102.
216.	102.2
217.	102.8
217.4	103.
218.	103.3
219.	103.9
219.2	104.
220.	104.4
221.	105.
222.	105.6
222.8	106.
223.	106.1
224.	106.7
224.6	107.
225.	107.2
226.	107.8
226.4	108.
227.	108.3
228.	108.9
228.2	109.
229.	109.4
230.	110.
231.	110.6
231.8	111.
232.	111.1
233.	111.7
233.6	112.
234.	112.3
235.	112.8
235.4	113.
236.	113.3
237.	113.9
237.2	114.
238.	114.4
239.	115.
240.8	115.6
241.	116.1
242.	116.7
242.6	117.
243.	117.2
244.	117.8
244.4	118.
245.	118.3
246.	118.9
246.2	119.
247.	119.4
248.	120.
249.	120.6
249.8	121.

Periodic Table of the Elements*

*From *Handbook of Chemistry and Physics*, 58th ed., CRC Press, Boca Raton, FL, 1971.

Periodic Table of the Elements

Key to chart:

Atomic Number →	**50** +2 +4 ← Oxidation States
Symbol →	**Sn**
Atomic Weight →	118.69
	-18-18-4 ← Electron Configuration

1a	2a	3b	4b	5b	6b	7b	8	8	8	1b	2b	3a	4a	5a	6a	7a	0	Orbit
1 +1 -1 **H** 1.00797 1																	**2** 0 **He** 4.0026 2	K
3 +1 **Li** 6.939 2-1	**4** +2 **Be** 9.0122 2-2											**5** +3 **B** 10.811 2-3	**6** +2 +4 -4 **C** 12.01115 2-4	**7** +1 +2 +3 +4 +5 -1 -2 -3 **N** 14.0067 2-5	**8** -2 **O** 15.9994 2-6	**9** -1 **F** 18.9984 2-7	**10** 0 **Ne** 20.183 2-8	K-L
11 +1 **Na** 22.9898 2-8-1	**12** +2 **Mg** 24.312 2-8-2											**13** +3 **Al** 26.9815 2-8-3	**14** +4 **Si** 28.086 2-8-4	**15** +3 +5 -3 **P** 30.9738 2-8-5	**16** +4 +6 -2 **S** 32.064 2-8-6	**17** +1 +5 +7 -1 **Cl** 35.453 2-8-7	**18** 0 **Ar** 39.948 2-8-8	K-L-M
19 +1 **K** 39.102 -8-8-1	**20** +2 **Ca** 40.08 -8-8-2	**21** +3 **Sc** 44.956 -8-9-2	**22** +3 +4 **Ti** 47.90 -8-10-2	**23** +3 +4 +5 **V** 50.942 -8-11-2	**24** +2 +3 +6 **Cr** 51.996 -8-13-1	**25** +2 +3 +4 +6 +7 **Mn** 54.9380 -8-13-2	**26** +2 +3 **Fe** 55.847 -8-14-2	**27** +2 +3 **Co** 58.9332 -8-15-2	**28** +2 +3 **Ni** 58.71 -8-16-2	**29** +1 +2 **Cu** 63.546 -8-18-1	**30** +2 **Zn** 65.37 -8-18-2	**31** +3 **Ga** 69.72 -8-18-3	**32** +4 **Ge** 72.59 -8-18-4	**33** +3 +5 -3 **As** 74.9216 -8-18-5	**34** +4 +6 -2 **Se** 78.96 -8-18-6	**35** +1 +5 -1 **Br** 79.904 -8-18-7	**36** 0 **Kr** 83.80 -8-18-8	-L-M-N
37 +1 **Rb** 85.47 -18-8-1	**38** +2 **Sr** 87.62 -18-8-2	**39** +3 **Y** 88.905 -18-9-2	**40** +4 **Zr** 91.22 -18-10-2	**41** +3 +5 **Nb** 92.906 -18-12-1	**42** +6 **Mo** 95.94 -18-13-1	**43** +6 +7 **Tc** (97) -18-13-2	**44** +3 **Ru** 101.07 -18-15-1	**45** +3 **Rh** 102.905 -18-16-1	**46** +2 +4 **Pd** 106.4 -18-18-0	**47** +1 **Ag** 107.868 -18-18-1	**48** +2 **Cd** 112.40 -18-18-2	**49** +3 **In** 114.82 -18-18-3	**50** +2 +4 **Sn** 118.69 -18-18-4	**51** +3 +5 **Sb** 121.75 -18-18-5	**52** +4 +6 -2 **Te** 127.60 -18-18-6	**53** +1 +5 +7 -1 **I** 126.904 -18-18-7	**54** 0 **Xe** 131.30 -18-18-8	-M-N-O
55 +1 **Cs** 132.905 -18-8-1	**56** +2 **Ba** 137.34 -18-8-2	**57*** +3 **La** 138.91 -18-9-2	**72** +4 **Hf** 178.49 -32-10-2	**73** +5 **Ta** 180.948 -32-11-2	**74** +6 **W** 183.85 -32-12-2	**75** +6 +7 **Re** 186.2 -32-13-2	**76** +3 +4 **Os** 190.2 -32-14-2	**77** +4 +4 **Ir** 192.2 -32-15-2	**78** +2 +4 **Pt** 195.09 -32-16-2	**79** +1 +3 **Au** 196.967 -32-18-1	**80** +1 +2 **Hg** 200.59 -32-18-2	**81** +1 +3 **Tl** 204.37 -32-18-3	**82** +2 +4 **Pb** 207.19 -32-18-4	**83** +3 +5 **Bi** 208.980 -32-18-5	**84** +2 +4 **Po** (210) -32-18-6	**85** +1 **At** (209) -32-18-7	**86** 0 **Rn** (222) -32-18-8	-N-O-P
87 +1 **Fr** (223) -18-8-1	**88** +2 **Ra** (226) -18-8-2	**89**** +3 **Ac** (227) -18-9-2																-O-P-Q

Transition Elements — Group 8

*Lanthanides:

58 +3 +4 **Ce** 140.12 -19-9-2	**59** +3 **Pr** 140.907 -20-9-2	**60** +3 **Nd** 144.24 -22-8-2	**61** +3 **Pm** (145) -23-8-2	**62** +3 **Sm** 150.35 -24-8-2	**63** +2 +3 **Eu** 151.96 -25-8-2	**64** +3 **Gd** 157.25 -25-9-2	**65** +3 **Tb** 158.924 -26-9-2	**66** +3 **Dy** 162.50 -28-8-2	**67** +3 **Ho** 164.930 -29-8-2	**68** +3 **Er** 167.26 -30-8-2	**69** +3 **Tm** 168.934 -31-8-2	**70** +2 +3 **Yb** 173.04 -32-8-2	**71** +3 **Lu** 174.97 -32-9-2

Orbit: -N-O-P

**Actinides:

90 +4 **Th** 232.038 -19-9-2	**91** +5 **Pa** (231) -20-9-2	**92** +3 +4 +5 +6 **U** 238.03 -21-9-2	**93** +3 +4 +5 +6 **Np** (237) -22-9-2	**94** +3 +4 +5 +6 **Pu** (244) -23-9-2	**95** +3 +4 +5 +6 **Am** (243) -24-9-2	**96** +3 **Cm** (247) -25-9-2	**97** +3 +4 **Bk** (247) -26-9-2	**98** +3 **Cf** (251) -28-9-2	**99** +3 **Es** (254) -29-8-2	**100** **Fm** (257) -30-8-2	**101** **Md** (256) -31-8-2	**102** (254) -32-8-2	**103** +3 **Lw** (254) -32-9-2

Orbit: -O-P-Q

Numbers in parentheses are mass numbers of most stable isotope of that element.

Basic Property of Water and Air

COMPONENTS OF ATMOSPHERIC AIR (APPROXIMATE)

Constituent	Content (%) by Volume	Content (ppm) by Volume
N_2	78.08	—
O_2	20.95	—
CO_2	0.33	—
Ar	—	18.18
He	—	5.24
CH_4	—	2.0
H_2	—	0.5
NO_2	—	0.5

MASS DENSITY AND VISCOCITY OF AIR

Temperature (°C)	Density (kg/m³)	Kinematic Viscosity (m²/s)
−20	1.392	1.152×10^{-5}
−10	1.340	1.245×10^{-5}
0	1.293	1.319×10^{-5}
10	1.247	1.412×10^{-5}
20	1.201	1.486×10^{-5}
30	1.160	1.607×10^{-5}
40	1.129	1.691×10^{-5}

Note: For conversion to ft²/s, use 1 m²/s = 10.764 ft²/s; for conversion to lb/ft³, use 1 kg/m³ = 0.06243 lb/ft³

MASS DENSITY OF WATER

	0°C	5°C	10°C	15°C	20°C	25°C	30°C
ρ (kg/m³)	999.9	1000.0	999.7	999.1	998.2	997.1	995.7

KINEMATIC VISCOSITY OF WATER (v): $v = \mu/\rho$

	0°C	5°C	10°C	15°C	20°C	25°C	30°C
(centistokes)							
cm²/s ($\times$ 10^{-2})	1.792	1.519	1.31	1.146	1.011	0.898	0.803
m²/s ($\times$ 10^{-6})	1.792	1.519	1.31	1.146	1.011	0.898	0.803
ft²/s ($\times$ 10^{-5})	1.926	1.633	1.408	1.232	1.087	0.965	0.864

Note: To convert centistokes to ft²/s, multiply by 1.075×10^{-3}.

ABSOLUTE VISCOSITY OF WATER (μ)

	0°C	5°C	10°C	15°C	20°C	25°C	30°C
(centipoises)							
g/cm·s ($\times$ 10^{-2})	1.792	1.519	1.310	1.145	1.009	0.895	0.800
kg/m·s ($\times$ 10^{-3})	1.792	1.519	1.310	1.145	1.009	0.895	0.800
lb·s/ft² ($\times$ 10^{-5})	3.74	3.17	2.73	2.39	2.11	1.87	1.67
newton·s/m² ($\times$ 10^{-3})	1.83	1.55	1.336	1.17	1.009	0.913	0.816

Notes: To convert centipoises to lb·s/ft², multiply by 2.088×10^{-5}. Newton $= $ kg·m/s² $= 10^5$ dyne.

Recommended Jar Test Procedure

This procedure is not applicable with the magnetic type of stirrer unit. The following jar test procedure is recommended for use with the Phipps and Bird laboratory stirrer (Figure A7-1).

REGULAR JAR TEST PROCEDURE

1. Jar Test Procedure. Place 1000 mL (1 L) of raw water in each 1 L nominal size beaker and set them on the jar tester. There should be a total of six beakers.

Flash Mixing: To simulate flash mixing, begin by initiating mixing for all six beakers at 100 rpm. Then pipette a predetermined amount of alum solution to each beaker as quickly as possible. The dosage

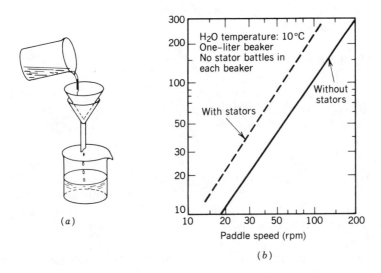

(a)

(b)

Figure A7-1 Mixing speed versus *G*-value and filterability test. (a) Filterability test. (b) Velocity gradient versus mixing speed.

rate is 1 mg/L because 1 mL of 0.1% alum solution is added to 1 L of sample. Remember to adjust the mixing time of each beaker to compensate for the time delay that results from pipetting beakers number 2 through 6.

Note: If an additional chemical such as lime or polymer is to be added, the chemical should be added separately from the alum since the two chemicals will immediately inter-react. The proper procedure is to first add alum to all the beakers as described above, then repeat the flash mixing step using the second chemical.

Flocculation: The flocculation process is simulated by applying the following "tapered mixing" immediately after flash mixing. Follow these instructions in the given order.

7.5 min at a mixing speed of 70 rpm	$G \times t = (7.5 \times 60) \times 60 = 2.7 \times 10^4$
7.5 min at a mixing speed of 40 rpm	$G \times t = (7.5 \times 60) \times 30 = 1.35 \times 10^4$
5 min at a mixing speed of 25 rpm	$G \times t = (5 \times 60) \times 15 = \underline{4.5 \times 10^3}$
A total of 20 min	$G \times t \text{ (approximately)} = 4.5 \times 10^4$

Settling: Allow a standing time of 20 min for settling to occur. Avoid exposing the beakers to direct sunlight.

Sampling: Carefully decant approximately 150 mL of the supernatant from each beaker. Avoid disturbing the settled particles and the scum on the surface of the water.

2. *Measurements and Observations*

Time: Record the minimum time (in minutes) that is required to form "pinpoint" floc during the flocculation step.

Size: At the end of the flocculation step record the size of the (majority) floc by using the "yard stick" method provided in Figure A7-2.

Water Quality: Measure the following water quality characteristics for the finished water (the supernatant): water temperature, turbidity, and pH. Optionally, the color and alkalinity of the water may also be recorded. Refer to Table A7-1 for an example of the jar test data sheet.

Filterability Test: The filterability of the finished water (supernatant) is determined by filtering 70 mL of the selected sample through Whatman No. 1 or similar filter paper, or an equivalent type with a pore

Floc size   Description

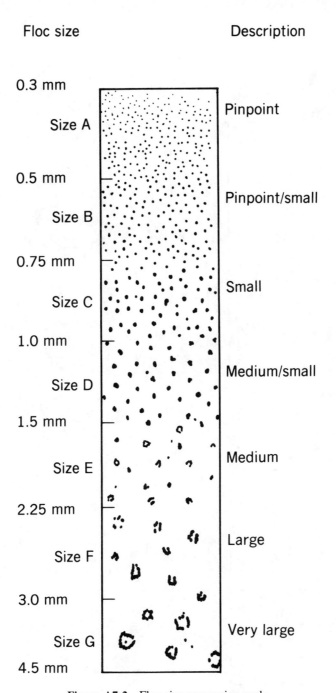

Figure A7-2   Floc size measuring scale.

TABLE A7-1 Jar Test Results

Jar Test Conditions

Rapid mixing: 100 rpm for 10 s ($G \times t = 1500$)

Flocculation: 60 rpm for 7.5 min
40 rpm for 7.5 min ($G \times t = 4.7 \times 10^4$)
20 rpm for 5.0 min

Settling time: 20 min

Filterability test: 50 mL of settled water to be filtered through Whatman No. 1 filter paper

Beaker Number	Chemicals Alum (mg/L)	(mg/L)	Floc Time (min)	Size	Settled Water Turbidity (NTU)	pH	Filterability Time (min)	Turbidity (NTU)	Remark
1	0	—	—	—	66	7.5	6.3	15	Raw water
2	15	—	8	C	13	7.2	8.4	4.4	Alum
3	20	—	2.5	C	7	7.1	5.3	2.3	
4	25	—	1.5	C	3.2	7.1	3.5	1.7	
5	30	—	1.5	C	2.5	6.9	3.3	1.4	
		$FeCl_3$							
6	—	6	3.5	C	11	7.1	8.1	3.8	$FeCl_3$
7	—	10	0.8	D	4	6.9	7.2	1.5	
8	—	14	0.5	D	2.7	6.7	5.1	1.2	
9	—	18	0.5	D	2	6.5	4.2	1	
		Chitosan							
10	—	1	2	C	12	7.4	6.3	1	Chitosan
11	—	2	1.5	C	6	7.4	—	—	
12	—	4	1.5	C	5.3	7.4	3.5	0.7	
13	—	8	1.5	C	4.9	7.3	4.5	0.6	

1	0	—	—	—	11	8.2	6.5	3.8	Raw water
2	2	—	—	—	—	—	—	—	Alum
3	5	—	7	A	6	8.1	6	1.2	
4	10	—	3.5	C	1.8	8.0	3.1	0.6	
5	5	Cationic floc 1	5	C	2.8	8.1	3.0	0.3	Alum and cationic polymer
6	10	1	2.8	D	1.3	8.0	2.8	0.23	
7	—	FeCl$_3$ 5	6.5	B	4	8.0	5.6	1.2	FeCl$_3$
8	—	10	3.1	C	1.3	7.9	3	0.35	
9	—	3	—	B	3.1	8.2	3	0.35	FeCl$_3$ and cationic polymer (0.25 mg/L)
10	—	5	—	C	2.8	8.1	2.8	0.19	

size of about 1–2 μm, and recording the time required to filter 50 mL of the sample. Lastly, measure the turbidity of the filtrate.

Notes: By having stators in each beaker, both the mixing gradient and the velocity gradient are improved. Rectangular jars are preferred over the regular beakers because their corners create more eddies and turbulences. However, the rectangular jars are special order items and are therefore more expensive. (See H. E. Hudson, Jr., *Water Clarification Processes*, Van Nostrand Reinhold, New York, 1981.)

Item 3 is not a routine jar test. Yet, they are basic water treatability evaluation tests.

3. Coagulation and Flocculation Control Tests

Optimization of Coagulant: Alum or ferric chloride should be chosen as a main coagulant. Due to economic reasons, the cationic polymers are usually considered to be a coagulant aid.

Proper Dosage: The proper dosage of main coagulant is evaluated by a two-step process. First, a wide range of dosages is covered by adding 2.5, 5, 10, 20, 40, and 80 mg/L of coagulant such as alum. The second step is a refinement of step one. For example, if the first test yields a good result at 20 mg/L of alum, the second step should evaluate the following dosages: 12.5, 15, 17.5, 20, 22.5, 25, and 27.5 mg/L.

Optimization of Cationic Polymer: This should follow the same concept as optimization of alum dosage. However, the maximum cationic polymer dosage is generally limited to 5 mg/L. The suggested dosages are 0.25, 0.5, 1.2, and 4 mg/L.

In cases where alum and cationic polymer are fed together, the following dosages may be tried:

Alum	(mg/L)	6	12	20	6	12	20
Cationic polymer	(mg/L)	0.25	0.25	0.25	0.5	0.5	0.5

Optimization of Mixing Energy and Time

A Phipps and Bird jar tester, or an equivalent unit, is capable of adjusting G-values of 15 to 100 s^{-1}.

Mixing energy levels of 15, 30, 45, and 60 s^{-1} and mixing times of 10, 20, 30, and 40 minutes should be evaluated under the optimized coagulant dosage. As a rule of thumb, the target value is $G \times t = 5 \times 10^4$ (t in seconds) and the minimum mixing time is 15 minutes.

Floc Settling Velocity Measurement

The jar test should be performed with 2-liter size beakers, rather than 1-liter beakers, under a particular set of conditions: whenever the floc settling velocity is related to the performance of the actual clarifier. When the flocculation process is completed—after the commencement of the settling period—samples of 50 ml or less are collected at 1, 2, 4, 8, and 16 minutes by pipetting or siphoning from a fixed depth of 10 cm from the original water level. The fixed depth represents settling velocities of 10, 5, 2.5, 1.25, and 0.63 cm/min respectively. A settling velocity profile curve is then derived by plotting the turbidity of the settled water samples (at the designated times) divided by the raw water turbidity against the corresponding settling velocity. The settling velocities may be converted to the hydraulic surface loading of the sedimentation tanks. For example, settling velocities of 2, 4, and 8 cm/min are approximately equivalent to 0.5, 1, and 2 gpm/sf (1.2, 2.4, and 4.8 m/h).

Evaluation of the Chemical Application Sequence

The supplemental chemicals should be evaluated under the optimized alum dosage and mixing conditions at the following times: when fed two minutes prior to the addition of alum; when fed at the same time as alum; and 5 to 10 minutes after alum flocculation has begun.

Whenever the filter washwaste is recycled, the recycling point should also be evaluated both prior and after the alum application point.

Core Sampling of the Filter Bed

1. Preparation. In order to obtain core samples, the engineer or plant operator should have a box of small plastic bags, such as sandwich bags, a non-water-soluble marker (permanent ink marker), and core sampling pipes similar to those illustrated below. A minimum of 12 bags will be required per sampling point if samples are to be obtained before and after backwashing. The interior diameter of the pipe should be approximately 1.5 in. and lines should be lightly sawed on the outside of the pipe to indicate the depth of the core sample (Figure A8-1). Arrangements should be made in advance with the chief plant operator to drain the filter water below the filter bed prior to sampling.

2. Core Sampling Technique. First make sure that there is no standing water on the surface of the filter bed. Select three representative points in the filter box as the core sampling spots. Gently push the sampling pipe into the bed in a circular motion until the first marker line (1 in.) is aligned with the surface of the filter bed. Slowly pull the pipe up in a whirling motion enlarging the size of the hole made by the sampling pipe. Next, gently blow the sample into a sampling backet and transfer to a plastic bag. Label the bag "0–1 in. sample" with the indelible marker. Since a minimum of 200 mL of sample is required, repeat the process if necessary; obtain the subsequent samples next to the original hole.

Continue obtaining samples by carefully inserting the sampling pipe into the hole that was created by the first coring. Do not scrape medium off the sides of the hole as this will contaminate the core sample. Carefully push the tube into the filter bed in the previously described manner until the second mark (2 in.) is aligned with the surface of the filter bed. Slowly remove the pipe while enlarging the hole with a circular motion. Blow the sample into the backet and transfer it to a plastic bag marked "1–2 in." Repeat this process until samples from the various depths are collected from the same spot.

When the sampling tube suddenly becomes difficult to push down, the top of the gravel bed has been reached. Operators and engineers will also know when the gravel bed has been reached because the gravel will not remain in the sampling pipe as the pipe is removed from the

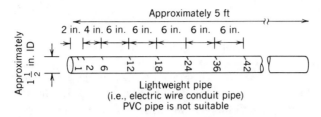

Figure A8-1   Core sampling pipe.

hole. Once the top of the gravel bed has been reached, measure and record this depth; this is the depth of the filter bed.

3. Sieve Analysis and Specific Gravity of the Various Samples. These two qualities should be determined in accordance with the methods specified by Standard B-100 of AWWA.

4. Sludge Retention Analysis. Prepare 50 mL of test sample from each of the sample bags by lightly tamping the core samples into a 50 mL graduated cylinder until the volume of the sample becomes settled and is shown as 50 mL. Transfer the 50 mL sample from the graduated cylinder into a 500 mL wide mouth bottle or flask. Add 100 mL of tap water and shake vigorously for 30 s. Drain the turbid water into another 500 mL beaker. Repeat this washing procedure four more times, using a total of 500 mL of water to wash out the sludge that is adhered to the filter medium. Measure and record the turbidity of the last 500 mL washing.

Multiply the recorded turbidity by 2 so that the final tabulations for each depth will list the turbidity for 100 mL of sample, instead of 50 mL of sample. Plot these results as the bed depth versus amount of sludge (turbidity), as in Figures 3.2.6-10 and 3.2.6-11.

5. Evaluation of the Results. The three features that should be analyzed when evaluating the size characteristic of the filter bed are the size of the medium in the top 2 in. (5 cm), the medium size profile across the entire depth of the bed, and the media interface of multimedia beds.

The first feature to be studied is the size of the medium in the upper 2 in. of the filter bed. This layer is composed of small-sized medium grains. The effective size of these grains should be no smaller than 90% of the effective size specified for the medium type. If the medium is less than 80% of the specified size, this top layer should be scraped off in order to maintain a reasonable filter run length.

The second feature to be analyzed is the medium size profile across the entire depth of the filter bed. The size and uniformity coefficient of the medium in the upper 6 in. (15 cm) of the bed should be nearly equal. If the size of the medium becomes progressively larger toward the bottom of the filter bed, the chance for early turbidity breakthrough is increased. This problem may be avoided by specifying uniformity coefficient that is less than 1.5, preferably 1.4.

The last feature to be examined is the interface zone of dual media or trimedia filter beds. An interface zone that is over 6 in. will have a smaller void ratio and therefore promote and accumulation of floc in the interface. This problem may be minimized or eliminated by specifying media with complementary characteristics (complementary with respect to fluidization during backwash) for each layer and by providing an effective auxiliary scouring system such as subsurface washing. Under these circumstances, air-scour wash is not necessarily the most effective system because the vigorous boiling action, generated by the compressed air, only occurs in the top 6–9 in. (15–22.5 cm) of the filter bed; thus, the air bubbles short-circuit to the surface of the bed with a "pulse-collapse" motion.

Cleanliness of filterbed: can be properly evaluated by the mud profile chart based on the sludge retention analysis. Detailed discussions on this subject are found on page 253.

Tracer Test

TRACER TEST STUDY

Tracer tests are commonly performed by injecting a slug dose of dye, radio-active substance, or a salt solution into the inlet of a basin or tank and then measuring the concentration of the tracer in the effluent at various time intervals. The effluent monitoring should be carried on until substantially all the tracer has passed through the basin. The type of tracer that can be used for water treatment and water supply systems is limited to lithium chloride and fluoride because of concern for the safety of public health.

Chlorides such as sodium chloride, which is readily available as well as cost effective, is not an effective tracer because of its high specific gravity as a solution produces a density flow in the tank or basin. Often, lithium chloride is used as a tracer since it is safe to handle and easy to make solution due to its high solubility (63 g/100 mL at 0°C, exothermic reaction). However, lithium chloride is rather expensive (3 $/lb). Lithium chloride is available from Lithco, a subsidiary of FMC Corp., Gastonia, NC (Tel. 704-868-5350). Commonly available fluoride compounds are sodium fluoride, sodium silico flouride, and hydrofluosilicic acid. The first two are available in a dry form but the last one is available only as a liquid. The dry compounds have very low solubility in water (4 and 0.76%, respectively, at 25°C) and dust from these compounds can be hazardous to health. The fluosilicic acid is available as a liquid (23% solution) and it has an infinite solubility but is extremely acidic and therefore has a problem of safety in handling. Although the cost of these fluoride chemicals is about 10–30% of that for lithium chloride, the cost benefit is often offset by the safety concerns and troublesome preparation of the tracer solution. One of the important characteristics of an effective tracer is that it cannot be removed or consumed by the treatment process. Fluoride can be removed by alum flocculation and the lime softening process in significant degree. Lithium, however, is much more stable than fluoride.

Tracer Test Methods

The two basic methods of tracer addition are the step-dose method and the slug-dose method. The former method doses the tracer at a constant rate until the concentration at the desired endpoint reaches a steady-state level.

The latter method applies a large and instantaneous dose of tracer at an entrance to a tank where adequate mixing is provided, and the tracer is recorded as it passes through the tank by taking samples at the effluent of the tank at certain intervals.

Amount of Tracer to Be Used

Regardless of the type of tracer, the amount of the tracer chemical to be fed should be adequate so that the residual of the tracer in the water samples at the outlet of the tank is sufficiently high to monitor the flowthrough time very clearly. A rule of thumb for lithium chloride is about 2–3 mg/L of lithium of peak concentration with the slug-dose method. When fluoride compound is used, the maximum concentration of fluoride ion in the sample water should be 1–1.5 mg/L in order not to exceed the MCL of fluoride of the drinking water standards. Often, the background level of fluoride ion in most waters is on the order of 0.5 mg/L and therefore not much additional fluoride can be added. The background level of lithium in most waters for water supply is negligibly small and generally does not fluctuate.

It should be realized that the flow pattern in a tank or basin is never an ideal plug flow and substantial degrees of flow short-circuiting, deadzone, and mixing may exist. Therefore, the amount of tracer chemical required is often significantly over the amount computed based on the assumption that the flow in a tank is plug flow. When a tracer test using the slug-dose method is conducted for a flocculation tank with flocculators, the tracer chemical added is dispersed in the entire flocculation tank. Therefore, the amount of tracer to be added is based on the entire flocculation tank volume and desired lithium ion concentration in the sampled water.

Sampling

Sampling at the outlet of the tank should include the water before tracer addition in order to establish the background. After the tracer addition, the sampling should be started at 5% of the computed mean detention time and sampling should be continued until the tracer level drops back to the background level. For the slug-dose method, the testing period is usually 2–2.5 times longer than the computed detention time. The frequency of sampling is dependent on the test method, the anticipated detention time, the cost of tracer analysis, and the total budget of the tracer study. For the slug-dose method, where the calculated detention time is less than 30 min, 2.5–3 min sampling intervals are preferable. It should be carried out until the tracer concentration goes down a level that is about 25% higher than the background level. After this point, the sampling frequency can be every 5–10 min until the tracer level is no longer detected. The initial sampling intervals can be 5–10 min when sedimentation tanks or clearwells are tested because their mean detention time is normally 2–5 h. In the case of the step-dose method, the sampling should also be started at 5% of the computed mean time detention time of the tank and the same sampling frequency as the slug-dose method should be applied. However, the sampling should be terminated after the measured tracer residual concentration reaches a steady-state value.

Notes of the field test should include the test method adopted, time of tracer addition, time of each sampling, plant flow rate during the test, water level if it is fluctuating, water temperature, and ambient temperature. In addition, the time at which tracer feed is stopped should also be recorded if the step-dose method is used.

Data Evaluation

Slug-Dose Method   The test results should be tabulated and the results organized by the values of C/C_0 versus t/T [where C_0 is the background concentration, C is the tracer concentration in sample water, (concentration in sample minus background concentration), t is the time of sample collection, and T is the calculated detention time (which is the volume of water in the tank divided by the flow rate)]. The results are then plotted on a graph as shown in Figure A9-1 with the values of C/C_0 on the vertical axis and t/T on horizontal axis. If so desired, the actual sampling time after the tracer injection can be used instead of t/T on the horizontal axis. Case 1 as noted on the figure is a sedimentation tank with an intermediate diffuser wall at the middle of the tank length and an inlet diffuser wall. The samples were collected at three points along the tank effluent. The graph shows that three different levels of peaks, different shapes (tracer peak at location A is a flat shape), and the peak of all three lines are located at different t/T scales. The results show that the flow in the tank is not uniform: the flow is skewed to the west side of the tank and a flow stagnation occurs on the eastern side of the wall, with an average flowthrough time of about 50% of the calculated detention time. The skewed flow is due to a east to west direction of flow in the influent channel, which is perpendicular to the flow direction in the tank. The diffuser walls have too much opening (15% of cross section) so that they do not provide a controlled headloss to redistribute the flow into the tank. Case 2 shows the tracer curves for two common rectangular sedimentation tanks constructed without effective influent diffuser walls. Samples were taken only at a middle point of the tank effluent and therefore the evaluation of the flow characteristics in the tank is limited. However, it is obvious that the actual flow through time is only about 30% of the computed detention time.

Step-Dose Method   Table A9-1 shows results of the step-dose tracer test for a poorly compartmentalized flocculation tank that has $T = 30$ min using lithium chlorine as a tracer. Here, the background lithium concentration was 0.1 mg/L and a constant dosage of lithium chlorine at 3 mg/L as lithium was added to the inlet of the tank.

Figure A9-2 is the plot from the table with the C/C_0 figures plotted against

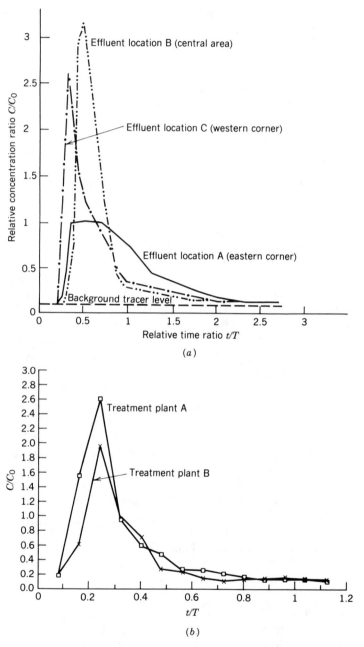

Figure A9-1   Tracer curves of conventional sedimentation basins (slug-dose method).

TABLE A9-1   Flocculation Tank Tracer Test Data

Time (A) (min)	Total Lithium Level (mg/L)	C^a (mg/L)	$C/C_0^{\ b}$
0	0.01	0	0
3	0.01	0	0
6	0.31	0	0.1
9	1.03	1.02	0.34
12	1.76	1.75	0.58
15	1.91	1.90	0.63
18	2.24	2.23	0.73
21	2.15	2.14	0.71
24	2.68	2.67	0.89
27	2.72	2.71	0.90
30	2.66	2.65	0.88
33	2.69	2.68	0.90
36	2.73	2.72	0.91
39	2.84	2.83	0.94
42	2.86	2.85	0.95
45	2.85	2.84	0.95

^{a}Net increase of lithium concentration (0.01 mg/L is background level).
$^b C_0$ = applied tracer dosage = 3 mg/L.

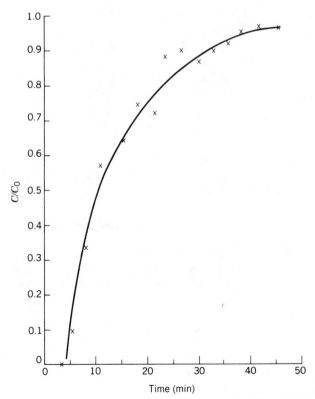

Figure A9-2   Tracer curve of a poorly compartmentalized flocculation tank (step-dose method).

the sampling time on the horizontal axis. The figure shows a severe flow short-circuiting in which it took only 6 min for the first tracer to get to the outlet of the tank. A C/C_0 value of 0.5 corresponds to $t = 11$ min so that one-half of the total flow is flowing out after 11 min of detention time. After 30 min, almost 90% of flow has passed through the tank. If the disinfectant contact time (t for $C \times t$) is determined through this tracer test, t_{10} should be used as the t. The t_{10} value is obtained from Figure A9-2 at $C/C_0 = 0.1$ by first drawing a horizontal axis to give the time t. In this particular case, t_{10} is 5 min.

Calculation Example

A tracer test is planned for one of three flocculation tanks (flocculator system) which has a 20 min mean time (calculated time) and also for chlorine contact tank (effluent weir control) which has a 15 min contact time at a normal plant flow rate of 30 mgd. Select the test method and determine the amount of chemical required as well as the application method.

Tracer Test for the Flocculation Tank   Set the test conditions as follows:

Method	Slug-dose method
Tracer	Lithium chloride
Level of tracer in sampled water	Approximately 3 mg/L as lithium ion
Flocculation tank	Tracer test for only one tank out of the three tanks
Tracer feeding	Prepare 30% LiCl solution on site using several plastic buckets and pour into the flocculation tank inlet where turbulence is greatest

We now calculate the amount of lithium chloride required for the flocculation tank. When we treat three tanks with a detention time of 20 min each at 30 mgd, the volume of one tank is

$$[(30 \div 3) \times 1.547] \times (60 \times 20) = 18,564 \text{ ft}^3$$

The amount of chemical required to obtain 3 mg/L residual should be computed based on the entire volume of the tank, since the chemical will be completely dispersed into the entire tank by the mechanical flocculators.

$$\frac{3}{1,000,000} = \frac{x}{18,564 \times 62.4}$$

$$x = 3.48 \text{ lb}$$

The molecular weight of lithium chloride is 42.5 and the molecular weight of lithium is only 7 or 16.5% of the chemical, assuming almost 100% purity.

The 3 mg/L of residual should be as lithium so that the amount of lithium chloride to be purchased is

3.48 divided by 0.16 = 21.75 lb,   say 22 lb   (costing approximately $70)

Make approximately 20% solution (2.2 lb/gal), providing five 4-gal capacity buckets. Place approximately 2 gal of water in each of the five buckets and then add 4.4 lb of the chemical to each, slowly mixing with a paddle. Heat will be generated during this dissolving process, but it will not be high enough to melt the plastic buckets.

Tracer Test for Chlorine Contact Tank   In general, chlorine is added at the tank inlet by means of a diffuser pipe. Therefore, the tracer may be added through the same diffuser if possible. If not, additional diffuser may be installed next to the chlorine diffuser.

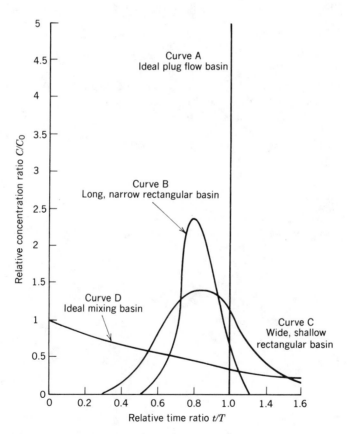

Figure A9-3   Generalized tracer curves under certain conditions (based on the slug-dose tracer test method).

Set the test conditions as follows:

Method	Step-dose method
Tracer	Lithium chloride
Level of tracer in sampled water	2 mg/L as lithium ion
Tracer feeding	Lithium chloride solution is metered by a metering pump and it is fed through the diffuser pipe
Duration of tracer feeding	15–20 min of a continuous feeding

The amount of lithium chloride required is

$$(2 \times 8.34 \times 30) \times \frac{0.333}{24} = 6.95 \text{ lb}$$

Since lithium is 16.5% of lithium chlorine (100% purity), the actual amount of lithium chloride needed is

$6.95 \div 0.16 = 43.44$ lb,   say 45 lb   (cost approximately \$140)

The capacity of the chemical feeder is calculated as follows. A 20% lithium chloride solution is to be prepared. Since 20% solution should contain 2.2 lb of the chemical per gallon of solution, 18.75 gal of solution is prepared in a 55 gal drum or other suitable container. The feeder capacity is determined to feed 19.75 gal of solution in 20 min:

$$19.75 \div \frac{20}{60} = 59.25 \text{ gph}, \quad \text{say 60 gph capacity pump}$$

Figure A9-3 shows generalized tracer curves, under certain physical conditions of the process unit, as a reference.

Gas Law

Generalized gas law: $PV = nRT$

where P = atmospheric pressure (P = 1 as standard pressure),
$\quad\;\; V$ = volume of a gas in liter,
$\quad\;\; n$ = number of moles of gas,
$\quad\;\; R$ = molar gas constant,
$\quad\;\; T$ = absolute temperature (K) (0°C = 273.15°K).

Under standard temperature (20°C) and pressure (1 atm), R and n are given as follows:

$$R = \frac{PV}{nT} = \frac{1\ \text{atm} \times 22.4\ \text{L}}{1\ \text{mole} \times 273.15\ \text{K}} = 0.0821\ \frac{\text{L}}{\text{K·mole}} = 0.0821\ \text{L/K·mole}$$

$$n = \frac{g}{M}$$

where g = mass of the gas in grams,
$\quad\;\; M$ = molecular weight of the gas.

Under standard temperature and pressure (STP), the density, volume, and molecular weight of a gas are given as follows:

Density (g/L) of gas: $d = \dfrac{g}{V} = \dfrac{MP}{RT}$ or $d = \dfrac{M}{24.053}$

Volume (L) of gas: $V = \dfrac{gRT}{MP}$ or $V = \dfrac{g \times 24.053}{M}$

Molecular weight (g/mole) of gas: $M = \dfrac{gRT}{PV}$ or $M = \dfrac{g \times 24.053}{V}$

Ozone Concentration Conversions*

*Adapted from the EPA—Disinfection Manual.

English Unit Equivalents for Ozone Concentration

Standard Pressure .... 14.696 psi (101 kPa)
Standard Temperature .... 68°F (20°C)
Gram Molecular wt. of air .... 29 g/mole
Gram Molecular wt. of oxygen .... 32 g/mole
Gram Molecular wt. of ozone .... 48 g/mole
Molar Volume .... 0.08205 L/mole/°K

English

| Metric | Percent Weight | | Percent Volume | PPM Weight | | PPM |
Wt Ozone to Vol Gas (g/m³)	Air (%)	Oxygen (%)	(%)	Air (ppm)	Oxygen (ppm)	Volume (ppm)
2	0.17	0.15	0.10	1,658	1,503	1,002
4	0.33	0.30	0.20	3,313	3,004	2,004
6	0.50	0.45	0.30	4,967	4,503	3,007
8	0.66	0.60	0.40	6,618	6,001	4,009
10	0.83	0.75	0.50	8,267	7,498	5,011
12	0.99	0.90	0.60	9,914	8,993	6,013
14	1.16	1.05	0.70	11,559	10,486	7,015
16	1.32	1.20	0.80	13,201	11,978	8,018
18	1.48	1.35	0.90	14,842	13,469	9,020
20	1.65	1.50	1.00	16,480	14,958	10,022
22	1.81	1.64	1.10	18,116	16,446	11,024
24	1.98	1.79	1.20	19,750	17,932	12,026
26	2.14	1.94	1.30	21,382	19,417	13,029
28	2.30	2.09	1.40	23,012	20,900	14,031
30	2.46	2.24	1.50	24,640	22,381	15,033
32	2.63	2.39	1.60	26,265	23,862	16,035
34	2.79	2.53	1.70	27,889	25,340	17,038
36	2.95	2.68	1.80	29,510	26,818	18,040
38	3.11	2.83	1.90	31,129	28,294	19,042
40	3.27	2.98	2.00	32,746	29,768	20,044
42	3.44	3.12	2.10	34,362	31,241	21,046
44	3.60	3.27	2.20	35,974	32,712	22,049
46	3.76	3.42	2.31	37,585	34,182	23,051
48	3.92	3.57	2.41	39,194	35,651	24,053
50	4.08	3.71	2.51	40,801	37,118	25,055

Venturi Type Flow Meter Selections*

CAPACITY TABLE - MGD
(Millions of Gallons per Day)

A thru E - Venturi Tube
B thru D - Venturi Insert Nozzle
A thru D - Dall Flow Tube

Size Throat	Range Tube Size - - Inches of Water										*Exact Diff. for Flows with 80* Range Tube
	20	30	40	60	80*	120	160	240	276	320	
6A	----	----	0.23	0.28	0.32	0.40	0.45	0.56	0.60	0.65	78.23
B	----	0.32	0.37	0.45	0.52	0.65	0.72	0.90	0.95	1.00	83.32
C	----	0.52	0.60	0.72	0.85	1.00	1.20	1.40	1.50	1.70	81.02
D	----	0.85	0.98	1.20	1.40	1.70	2.00	2.40	2.50	2.80	84.35
E	----	1.10	1.20	1.50	1.70	2.20	2.40	2.90	3.00	3.40	77.02
8A	----	0.35	0.42	0.50	0.58	0.72	0.80	1.00	1.00	1.20	80.54
B	0.45	0.56	0.65	0.80	0.90	1.10	1.30	1.60	1.60	1.80	81.06
C	0.75	0.90	1.10	1.30	1.50	1.80	2.10	2.60	2.80	3.00	79.68
D	1.20	1.50	1.70	2.10	2.40	3.00	3.40	4.20	4.50	4.80	80.74
E	1.40	1.80	2.00	2.50	2.90	3.60	4.20	5.00	5.40	5.80	79.99
10A	0.45	0.55	0.65	0.80	0.90	1.10	1.30	1.60	1.70	1.80	79.69
B	0.75	0.90	1.00	1.30	1.50	1.80	2.10	2.60	2.70	3.00	82.60
C	1.20	1.50	1.70	2.10	2.40	3.00	3.40	4.20	4.50	4.80	78.05
D	2.00	2.40	2.80	3.40	4.00	4.80	5.50	6.80	7.20	8.00	86.08
E	2.40	2.90	3.40	4.20	5.00	5.80	6.60	8.50	8.80	9.50	83.71
12A	0.70	0.85	0.98	1.20	1.40	1.70	2.00	2.40	2.60	2.80	81.84
B	1.10	1.30	1.50	1.90	2.20	2.70	3.10	3.80	4.00	4.40	83.34
C	1.70	2.10	2.50	3.00	3.50	4.20	5.00	6.00	6.50	7.00	81.04
D	2.80	3.50	4.00	5.00	5.60	7.00	8.00	9.80	10.00	11.00	81.72
E	3.40	4.20	4.80	6.00	7.00	8.40	9.50	12.00	13.00	14.00	81.56
14A	0.90	1.10	1.30	1.60	1.80	2.30	2.60	3.20	3.40	3.70	76.01
B	1.40	1.80	2.00	2.50	2.90	3.50	4.00	5.00	5.20	5.80	83.70
C	2.40	2.90	3.40	4.20	4.80	5.80	6.80	8.50	8.80	9.50	81.95
D	3.80	4.80	5.50	6.80	7.50	9.50	11.00	13.00	14.00	15.00	78.14
E	4.80	5.60	6.60	8.00	9.50	12.00	13.00	16.00	17.00	20.00	78.62
16A	1.20	1.50	1.70	2.10	2.40	3.00	3.40	4.20	4.50	5.00	78.30
B	2.00	2.40	2.80	3.40	4.00	4.80	5.50	6.80	7.20	8.00	85.80
C	3.20	4.00	4.50	5.50	6.30	8.00	9.00	11.00	12.00	13.00	78.65
D	5.00	6.00	7.00	8.50	10.00	12.00	14.00	17.00	18.00	20.00	84.09
E	6.50	8.00	9.00	11.00	13.00	16.00	18.00	22.00	24.00	26.00	82.53
18A	1.50	1.80	2.10	2.60	3.00	3.70	4.20	5.20	5.60	6.00	80.05
B	2.40	2.90	3.30	4.20	4.80	5.80	6.80	8.00	8.50	9.50	85.42
C	3.80	4.80	5.50	6.80	8.00	9.50	11.00	13.00	14.00	16.00	85.25
D	6.50	8.00	9.00	11.00	13.00	16.00	18.00	22.00	23.00	25.00	86.83
E	8.00	9.50	11.00	13.00	16.00	19.00	22.00	26.00	29.00	33.00	81.83
20A	1.80	2.30	2.80	3.20	3.70	4.50	5.20	6.50	6.80	7.50	80.18
B	3.00	3.70	4.20	5.20	6.00	7.50	8.50	10.00	11.00	12.00	82.59
C	5.00	6.00	7.00	8.50	10.00	12.00	14.00	17.00	18.00	20.00	82.62
D	8.00	10.00	11.00	14.00	16.00	20.00	23.00	28.00	30.00	33.00	76.88
E	9.80	12.00	14.00	17.00	19.00	24.00	28.00	34.00	36.00	40.00	75.54
24A	2.70	3.30	3.70	4.50	5.40	6.50	7.50	9.50	9.80	11.00	82.60
B	4.30	5.40	6.20	7.50	8.50	10.00	12.00	15.00	16.00	17.00	79.66
C	7.00	8.50	9.80	12.00	14.00	17.00	20.00	24.00	26.00	28.00	81.31
D	11.00	14.00	16.00	19.00	22.00	27.00	32.00	40.00	40.00	45.00	79.49
E	13.00	17.00	19.00	23.00	28.00	32.00	38.00	48.00	50.00	54.00	73.44
30A	4.20	5.20	6.00	7.50	8.50	10.00	12.00	15.00	15.00	17.00	78.45
B	7.00	8.50	9.80	12.00	14.00	17.00	20.00	24.00	26.00	28.00	82.87
C	11.00	13.00	16.00	19.00	22.00	27.00	31.00	38.00	40.00	44.00	80.29
D	17.00	21.00	24.00	30.00	35.00	42.00	50.00	60.00	63.00	70.00	82.41
E	22.00	27.00	31.00	38.00	44.00	54.00	62.00	75.00	80.00	88.00	80.03
36A	6.50	8.00	9.00	11.00	13.00	16.00	18.00	22.00	24.00	25.60	83.53
B	9.80	12.00	14.00	17.00	20.00	24.00	28.00	34.00	36.00	40.00	82.77
C	16.00	20.00	23.00	28.00	32.00	40.00	45.00	56.00	60.00	65.00	80.30
D	25.00	31.00	36.00	44.00	51.00	62.00	72.00	88.00	90.00	100.00	84.39
E	31.00	38.00	44.00	54.00	62.00	80.00	88.00	110.00	110.00	130.00	76.63
42A	8.00	10.00	11.00	14.00	16.00	20.00	23.00	28.00	30.00	32.00	78.78
B	13.00	16.00	19.00	23.00	27.00	33.00	38.00	45.00	50.00	54.00	82.09
C	22.00	27.00	31.00	38.00	44.00	54.00	62.00	75.00	80.00	88.00	81.95
D	35.00	42.00	50.00	60.00	70.00	85.00	98.00	120.00	120.00	140.00	85.81
E	43.00	54.00	62.00	75.00	85.00	110.00	120.00	150.00	160.00	170.00	77.94

*Adapted from BIF Catalog.

Size Throat	Range Tube Size - - Inches of Water										*Exact Diff. for Flows with 80" *Range Tub
	20	30	40	60	80*	120	160	240	276	320	
48A	11.00	13.00	16.00	19.00	22.00	27.00	31.00	38.00	40.00	44.00	80.20
B	17.00	21.00	24.00	30.00	35.00	42.00	50.00	60.00	65.00	69.00	80.86
C	29.00	35.00	40.00	50.00	58.00	70.00	80.00	100.00	100.00	120.00	82.25
D	45.00	55.00	65.00	80.00	90.00	110.00	130.00	150.00	160.00	180.00	83.15
E	56.00	70.00	80.00	98.00	110.00	140.00	160.00	200.00	210.00	230.00	76.49
54A	14.00	17.00	20.00	24.00	28.00	34.00	38.00	48.00	51.00	56.00	81.10
B	22.00	27.00	31.00	38.00	44.00	54.00	62.00	75.00	80.00	88.00	79.78
C	36.00	45.00	51.00	63.00	72.00	90.00	100.00	130.00	130.00	150.00	78.99
D	58.00	70.00	80.00	98.00	110.00	140.00	160.00	200.00	210.00	230.00	77.54
E	70.00	85.00	98.00	120.00	140.00	170.00	200.00	240.00	260.00	280.00	81.45
60A	17.00	21.00	25.00	30.00	35.00	42.00	48.00	60.00	63.00	70.00	82.35
B	27.00	33.00	38.00	48.00	54.00	66.00	80.00	95.00	100.00	110.00	78.84
C	44.00	54.00	62.00	75.00	88.00	110.00	130.00	150.00	160.00	180.00	78.70
D	70.00	85.00	98.00	120.00	140.00	170.00	200.00	240.00	260.00	280.00	82.41
E	85.00	110.00	120.00	150.00	170.00	210.00	240.00	300.00	320.00	350.00	74.66
66A	21.00	25.00	29.00	36.00	42.00	51.00	60.00	72.00	75.00	85.00	80.99
B	33.00	40.00	48.00	58.00	66.00	80.00	95.00	110.00	120.00	130.00	80.44
C	54.00	66.00	75.00	95.00	110.00	130.00	150.00	190.00	200.00	220.00	83.10
D	80.00	100.00	110.00	140.00	160.00	200.00	230.00	280.00	300.00	320.00	78.69
E	100.00	130.00	150.00	180.00	210.00	250.00	290.00	360.00	380.00	420.00	82.10
72A	25.00	30.00	35.00	43.00	50.00	60.00	70.00	85.00	90.00	100.00	81.04
B	40.00	48.00	56.00	68.00	80.00	95.00	110.00	140.00	140.00	160.00	83.45
C	65.00	80.00	90.00	110.00	130.00	160.00	180.00	220.00	230.00	250.00	84.47
D	98.00	120.00	140.00	170.00	200.00	240.00	280.00	340.00	360.00	400.00	81.11
E	120.00	150.00	170.00	210.00	240.00	300.00	340.00	420.00	450.00	480.00	75.73
78A	29.00	35.00	42.00	50.00	58.00	70.00	85.00	100.00	110.00	120.00	79.18
B	48.00	58.00	65.00	80.00	95.00	110.00	130.00	160.00	170.00	180.00	83.48
C	75.00	95.00	110.00	130.00	150.00	180.00	210.00	260.00	280.00	300.00	80.06
D	120.00	140.00	160.00	200.00	230.00	280.00	330.00	400.00	430.00	480.00	77.88
E	140.00	180.00	200.00	250.00	290.00	350.00	400.00	500.00	540.00	580.00	79.96
84A	33.00	40.00	48.00	58.00	66.00	80.00	95.00	120.00	120.00	140.00	76.22
B	54.00	65.00	75.00	90.00	110.00	130.00	150.00	180.00	200.00	210.00	85.16
C	85.00	110.00	120.00	150.00	170.00	210.00	240.00	300.00	320.00	350.00	76.45
D	130.00	160.00	190.00	230.00	270.00	330.00	380.00	450.00	480.00	540.00	85.40
E	170.00	210.00	240.00	290.00	330.00	420.00	480.00	580.00	620.00	680.00	77.27
90A	38.00	48.00	54.00	66.00	75.00	95.00	110.00	130.00	140.00	150.00	76.62
B	65.00	80.00	90.00	110.00	130.00	160.00	180.00	220.00	240.00	250.00	83.44
C	98.00	120.00	140.00	170.00	200.00	240.00	280.00	340.00	360.00	400.00	80.30
D	160.00	190.00	220.00	270.00	310.00	380.00	440.00	540.00	580.00	620.00	79.82
E	200.00	240.00	280.00	340.00	400.00	480.00	560.00	680.00	720.00	800.00	81.65
96A	44.00	54.00	62.00	75.00	88.00	110.00	120.00	150.00	160.00	180.00	81.02
B	70.00	85.00	98.00	120.00	140.00	170.00	200.00	240.00	260.00	280.00	80.86
C	120.00	140.00	160.00	200.00	230.00	280.00	330.00	400.00	420.00	450.00	80.83
D	180.00	220.00	250.00	310.00	360.00	440.00	510.00	620.00	660.00	720.00	83.15
E	230.00	280.00	320.00	400.00	450.00	550.00	650.00	750.00	800.00	900.00	79.83
108A	55.00	68.00	80.00	95.00	110.00	140.00	160.00	190.00	200.00	220.00	79.03
B	85.00	110.00	120.00	150.00	170.00	210.00	240.00	300.00	320.00	350.00	76.18
C	140.00	180.00	200.00	250.00	290.00	350.00	420.00	500.00	540.00	580.00	81.42
D	230.00	280.00	320.00	400.00	450.00	550.00	650.00	750.00	800.00	900.00	81.11
E	280.00	350.00	400.00	500.00	580.00	700.00	800.00	980.00	1000.00	1100.00	82.79
120A	70.00	85.00	98.00	120.00	140.00	170.00	200.00	240.00	260.00	280.00	82.37
B	110.00	130.00	160.00	190.00	220.00	270.00	310.00	380.00	400.00	440.00	79.92
C	180.00	220.00	250.00	310.00	360.00	440.00	510.00	620.00	660.00	720.00	82.32
D	280.00	340.00	400.00	480.00	550.00	680.00	800.00	950.00	1000.00	1100.00	79.50
E	350.00	430.00	500.00	620.00	700.00	850.00	1000.00	1200.00	1300.00	1400.00	79.12

Water Treatment Chemicals

ALUMINUM SULFATE*

Properties

Chemical Formula.............$Al_2(SO_4)_3 \cdot xH_2O$ (approx. $14H_2O$)
Molecular Weight.............594 (For $14H_2O$ product)
Appearance......................White to cream color
Solubility, 32°F.................105 parts in 100 parts water
pH 1% Solution.................Approximately 3.5

Bulk Density:	Powdered	38 to 45 pounds per cubic foot
	Ground	63 to 71 pounds per cubic foot
	Rice	57 to 61 pounds per cubic foot

Angle of Repose:	Powdered	65°
	Ground	43°
	Rice	38°

Grades

Commercial Dry...................Powdered (97% thru 100 mesh)
Ground (90% thru 10 mesh)
Rice (100% thru 6 mesh)
(95% on 20 mesh)
Lump (90% thru 2.5 inch ring)

Commercial Liquid................8.3% Al_2O_3 Liquid

Iron-Free.............................Powdered
Ground
Lump

Iron-Free Liquid....................8.3% Al_2O_3 Liquid

	COM-MERCIAL DRY	COM-MERCIAL LIQUID	IRON-FREE DRY	IRON-FREE LIQUID
Total soluble Al_2O_3	17.1 %	8.3 %	17.2 %	8.3 %
Free Al_2O_3	0.2 %	0.1 %	0.25 %	0.1 %
Total Fe (as Fe_2O_3)	0.4 %	0.2 %	0.002%	0.004%
Actual Fe_2O_3	0.04%	0.03%	—	—
Insoluble in water	0.05%	0.03%	0.01 %	0.005%

GRAVITY CONVERSIONS & WEIGHT EQUIVALENTS OF COMMERCIAL ALUMINUM SULFATE SOLUTIONS

These are typical values, varying slightly from one producing point to another because of variations in raw materials.

°Be at 60° F	Sp. Gr.	% Al_2O_3	% $Al_2(SO_4)_3$	Lbs./Gal.	% (Dry) 17% Alum	Pounds Dry Alum per Gallon	Grains Dry Alum per Gallon
32.16	1.2850	7.37	24.73	10.713	43.34	4.64	32,480
32.38	1.2875	7.42	24.90	10.734	43.63	4.68	32,760
32.63	1.2900	7.47	25.07	10.755	43.92	4.72	33,040
32.81	1.2925	7.52	25.24	10.776	44.22	4.77	33,390
33.03	1.2950	7.57	25.40	10.796	44.51	4.81	33,670
33.27	1.2975	7.61	25.54	10.817	44.75	4.84	33,880
33.46	1.3000	7.69	25.81	10.838	45.22	4.90	34,300
33.68	1.3025	7.74	25.97	10.859	45.51	4.94	34,580
33.89	1.3050	7.80	26.18	10.880	45.86	4.99	34,930
34.10	1.3075	7.86	26.38	10.901	46.22	5.04	35,280
34.31	1.3100	7.92	26.58	10.921	46.57	5.09	35,630
34.52	1.3125	7.97	26.75	10.942	46.86	5.13	35,910
34.73	1.3150	8.03	26.95	10.963	47.22	5.18	36,260
34.94	1.3175	8.08	27.11	10.984	47.51	5.22	36,540
35.15	1.3200	8.14	27.32	11.005	47.86	5.27	36,890
35.36	1.3225	8.19	27.48	11.026	48.16	5.31	37,170
35.57	1.3250	8.25	27.69	11.047	48.51	5.36	37,520
35.77	1.3275	8.30	27.85	11.067	48.80	5.40	37,800
35.98	1.3300	8.35	28.02	11.088	49.10	5.44	38,080
36.18	1.3325	8.40	28.19	11.109	49.39	5.49	38,430
36.39	1.3350	8.46	28.39	11.130	49.74	5.54	38,780
36.59	1.3375	8.52	28.59	11.151	50.10	5.59	39,130
36.79	1.3400	8.58	28.79	11.172	50.45	5.64	39,480
36.99	1.3425	8.64	28.99	11.192	50.80	5.69	39,830
37.19	1.3450	8.70	29.20	11.213	51.16	5.74	40,180
37.39	1.3475	8.75	29.36	11.234	51.45	5.78	40,460
37.59	1.3500	8.80	29.53	11.255	51.74	5.82	40,740
37.79	1.3525	8.85	29.70	11.276	52.04	5.87	41,090
37.99	1.3550	8.91	29.90	11.297	52.39	5.92	41,440
38.19	1.3575	8.97	30.10	11.317	52.74	5.97	41,790
38.38	1.3600	9.04	30.34	11.338	53.16	6.03	42,210

*Adapted from Aluminum Sulfate Bulletins by Allied Chemical (1968) and Stauffer Chemical (1967).

CRYSTALLIZATION TEMPERATURE OF SULFURIC ACID AND ALUM

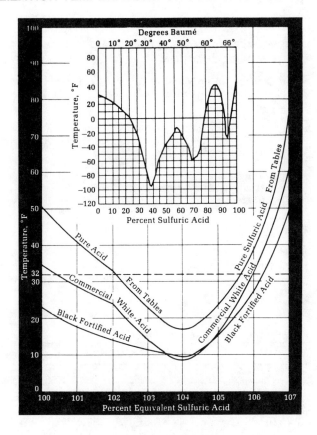

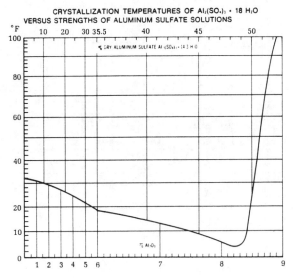

Figure A13-1 Crystallization temperature of sulfuric acid and alum.

FERRIC CHLORIDE*

PROPERTIES OF 42 BAUMÉ FERRIC CHLORIDE SOLUTION

Description of Material	Red brown
At Room Temperature	Solution
Formula	FeCl₃ in solution
Molecular Weight	162.22
FeCl₃	39%
Acidity as HCl	Less than 0.50%
Specific Gravity, 25°/4°C.	1.40
Melting Point	—6°C.
Boiling Point	109°C.
Viscosity, 25°C., Centistokes	6
Surface Tension, 25°C., dynes/cm.	60
Specific Heat, 0-98°C.	0.6

FERRIC CHLORIDE SOLUTIONS						
Concentration-Sp. Gr. Data at 68-77° F.					Freezing Range °F.	
% FeCl₃	Sp. Gr.	° Bé	Lb. Sol. / Gal. Sol.	Lb. FeCl₃ / Gal. Sol.	Undis-turbed	Agitated or Seeded
0	1.000	0.0	8.32	0.0	32	32
5	1.042	5.9	8.67	0.43	27	28
10	1.088	11.8	9.12	0.91	20	24
15	1.134	17.0	9.44	1.42	10	16
20	1.185	22.8	9.87	1.98	− 4	5
25	1.238	27.9	10.30	2.58	−22	−10
30	1.298	33.4	10.80	3.24	−50	−36
35	1.360	38.8	11.32	3.97	−42	−12
37	1.391	40.8	11.59	4.28	−15	15
38	1.402	41.8	11.69	4.44	− 5	21
39	1.415	42.6	11.78	4.59	4	23
40	1.430	43.7	11.90	4.77	10	25
41	1.445	44.8	12.03	4.93	15	28
42	1.459	45.5	12.14	4.98	20	31
43	1.472	46.6	12.24	5.27	24	35
44	1.485	47.4	12.36	5.44	28	40
45	1.500	48.6	12.48	5.62	32	45
46	1.514	49.3	12.60	5.79	35	50
47	1.530	50.3	12.73	5.98	39	58
48	1.545	51.1	12.86	6.17	43	68

*Adapted from Dow Chemical's Ferric Chloride Bulletin (1965).

CAUSTIC SODA*

density of caustic soda solutions

AT 60° F. (15.5° C.)

% NaOH	% Na2O	Specific Gravity 60/60°F.	Baumé Am. Std.	Twaddell 60°F.	Grams NaOH Per Liter	Pounds NaOH		Pounds of Solution	
						Per Gal.	Per Cu. Ft.	Per Gal.	Per Cu. Ft.
2	1.55	1.023	3.26	4.6	20.51	.17	1.28	8.53	63.80
4	3.10	1.045	6.24	9.0	41.65	.35	2.60	8.71	65.18
6	4.65	1.067	9.10	13.4	63.92	.53	3.99	8.90	66.55
8	6.20	1.090	11.97	18.0	87.15	.73	5.44	9.09	67.98
10	7.75	1.112	14.60	22.4	111.18	.93	6.94	9.27	69.36
12	9.30	1.134	17.13	26.8	136.01	1.13	8.49	9.45	70.74
14	10.85	1.156	19.57	31.2	161.64	1.35	10.09	9.63	72.10
16	12.40	1.178	21.91	35.6	188.40	1.57	11.76	9.82	73.47
18	13.95	1.201	24.27	40.2	215.31	1.80	13.44	10.01	74.91
20	15.50	1.223	26.44	44.6	244.47	2.04	15.26	10.20	76.28
22	17.05	1.245	28.53	49.0	273.62	2.28	17.08	10.38	77.65
24	18.60	1.267	30.56	53.4	303.74	2.53	18.96	10.56	79.02
26	20.15	1.289	32.51	57.8	334.82	2.79	20.90	10.75	80.39
28	21.69	1.310	34.31	62.0	366.54	3.06	22.88	10.92	81.70
30	23.24	1.332	36.14	66.4	399.22	3.33	24.92	11.11	83.08
32	24.79	1.353	37.83	70.6	432.54	3.61	27.00	11.28	84.39
34	26.34	1.374	39.47	74.8	466.82	3.89	29.14	11.46	85.71
36	27.89	1.394	40.98	78.8	501.43	4.18	31.30	11.62	86.95
38	29.44	1.415	42.53	83.0	537.15	4.48	33.53	11.80	88.25
40	30.99	1.434	43.88	86.8	573.20	4.78	35.78	11.96	89.45
42	32.54	1.454	45.28	90.8	610.20	5.09	38.09	12.12	90.70
44	.34.09	1.473	46.56	94.6	647.53	5.40	40.42	12.28	91.87
46	35.64	1.492	47.82	98.4	685.82	5.72	42.81	12.44	93.06
48	37.19	1.511	49.04	102.2	724.74	6.04	45.24	12.60	94.24
50	38.74	1.530	50.23	106.0	764.47	6.38	47.72	12.76	95.43
52	40.29	1.549	51.39	109.8	804.84	6.71	50.24	12.92	96.62

approximate freezing temperatures of caustic soda solutions

% NaOH Solution	Approx. F.P. °F	% NaOH Solution	Approx. F.P.°F	% NaOH Solution	Approx. F.P. °F	% NaOH Solution	Approx. F.P. °F
2	30	22	−11	42	57	62	134
4	26	24	−2	44	51	64	140
6	22	26	+14	46	43	66	145
8	18	28	25	48	49	68	148
10	13	30	34	50	53	70	148
12	7	32	42	52	71	72	146
14	+1	34	51	54	88	73	144
16	−5	36	57	56	102	74	143
18	−13	38	60	58	116	74.2	140
20	−15	40	60	60	126	74.5	144

*Adapted from Caustic Soda Bulletin by Stauffer Chemical (1965).

AQUA AMMONIA*

AQUA AMMONIA
Neutralization Grade

CHEMICAL PROPERTIES

Property	Product Specification	Typical Properties
Ammonia Content	24.5% Min.	24.9%
Water Content	75.5% Max.	

PHYSICAL PROPERTIES

Chemical Formula	NH_4OH
Specific Gravity @ 60°F	0.910
Pounds Per Gallon @ 60°F	7.58
Color	Colorless

Vapor Pressure

°F	80	100	120	140
psig	-	0.9	8.3	19.3

Corrosivity — Corrosive to copper, copper alloy, aluminum and galvanized surfaces.

AQUA AMMONIA / 26° Be'
Industrial Grade

CHEMICAL PROPERTIES

Property	Product Specification	Typical Properties [1]
Ammonia Content	29.4% Min.	29.6%
Carbon Dioxide		20.0 ppm Max.
Chlorides		0.5 ppm Max.
Phosphates		0.5 ppm Max.
Sodium		10.0 ppm Max.
Pyridine		Passes A.C.S. Test
Sulfur (As Sulfate)		3.0 ppm Max.
Iron		0.5 ppm Max.
Heavy Metals (As Lead)		1.0 ppm Max.
Organic		Passes A.C.S. Test
Residue After Ignition		5.0 ppm

PHYSICAL PROPERTIES

Chemical Formula	NH_4OH
Specific Gravity @ 60°F	0.897
Pounds Per Gallon @ 60°F	7.47
Color	Water White

Vapor Pressure

°F	50	80	100	120	140
psig	-	0	7.8	18.9	34.3

Corrosivity — Corrosive to copper, copper alloy, aluminum and galvanized surfaces.

*Adapted from UNOCAL 76 Aqua Ammonia Bulletin (1984).

CHLORINE

Properties of Chlorine Gas

Symbol: Cl_2
Atomic wt.: 35.457
Atomic number: 17
Isotopes: 33, 34, 35, 36, 37, 38, 39
Density (see appendix) at 34°F and 1 atm: 0.2006 lb/ft³
Specific gravity at 32°F and 1 atm: 2.482 (air = 1)
Liquefying point at 1 atm: −30.1°F (−34.5°C)
Viscosity (see Appendix) at 68°F: 0.01325 centipoise (approximately the same as saturated
  steam between 1 and 10 atm)
Specific heat at constant pressure of 1 atm and 59°F: 0.115 Btu/lb/°F
Specific heat at constant volume at 1 atm pressure and 59°F: 0.085 Btu/lb/°F
Thermal conductivity at 32°F: 0.0042 Btu/hr/ft²/ft
Heat of reaction with NaOH: 626 Btu/lb Cl_2 gas
Solubility in water at 68°F and 1 atm: 7.29 g/L.

Combining Quantities

1 lb chlorine gas combines with
  1.10 lb commercial hydrated lime (95% $Ca(OH)_2$
$$2Ca(OH)_2 + 2 Cl_2 = Ca(OCl)_2 + Cl Cl_2 + 2H_2O$$
  0.83 lb commercial quicklime (95% CaO)
$$2Ca + H_2O + Cl_2 = Ca(OCl)_2 + CaCl_2 + 2H_2O$$
  1.13 lb caustic soda (100% NaOH)
$$2NaOH + Cl_2 = NaOCl + NaCl + H_2O$$
  2.99 lb soda ash
$$2 Na_2CO_3 + Cl_2 + H_2O = NaOCL + NaCl + 2NaHCO_3$$

Properties of Liquid Chlorine

Critical temperature	144°C; 291.2°F
Pretical pressure	1118.36 psia
Critical density	573 g/1; 35.77 lb/ft³
Compressibility	0.0118% per unit vol per atm increase at 68°F
Density (see Appendix) at 32°F	91.67 lbs/ft³
Specific gravity at 68°F	1.41 (water = 1)
Boiling point (liquefaction point) 1 atm	−34.5°C; −30.1°F
Freezing point	−100.98°C; −149.76°F
Viscosity (see Appendix) at 68°F	.345 centipoise (approx. 0.35 × water at 68°F)
1 volume liquid at 32°F and 1 atm	457.6 vol gas
1 lb liquid at 32°F and 1 atm	4.98 ft³ gas
Specific heat	0.226 Btu/lb/°F
Latent heat of vaporization	123.8 Btu/lb at −29.3°F
Heat of fusion	41.2 Btu/lb at −150.7°F

CHLORINE GAS SCRUBBING SYSTEM

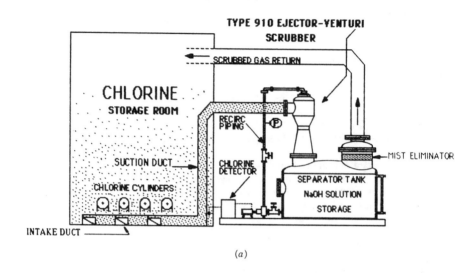

(a)

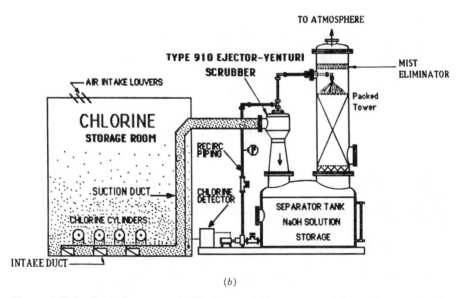

(b)

Figure A13-2 Typical emergency chlorine scrubbing systems: (a) recycle system and (b) once-through system.

*Adapted from Environmental Systems Technology Corp.

Hydraulics: Basic Data and Formulas

USEFUL CONVERSION FACTORS

1 Million gallons per day (mgd) = 1.547 cfs = 694.5 gpm
1 Cubic foot per second (cfs) = 448.8 gpm = 1.98 AF/day
1 Miner's inch = 1.5 cfm
1 Cubic meters per second (m³/s) = 35.3 cfs = 22.8 mgd
1 Megaliters per day (ML/d) = 0.264 mgd
1 Acre foot per day = 0.326 mgd = 0.5 cfs
1 cubic foot = 7.48 gallons
1 psi = 0.07 kg/cm² = 6.9 Pa
1 hp = 7.46 kW
1 cubic foot of water = 62.4 lb
1 gallon of water = 8.34 lb
1 psi = 2.307 ft of water
Earth's gravitational acceleration g = 32.17 ft/s² = 9.806 m/s²
Standard atmospheric pressure = 14.7 psi = 1013 millibar = 1 kg/cm²
    = 760 mm mercury

FUNDAMENTAL EQUATIONS

Equation of mass: $M = W/g$

Manning's formula for open channel: $v = \dfrac{1.486}{n} R^{0.667} S^{0.5}$

where V = mean velocity (fps),
    R = hydraulic radius (ft) $(R = A/p)$,
    S = hydraulic gradient,
    n = coefficient of roughness (0.01–0.02).

From Manning's formula: $S = \dfrac{nv}{1.486\ R^{0.667}}$

Chezy's formula for open channel: $v = C(RS)^{0.5}$

635

where V, R, S = same as above,
C = coefficient.

$$C = \frac{1.486}{n} R^{0.1667}$$

Darcy–Weisback formula for pipe line: $h_f = f\left(\frac{l}{d}\right)\left(\frac{v^2}{2g}\right)$

where h_f = friction loss (ft),
l = length of pipeline (ft),
d = inside diameter of the pipe (ft),
v = mean flow velocity (fps),
g = acceleration of gravity (32.2 ft/s²),
f = friction coefficient (0.0002–0.002).

Hazen–William formula for pipe line: $v = 1.318CR^{0.63}S^{0.54}$

where v = mean velocity (fps),
S = hydraulic gradient,
R = hydraulic radius (ft) ($R = A/p$),
A = cross-sectional area (ft²),
P = wetted perimeter (ft),
C = friction coefficient (80–140).

RELATION BETWEEN SURFACE CONDITIONS AND FRICTION COEFFICIENT

Condition	Friction Coefficient			Notes
	n (Manning's)	C (Hazen's)	f (Darcy's)	
Very smooth surface	0.010	140	0.0002	PVC pipe Clean cement-lined pipe
Fair condition surface	0.013	120	0.0012	Unlined pipe
Rough surface	0.016	100	0.0025	Rusted pipe

HEADLOSS IN OPEN CHANNEL

1. Sudded Contraction (Including Inlet Loss)

$$h = C_1 \frac{v_2^2 - v_1^2}{2g}$$

where C_1 = 0.5 for sharp cornered entrance
   = 0.25 for rounded cornered entrance
   = 0.05–0.1 for bellmouthed entrance (use 0.1 for design),
  v_1 = flow velocity in upstream (fps),
  v_2 = flow velocity in downstream (fps).

2. Sudded Enlargement (Including Outlet Loss)

$$h = C_2 \frac{v_1^2 - v_2^2}{2g}$$

where C_2 = 1.0 for sudded enlargement
   = 0.1–0.2 for bellmouthed outlet (use 0.2 for design),
  v_1 = flow velocity in upstream (fps),
  v_2 = flow velocity in downstream (fps).

3. Turns in the Baffled Channel (Plug Flow Condition Only)

$$h = K \cdot \frac{v^2}{2g}$$

where K = 1.7 for 90° turn
   = 3.2 for 180° turn,
 v = mean flow velocity (fps).

4. Flowing Through Bar Screen

$$h = \frac{v^2 - v_a^2}{2g}\left(\frac{1}{0.7}\right)$$

where v = flow velocity in-between bars (fps),
  v_a = approaching flow velocity (fps).

5. Friction Loss in Open Channel

$$h_f = \frac{2gn^2}{R^{0.333}}\frac{L}{R}\frac{v^2}{2g}, \qquad R = \frac{A}{p}$$

where L = length of the channel (ft).

HEADLOSS FOR PIPE FLOW

Values of K for the basic formula: $\quad h = K\dfrac{v^2}{2g}$

Item	K	Item	K
Butterfly valve (full open)	0.25 (>16 in. diameter) 0.35 (<12 in. diameter)	Return bend	
		Screwed ends	2.2
		Flanged ends	
Gate valve (full open)	0.12 (>16 in. diameter) 0.20 (<12 in. diameter)	Regular	0.38
		Long radius	0.25
		Regular Y branches (flow in the branches)	0.5–0.75
Plug valve (screwed end)	0.85	Sudden contraction	
		$d/D = \frac{1}{4}$	0.42
		$d/D = \frac{1}{2}$	0.35
Diaphragm valve	2.3	$d/D = \frac{3}{4}$	0.19
Check valve			
Swing type	0.6–2.3	*Note: v* to be used is in branches	
Silent check	2.5–3.3	Sudden enlargement	
Glove valve	4.0	$d/D = \frac{1}{4}$	0.92
Ball valve	0.15	$d/D = \frac{1}{2}$	0.35
Foot valve (excluding strainer)	0.8	$d/D = \frac{3}{4}$	0.19
90° Elbow (flanged ends)		*Note: V* is in smaller pipe	
Regular	0.2–0.3	Entrance loss	
Long radius	0.15–0.25	Inward projecting	0.75
90° Elbow (screwed ends)		Sharp cornered	0.5
		Rounded	0.25
Short	0.9	Bellmouthed	0.10
Medium	0.75	Outlet loss	
45° Elbow		Sudden enlargement	1.0
Regular (screwed)	0.3–0.45	Bellmouthed	0.1–0.2
Long radius (flanged)	0.18–0.2	Tee (same sizes, same flow rate)	
22.5° Elbow (flanged)	0.10–0.12	90° turn	1.1–1.5
Mitter bends (large size, welded ends)		Run of tee	0.2–0.5
		Siphon	2.8
90	0.25–0.36	Increaser	0.25
45	0.15–0.20		
22.5	0.07–0.12	*Note:* $h = 0.25\dfrac{v_1^2 - v_2^2}{2g}$	

where v_1 = velocity in smaller pipe,
v_2 = velocity in larger pipe

Reducer
Bushing	1.4
Regular	0.25
Regular to large size pipes (36 in.)	0.05

Note: v is in the smaller size pipe

Various "tee-flow" conditions. (Adapted from Degremont, *Water Treatment Handbook*, Halstead Press, New York, 1979.) *Note*: Pipe size of branch is the same as main pipeline.

1. Outing Flow

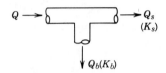

Q_0/Q	0.3	0.5	0.7	1.0
K_b	1.0	1.1	1.25	1.45
K_s	0.05	0.1	0.2	—

2. Merging Flow

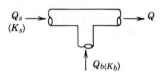

Q_m/Q	0.3	0.5	0.7	1.0
K_b	0.1	0.55	0.8	1.2
K_s	0.35	0.45	0.6	—

3. Splitting Flow

$$K_a = 1 + 0.3 \left(\frac{Q_a}{Q}\right)^2$$

$$k_b = 1 + 0.3 \left(\frac{Q_b}{Q}\right)^2$$

4. Converging Flow

$$K_a = 2 + 3 \left[\left(\frac{Q_b}{Q}\right)^2 - \frac{Q_2}{Q}\right]$$

$$K_b = 2 + 3 \left[\left(\frac{Q_a}{Q}\right)^2 - \frac{Q_b}{Q}\right]$$

ORIFICE

$$Q = Ca(2gh)^{0.5} \quad \text{or} \quad h = \frac{1}{2g}\left(\frac{Q}{Ca}\right)^2$$

where Q = discharge rate (cfs),
a = orifice area (ft^2),
C = coefficient, 0.65–0.8,
h = headloss across the orifice (ft).

WEIRS (FREE DISCHARGE CONDITION)

1. *Sharp Crested Rectangular Weir*

$$Q = 3.33 \, LH^{1.5}$$

2. *Broad Crested Rectangular Weir*

$$Q = 3.09L(H + h_v)^{1.5}$$

3. *90° V-Notch Weir*

$$Q = 2.54H^{2.5}$$

4. *60° V-Notch Weir*

$$Q = 1.45H^{2.5}$$

5. *45° V-Notch Weir*

$$Q = 1.07H^{2.5}$$

6. *22.5° V-Notch Weir*

$$Q = 0.5H^{2.5}$$

where Q = discharge rate (cfs),
L = effective weir length (ft),
H = head (water height) over the crest (ft),
h_v = velocity head of approach (ft).

HYDRAULIC ELEMENTS FOR CIRCULAR PIPE

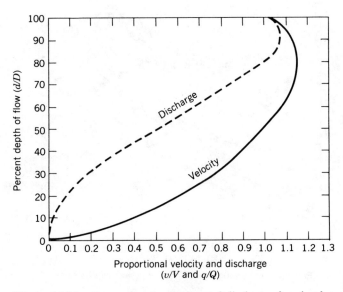

Figure A14-1 Proportional velocity and discharge for circular pipe.

Tertiary Filter Performance

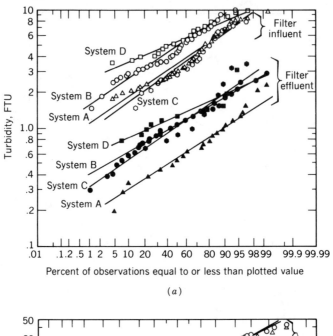

(a)

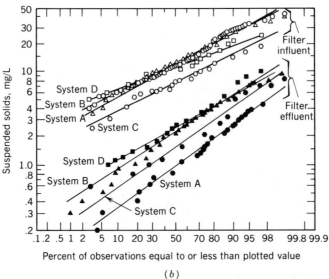

(b)

Figure A15-1 Frequency distributions of (a) filter influent and effluent turbidity and (b) filter influent and effluent suspended solids. System A: Flocculation, sedimentation, and filtration of secondary effluent with 50 mg/L alum and dual-media filter at 5 gpm/ft². System B: Coagulation and filtration (in-line filtration) of secondary effluent with 5 mg/L alum and dual-media filter at 5 gpm/ft². System C: Two-stage granular carbon adsorption process. Both beds have 10 min EBCT each. System D: Coagulation and filtration (in-line filtration) of nitrified secondary effluent with 5 mg/L alum and dual-media filter at 5 gpm/ft². (Adapted from Sanitation Districts of Los Angeles County Report.)

Nomographical Determination of pH$_s$ and Langelier Index

Graph and Nomograph designed by Mr. Ch. Hoover for the determination of the saturation pH and the water saturation index according to the LANGELIER formula (correct for pH values between 7 and 9·5).

Column 1

Temperature total salinity (pK'$_2$—pK$_s$) constant

Data necessary for the determination of the saturation pH value (pHs):

(a) Total alkalinity as calcium carbonate (CaCO$_3$) in p.p.m. (mg/litre).
(b) Calcium as Ca in p.p.m. (mg/litre).
(c) Total dissolved salts in p.p.m. (mg/litre).
(d) Temperature in °C (for which the saturation pH value is required).

How to use the graph

(a) For the known value of total dissolved salts mark off in column 1 the temperature total salinity value using th. appropriate temperature curve.

Total dissolved salts in p.p.m. (mg/litre)

Column 2
Pivot Line

Column 3
Calcium

Column 4
pHs

Column 5
Alkalinity p.p.m. as CaCO$_3$

(b) Draw a straight line to connect this value to the known calcium concentration in column 3, and note the point at which this line crosses column 2.
(c) Draw a straight line to connect this point on the pivot line (column 2) with the point on column 5 corresponding to the known alkalinity. Read off the saturation pH value at the point where this line crosses column 4.
The saturation index is the difference between the measured pH of the water and the saturation pH value.

Example:

pH of the water	Saturation pH	Saturation index
7·6	8·1	− 0·5 (corrosive)
8·4	7·8	+ 0·6 (scale-forming)

Marston's Soil Coefficient for Pipe Trench Conditions

1. Trench condition:

$$W_d = C_d W B_d^2$$

where
W_d = fill load in pounds per linear foot of pipe
C_d = load coefficient (a function of H/B_d and $K\mu$)
W = unit weight of fill material in pounds per cu ft
B_d = width of trench at the level of the top of pipe, in ft.
H = backfill depth, ft

2. Embankment condition:

$$W_c = C_c W B_c^2$$

where
W_c = load per unit length of pipe
C_c = load coefficient (a function of H/B_c and $r_{sd}p$)
W = unit weight of fill material
B_c = outside diameter of the pipe, in feet

MARSTON SOIL COEFFICIENTS (C_d) FOR TRENCH CONDUITS

A = $K\mu$ = •1924 Granular materials without cohesion
B = $K\mu$ = •165 Maximum for sand and gravel
C = $K\mu$ = •150 Maximum for saturated top soil
D = $K\mu$ = •130 Ordinary maximum for clay
E = $K\mu$ = •110 Maximum for saturated clay

H/B_d	A	B	C	D	E	H/B_d	A	B	C	D	E
0.05	0.050	0.050	0.050	0.050	0.050	3.00	1.780	1.904	1.978	2.083	2.196
0.10	0.098	0.098	0.099	0.099	0.099	3.10	1.810	1.941	2.018	2.128	2.247
0.15	0.146	0.146	0.147	0.147	0.148	3.20	1.840	1.976	2.057	2.172	2.297
0.20	0.192	0.194	0.194	0.195	0.196	3.30	1.869	2.010	2.095	2.215	2.346
0.25	0.238	0.240	0.241	0.242	0.243	3.40	1.896	2.044	2.131	2.257	2.394
0.30	0.283	0.286	0.287	0.289	0.290	3.50	1.923	2.076	2.167	2.298	2.441
0.35	0.327	0.331	0.332	0.335	0.337	3.60	1.948	2.107	2.201	2.338	2.487
0.40	0.371	0.375	0.377	0.380	0.383	3.70	1.973	2.137	2.235	2.376	2.531
0.45	0.413	0.418	0.421	0.425	0.428	3.80	1.997	2.166	2.267	2.414	2.575
0.50	0.455	0.461	0.464	0.469	0.473	3.90	2.019	2.194	2.299	2.451	2.618
0.55	0.496	0.503	0.507	0.512	0.518	4.00	2.041	2.221	2.329	2.487	2.660
0 60	0.536	0.544	0.549	0.555	0.562	4.10	2.062	2.247	2.359	2.522	2.701
0.65	0.575	0.585	0.591	0.598	0.606	4.20	2.082	2.273	2.388	2.556	2.741
0.70	0.614	0.625	0.631	0.640	0.649	4.30	2.102	2.297	2.416	2.589	2.780
0.75	0.651	0.664	0.672	0.681	0.691	4.40	2.121	2.321	2.443	2.621	2.819
0.80	0.689	0.703	0.711	0.722	0.734	4.50	2.139	2.344	2.469	2.652	2.856
0.85	0.725	0.741	0.750	0.763	0.775	4.60	2.156	2.366	2.495	2.683	2.893
0.90	0.761	0.779	0.789	0.802	0.817	4.70	2.173	2.388	2.520	2.713	2.929
0.95	0.796	0.816	0.827	0.842	0.857	4.80	2.189	2.409	2.543	2.742	2.964
1.00	0.830	0.852	0.864	0.881	0.898	4.90	2.204	2.429	2.567	2.770	2.999
1.05	0.864	0.887	0.901	0.919	0.938	5.00	2.219	2.448	2.590	2.798	3.032
1 10	0.897	0.922	0.937	0.957	0.977	5.10	2.234	2.467	2.612	2.825	3.065
1 15	0.929	0.957	0.973	0.994	1.016	5.20	2.247	2.486	2.633	2.851	3.098
1.20	0.961	0.991	1.008	1.031	1.055	5.30	2.261	2.503	2.654	2.877	3.129
1.25	0.992	1.024	1.042	1.067	1.093	5.40	2.273	2.520	2.674	2.901	3.160
1.30	1.023	1.057	1.076	1.103	1.131	5.50	2.286	2.537	2.693	2.926	3.190
1.35	1.053	1.089	1.110	1.139	1.168	5.60	2.298	2.553	2.712	2.949	3.220
1.40	1.082	1.121	1.143	1.173	1.205	5.70	2.309	2.568	2.730	2.972	3.248
1.45	1.111	1.152	1.176	1.208	1.241	5.80	2.320	2.583	2.748	2.995	3.277
1.50	1.140	1.183	1.208	1.242	1.278	5.90	2.330	2.598	2.766	3.017	3.304
1.55	1.167	1.213	1.240	1.276	1.313	6.00	2.340	2.612	2.782	3.038	3.331
1.60	1.195	1.243	1.271	1.309	1.349	6.20	2.360	2.639	2.814	3.079	3.383
1.65	1.221	1.272	1.301	1.342	1.384	6.40	2.377	2.664	2.845	3.118	3.433
1.70	1.248	1.301	1.332	1.374	1.418	6.60	2.394	2.687	2.873	3.155	3.481
1.75	1.273	1.329	1.361	1.406	1.452	6.80	2.409	2.709	2.900	3.190	3.527
1.80	1.299	1.357	1.391	1.437	1.486	7.00	2.423	2.730	2.925	3.223	3.571
1.85	1.323	1.385	1.420	1.469	1.520	7.20	2.436	2.749	2.949	3.255	3.613
1.90	1.348	1.412	1.448	1.499	1.553	7.40	2.448	2.767	2.971	3.285	3.653
1.95	1.372	1.438	1.476	1.530	1.586	7.60	2.459	2.784	2.992	3.313	3.691
2.00	1.395	1.464	1.504	1.560	1.618	7.80	2.470	2.799	3.012	3.340	3.728
2.10	1.440	1.515	1.558	1.618	1.682	8.00	2.479	2.814	3.031	3.366	3.763
2.20	1.484	1.564	1.610	1.675	1.744	8.50	2.500	2.847	3.073	3.424	3.845
2.30	1.526	1.612	1.661	1.731	1.805	9.00	2.517	2.875	3.109	3.476	3.918
2.40	1.567	1.658	1.711	1.785	1.865	9.50	2.532	2.898	3.141	3.521	3.983
2.50	1.606	1.702	1.759	1.838	1.923	10.0	2.543	2.919	3.167	3.560	4.042
2.60	1.643	1.745	1.805	1.890	1.980	15.0	2.591	3.009	3.296	3.768	4.378
2.70	1.679	1.787	1.850	1.940	2.036	20.0	2.598	3.026	3.325	3.825	4.490
2.80	1.714	1.827	1.894	1.989	2.090	30.0	2.599	3.030	3.333	3.845	4.539
2.90	1.747	1.867	1.937	2.037	2.144	40.0	2.599	3.030	3.333	3.846	4.545

Load Coefficient Chart for Positive Projecting Embankment Condition

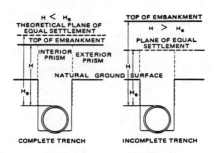

COMPLETE AND INCOMPLETE
TRENCH CONDITIONS

LOAD COEFFICIENT CHART FOR POSITIVE PROJECTING EMBANKMENT CONDITION

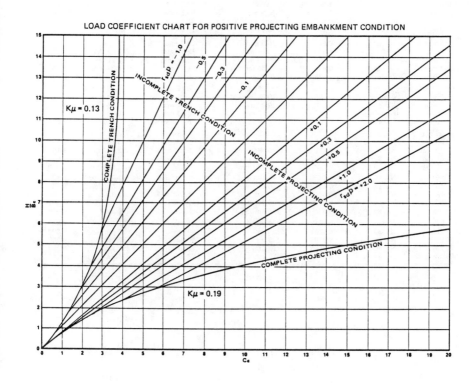

USA Standard Sieves

Nominal Dimensions, Permissible Variations for Wire Cloth of Standard Test Sieves (USA Standard Series)

Sieve Designation Standard†	Sieve Designation Alternative	Nominal Sieve Opening in.	Permissible Variation of Average Opening from the Standard Sieve Designation	Maximum Opening Size for not More Than 5 percent of Openings	Maximum Individual Opening	Nominal Wire Diameter mm
125 mm	5 in.	5	±3.7 mm	130.0 mm	130.9 mm	8.0
106 mm	4.24 in.	4.24	±3.2 mm	110.2 mm	111.1 mm	6.40
100 mm	4 in.	4	±3.0 mm	104.0 mm	104.8 mm	6.30
90 mm	3¹/₂ in.	3.5	±2.7 mm	93.6 mm	94.4 mm	6.08
75 mm	3 in.	3	±2.2 mm	78.1 mm	78.7 mm	5.80
63 mm	2¹/₂ in.	2.5	±1.9 mm	65.6 mm	66.2 mm	5.50
53 mm	2.12 in.	2.12	±1.6 mm	55.2 mm	55.7 mm	5.15
50 mm	2 in.	2	±1.5 mm	52.1 mm	52.6 mm	5.05
45 mm	1³/₄ in.	1.75	±1.4 mm	46.9 mm	47.4 mm	4.85
37.5 mm	1¹/₂ in.	1.5	±1.1 mm	39.1 mm	39.5 mm	4.59
31.5 mm	1¹/₄ in.	1.25	±1.0 mm	32.9 mm	33.2 mm	4.23
26.5 mm	1.06 in.	1.06	±0.8 mm	27.7 mm	28.0 mm	3.90
25.0 mm	1 in.	1	±0.8 mm	26.1 mm	26.4 mm	3.80
22.4 mm	⁷/₈ in.	0.875	±0.7 mm	23.4 mm	23.7 mm	3.50
19.0 mm	³/₄ in.	0.750	±0.6 mm	19.9 mm	20.1 mm	3.30
16.0 mm	⁵/₈ in.	0.625	±0.5 mm	16.7 mm	17.0 mm	3.00
13.2 mm	0.530 in.	0.530	±0.41 mm	13.83 mm	14.05 mm	2.75
12.5 mm	¹/₂ in.	0.500	±0.39 mm	13.10 mm	13.31 mm	2.67
11.2 mm	⁷/₁₆ in.	0.438	±0.35 mm	11.75 mm	11.94 mm	2.45
9.5 mm	³/₈ in.	0.375	±0.30 mm	9.97 mm	10.16 mm	2.27
8.0 mm	⁵/₁₆ in.	0.312	±0.25 mm	8.41 mm	8.58 mm	2.07
6.7 mm	0.265 in.	0.265	±0.21 mm	7.05 mm	7.20 mm	1.87
6.3 mm	¹/₄ in.	0.250	±0.20 mm	6.64 mm	6.78 mm	1.82
5.6 mm	No. 3¹/₂	0.223	±0.18 mm	5.90 mm	6.04 mm	1.68
4.75 mm	No. 4	0.187	±0.15 mm	5.02 mm	5.14 mm	1.54
4.00 mm	No. 5	0.157	±0.13 mm	4.23 mm	4.35 mm	1.37
3.35 mm	No. 6	0.132	±0.11 mm	3.55 mm	3.66 mm	1.23
2.80 mm	No. 7	0.111	±0.095 mm	2.975 mm	3.070 mm	1.10
2.36 mm	No. 8	0.0937	±0.080 mm	2.515 mm	2.600 mm	1.00
2.00 mm	No. 10	0.0787	±0.070 mm	2.135 mm	2.215 mm	0.900
1.70 mm	No. 12	0.0661	±0.060 mm	1.820 mm	1.890 mm	0.810
1.40 mm	No. 14	0.0555	±0.050 mm	1.505 mm	1.565 mm	0.725
1.18 mm	No. 16	0.0469	±0.045 mm	1.270 mm	1.330 mm	0.650
1.00 mm	No. 18	0.0394	±0.040 mm	1.080 mm	1.135 mm	0.580
850 μm	No. 20	0.0331	±35 μm	925 μm	970 μm	0.510
710 μm	No. 25	0.0273	±30 μm	774 μm	815 μm	0.450
600 μm	No. 30	0.0234	±25 μm	660 μm	695 μm	0.390
500 μm	No. 35	0.0197	±20 μm	550 μm	585 μm	0.340
425 μm	No. 40	0.0165	±19 μm	471 μm	502 μm	0.290
355 μm	No. 45	0.0139	±16 μm	396 μm	425 μm	0.247

Standard Electric Motors and Their Characteristics

Synchronous and Full Load Speeds of
Standard A.C. Induction Motors

NUMBER OF POLES	60 CYCLE RPM		50 CYCLE RPM	
	SYNC.	F.L.	SYNC.	F.L.
2	3600	3500	3000	2900
4	1800	1770	1500	1450
6	1200	1170	1000	960
8	900	870	750	720
10	720	690	600	575
12	600	575	500	480
14	515	490	429	410
16	450	430	375	360
18	400	380	333	319
20	360	340	300	285
22	327	310	273	260
24	300	285	240	230
26	277	265	231	222
28	257	245	214	205
30	240	230	200	192

Full Load Amperes For All Speeds and Frequencies

MOTOR HP	SINGLE-PHASE A-C		THREE-PHASE A-C INDUCTION TYPE SQUIRREL CAGE & WOUND ROTOR			DIRECT CURRENT	
	115 VOLTS	230 VOLTS*	230 VOLTS*	460 VOLTS	575 VOLTS	120 VOLTS	240 VOLTS
½	9.8	4.9	2.0	1.0	.8	5.2	2.6
¾	13.8	6.9	2.8	1.4	1.1	7.4	3.7
1	16	8	3.6	1.8	1.4	9.4	4.7
1½	20	10	5.2	2.6	2.1	13.2	6.6
2	24	12	6.8	3.4	2.7	17	8.5
3	34	17	9.6	4.8	3.9	25	12.2
5	56	28	15.2	7.6	6.1	40	20
7½	80	40	22	11	9	58	29
10	100	50	28	14	11	76	38
15			42	21	17	112	55
20			54	27	22	148	72
25			68	34	27	184	89
30			80	40	32	220	106
40			104	52	41	292	140
50			130	65	52	360	173
60			154	77	62	430	206
75			192	96	77	536	255
100			240	120	96		350
125			296	148	118		440
150			350	175	140		530
200			456	228	182		710
250			558	279	223		

* For full-load currents of 208 and 200 volt motors, increase the corresponding 230 volt motor full-load current by 10 and 15 per cent respectively.

■ INDEX